INTERNATIONAL CENTRE FOR MECHANICAL SCIENCES

COURSES AND LECTURES - No. 297

ROTORDYNAMICS 2

PROBLEMS IN TURBOMACHINERY

EDITED BY

N. F. RIEGER
STRESS TECHNOLOGY INCORPORATED

SPRINGER-VERLAG WIEN GMBH

Le spese di stampa di questo volume sono in parte coperte da contributi del Consiglio Nazionale delle Ricerche.

This volume contains 354 illustrations.

Originally published by Springer-Verlag Wien New York in 1988

ISBN 978-3-211-82091-9 **ISBN 978-3-7091-2846-6 (eBook)**
DOI 10.1007/978-3-7091-2846-6

PREFACE

Current attention in turbomachinery design and operation is sharply focused on the achievement of higher levels of availability and reliability. The usual trends towards minimum weight and low vibration have become standard well-recognized criteria for turbomachines of all types. The thrust of new work is towards increasing the number of operating hours per year and to minimize outage and maintenance periods.

It is hoped that the technology presented in this volume, "Rotordynamics II – Problems of Turbomachinery" will contribute towards the above objectives by its consolidated presentation of existing material, and with the inclusion of newly prepared material on a range of new and important topics. This latter is not elsewhere available in a single volume. The subject of rotordynamics now covers a large number of contributing disciplines, and it is now soundly based on long established principles. Nonetheless, rotordynamics continues to evolve new and fast-developing aspects, such as the recent surge in instrumentation development, analytical equipment, and in monitoring devices, all of which are part of the current revolution in micro-electronics. Certain new devices such as magnetic bearings also offer great promise for future application in space, computer, and medical applications. These devices have themselves given rise to sub-sets of rotordynamics technology. In more conventional areas, new problems such as torsional transient effects on machine survivability, and the recent emphasis on blading life improvement have contributed to increased demands for structural component reliability, and have required additional technology involving life prediction and component life extension.

Rotordynamics research is therefore much different, more detailed, and perhaps more exacting than the situation which existed ten or fifteen years ago. It is therefore timely to focus these newer developments into a single volume, written by many experts involved in developing this new technology. Such a volume should provide designers and operators with a concentration of new material which will, in turn, further contribute to the development of more reliable and better functioning turbomachinery.

The initial chapters of this volume discuss the operating properties of fluid-film bearings, and the analysis of rotor behavior in such bearings. This fundamental material is utilized in Chapter Four, where a discussion of problems of rotor balancing is presented. This section covers balancing machines, rigid rotor balancing, flexible rotor balancing, and several balancing case histories from practice. Chapter Five deals with techniques for identification of stiffness, damping, and inertia coefficients for seals, for instances where the dynamic behavior of fluid-moving turbomachines requires the contributions of the interstage and gland seals to be included in the analysis. Chapter Six is a discussion of the stability of rotors in two oil-film bearings. The prediction of instability threshold speeds is demonstrated, and verified by experimental studies.

The technology presented in the preceding six chapters is coordinated in Chapter Seven, which deals with the computer analysis of rotor-bearing systems. A specific large computer code called PALLA for the dynamic analysis of rotor structural systems is described. The functioning of this code is demonstrated using several examples from practice. This code analyzes several rotating shafts in fluid-film bearings and seals, mounted on a flexible foundation. Although codes of this type have existed since Prohl's critical speed analysis in 1945, to this day it is still difficult to decide on the degree of complexity which the ideal or optimum rotordynamic computer code should contain. PALLA provides the analyst with a comprehensive tool for complex rotor-structure systems, to obtain response and stability information in using well defined support properties.

Machine-soil interactions are discussed in Chapter Nine. The theory of such interactions is established through the use of viscoelastic field equations, and applied, with experimental verification, to the case of a turbomachine frame on a soil foundation.

Experimental methods for the study of rotor behavior in bearings are discussed in Chapter Eight. Sensor technology is presented first, and when the modern principles of this science have been described, several experimental techniques for analysis of rotors in bearings and seals are discussed with examples. Chapter Eight describes certain aspects of possible interactions which may occur beteen a machine and its foundation. The practical aspects of this presentation throw valuable light upon a complex subject which faces both analysts and experimentalists, and complements the discussion of the PALLA code in Chapter Seven.

The remaining three chapters deal with special topics which have now developed into sophisticated sub-technologies. Chapter Ten discusses problems of torsional shaft systems. The analysis of turbine-generator torsional transient vibrations has recently undergone considerable development due to major torsional transient problems which have occurred in units around the world. Analytical techniques for torsional systems using modal analysis are discussed in the second section of Chapter Ten, and the complications introduced by the presence of gears in a drive system are discussed in section three of this chapter. Problems of turbine blades are discussed in Chapter Eleven. The first presentation deals with free vibrations and forced vibrations of blades, and the second section examines the current state-of-the-art for blade excitation and damping. Case studies involving problems of turbine blades are discussed in section three of this Chapter, and a consolidation of existing approaches for life evaluation of blades is discussed in section four. Experimental and analytical studies on the damping properties of steam turbine blades are described in part five of Chapter Eleven.

Chapter Twelve describes several special topics. The first of these is magnetic bearings for which a comprehensive introduction to this subject with applications is presented in the first section. The technology of magnetic bearings is currently being advanced by needs in space technology, computers, and elsewhere. This section deals with control systems of magnetic bearings, and the details presented for application of such devices should provide designers and users with much valuable guidance. The final section of Chapter Twelve deals with vibrations in variable speed machines. This topic is of interest with all machines which much traverse one or more critical speeds during runup and rundown, and machines which operate over a broad range of operating speed such as gas turbines, utility steam turbines, and pumps. The importance of the rate at which a critical speed is traversed has been recognized since Lewis's work in 1932. The technology of this chapter should find further application in life evaluation techniques discussed above for such machines.

As Editor it is my pleasure to thank all authors who have contributed their labors, creativity, and valuable time to prepare their sections of this volume. Warm thanks will doubtless also be expressed by the many readers who scan these pages. Thanks are also due to the diligent staff at CISM, in particular Professor Giovanni Bianchi of the Politecnico di Milano whose idea it was to publish these proceedings; to Professor Carlo Tasso who supervised the preparation of the volume itself, and Signora Bertozzi who had the delicate task of guiding these many authors towards a common goal. Our publisher Springer-Verlag also deserves our grateful thanks for producing such a fine manuscript, and we also record our indebtedness to all the secretaries who worked long and hard to ensure that the quality of the manuscript was achieved to the satisfaction of the authors. My own thanks go to Ms. Candace Rogers of Stress Technology who did an outstanding job of personally checking all the manuscripts that went into this volume. I extend my grateful thanks to all these people for their contributions.

It is the hope of all those who have contributed to this volume that their labors have produced a new state-of-the-art document which will serve as a guide for the creators of new turbomachines and for those who must maintain the present machines.

Neville F. Rieger
Rochester, New York

CONTENTS

CHAPTER I

INTRODUCTION

N.F. Rieger
Stress Technology Incorporated, Rochester, New York, USA

ABSTRACT

The major problem areas of rotordynamics are identified. The manner in which these problems are addressed as Chapters of this book is described. A number of commonly-used rotordynamic terms are defined for reference purposes. The orbits of several frequently observed whirl motions are discussed and correlated with their causes, for convenience in problem diagnosis. Literature sources for further information on specific problems including books, papers, and conference proceedings, with a reference listing are presented. Prominent references in the historical development of rotordynamics are cited.

1.1 Problems and Scope of Rotordynamics

Rotor vibration problems may arise from a number of sources, of which the most important are residual rotor unbalance and rotor instability. Rotor unbalance causes a rotating force in synchronism with running speed (synchronous unbalance), whereas rotor instability is a self-excited vibration which may arise from bearing fluid-film effects, electromagnetic effects, flow effects, or from some combination of these factors. Several other mechanisms have also led to rotor vibration problems in the past. For example, hysteresis (dry friction) effects have lead to whirl in lightly damped machinery; different shaft lateral stiffnesses are a known cause of unstable whirl in electrical machinery; and non-linear foundation effects have led to vibrations in certain speed ranges. In most instances, practical 'fixes' can be applied to reduce or eliminate these undesirable vibrations once the problem is correctly diagnosed. The design of smooth running rotating machinery requires that each of these problem areas should be carefully eliminated from the machine system specifications. In most instances this can be done quite readily if adequate allowance is made for the machine dynamic characteristics as part of the design.

The dynamic characteristics of greatest interest in rotating machinery are:

a) Rotor critical speeds in the operating range.

b) Unbalance response amplitudes at critical speeds.

c) Threshold of resonant whip instability.

d) Bearing transmitted force.

e) System torsional critical speeds.

f) Gear dynamic loads.

g) Disk natural frequency (compressor, turbine, gear).

h) Bucket, blade, and impeller natural frequencies and modes.

i) Blade flutter frequencies.

j) Rotating stall and surge thresholds.

Many other important vibration topics such as noise and structural vibration of rotating machinery could be added to this list. Each of the above subjects has an extensive published literature. In this book, the fundamentals of each topic are presented with comments and discussions on important recent contributions to the literature of each. No single book presently covers the entire subject of rotor bearing dynamics, but a thorough appreciation may be gathered from those publications listed in the references to each chapter.

The properties of hydrodynamic bearings are discussed in Chapter 2. The plain cylindrical bearing is considered first, and expressions for its static performance are derived, using the short bearing form of Ockvirk for convenience. This theory is then extended to the dynamic case, and expressions are developed for the linearized theory of short bearings. Properties of special bearing types, such as partial, tilting pad, and multi-arc bearings are discussed. This chapter provides a basis upon which the theory and experience of rotors in flexible supports can be developed in the chapters which follow.

Chapter 3 contains an introduction to the analysis of rotors in bearings. The basic theory of the Jeffcott-Foppl rotor is developed in detail with emphasis on the principles involved. This leads to the well-known expressions for critical speed and unbalance response.

The principles of rotor balancing are discussed in Chapter 4. The fundamentals of rigid rotor balancing using two correction planes are first presented and the development of modern balancing machines of the Lawaczek-Heyman type is discussed. Case histories of certain balancing problems which have been encountered in practice are described.

Chapter 5 discusses the principles of hydrodynamic seals and their influence on rotor performance. The representation of the performance of certain types of seals in terms of stiffness, damping, and inertia coefficients is discussed, with examples which compare seal performance with test and experimental data.

Chapter 6 deals with stability problems of rigid rotors in bearings. The relationship between the eight linearized coefficients and the instability threshold speed is developed. Experience with instability in practice is presented, and experimental comparisons with predicted threshold data based on matrix eigenvalue extraction procedures is presented.

In Chapter 7 a general-purpose computer program for the behavior of a rotor on a flexible foundation is discussed. The structure of the code is described, with an example of its use on a practical problem. The properties of several computer programs which are available from various sources for rotordynamic analysis are discussed.

Practical procedures for obtaining test data on rotordynamic performance of rotors in supports are discussed in Chapter 8. The technology and application of several types of measuring sensors (capacitance, inductance, light sensors, etc.) and their readout equipment are presented. The use of such equipment for rotor performance measurements is discussed with several practical test examples is also described.

The problem of interaction between a rotor and its supporting foundation is discussed in Chapter 9. Many types of such foundations exist, and their dynamic influence on rotor performance is discussed, based on generalized findings from a program of experiments, and correlation of results with supporting theory.

Torsional dynamics of drive trains is discussed in Chapter 10. The systems discussed include turbine-generator sets for which an appropriate modal theory is derived and presented with experimental correlation, and a program of tests on practical turbine-generator sets with an analysis of test results compared with transient response theory. Several practical examples of geared industrial drive problems which have involved torsional dynamic analysis for their solution are discussed.

Chapter 11 provides a comprehensive discussion of turbomachine blade problems. Principles of vibration of such blades are first discussed, followed by a discussion of the types of excitation and damping forces which may be encountered in practice, together with their most probable sources. Case studies of certain classical problems are given, together with a discussion of current procedures for blade life evaluation.

Certain special topics of current interest in the vibration of rotating machinery are presented in Chapter 12. The principles and application of magnetic bearings to rotating machinery is an important topic which holds considerable promise for future developments in high speed equipment. Selected theoretical developments and supporting test developments are described for both passive and active magnetic rotor supports. A second topic is the problem of the transient passage of a mechanical component through resonance during runup and rundown operation. The principles of this problem for a rotor passing through a critical speed, and for a blade passing through an excitation harmonic are described, with sample applications.

The extensive literature which deals with vibration problems of rotors in bearings indicates the scope and sophistication of the modern technology of this subject. This book provides both a broad and up-to-date review of today's technology of rotordynamics. The specialists who have written the various chapters of this book are experts in each area. It is hoped that this focus of timely expertise will be of value and guidance to users of this book.

1.2 Definitions

Disk - A wheel, usually solid and axially slim, on which mechanical work is performed or from which work is extracted. Examples: turbine disk, compressor wheel.

Shaft - A beam, usually axisymmetric (commonly circular) on which disks may be carried, either integrally or by shrink fitting.

<u>Rotor</u> - An assembly of disks on a shaft or simply a massive shaft, mounted in supporting bearings. Two special classes of rotors are: rigid rotor, flexible rotor.

<u>Rigid rotor</u> - A rotor which operates substantially below its first bending critical speed. A rigid rotor can be brought into, and will remain in, a state of satisfactory balance at all operating speeds when balanced on any two arbitrarily selected correction planes.

<u>Flexible rotor</u> - A rotor which operates close enough to, or beyond its first bending critical speed for dynamic effects to influence rotor deformations. Rotors which cannot be classified as rigid rotors are considered to be flexible rotors.

<u>Bearing</u> - Any low friction support which carries the rotor and provides dynamic constraint in the transverse and/or axial directions. The two main categories are fluid-film bearings and rolling-element bearings. Types: journal bearing, thrust bearing.

<u>Fluid-film bearing</u> - A bearing whose low friction property derives from the thin fluid layer between the rotor surface and the bearing metal. The fluid-film layer may be generated by journal rotation (hydrodynamic bearing) or by externally pressurized pumped fluid under pressure (hydrostatic bearing).

<u>Rolling-element bearing</u> - The low friction property derives from mechanical rolling with marginal lubrication, using ball or roller elements.

<u>Journal</u> - Specific portions of shaft surface from which rotor applied loads are transmitted to bearing supports.

<u>Pedestal</u> - Bearing support possessing mass elastic properties, mounted on the machine foundation.

<u>Foundation</u> - Machine support. May be rigid or possess mass elastic properties.

<u>System</u> - The interacting combination of rotor, bearings, pedestals and foundation which responds as a complex to dynamic excitation.

<u>Whirl</u> - Rotor transverse orbital motion about the static equilibrium position, at any axial location.

<u>Natural frequency</u> - Any frequency of free vibration at which a natural mode of the system assumes its maximum amplitude.

Critical speed - Rotor speed at which local maximum amplitude whirling occurs. Where the rotor has no significant gyroscopic effects, a critical speed occurs whenever the rotor speed coincides with a system natural frequency.

Unbalance - Eccentricity of local c.g. of rotor from undisturbed axis of rotation. Product of rotor local mass times eccentricity of c.g. from shaft elastic axis. Also expressed in terms of eccentricity along, e.g., in micro-inches of eccentricity.

Unbalance response - Whirl maximum amplitude at a given speed, caused by dynamic forcing action of rotating unbalance on the system.

Sub-critical speed - System vibration occurring at integer sub-multiple of main system critical speed ($1/2\ \omega_c$, $1/3\ \omega_c$, $1/4\ \omega_c$), ... $1/10\ \omega_c$) arising from stiffness non-linearity, such as 'flat' shaft.

Whirl instability - Condition in which whirl radius increases with time. May arise from variety of causes, e.g., bearings, shaft hysteresis or geometry, foundation, torque or speed fluctuation. Instability may grow indefinitely in time, or become bounded by new constraints.

Half-frequency whirl - Instability of dynamically rigid rotor in hydrodynamic fluid-film bearings, arising from rotor fluid interaction. This instability commences near the first system critical speed and occurs with a frequency somewhat less than half the rotational speed. Associated with rigid rotor systems where bearing properties predominate.

Resonant whip - System instability involving rotor flexure, occurring in fluid-film bearing systems. It arises from flexible rotor fluid-film interaction. It is similar in nature to half-frequency whirl, except for presence of rotor flexibility effect.

Fractional frequency whirl - Instability of dynamically rigid rotor in externally pressurized fluid-film bearings. Otherwise similar to half-frequency whirl, except that whirl frequency is usually much less than half rotational speed, e.g., 0.25ω to 0.45ω whirl frequency range.

1.3 Nature of Whirl Motions

A rotor is said to whirl when the c.g. of any cross section traces out an orbit in time, instead of remaining at a fixed point. If identical whirl orbits are traced out with successive shaft rotations, the whirl is said to be stable. If the orbit increases in size with successive rotations, the whirl is then unstable and may subsequently grow until the orbit becomes bounded either by the internal forces of the system, or by some external constraint, e.g., bearing rub, guard ring, shutdown, etc. Smooth machine operation is characterized by small, stable rotor whirl orbits, and by the absence of any instabilities within the machine operating range.

Some typical whirl orbits are shown in Figure 1. The circular orbit in Figure 1 represents the synchronous whirling of a rotor in isotropic radial supports. The frequency of the excitation is synchronous as shown by the absence of loops within the orbit. Excitation in such a case usually arises from rotor unbalance.

An elliptical orbit, Figure 2, may arise from dissimilar bearing stiffnesses or support stiffnesses in the horizontal and vertical directions. These stiffnesses give different displacements for the same rotor centrifugal force. Where complex (i.e., cross-coupled) stiffness and damping properties exist as in fluid-film bearings, the major and minor axes of the orbit then occur at some angle with respect to the x and y coordinate directions, as shown. Where cross-coupled effects are absent, the axes of the ellipse coincide with the x and y coordinate directions.

If the whirl is non-synchronous, i.e., the rotor whirls at a frequency ν, other than the rotational frequency ω, the orbit will contain a loop, frequently like that shown in Figure 3 for half-frequency whirl. The loop is internal indicating that the whirl is in the direction of rotation.

Other non-synchronous excitations may occur at several times rotational frequency, e.g., multi-pole electrical stimuli give rise to multi-lobe whirl orbits such as that shown in Figure 4.

Instabilities such as half-frequency whirl are frequently bounded. In such a case the whirl would be initiated by crossing the instability threshold speed. The whirl then becomes a growing transient whose radius increases until a new equilibrium position is found. Frequently this is established by the fluid forces themselves and no damage is done. The whirl then continues at a larger radius. Otherwise the whirl increases until some structural constraint such as the bearing surface or the machine casing is encountered. A bounded instability whirl is shown in Figure 5, including the transient motion from the first condition of equilibrium to the second equilibrium condition.

Another type of transient condition is shown in Figure 6. The rotor is initially operating in a small stable unbalance whirl condition. The rotor system then receives a transverse shock, and the journal displaces abruptly in a radial direction within the bearing clearance, but without contacting the bearing surface. Following the shock, the rotor motion is a damped decaying transient, as it returns to its original small unbalance whirl condition.

Many other interesting types of whirl orbits have been observed, such as those associated with system non-linearities. Further comments on whirl motions are given in detail in the chapters dealing with this and related topics.

1.4 Rotordynamics Information Sources

The rapid development of rotor-bearing dynamics has resulted in a very extensive published literature and several books. The best known books on this subject are listed in Table 1.1: see references [1.1] through [1.13] of this Table. Details are given in Section 1.6 at the end of this chapter. Most books deal with some selected aspect of the subject, e.g., stability or balancing, rather than a comprehensive treatment.

The great practical relevance of rotordynamics studies has also led to several major conferences, at which the current technical developments have been presented. Vibrations conferences featuring rotordynamics sessions were initiated by the American Society of Mechanical Engineers in 1967. Since then conferences have been held in the U.S.A. every two years, and papers are published in the ASME Transactions. An international conference on Vibrations in Rotating Machinery has been held every four years since 1972 by the Institution of Mechanical Engineers of London. Bound volumes of proceedings are available from 1972 through 1984. An international Symposium on Rotordynamics was also held by the International Union of Theoretical and Applied Mechanics (IUTAM) in Lyngby, Denmark in 1974. A book of proceedings is available from this meeting. In recent years symposia on rotordynamics have been organized by the International Federation for Theory of Machines and Mechanisms (IFToMM), in 1979, 1983, and 1987. Published proceedings from these symposia containing state-of-the-art papers are also available. An international Rotordynamics Conference has been organized by IFToMM every four years, in 1982 (Rome), and in 1986 (Tokyo).

Beginning in 1976 the Vibration Institute in the U.S.A. has organized annual seminar meetings on vibration problems, with an emphasis on problems of rotating machinery. Other conferences held by the Vibration Institute which emphasized instrumentation and test measurement and balancing of rotors, are also listed. Proceedings are available for these meetings. In each instance these proceedings consist of a collection of original papers.

1.5 The Classical Literature of Rotor-Bearing Dynamics

The critical speeds of a uniform elastic shaft were first investigated by Rankine [1.14] in 1869, who devised the term 'critical speed.' This phenomenon arose in factory overhead pulley shafting at that time. It was incorrectly thought by certain investigators to be an unstable condition. Dunkerley [1.15] in 1894 presented an excellent collection of critical speed studies related to pulley shafting, and gave his well-known method with its experimental verification. Kerr [1.16] in 1916 precipitated an extended discussion of turbine critical speed problems and the mechanics of whirling. Jeffcott [1.17] in 1919 resolved the evident confusion surrounding the mechanics of rotor unbalance whirl in his classical analysis of this problem. Stodola [1.18] proposed the linear coefficient representation of fluid film bearing properties in 1917. Hysteretic whirl was first investigated by Newkirk [1.19] in 1924 during studies of blast furnace compressor vibrations, and explained by Kimball [1.20]. Newkirk and Taylor [1.21] observed oil film whirl and resonant whipping for the first time in 1925. An explicit review of basic rotordynamics problems was presented in 1933 by Smith [1.22]. Robertson [1.23], [1.24], [1.25], [1.26], [1.27] presented a series of important rotordynamics papers between 1932 and 1935 on the subjects of bearing whirl, rotor transient whirl, and hysteretic whirl.

Demands for larger and faster rotors, together with development and application of the computer, had led to sustained development of rotor-bearing technology since World War II. Hagg [1.28] conducted experiments on half-frequency whirl in 1946. The mechanism of bearing instability was first explained by Poritsky [1.29] in 1952, and experimentally confirmed by Boeker and Sternlicht [1.30] in 1955. The linear theory of rotor instability was established by Lund [1.31] in 1963. Prohl [1.32] introduced discrete mass numerical calculation of rotors in 1946. Hagg [1.33] studied unbalance response of rotors in bearings in 1948. Hagg and Sankey [1.34], Sternlicht [1.35], Lund [1.36] and others developed numerical hydrodynamical analysis procedures for bearings between 1955 and 1965. Procedures for flexible rotor balancing were suggested by Linn [1.37] in 1928 and by Thearle [1.38] in 1935. Their application was demonstrated by Grobel [1.39] and developed into practical procedures by Goodman [1.40], Bishop and Gladwell [1.41], Lund [1.42] and others since 1952.

1.6 References

1.1 Dimentberg, F. M., Flexural Vibrations of Rotating Shafts, Butterworth and Company, Ltd., London, England, 1961.

1.2 Tondl, A., Some Problems of Rotor Dynamics, Publishing House, Czechoslovakian Academy of Sciences, Prague, 1965.

1.3 Gunter, E. J., Jr., 'Dynamic Stability of Rotor-Bearing Systems,' NASA Report SP-112, 1966.

1.4 Rieger, N. F., Poritsky, H., Lund, J. W., et. al., 'Rotor-Bearing Dynamics Design Technology,' Volumes 1-9, Wright-Patterson Air Force Base Aero Propulsion Laboratory Reports, 1965-1968.

1.5 Wilcox, J. B., Dynamic Balancing of Rotating Machinery, Sir Isaac Pitman and Sons, Ltd., London, England, 1967.

1.6 Smith, D. M., Journal Bearings in Turbomachinery, Chapman and Hall Ltd., London, England, 1969.

1.7 Eshleman, R., Shapiro, W., Rumbarger, J., Rieger, N. F., 'Flexible Rotor-Bearing System Dynamics,' Volume I: Critical Speeds, Volume II: Bearing Properties, Volume III: Unbalance Response and Balancing, ASME Design Division Monographs, 1973.

1.8 Wilson, D., Pan, C., Allaire, P., 'Rotor-Bearing Dynamics Design Technology,' Second Edition, WPAFB Fuels and Lubricants Division, 1977.

1.9 Federn, K., Auswuchttechnik, Bd. 1, Berlin, Springer, 1977.

1.10 Rieger, N. F., Vibrations of Rotating Machinery, Vibration Institute, Clarendon Hills, Illinois, 2nd Edition, 1982.

1.11 Rao, J. S., Dynamics of Rotors, Wiley-Eastern Publishers, New Delhi, India, 1983.

1.12 Rieger, N. F., Balancing of Rigid and Flexible Rotors, Shock and Vibration Information Center, Naval Research Laboratory, Washington, DC, 1987.

1.13 Kellenberger, W., Elastisches Wuchten, Springer-Verlag, 1987.

1.14 Rankine, W. J. McQ., 'On the Centrifugal Force of Rotating Shafts,' Engineer London, Volume 27, p. 249, 1869.

1.15 Dunkerley, S., 'On the Whirling and Vibration of Shafts,' Phil. Transactions of the Royal Society, Series A, Vol. 185, p. 229, 1895.

1.16 Kerr, W., 'On the Whirling Speed of Loaded Shafts,' Engineering, pp. 150, 296, 386, 410, and 420, February 18, 1916. Discussions by C. Chree, A. Morley, A. Stodola, H. Naylor, H. Jeffcott, J. Danus, W. Kerr.

1.17 Jeffcott, H. H., 'The Lateral Vibration of Loaded Shafts in the Neighborhood of a Whirling Speed- The Effect of Want of Balance,' Phil. Magazine, Series 6, Vol. 37, p. 304, 1919.

1.18 Stodola, A., Steam and Gas Turbines, Vols. 1 and 2, Translated by L. C. Loewenstein, McGraw-Hill Book Company, Inc., New York, 1927.

1.19 Newkirk, B. L., 'Shaft Whipping,' General Electric Review, Vol. 27, p. 169, 1924.

1.20 Kimball, A. L., Lovell, J., 'Internal Friction as a Cause of Shaft Whirling,' Trans. ASME, Vol. 48, 1926.

1.21 Newkirk, B. L., Taylor, H. D., 'Oil Film Whirl - An Investigation of Disturbances on Oil Films in Journal Bearings,' General Electric Review, Vol. 28, 1925.

1.22 Smith, D. M., 'The Motion of a Rotor Carried by a Flexible Shaft in Flexible Bearings,' Proceedings of the Royal Society, Series A., Vol. 142, p. 92, 1933.

1.23 Robertson, D., 'The Vibrations of Revolving Shafts,' Phil. Magazine, Series 7, Vol. 13, p. 862, 1932.

1.24 Robertson, D., 'The Whirling of Shafts,' The Engineer, Vol. 158, p. 216, 1934.

1.25 Robertson, D., 'Whirling of a Journal in a Sleeve Bearing,' Phil Magazine, Series 7, Vol. 15, p. 113, 1933.

1.26 Robertson, D., 'Transient Whirling of a Rotor,' Phil. Magazine, Series 7, Vol. 20; p. 793, 1935.

1.27 Robertson, D., 'Hysteretic Influences on the Whirling of Rotors,' Proceedings IMechE, Vol. 131, p. 513, 1935.

1.28 Hagg, A. C., 'Some Vibration Aspects of Lubrication,' Lubrication Engineering, pp. 166-169, August 1948.

1.29 Poritsky, H., 'Contribution to the Theory of Oil Whip,' Trans. ASME, Vol. 75, pp. 1153-1161, 1953.

1.30 Boeker, G. F., Sternlicht, B., 'Investigation of Translatory Fluid Whirl in Vertical Machines,' Trans. ASME, Vol. 78, 1956.

1.31 Lund, J. W., 'The Stability of an Elastic Rotor in Journal Bearing with Flexible Damped Supports,' Trans. ASME, Journal of Basic Engineering, Vol. 87, Series E, 1965.

1.32 Prohl, M. A., 'A General Method for Calculating Critical Speeds of Flexible Rotors,' Trans. ASME, Vol. 67, Journal of Applied Mechanics, Vol. 12, p. A-142, 1946.

1.33 Hagg, A. C., 'Some Vibration Aspects of Lubrication,' Lubrication Engineering, pp. 166-169, August 1948.

1.34 Hagg, A. C., Sankey, G. O., 'Elastic and Damping Properties of Oil Film Journal Bearings for Application to Unbalance Vibration Calculations,' Trans. ASME, Journal of Applied Mechanics, Vol. 25, p. 141, 1958.

1.35 Sternlicht, B., 'Elastic and Damping Properties of Cylindrical Journal Bearings,' Trans. ASME, Journal of Applied Mechanics, Series D., Vol. 81, p. 101, 1959.

1.36 Lund, J. W., Sternlicht, B., 'Rotor-Bearing Dynamics with Emphasis on Attenuation,' Trans. ASME, Journal of Basic Engineering, Series D, 1962.

1.37 Linn, F. C., 'Method of and Means for Balancing Rotors,' U.S. Patent No. 1,776,125, September 16, 1930.

1.38 Thearle, E. L., 'Dynamic Balancing of Rotating Machinery in the Field,' General Electric Company, APM-56-19, Schenectady, NY, 1935.

1.39 Grobel, L. P., 'Balancing Turbine-Generator Rotors,' General Electric Review, Vol. 56, No. 4, P. 22, 1953.

1.40 Goodman, T. P., 'A Least-Squares Method for Computing Balancing Corrections,' ASME Publication No. 63-WA-295, September 1964.

1.41 Bishop, R. E. D., Gladwell, G. M. L., 'The Vibration and Balancing of an Unbalanced Flexible Rotor,' Journal of Mechanical Engineering Science, Vol. 1, No. 1, p. 66, 1959.

1.42 Lund, J. W., Computer Program, See: Rieger, N. F., 'Computer Program for Balancing of Flexible Rotors,' Mechanical Technology Incorporated Report 67TR68, September 1967.

Table 1.1 Books on Rotordynamics

REFERENCE	AUTHOR	PUBLISHER	TOPICS EMPHASIZED	EDITION	DATE
1.1	Dimentberg	Pergammon	Classical shaft dynamics for turbine generator applications.	1	1961
1.2	Tondl	Publishing House CSSR	Instability problems of rotors in bearings.	1	1965
1.3	WP/MTI	WPAFB/AFAPL	Rotor-bearing dynamics technology.	1	1965
1.4	Wilcox	Pitman	Balancing of rotors.	1	1966
1.5	Gunter	NASA	Instability of simple rotors.	1	1966
1.6	Smith	Pitman	Bearing analysis.	1	1970
1.7	Eshleman, Shapiro, Rumbarger, Rieger	ASME	Rotor-bearing dynamics literature.	1	1973
1.8	WP/SRI	WPAFB/APAFL	Rotor-bearing dynamics technology update.	1	1973
1.9	Federn	Springer	Balancing of rigid rotors.	1	1978
1.10	Rieger	V.I.	Vibrations of rotating machinery.	2	1982
1.11	Rao	Wiley-Eastern	Dynamics of rotors.	1	1983
1.12	Rieger	NRL	Balancing of rigid and flexible rotors.	1	1987
1.13	Kellenberger	Springer	Rotordynamics.	1	1987

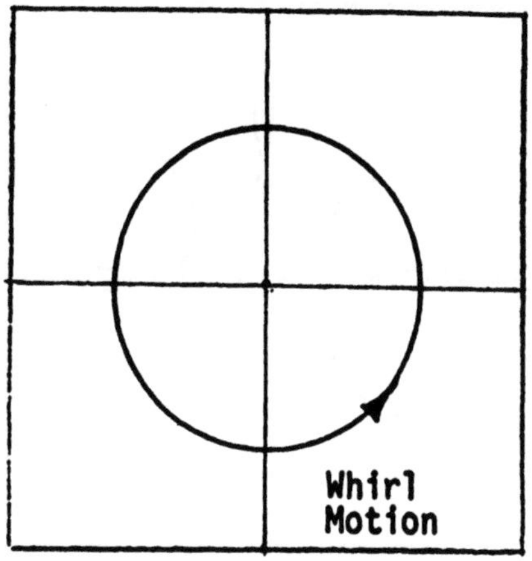

Figure 1 Circular Orbit

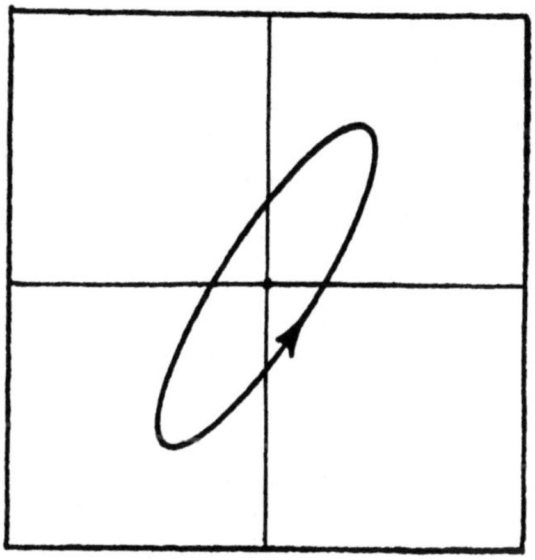

Figure 2 Elliptical Orbit

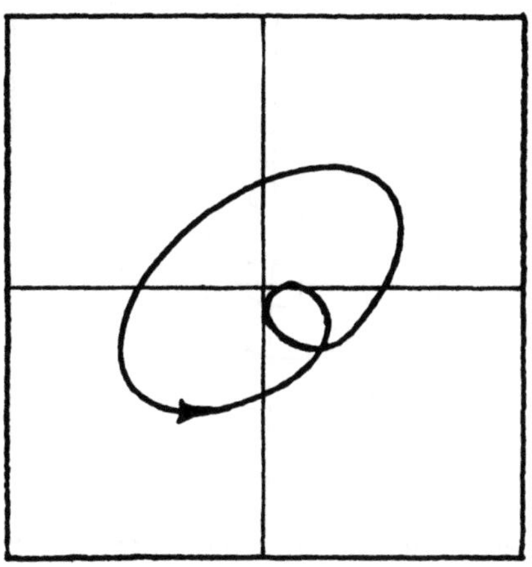

Figure 3 Non-Synchronous Whirl Half-Frequency Bearing Excitation

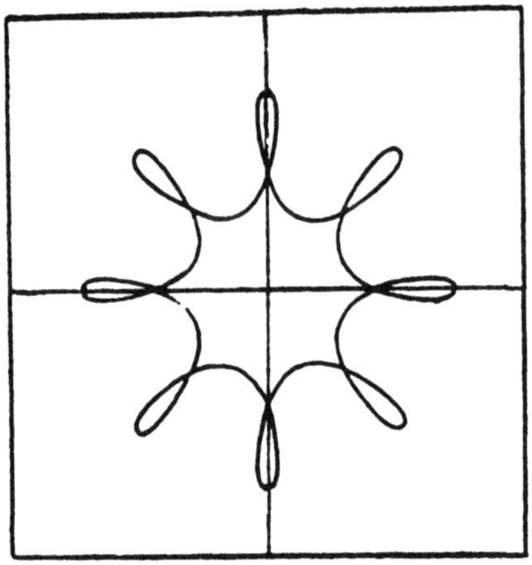

Figure 4 Non-Synchronous Whirl Bearing Rub or Electrical Excitation Determines Number of Cusps

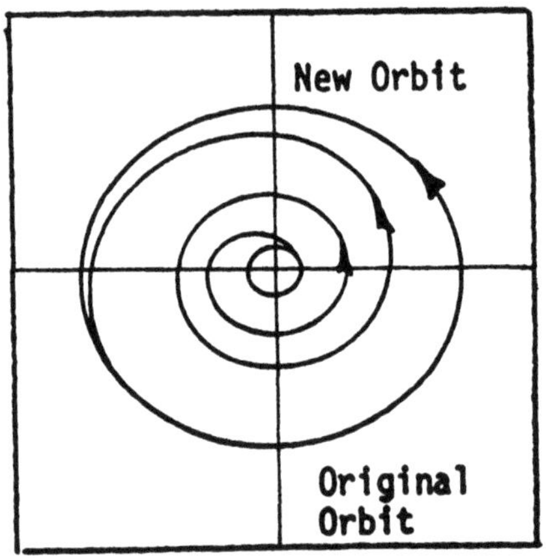

Figure 5 Transient Growth to New Stable Orbit

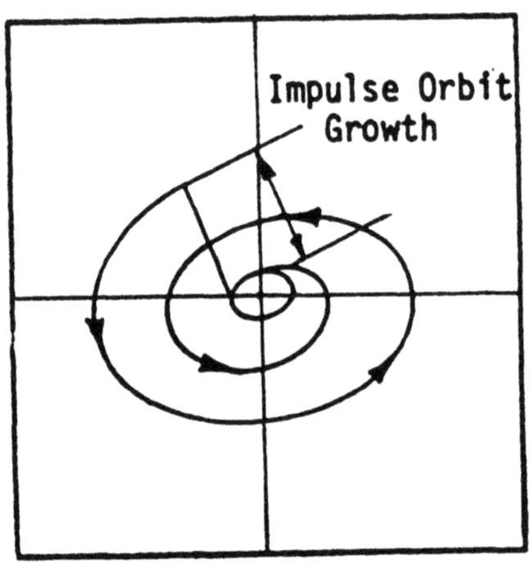

Figure 6 Transient Whirl Decay to Stable Orbit Following Impulse

CHAPTER 2

BEARING PROPERTIES

E. Krämer
Technische Hochschule Darmstadt, Darmstadt, FRG

ABSTRACT

The fundamental analysis of hydrodynamic lubricated journal bearings is presented. It is shown how displacements at static load, stiffness- and damping coefficients may be computed. The formulas of the short circular bearing are presented. Finally some remarks on special bearing types are made.

Symbols

F	Journal load
F_s	Static journal load
L	Bearing length
R	Bore radius
S	Duty parameter
So	Sommerfeld number
d_{ik}	Damping coefficient
e	Eccentricity
h	Film thickness
h_0	Film thickness at centric journal

k_{ik}	Stiffness coefficient
p	Film pressure
$\bar{p}$	Mean bearing pressure
r	Journal radius
t	Time
x_1, x_2	Displacements
Ω	Angular velocity of journal
α	Attitude angle
β	Width ratio
β_{ik}	Damping coefficient, dimensionless
γ	Angle between eccentricity and X-axis
γ_{ik}	Stiffness coefficient, dimensionless
δ	Radial clearance
ε	Eccentricity ratio
η	Dynamic viscosity
$\bar{\eta}$	Mean dynamic viscosity
ϑ	Angle of resulting film force

Rotors are supported by two or more radial bearings. These are mounted in bushings carried by pedestals. Two types of bearings are used, journal bearings and ball bearings. Journal bearings are lubricated by liquid or gas, normally oil is used.

The flexural vibrations of rotors essentially are influenced by the dynamic properties of the bearings. This chapter deals with the main results of the theory of oil lubricated journal bearings.

A review for gas bearings is given in /1/. Some results for ball bearings are published in /2, ...,8/.

2.1 Reynolds Equation and Resulting Pressure Force

We consider the system journal, lubricant film, bushing. The journal rotates and it is loaded by static and dynamic forces, leading to appropriate motion. This behaviour may be studied by measurements or by computations. In the following the basic analytical solution is presented. More details are given in /9/, /10/, /11/ among others.

The system is shown in figure 2.1. The bore of the bushing is assumed cylindrical with arbitrary, but nearly circular cross-section with radius R. We assume a fixed coordinate system X, Y, Z with the origin C_B. The journal is cylindrical (radius r, center C_J) rotating with angular velocity Ω. It is assumed, that it moves only translatory in X- and Y-direction, that means the axis remains parallel to Z-direction. The position of the journal

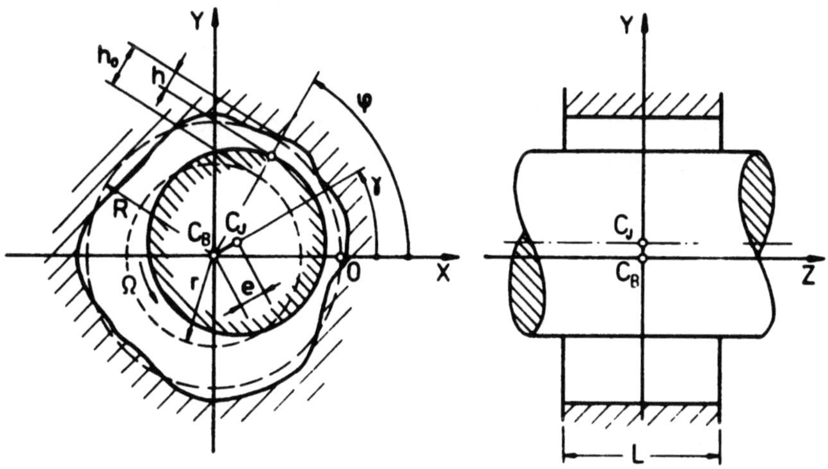

Fig. 2.1 Geometry of a journal bearing.
C_B Bearing center, C_J Journal center

center C_J is determined by the eccentricity e and the angle γ. In case of centric position (e=0) the film thickness is denoted by h_0 which is generally $h_0=h_0(\varphi)$. With e<<r in case of arbitrary journal position the film thickness is approximately

$$h(\varphi) = h_0(\varphi) - e\cos(\varphi-\gamma). \tag{2.1}$$

If the journal moves transversally, e=e(t) and $\gamma=\gamma(t)$, (2.1) becomes

$$h(\varphi,t) = h_0(\varphi) - e(t)\cos[\varphi-\gamma(t)], \tag{2.2}$$

which is called "gap function".

Furthermore the following additional assumptions are made:

1. The lubricant is massless, incompressible and sticking at the walls.
2. The lubricant is Newtonian.
3. Flow is laminar.
4. The lubricant pressure is constant in radial direction.
5. Flow velocities in radial direction are neglected.
6. Velocity gradients in radial direction are large compared to those in tangential and axial direction.
7. The film thickness is small compared to journal radius.
8. The curvature of lubrication gap is neglected, i.e. $R=\infty$.
9. The surfaces of journal and bore are smooth and rigid.

According to assumption 8 the lubricant gap now is shaped as shown in figure 2.2, where

x, y, z	coordinates
u, v, w	flow velocities
$U=\Omega r$	circumferential velocity of journal.

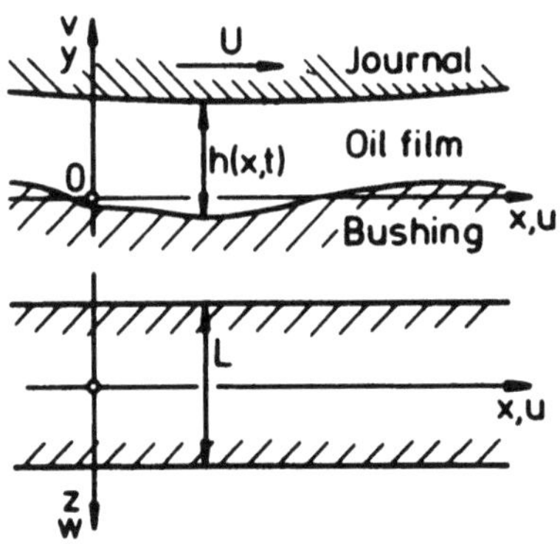

Fig. 2.2 Cross-section as basis for theory

For a small fluid element of sides dx, dy, dz the condition of force equilibrium yields

$$\frac{\partial \tau_{xy}}{\partial y} = \frac{\partial p}{\partial x} \qquad \text{and} \qquad \frac{\partial \tau_{zy}}{\partial y} = \frac{\partial p}{\partial z} \tag{2.3}$$

with p pressure of fluid

τ_{xy}, τ_{zy} shearing stresses.

According to assumption 2 the shearing stresses are

$$\tau_{xy} = \eta\frac{\partial u}{\partial y} \quad \text{and} \quad \tau_{zy} = \eta\frac{\partial w}{\partial y} , \tag{2.4}$$

where η denotes the dynamic viscosity.

Equations (2.3) and (2.4) result

$$\frac{\partial^2 u}{\partial y^2} = \frac{1}{\eta}\frac{\partial p}{\partial x} \quad \text{and} \quad \frac{\partial^2 w}{\partial y^2} = \frac{1}{\eta}\frac{\partial p}{\partial z} . \tag{2.5}$$

Integration of (2.5) yields

$$u = \frac{1}{2\eta}\frac{\partial p}{\partial x}y^2 + C_1 y + C_2 , \qquad w = \frac{1}{2\eta}\frac{\partial p}{\partial z}y^2 + C_3 y + C_4 , \tag{2.6}$$

and finally with the boundary conditions

$u=0$, $w=0$ at $y=0$ (fixed bushing)

and $u=U$, $w=0$ at $y=h$:

$$\left.\begin{aligned} u &= \frac{1}{2\eta}\frac{\partial p}{\partial x}(y^2-hy) + \frac{U}{h}y \\ w &= \frac{1}{2\eta}\frac{\partial p}{\partial z}(y^2-hy) \quad . \end{aligned}\right\} \tag{2.7}$$

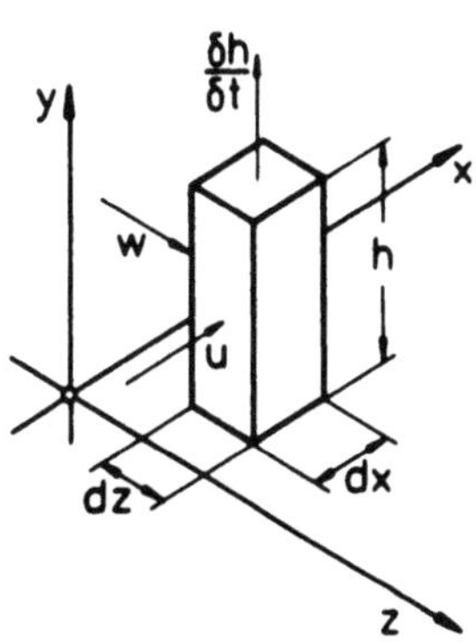

Fig. 2.3 Continuity of flow in a column

Furthermore the condition for continuity of flow is needed. For this we consider a column of fluid of height h and base dx·dz (figure 2.3).

With the flow rates Q'_x, Q'_z in x- and z-direction the condition of continuity is

$$\frac{\partial Q'_x}{\partial x}dxdzdt + \frac{\partial Q'_z}{\partial z}dzdxdt + \frac{\partial h}{\partial t}dtdzdx = 0$$

respectively

$$\frac{\partial Q'_x}{\partial x} + \frac{\partial Q'_z}{\partial z} + \frac{\partial h}{\partial t} = 0 \quad . \tag{2.8}$$

With (2.7) the flow rates are

$$\left.\begin{aligned} Q'_x &= \int_0^h u\,dy = \frac{-h^3}{12\eta}\frac{\partial p}{\partial x} + \frac{1}{2}Uh \\ Q'_z &= \int_0^h w\,dy = \frac{-h^3}{12\eta}\frac{\partial p}{\partial z} \end{aligned}\right\} \tag{2.9}$$

and (2.8) leads to

$$\frac{\partial}{\partial x}\left(\frac{h^3}{\eta}\frac{\partial p}{\partial x}\right) + \frac{\partial}{\partial z}\left(\frac{h^3}{\eta}\frac{\partial p}{\partial z}\right) = 6U\frac{\partial h}{\partial x} + 12\frac{\partial h}{\partial t} \tag{2.10}$$

which is called Reynolds equation.

Now we return to the original cross-section of figure 2.1. With $x=\varphi R$ and (2.2) the needed derivatives on the right hand side of (2.10) are

$$\frac{\partial h}{\partial x} = \frac{1}{R}\left[\frac{\partial h_o}{\partial \varphi} + e\sin(\varphi-\gamma)\right]$$

$$\frac{\partial h}{\partial t} = -e\frac{d\gamma}{dt}\sin(\varphi-\gamma) - \frac{de}{dt}\cos(\varphi-\gamma) \quad .$$

With $U=\Omega R$, assuming constant viscosity $\eta=\bar{\eta}$ and neglecting the small difference between r and R the following form of the Reynolds equation is obtained

$$\frac{1}{R^2}\frac{\partial}{\partial\varphi}\left(h^3\frac{\partial p}{\partial\varphi}\right) + h^3\frac{\partial^2 p}{\partial z^2} = 6\bar{\eta}\left[\Omega\frac{\partial h_o}{\partial\varphi} + e(\Omega-2\dot{\gamma})\sin(\varphi-\gamma) - 2\dot{e}\cos(\varphi-\gamma)\right]. \tag{2.11}$$

The solution of (2.11) is the film pressure $p(\varphi, z, t)$. It depends on one hand upon the viscosity $\bar{\eta}$ and the angular velocity Ω, on the other hand upon the position e, γ and the velocities $\dot{e}$ and $\dot{\gamma}$ of the journal center. In case of noncircular bore the term $\partial h_0/\partial\varphi \neq 0$, thus it generates film pressure.

To obtain the resulting force firstly the pressure is integrated in axial direction:

$$dF = [\int_{-L/2}^{+L/2} p(\varphi,z,t)dz]Rd\varphi \quad , \tag{2.12}$$

where dF is the force on a small area $LRd\varphi$ at arbitrary angle with the components

$$dF_1 = dF\cos\varphi \quad , \quad dF_2 = dF\sin\varphi \quad . \tag{2.13}$$

A further integration in circumferential direction yields to the components of the resultant force F:

$$F_1 = \int_{\varphi=0}^{2\pi} dF_1 \quad , \quad F_2 = \int_{\varphi=0}^{2\pi} dF_2 \quad . \tag{2.14}$$

The resultant is

$$F = \sqrt{F_1^2 + F_2^2} \quad . \tag{2.15}$$

Its angle to X-coordinate (figure 2.4) is given by

$$\tan\vartheta = \frac{F_2}{F_1} \quad . \tag{2.16}$$

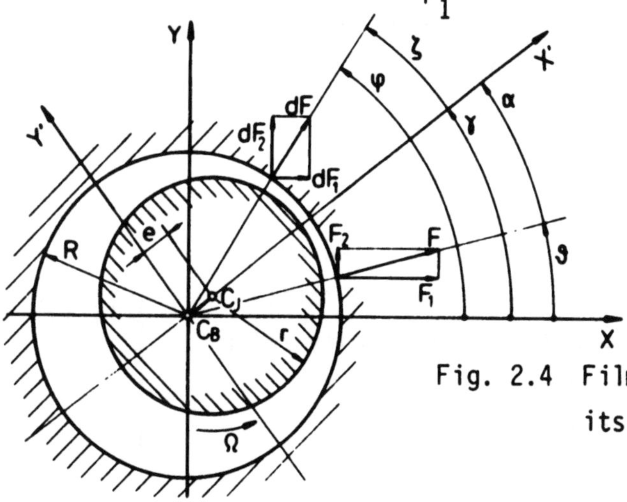

Fig. 2.4 Film forces dF, F and its components

In the following we consider the journal load which is equal to the film force by amount and angle.

2.2 Static Load

Consider a horizontal rotor which is only loaded by its weight. The components of the journal load are

$$F_1 = 0 \quad , \quad F_2 = -F_s \quad ,$$

where F_s is the weight force. Here the journal center is located in the fourth quadrant of the X, Y system (figure 2.5). With F_I, F_{II} as components of F_s in the X', Y' system the attitude angle α is given by

$$\tan\alpha = \frac{-F_{II}}{F_I} \quad . \tag{2.17}$$

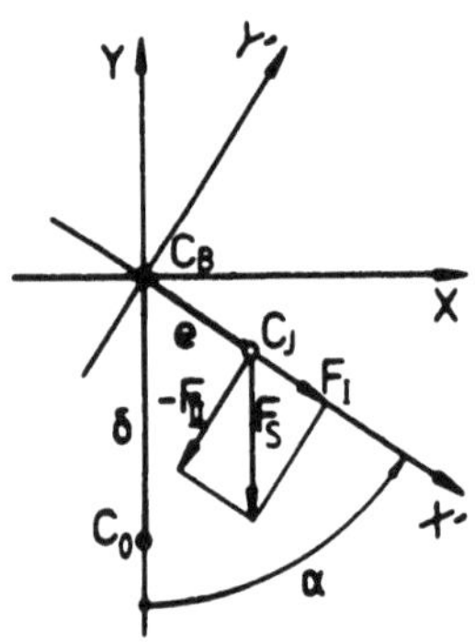

Fig. 2.5 Location of journal center in case of static load

Conventionally the Sommerfeld number is used as a dimensionless form of the static journal force, i.e.

$$So = \frac{\bar{p}\psi^2}{\eta\Omega} \quad , \tag{2.18}$$

where

$$\left.\begin{array}{lll} \delta = R-r & & \text{radial clearance} \\ \psi = \delta/r & & \text{clearance ratio} \\ \varepsilon = e/\delta & & \text{eccentricity ratio} \\ \bar{p} = \dfrac{F_s}{2RL} & & \text{mean pressure.} \end{array}\right\} \tag{2.19}$$

The Sommerfeld number and the attitude angle are functions of eccentricity ratio: $So(\varepsilon)$, $\alpha(\varepsilon)$. For example see (2.45), (2.43) in chapter 2.4.2. Hence the location ε,α of journal center is a function of Sommerfeld number. The example given by figure 2.8 is typical: the plot is nearly semicircular. Sometimes it is called displacement orbit, since it describes the journal position at different angular speeds (C_0 at $\Omega=0$, C_B at $\Omega=\infty$).

The Sommerfeld number sometimes is substituted by its reciprocal value. D.M. Smith introduced in /9/ the duty parameter, which is defined under using the above symbols by

$$S = \frac{1}{2\pi}\frac{\bar{\eta}\Omega}{\bar{p}\psi^2} = \frac{1}{2\pi So} \quad . \tag{2.20}$$

2.3 Stiffness and Damping

Initially static load as considered in chapter 2.2 is assumed. In case of small displacements x_1, x_2 and small velocities $\dot{x}_1$, $\dot{x}_2$ (figure 2.6) the components of the journal force are in first approximation

$$\left.\begin{aligned} F_1 &= \frac{\partial F_1}{\partial x_1}x_1 + \frac{\partial F_1}{\partial x_2}x_2 + \frac{\partial F_1}{\partial \dot{x}_1}\dot{x}_1 + \frac{\partial F_1}{\partial \dot{x}_2}\dot{x}_2 \\ F_2 &= -F_s + \frac{\partial F_2}{\partial x_1}x_1 + \frac{\partial F_2}{\partial x_2}x_2 + \frac{\partial F_2}{\partial \dot{x}_1}\dot{x}_1 + \frac{\partial F_2}{\partial \dot{x}_2}\dot{x}_2 \quad . \end{aligned}\right\} \tag{2.21}$$

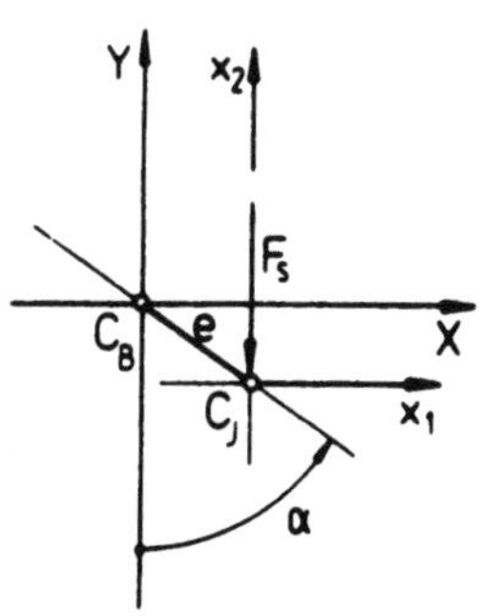

Fig. 2.6 Displacements x_1, x_2 of journal center from static position

The additional forces are stiffness forces

$$\left.\begin{aligned} F_{S1} &= k_{11}x_1 + k_{12}x_2 \\ F_{S2} &= k_{21}x_1 + k_{22}x_2 \end{aligned}\right\} \qquad (2.22)$$

with stiffness coefficients $$k_{ik} = \frac{\partial F_i}{\partial x_k} \qquad (2.24)$$

and damping forces

$$\left.\begin{aligned} F_{D1} &= d_{11}\dot{x}_1 + d_{12}\dot{x}_2 \\ F_{D2} &= d_{21}\dot{x}_1 + d_{22}\dot{x}_2 \end{aligned}\right\} \qquad (2.25)$$

with damping coefficients $$d_{ik} = \frac{\partial F_i}{\partial \dot{x}_k} \quad . \qquad (2.26)$$

The coefficients may be expressed by

$$k_{ik} = \gamma_{ik}\frac{F_s}{\delta} \quad \text{and} \quad d_{ik} = \beta_{ik}\frac{F_s}{\delta\Omega} \quad , \qquad (2.27)$$

where the dimensionless terms γ_{ik}, β_{ik} depend on the bearing parameters and are functions of eccentricity ratio ε. For the short circular bearing they are derived in chapter 2.4.3.

2.4 Short Bearing Theory

Many papers are published on the solution of the Reynolds equation and the estimation of stiffness and damping coefficients. Exemplary see /12, 13/, a review is given by J.W. Lund in /14/. The computations are somewhat extensive and normally only numerical solutions are possible. For bearings with circular bore G.B. Dubois and F.W. Ocvirk showed in /15/, that in case of short bearings, L<<R, the derivate of film pressure in circumferential direction may be neglected. This means that in (2.11) the first term may be set to zero since the term $h^3\partial^2p/\partial z^2$ dominates. The computation of stiffness and damping coefficients by this theory is published in /15, 9, 10/ and is summarized in the following.

2.4.1 Equation of Film Pressure and of Resulting Force

With circular bore the gap h_0 is constant and equal to the clearance

$$h_0 = R-r = \delta,$$

whereby (2.2) becomes

$$h(\varphi,t) = \delta-e(t)\cos[\varphi-\gamma(t)] \quad . \tag{2.28}$$

Neglecting the first term in (2.11) leads with $\partial h_0/\partial\varphi=0$ to

$$\frac{d^2\bar{p}}{dz^2} = \frac{6\bar{\eta}}{h^3}[e(\Omega-2\dot{\gamma})\sin(\varphi-\gamma)-2\dot{e}\cos(\varphi-\gamma)] \quad , \tag{2.29}$$

by integration to

$$p(\varphi,z,t) = \frac{6\bar{\eta}}{h^3}[\,]\,\frac{z^2}{2} + C_1 z + C_2$$

and by the conditions $\partial p/\partial z=0$ at $z=0$ and $p=0$ at $z=\pm L/2$ to the film pressure function

$$p(\varphi,z,t) = \frac{3\bar{\eta}}{h^3}[e(\Omega-2\dot{\gamma})\sin(\varphi-\gamma)-2\dot{e}\cos(\varphi-\gamma)]\,(z^2-\frac{L^2}{4}) \quad . \tag{2.30}$$

Integration within the limits $-L/2\leqq z\leqq +L/2$ gives with (2.28) the film load per unit length

$$q(\varphi,t) = -\frac{\bar{\eta}L^3\Omega}{2\delta^2}\,\frac{\varepsilon(1-2\frac{\dot{\gamma}}{\Omega})\sin(\varphi-\gamma)-2\frac{\dot{\varepsilon}}{\Omega}\cos(\varphi-\gamma)}{[1-\varepsilon\cos(\varphi-\gamma)]^3} = -\frac{\bar{\eta}L^3\Omega}{2\delta^2}\,q'(\varphi,\dot{t}). \tag{2.31}$$

In static case, $\dot{\gamma}=0$, $\dot{\varepsilon}=0$, the load becomes

$$q(\varphi) = -\frac{\bar{\eta}L^3\Omega}{2\delta^2}\,\frac{\varepsilon\sin(\varphi-\gamma)}{[1-\varepsilon\cos(\varphi-\gamma)]^3} \quad . \tag{2.32}$$

This theoretical load (2.32) is plotted in figure 2.7, showing regions with positive and negative loads. In reality negative loads are impossible and therefore they will be assumed as zero.

With this assumption and with (2.12), (2.13), (2.14) and (2.31) the components of the resulting force are

$$F_1 = -K \int_{-\tau+\gamma}^{\gamma} q'(\varphi,t)\cos\varphi d\varphi \ , \ F_2 = -K \int_{-\tau+\gamma}^{\gamma} q'(\varphi,t)\sin\varphi d\varphi \tag{2.33}$$

with

$$K = \frac{\bar{\eta} L^3 R \Omega}{2\delta^2} \ . \tag{2.34}$$

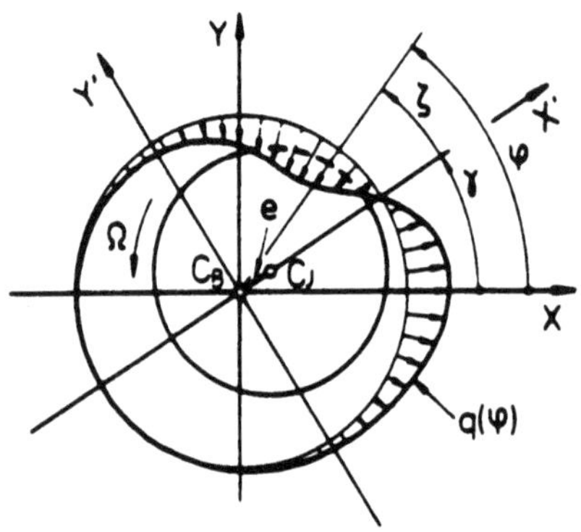

Fig. 2.7 Theoretical film load of short bearing

The equations (2.33) may be integrated in closed form as shown in /11/. With the dimensionless force

$$\bar{F}_i = \frac{F_i}{2RL} \frac{\psi^2}{\bar{\eta}\Omega} \frac{1}{\beta^2} \ ; \ i = 1,2,I,II \ , \tag{2.35}$$

where

$$\beta = \frac{L}{2R} \quad \text{is the width ratio} \ , \tag{2.36}$$

the components are

$$\left.\begin{aligned} \bar{F}_1 &= \bar{F}_I \cos\gamma - \bar{F}_{II} \sin\gamma \\ \bar{F}_2 &= \bar{F}_I \sin\gamma + \bar{F}_{II} \cos\gamma \ , \end{aligned}\right\} \tag{2.37}$$

where

$$\left.\begin{aligned} \bar{F}_I &= (1-2\frac{\dot{\gamma}}{\Omega}) \frac{2\varepsilon^2}{(1-\varepsilon^2)^2} + \tau \frac{\dot{\varepsilon}}{\Omega} \frac{1+2\varepsilon^2}{(1-\varepsilon^2)^{5/2}} \\ \bar{F}_{II} &= -\frac{\tau}{2}(1-2\frac{\dot{\gamma}}{\Omega}) \frac{\varepsilon}{(1-\varepsilon^2)^{3/2}} - \frac{\dot{\varepsilon}}{\Omega} \frac{4\varepsilon}{(1-\varepsilon^2)^2} \end{aligned}\right\} \tag{2.38}$$

are the components with respect to the X', Y' system. The latter may be splitted in a static part which depends on the eccentricity ratio ε and in a dynamic part which is a function of the velocities $\dot{\varepsilon}, \dot{\gamma}$ and also of ε:

$$\bar{F}_I(\varepsilon,\dot{\varepsilon},\dot{\gamma}) = \bar{F}'_I(\varepsilon) + \bar{F}''_I(\varepsilon,\dot{\varepsilon},\dot{\gamma})$$
$$\bar{F}_{II}(\varepsilon,\dot{\varepsilon},\dot{\gamma}) = \bar{F}'_{II}(\varepsilon) + \bar{F}''_{II}(\varepsilon,\dot{\varepsilon},\dot{\gamma}) \; . \qquad (2.39)$$

2.4.2 Static Load

The static case is determined by

$$\bar{F}_I = \bar{F}'_I = \frac{2\varepsilon^2}{(1-\varepsilon^2)^2} \; , \quad \bar{F}_{II} = \bar{F}'_{II} = -\frac{\pi}{2}\frac{\varepsilon}{(1-\varepsilon^2)^{3/2}} \; . \qquad (2.40)$$

Assuming the same force as in chapter 2.2 with the components

$$\bar{F}_1 = 0, \quad \bar{F}_2 = -\bar{F}_s \quad ,$$

leads with (2.35), (2.18) and (2.36) to

$$\bar{F}_s = \frac{So}{\beta^2} \qquad (2.41)$$

The angle γ is determined with the first equation (2.37) and $\bar{F}_1=0$ by

$$0 = \bar{F}'_I\cos\gamma - \bar{F}'_{II}\sin\gamma$$

and with (2.40) by

$$\tan\gamma = -\frac{4}{\pi}\frac{\varepsilon}{(1-\varepsilon^2)^{1/2}} \quad . \qquad (2.42)$$

Finally the attitude angle α results with (2.17) and (2.40) from

$$\tan\alpha = \frac{\pi}{4}\frac{(1-\varepsilon^2)^{1/2}}{\varepsilon} \quad . \qquad (2.43)$$

In our case the second equation (2.37) is

$$\bar{F}_2 = -\bar{F}_s = -\frac{So}{\beta^2} = \bar{F}'_I\sin\gamma + \bar{F}'_{II}\cos\gamma \quad , \qquad (2.44)$$

whereby $\sin\gamma$ and $\cos\gamma$ may be expressed by (2.42). This leads to the following expression for the Sommerfeld number

$$So = \frac{\pi}{2}\beta^2 \frac{\varepsilon}{(1-\varepsilon^2)^2}\sqrt{1+[(\frac{4}{\pi})^2-1]\varepsilon^2} \quad . \qquad (2.45)$$

Notice that the Sommerfeld number is proportional to the squared width ratio.

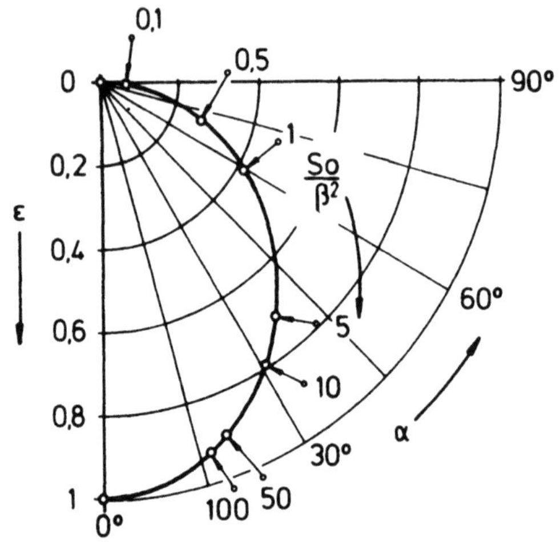

Fig. 2.8 Displacement orbit of short bearing

The displacement orbit of journal center is determined by (2.45), (2.43) and plotted in figure 2.8. For numerical evaluations $So/\beta^2(\varepsilon)$ and $\alpha(\varepsilon)$ are plotted in figure 2.9.

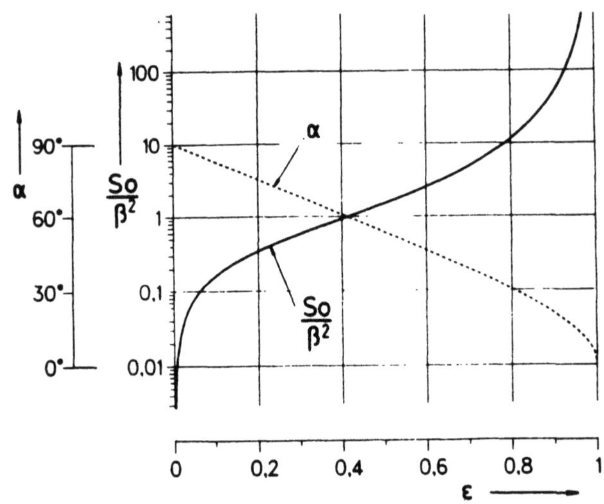

Fig. 2.9 Sommerfeld number and attitude angle of short circular bearing

2.4.3 Stiffness and Damping Coefficients

Stiffness coefficients are derivatives of the force components with respect to the displacements x_1 and x_2, (2.24). In static case the components are given by (2.37)

$$\left.\begin{aligned} F_1 &= K(\bar{F}_I'\cos\gamma - \bar{F}_{II}'\sin\gamma) \\ F_2 &= K(\bar{F}_I'\sin\gamma + \bar{F}_{II}'\cos\gamma) \ , \end{aligned}\right\} \quad (2.46)$$

considering (2.34), (2.35) and (2.40). F_1 and F_2 are functions of ε and γ, hence

$$\frac{\partial F_i}{\partial x_k} = \frac{\partial F_i}{\partial \varepsilon}\frac{\partial \varepsilon}{\partial x_k} + \frac{\partial F_i}{\partial \gamma}\frac{\partial \gamma}{\partial x_k} \ . \quad (2.47)$$

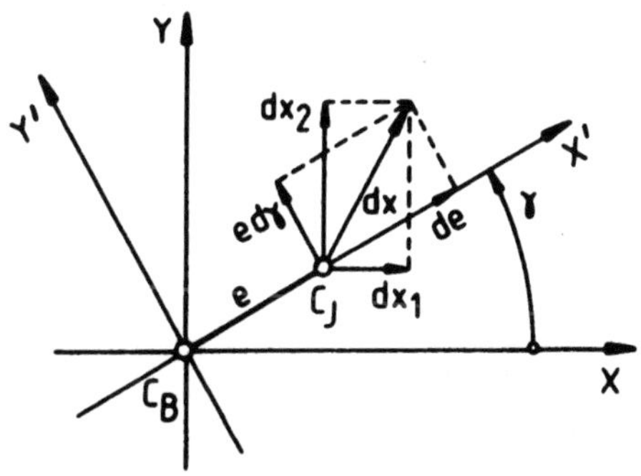

Fig. 2.10 Coordinates of small displacement dx of journal center

Referring to figure 2.10

$$\left.\begin{aligned} \frac{\partial \varepsilon}{\partial x_1} &= \frac{1}{\delta}\cos\gamma \ , &\quad \frac{\partial \varepsilon}{\partial x_2} &= \frac{1}{\delta}\sin\gamma \\ \frac{\partial \gamma}{\partial x_1} &= \frac{-1}{e}\sin\gamma \ , &\quad \frac{\partial \gamma}{\partial x_2} &= \frac{1}{e}\cos\gamma \end{aligned}\right\} \quad (2.48)$$

leads finally to

$$k_{ik} = \frac{\beta^2}{So} f_{ik}(\varepsilon)\frac{F_s}{\delta} = \gamma_{ik}\frac{F_s}{\delta}$$

with

$$\gamma_{ik} = \frac{\beta^2}{So} f_{ik}(\varepsilon) \quad (2.49)$$

and

$$\left.\begin{aligned} f_{11}(\varepsilon) &= \frac{2\varepsilon}{(1-\varepsilon^2)^2}[1+a(\varepsilon)] \\ f_{12}(\varepsilon) &= \frac{\pi}{2}\frac{1}{(1-\varepsilon^2)^{5/2}}\{1-[(\tfrac{4}{\pi})^2 a(\varepsilon)+1]\varepsilon^2\} \\ f_{21}(\varepsilon) &= -\frac{\pi}{2}\frac{1}{(1-\varepsilon^2)^{5/2}}\{1+[(\tfrac{4}{\pi})^2 a(\varepsilon)+2]\varepsilon^2\} \\ f_{22}(\varepsilon) &= \frac{2\varepsilon}{(1-\varepsilon^2)^3}\{1+[(\tfrac{4}{\pi})^2 a(\varepsilon)+2]\varepsilon^2\} \\ \text{where} \quad a(\varepsilon) &= \frac{1}{1+[(\tfrac{4}{\pi})^2-1]\varepsilon^2} \ . \end{aligned}\right\} \quad (2.50)$$

Figure 2.11 shows the dimensionless stiffness coefficients γ_{ik} as a function of So/β^2 or ε.

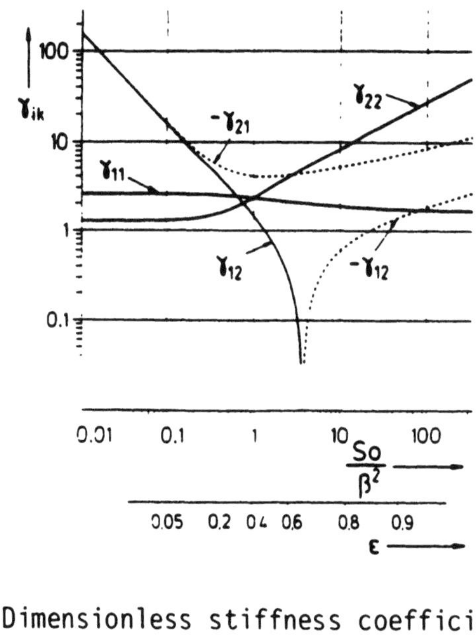

Fig. 2.11 Dimensionless stiffness coefficients of short bearing

Damping coefficients are derivatives of the force components with respect to the velocities $\dot{x}_1$, $\dot{x}_2$ (2.26). The derivatives of the static terms $\bar{F}'_I$, $\bar{F}'_{II}$ are zero, therefore the components F_1, F_2 are given by (2.46) if $\bar{F}'_I$, $\bar{F}'_{II}$ are replaced by the dynamic terms $\bar{F}''_I$, $\bar{F}''_{II}$ (see (2.39), (2.38)). The derivatives are given by

$$\frac{\partial F_i}{\partial \dot{x}_k} = \frac{\partial F_i}{\partial \dot{\varepsilon}} \frac{\partial \dot{\varepsilon}}{\partial \dot{x}_k} + \frac{\partial F_i}{\partial \dot{\gamma}} \frac{\partial \dot{\gamma}}{\partial \dot{x}_k} \quad ,$$

which leads in similar way as before to the damping coefficients

$$d_{ik} = \frac{\beta^2}{So} g_{ik}(\varepsilon) \frac{F_s}{\delta\Omega} = \beta_{ik} \frac{F_s}{\delta\Omega}$$

with

$$\beta_{ik} = \frac{\beta^2}{So} g_{ik}(\varepsilon) \tag{2.51}$$

and

$$\left.\begin{aligned} g_{11}(\varepsilon) &= \frac{\pi}{(1-\varepsilon^2)^{3/2}}\{1-[2(\tfrac{4}{\pi})^2-3]\varepsilon^2 a(\varepsilon)\} \\ g_{12}(\varepsilon) &= g_{21}(\varepsilon) = -\frac{4\varepsilon}{(1-\varepsilon^2)^2}\{1-[2(\tfrac{4}{\pi})^2-3]\varepsilon^2 a(\varepsilon)\} \\ g_{22}(\varepsilon) &= \frac{\pi}{(1-\varepsilon^2)^{3/2}}[1+(\tfrac{4}{\pi})^2 \frac{2+\varepsilon^2}{1-\varepsilon^2}\varepsilon^2 a(\varepsilon)] \quad , \end{aligned}\right\} \tag{2.52}$$

with $a(\varepsilon)$ from (2.50).

The dimensionless damping coefficients are shown in figure 2.12.

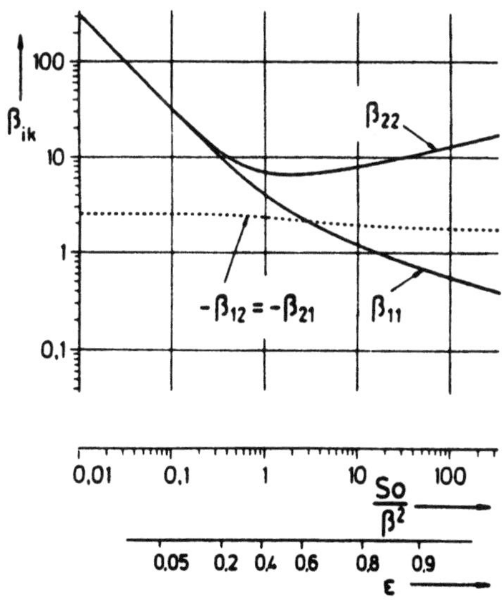

Fig. 2.12 Dimensionless damping coefficients of short bearing

2.4.4 Journal Orbit as a Result of Rotating Force

We consider a static load F_S and an additional load F_a which rotates slowly in XY-plane. The angular velocity of F_a is assumed to be small enough to neglect dynamic forces. For $F_a \ll F_s$ the displacement of journal center is determined by inversion of (2.22)

$$\left.\begin{aligned} x_1 &= h_{11}\,F_{S1} + h_{12}\,F_{S2} \\ x_2 &= h_{21}\,F_{S1} + h_{22}\,F_{S2} \end{aligned}\right\} \quad (2.53)$$

with $\quad [h_{ik}] = [k_{ik}]^{-1}$.

The components of the additional force are

$$F_{S1} = F_a \sin\omega t \quad , \quad F_{S2} = -F_a \cos\omega t \quad (\omega \to 0)$$

hence the orbit of C_J is elliptical.

As mentioned before equation (2.53) is limited to small forces F_a. Also for arbitrary large forces F_a a simple determination of the journal orbit is possible as explained in the following. The relation between static force and eccentricity is given by (2.45), and the attitude angle by (2.43). Therefore the position of C_J is given in case of $F_a=0$ by figure 2.13a and in case of $F_a>0, \omega t>0$ by figure 2.13b. The orbit is determined by $0 \leqq \omega t < 2\pi$.

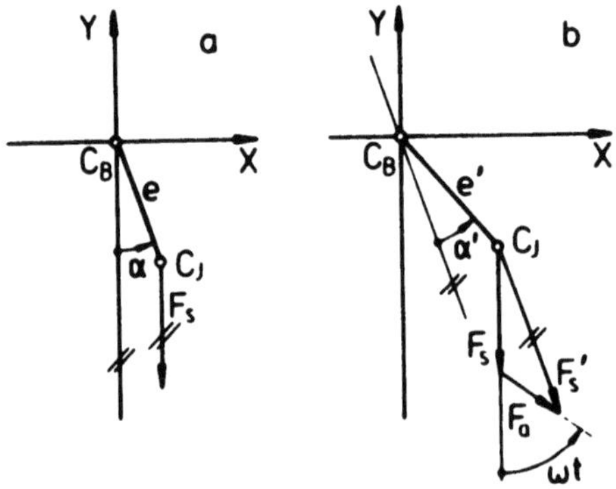

Fig. 2.13 Determination of orbit for arbitrary F_a/F_s

Figure 2.14 shows orbits for $F_a=0{,}5\ F_s$ at different eccentricities (ε=0,3; 0,6; 0,9). The solid lines show the exact solutions, the broken lines show the linearized one. The arrows indicate the direction of F_a. It may be stated that up to $F_a \approx 0{,}5\ F_s$ the linear solution is rather good.

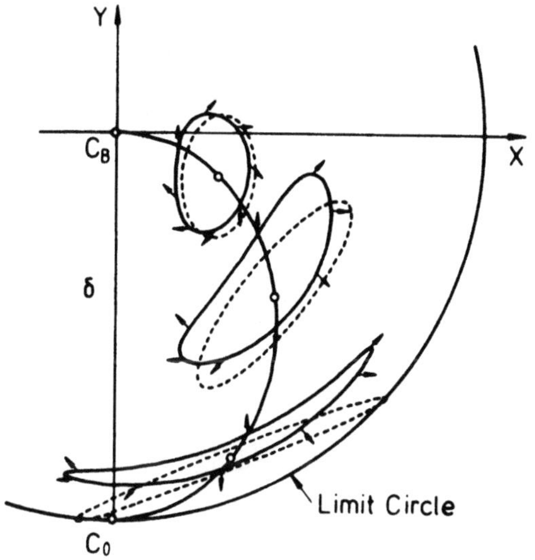

Fig. 2.14 Journal orbits in case of slowly rotating additional force $F_a = 0{,}5\ F_s$.
—— exact, ----- linearized

The figure shows that the orbit is nearly circular for small eccentricities and shaped like a banana for large eccentricities. For small eccentricities the bearing behaves nearly isotropic, but the displacements are nearly vertical to the additional force. At high eccentricities the bearing is stiff in radial direction and rather elastic in tangential direction. Roughly the stiffness is in the order of F_s/δ.

2.5 Properties of Special Bearing Types

The theory given in chapters 2.1 to 2.3 is not only valid for circular bore but also for other cross-sections. Furthermore partial film areas are possible. Special bearing types with different geometry and also with flexible pads are used with respect to stability, load capacity, fricton losses and other demands. Some types are plotted schematically in figures 2.15 a ... g.

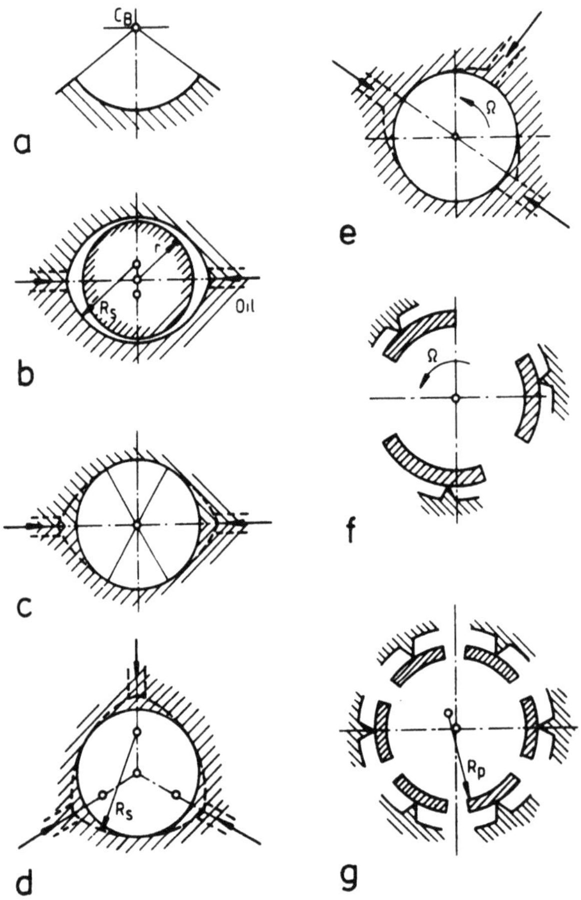

Fig. 2.15 Some special bearing types

a Short arc bearing

b Lemon bearing

c Pocket bearing

d Three-lobed bearing

e Special three-lobed bearing

f Tilting-pad bearing

g Tilting-pad bearing for vertical shaft

For bearings with arbitrary cross section and cylindrical bore T. Someya gives in /13/ a numerical solution for the static displacement orbit. J. Glienicke describes in /16,17 / results of comprehensive measurements for evaluation of stiffness and damping coefficients for several bearing types. The displacement orbit of a special three-pad-bearing is published by H.H. Ott /18/. Tilting-pad-bearings are investigated by Z.E. Varga /19/, H.J. Merker /20/ and H. Springer /21/. In /22/ E. Pollmann considers the influence of variable viscosity. These references are only some of the large number of published papers. More references are given in /14/.

References

/1/ Chandra, M., M. Malik and R. Sinhasan: Gas Bearings. Part 1: Dynamic Analysis and Solution Method. Wear 88 (3) pp 255-268.

/2/ Brändlein, J.: Die radiale Federsteifigkeit von Wälzlagern: Linearisierung der Kennlinien. Masch.-Markt 83 (1977) Nr. 61, S. 1182/84.

/3/ Klumpers, K.: Untersuchung des Dämpfungsverhaltens von Wälzlagern. FVA-Informationstagung Bad Homburg, Nov. 1977, 32 S. Forsch.vereinig. Antriebstechnik e.V., Cornelíusstr. 4, 6000 Frankfurt/M.

/4/ Klepzig, W.: Experimentelle Ermittlung der Dämpfungskonstante eines doppelreihigen Zylinderrollenlagers. Masch.-Bau-Techn. 26 (1977) Nr. 3, S. 112/15.

/5/ Voll, H.: Wälzgelagerte Spindeleinheiten: Einfluß der Lagergröße auf die statische Steifigkeit. Masch.-Markt Bd. 86 (1980) Nr. 29, S. 572/73.

/6/ Tamura, H., Tsuda, Y.: On the Spring Characteristics of a Ball Bearing. Bulletin of JSME, Vol. 23, No. 180, 1980.

/7/ Weck, M. u. L. Ophey: Experimentelle Ermittlung der Steifigkeit und Dämpfung radial belasteter Wälzlager. Ind. Anz. Bd. 103 (1981) Nr. 79, S. 32/35.

/8/ Fukata, S., E.H. Gad, T. Kondou, T. Ayage: On the Radial Vibration of Ball Bearings (Computer Simulation). Bull. JSME 28 (239), pp 899-904 (1985).

/9/ Smith, D.M.: Journal Bearings in Turbomachinery, Great Britain: Chapman and Hall, 1969.

/10/ Cameron, A.: Basic Lubrication Theory, London: Longman, 1971.

/11/ Lang, O.R., Steinhilper, W.: Gleitlager. Berlin, Heidelberg, New York: Springer 1978.

/12/ Someya, T.: Stabilität einer in zylindrischen Gleitlagern laufenden, unwuchtfreien Welle. Diss. TU Karlsruhe, 1962.

/13/ Someya, T.: Das dynamisch belastete Radial-Gleitlager beliebigen Querschnitts. Ing.-Arch. Band 34, Heft 1, S. 7-16, Berlin, Heidelberg, New York: Springer 1965.

/14/ Lund, J.W.: Evaluation of Stiffness and Damping Coefficients for Fluid-Film Bearings, Shock and Vibration Digest, 1979.

/15/ Dubois, G.B., Ocvirk, F.W.: Analytical Derivation and Experimental Evaluation of Short-Bearing Approximation for Full Journal Bearings. Report 1157, Cornell University, 1953

/16/ Glienicke, J.: Feder- und Dämpfungskonstanten von Gleitlagern für Turbomaschinen und deren Einfluß auf das Schwingungsverhalten eines einfachen Rotors. Diss. TU Karlsruhe, 1966.

/17/ Glienicke, J.: Experimental Investigation of the Stiffness and Damping Coefficients of Turbine Bearings and Their Application to Instability Prediction. Symposium in Nottingham. London: The Institution of Mechanical Engineers, 1966.

/18/ Ott, H.H.: Berechnung von Wellenlage und Reibung im Dreikeil-Traglager. Brown, Boveri Mitt. Bd. 46 (1959) Nr. 7, S. 395-406.

/19/ Varga, Z.E.: Wellenbewegung, Reibung und Öldurchsatz beim segmentierten Radialgleitlager von beliebiger Spaltform unter konstanter und zeitlich veränderlicher Belastung. Diss. ETH Zürich, 1971.

/20/ Merker, H.-J.: Über den nichtlinearen Einfluß von Gleitlagern auf die Schwingungen von Rotoren, Fortschrittsber. VDI Z, Reihe 11, Nr. 40 (1981).

/21/ Springer, H.: Stiffness and Damping Characteristics of Tilting-Pad Journal Bearings. CISM Courses and Lectures Nr. 273. Wien, New York: Springer Verlag, 1984.

/22/ Pollmann, E.: Das Mehrgleitflächenlager unter Berücksichtigung der veränderlichen Ölviskosität. Konstruktion 21 (1969) H. 3, S. 85-97.

CHAPTER 3

ANALYSIS OF ROTORS IN BEARINGS

E. Krämer
Technische Hochschule Darmstadt, Darmstadt, FRG

ABSTRACT

This part begins with the main facts about the simplest rotor model, the Jeffcott rotor. A short comment follows about the approximate computation of the first critical speed. Then the start-up and shut-down behaviour of a rotor is described. The influence of bearings also is studied using the Jeffcott model. A simple model shows the influence of bearing damping. In the last section the main aspects of unbalance behaviour of multi-bearing systems are discussed.

In general two kinds of rotors occur in practice, rotors for turbomachines and those for reciprocating machines. Both types vibrate in a rotatory and translatory manner. In this chapter translatory vibrations of turbomachine rotors are considered.

The aim of the designer is to develop rotors which operate safely during their whole lifetime (several decades), and do so with moderate vibrations. Many investigations about this problem have been published since the beginning of the century. A short review of several results is presented in the following.

3.1 Principles of Rotordynamic Analysis

Real rotors are rather complicated systems: their geometry is not simple, mostly they are supported static undetermined on more than two bearings. The oil film, the pedestals and the foundation may influence their dynamic behaviour. In spite of this many pheno mena may be illustrated by the simple rotor as described in the following chapter.

3.1.1 Jeffcott Rotor

The cross-section of a real rotor on two bearings is given in figure 3.1. The vibrations of such a rotor may be studied with good approximation by the model of figure 3.2. Investigations abou such a model, called Jeffcott rotor or Laval shaft, are given in /1, ..., 6/ and many other publications.

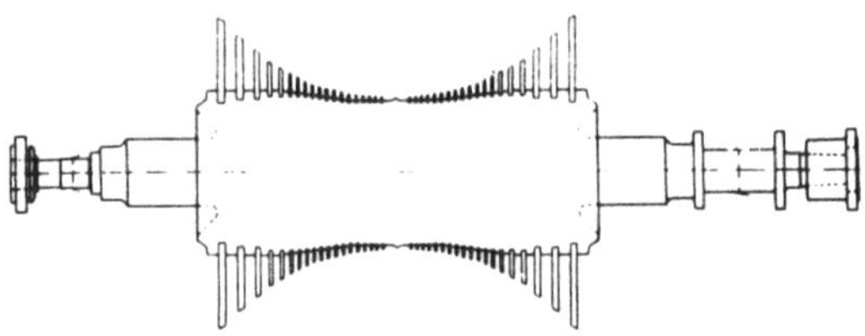

Fig. 3.1 Example of a real rotor

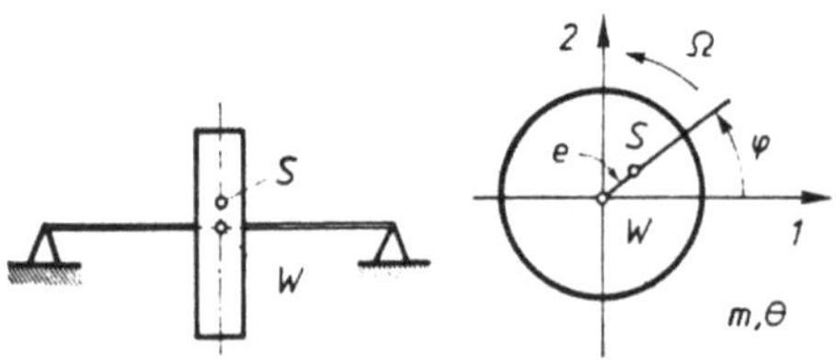

Fig. 3.2 Jeffcott Rotor Laval Shaft

The properties of this rotor are:

- massless shaft, symmetric to mid-span, stiffness k,
- rigid circular disc, mass m, polar mass moment Θ,
- rigid bearings,
- viscous damping d of translatory disc displacement.

W shaft center, displacements y_1, y_2
S mass center, displacements z_1, z_2.

With eccentricity e and angle φ

$$\left.\begin{aligned} z_1 &= y_1 + e\cos\varphi \\ z_2 &= y_2 + e\sin\varphi \end{aligned}\right\} \quad , \tag{3.1}$$

hence the system has three degrees of freedom.

By Newtons law the equations of motion are

$$\left.\begin{aligned} m\ddot{z}_1 &= -ky_1 - d\dot{y}_1 \\ m\ddot{z}_2 &= -ky_2 - d\dot{y}_2 - mg \\ \Theta\ddot{\varphi} &= -(ky_1 + d\dot{y}_1)e\sin\varphi - (ky_2 + d\dot{y}_2)e\cos\varphi + M_T \end{aligned}\right\} \tag{3.2}$$

where M_T denotes external torque.

Normally the radius of inertia $i=\sqrt{\Theta/m}$ is large compared to eccentricity e, therefore

$$\varphi = \Omega t \tag{3.3}$$

in case of external torque $M_T=0$ is a good approximation. With these assumptions and (3.1) the two equations

$$\left.\begin{aligned} m\ddot{y}_1 + d\dot{y}_1 + ky_1 &= me\Omega^2\cos\Omega t \\ m\ddot{y}_2 + d\dot{y}_2 + ky_2 &= me\Omega^2\sin\Omega t - mg \end{aligned}\right\} \tag{3.4}$$

for the displacements of shaft center remain. They are uncoupled, being the equations of a simple oscillator.

The homogenous equation

$$m\,\ddot{y}_i + d\dot{y}_i + ky_i = 0, \qquad i=1,\ 2 \tag{3.5}$$

has the solution

$$y_i = C_i e^{-\delta t}\cos(\omega_d t - \psi_i) \tag{3.6}$$

where $\omega_d = \sqrt{\omega_k^2 - \delta^2}$ (3.7) damped natural frequency

$\omega_k = \sqrt{\frac{k}{m}}$ (3.8) undamped natural frequency

$\delta = \frac{d}{2m}$ quantity of damping

C_i, ψ_i constants, determined by the initial conditions

By (3.6) the natural motion of the shaft is determined. This means that the shaft center describes a decreasing elliptical orbit (figure 3.3).

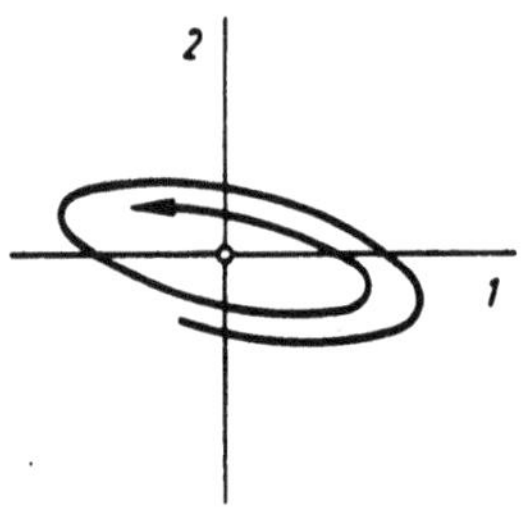

Fig. 3.3 Natural motion of shaft center

The inhomogenous equations (3.4) are valid for unbalance excitation of the weight-loaded rotor. Further external loads $F_1(t)$, $F_2(t)$ acting on the disc in directions 1 and 2 may be considered by adding them on the right hand side of (3.4). For the given case the steady state solution of (3.4) is

$$\left.\begin{aligned} y_1 &= \hat{y}\cos(\Omega t - \varepsilon) \\ y_2 &= \hat{y}\sin(\Omega t - \varepsilon) - y_0 \end{aligned}\right\} \quad (3.9)$$

where

$$\left.\begin{aligned} \hat{y} &= e\eta^2[(1-\eta^2)^2 + (2D\eta)^2]^{-1/2} = eV \\ \tan\varepsilon &= \frac{2D\eta}{1-\eta^2} \\ y_0 &= \frac{mg}{k} \quad \text{static displacement} \end{aligned}\right\} \quad (3.10)$$

$$\left.\begin{aligned} \eta &= \frac{\Omega}{\omega_k} \\ D &= \frac{d}{2\sqrt{km}} \quad \text{damping factor.} \end{aligned}\right\} \quad (3.11)$$

According to (3.9) for constant angular velocity Ω the shaft center describes a circular orbit with center $(0, -y_0)$ and with constant angle ε (figure 3.4). The relative amplitude $\hat{y}/e$ is given by the wellknown amplification function V [(3.10), figure 3.5)]. It shows that the amplitude $\hat{y}$ is maximum at $\Omega \approx \omega_k$. The corresponding speed

$$n_{cr} = \Omega_{cr}/2\pi \approx \omega_k/2\pi \quad (3.12)$$

is called critical speed.

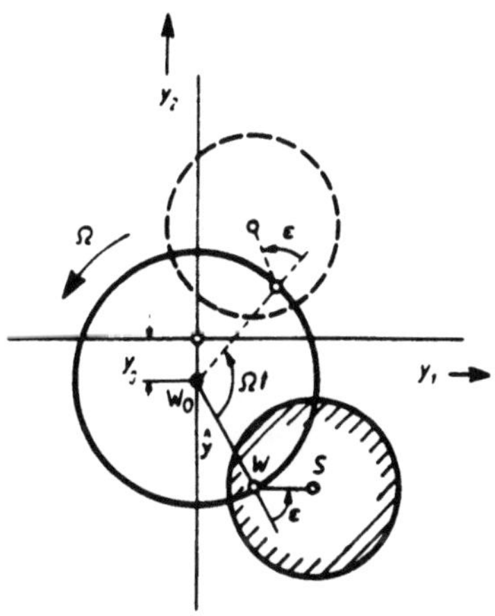

Fig. 3.4 Orbit of shaft center in case of unbalance excitation

Further information is given by the polar plot 3.6, which shows the position of shaft and mass center W, S at different relative velocities $\eta=\Omega/\omega_k$ and at t=0. In case of an undamped rotor two configurations are possible (figure 3.7), the mass center is outside ($\Omega<\omega_k$) or it is inside ($\Omega>\omega_k$) of the orbit. At $\Omega=\infty$ the mass center coincides with W_0, which is the static position of shaft center W (self-centering).

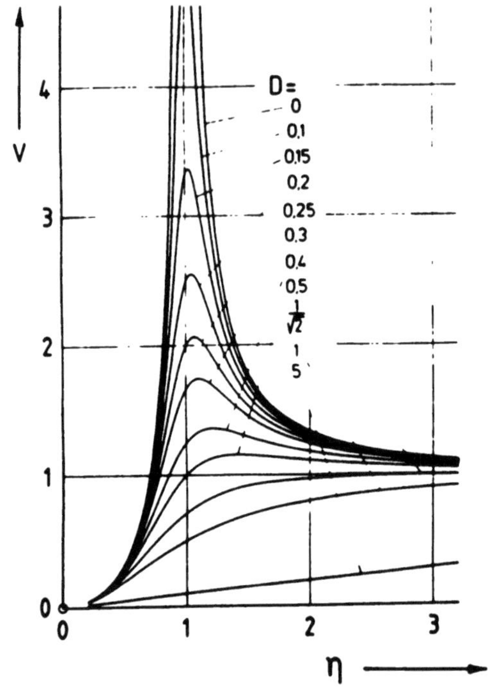

Fig. 3.5
Amplification function

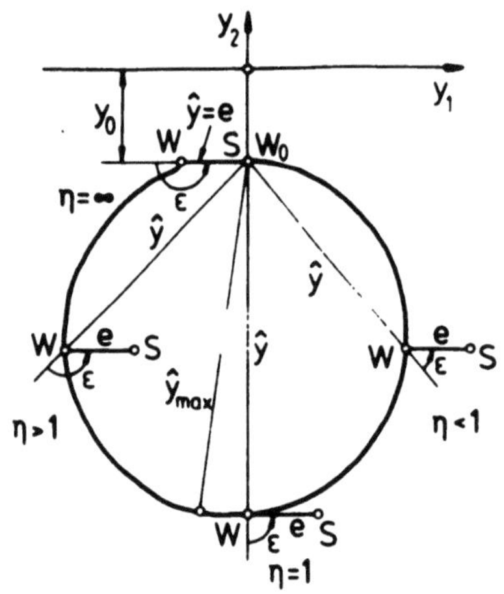

Fig. 3.6
Position of W and S for different angular velocities at t=0

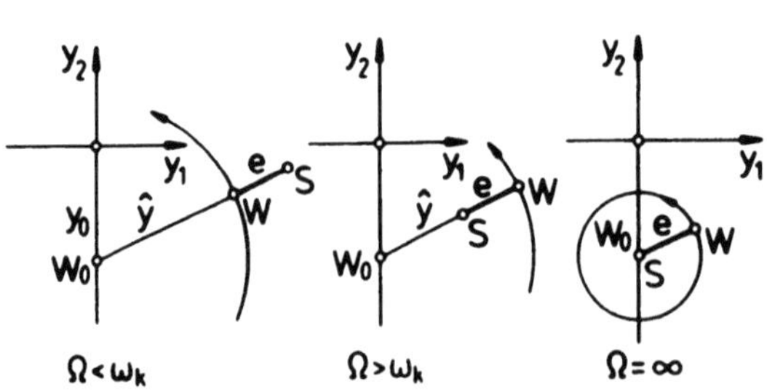

Fig. 3.7
Undamped rotor.
Possible configurations

3.1.2 Real Rotors with two Bearings

The first critical speed of real rotors, as shown for instance in figure 3.1, may be computed without using a computer accurately enough in the following manner.

As first approximation the critical speed for the Jeffcott rotor is according to (3.12)

$$n_{cr} \approx \frac{1}{2\pi}\sqrt{\frac{k}{m}}$$

and by (3.10) $$k = \frac{mg}{y_0} \quad ,$$

therefore $$n_{cr} \approx \frac{1}{2\pi}\sqrt{\frac{g}{y_0}} \quad , \qquad (3.13)$$

where g is the constant of gravity and y_0 the static displacement at mid-span, caused by rotor weight mg. The first natural frequency of a uniform bar supported in the same manner as the Jeffcott rotor is

$$f_1 = 1,57 \frac{1}{\ell^2}\sqrt{\frac{EI}{\rho A}} \quad ,$$

and its static displacement is

$$y_0 = \frac{5}{384}\frac{\rho A \ell^4 g}{EI} \quad ,$$

which leads with $n_{cr} \approx f_1$ finally to

$$n_{cr} \approx \frac{1,13}{2\pi}\sqrt{\frac{g}{y_0}} \qquad (3.14)$$

The geometry of real rotors lies between these two models. Therefore their first critical speed is limited by

$$0,159\sqrt{\frac{g}{y_0}} < n_{cr} < 0,179\sqrt{\frac{g}{y_0}} \quad . \qquad (3.15)$$

If y_0 is known, n_{cr} may be estimated accurately enough by (3.15).

Another approximation is given by the equation T+U=constant for undamped systems, which leads to

$$U_{max} = T_{max} \quad . \tag{3.16}$$

Assuming the first natural mode to be approximately equal to the static mode, given by displacements y_i, then the maximum potential energy

$$U_{max} = \frac{1}{2} g \Sigma m_i y_i \quad ,$$

where m_i are masses of the rotor concentrated at points i. The kinetic energy

$$T_{max} = \frac{1}{2}\Sigma m_i \dot{y}_i^2 = \frac{1}{2}\omega_n^2 \Sigma m_i y_i^2 \quad ,$$

where ω_n is the unknown natural frequency. Equation (3.16) leads to

$$\omega_n^2 = \frac{g}{y_o'} \; , \quad \text{where} \quad y_o' = \frac{\Sigma m_i y_i^2}{\Sigma m_i y_i} \tag{3.17}$$

and with $\omega_n \approx 2\pi n_{cr}$ and (3.13) finally to

$$n_{cr} \approx \frac{1}{2\pi}\sqrt{\frac{g}{y_o'}} \quad . \tag{3.18}$$

Usage of this equation needs only the static displacements y_i caused by the loads $m_i g$ at several points along the rotor axis. This procedure is also relevant to rotors with overhang, if the loads at overhang are assumed opposite to those between the bearings.

3.1.3 Acceleration through Critical Speed

In chapter 3.1.1 constant angular velocity of the rotor is assumed. During start-up and shut-down the angular velocity increases or decreases with time. Investigations about this problem are published in /7, ..., 12/ and in further papers. In the following the main results for simple models are described.

The typical behaviour during acceleration is shown in figure 3.8, which was computed acc. to /8/ by J. Greb in /13/. It shows the displacement as function of time for a damped single degree of freedom oscillator which is excited by the force

$$F(t) = \hat{F}\sin(\tfrac{1}{2}\alpha t^2) = \hat{F}\sin[\omega(t)t] \quad , \qquad (3.19)$$

where $\omega(t) = \frac{1}{2}\alpha t$, which means that the exciting frequency grows linearly. The figure shows that the envelope of amplitudes is slightly undulated.

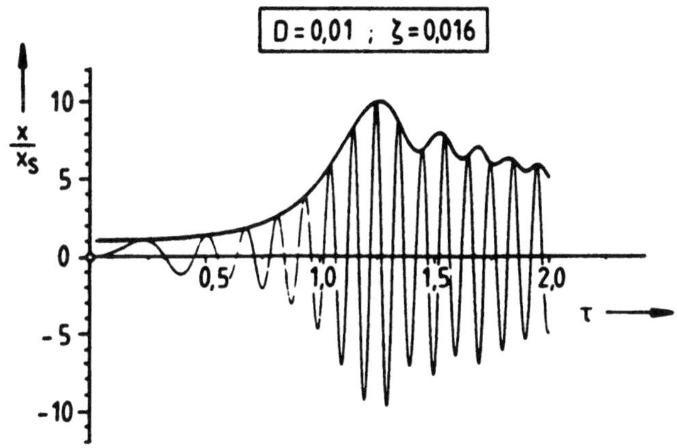

Fig. 3.8 Displacements of an oscillator, exited harmonically with linear increasing frequency.

$x_s = \frac{\hat{F}}{k}$ static displacement

$\tau = \frac{\alpha t}{\omega_k}$ dimensionless time

D damping factor

$\zeta = \frac{\alpha}{\omega_k^2}$ relative acceleration

$\omega_k = \sqrt{\frac{k}{m}}$ natural frequency

A Jeffcott rotor behaves during start-up similar to an accelerated oscillator. Figure 3.9 shows for example the envelope of amplitudes of shaft center for different values of acceleration.

With increasing acceleration the peaks decrease and are shifted to later times. During shut-down (figure 3.10) the peaks decrease with higher decelerations, they are also shifted to later times (shut-down: $\tau=10\to0$).

The figures show that the effect begins to be appreciable at $|\zeta|> 5\cdot10^{-4}$. To make this limit clear we define a starting time as follows. With $\tau=\alpha t/\omega_k$ and $\zeta=\alpha/\omega_k^2$ the real time is

$$t = \frac{\tau}{\zeta\omega_k} \quad .$$

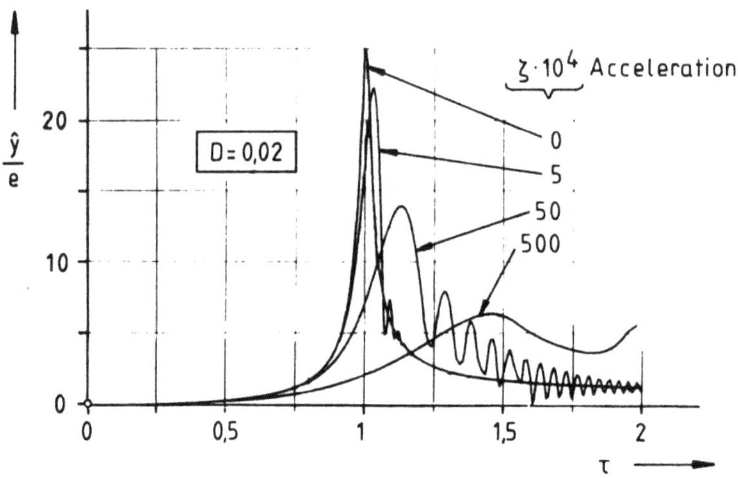

Fig. 3.9 Amplitudes of Jeffcott rotor during start-up

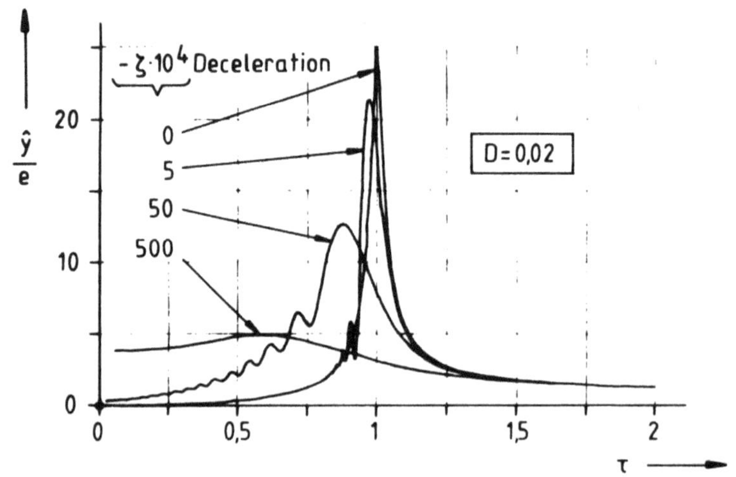

Fig. 3.10 Amplitudes of Jeffcott rotor during shut-down

For small accelerations the main peak is reached at $\tau \approx 1$, as the figures show. Therefore $\tau=1$ may determine the starting time

$$t_s = t(\tau=1) = \frac{1}{\zeta\omega_k} \quad . \tag{3.20}$$

By this definition the acceleration

$$\zeta = \frac{1}{t_s\omega_k} \quad ,$$

which is plotted in figure 3.11, where $n_{cr}=\omega_k/2\pi$ is the critical speed. The figure shows that $|\zeta| > 5\cdot10^{-4}$ means extremely short starting or stopping times.

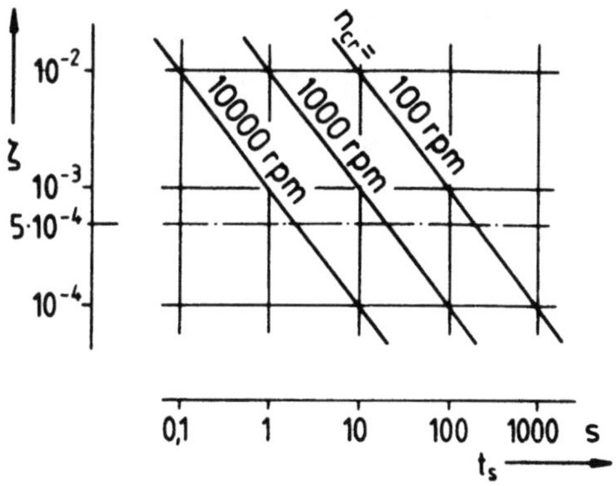

Fig. 3.11
Relative acceleration as function of starting time and critical speed

3.2 Theory of Rotor Bearing Systems

In chapter 3.1 only simple rotors supported on two rigid bearings are considered. In reality rotors are more complicated. They have more than two bearings, pedestals and a foundation. Sometimes the base also should be mentioned. This means that the two rigid bearings must be replaced by a more or less complicated dynamic system. This will be done step by step in the following chapters.

3.2.1 Jeffcott Rotor with Flexible Bearings

Firstly the model of figure 3.2 will be replaced by that of figure 3.12 which has two flexible bearings with stiffness k_{B1}, k_{B2} in horizontal and vertical direction. Denoting the stiffness of

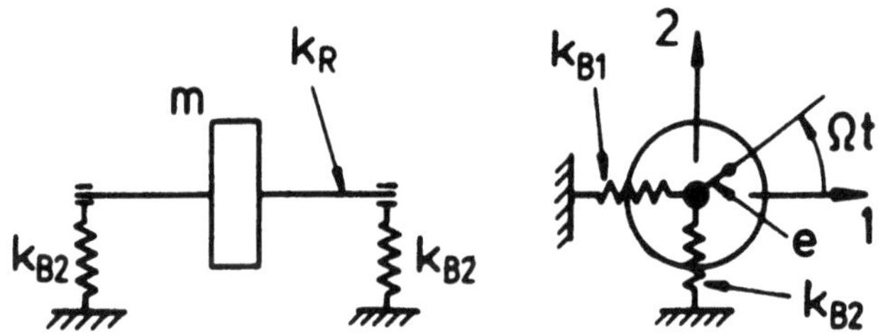

Fig. 3.12 Jeffcott Rotor with flexible bearings

rigidly supported rotor by k_R the new system has the stiffnesses

$$k_i = \frac{2k_R k_{Bi}}{k_R+2k_{Bi}} \quad , \qquad i = 1, 2 \quad . \tag{3.21}$$

The equations of motion may be obtained by substituting k_1 for k in the first and k_2 for k in the second equation of (3.4). The system has two undamped natural frequencies

$$\omega_i = \sqrt{\frac{k_i}{m}}$$

and two damped natural frequencies

$$\omega_{di} = \sqrt{\omega_i^2-\delta^2} \quad , \qquad i = 1, 2 \quad , \tag{3.22}$$

where δ=d/2m acc. to (3.7).

The particular solutions of the inhomogenous equations are

$$\left.\begin{aligned} y_1 &= \hat{y}_1 \cos(\Omega t-\varepsilon_1) \\ y_2 &= \hat{y}_2 \sin(\Omega t-\varepsilon_2)-y_{o2} \end{aligned}\right\} \tag{3.23}$$

where

$$\left.\begin{aligned} \hat{y}_i &= e\eta_i^2[(1-\eta_i^2)^2 + (2D_i\eta_i)^2]^{-1/2} \\ \tan\varepsilon_i &= \frac{2D_i\eta_i}{1-\eta_i^2} \\ y_{o2} &= \frac{mg}{k_2} \\ \eta_i &= \frac{\Omega}{\omega_i} \qquad D_i = \frac{d}{2\sqrt{k_i m}} \quad , \quad i = 1, 2 \ . \end{aligned}\right\} \tag{3.24}$$

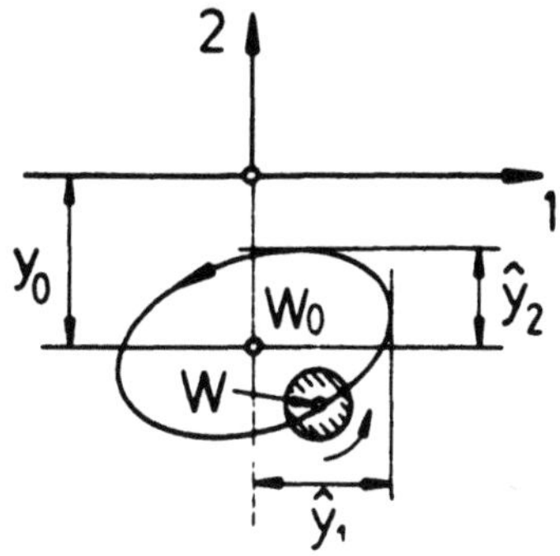

Fig. 3.13 Orbit of shaft center in case of unbalance excitation. Rotor with unequal bearing stiffnesses

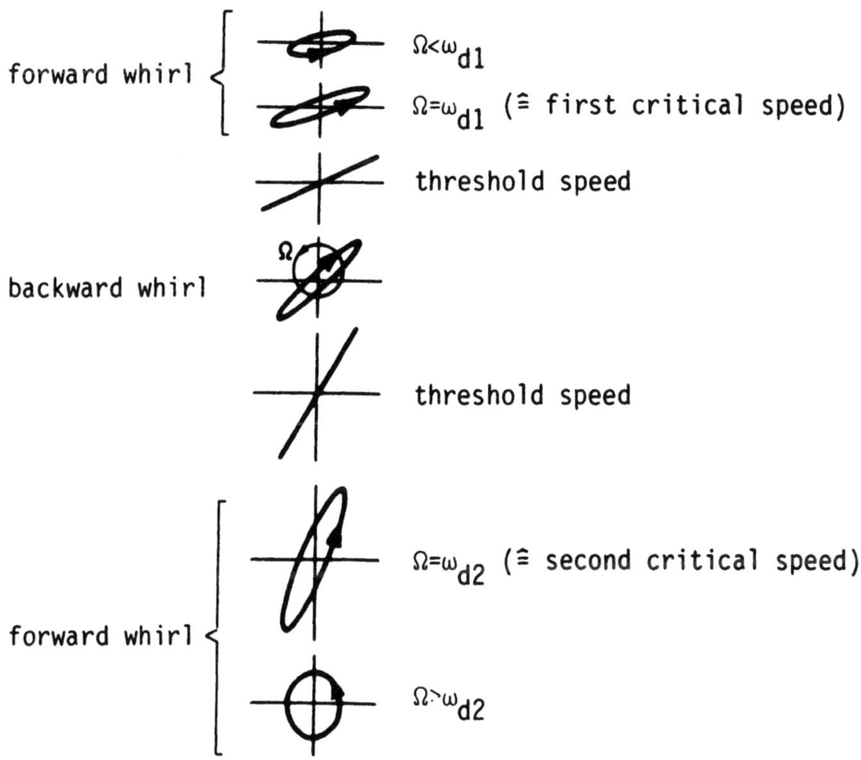

Fig. 3.14 Orbits of a rotor with unequal bearing stiffnesses at different angular velocities

The equations (3.23) describe the orbit of shaft center at unbalance excitation (figure 3.13). It is an ellipse (circle) if $k_{B2} \neq k_{B1}$ ($k_{B2} = k_{B1}$). The shaft center may whirl in the same direction as the rotor rotates (forward whirl) or in opposite direction (backward whirl). Figure 3.14 shows the typical behaviour at different angular velocities. The rotor whirls forward at low and at high velocities and it whirls backward in a region between ω_{d1} and ω_{d2}.

3.2.2 Remarks about Bearing Damping

In the previous chapter external damping at the disc is assumed. However rotors usually are damped at their bearings. To study this we consider a system shown in figure 3.15. Only displacements in one direction and damping acting only at the bearings are assumed.

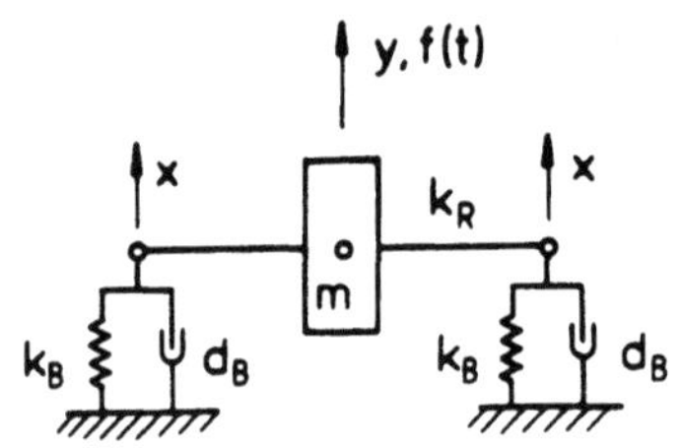

Fig. 3.15 Rotormodel with bearing damping

With the symbols of the figure the equations of motion are

$$\left.\begin{aligned} m\ddot{y} + k_R(y-x) &= f(t) \\ k_R(x-y) + 2(d_B\dot{x} + k_B x) &= 0 \quad . \end{aligned}\right\} \quad (3.24)$$

The homogenous system leads to the characteristic equation

$$\bar{d}_B\bar{\lambda}^3 + (0{,}5+\kappa)\bar{\lambda}^2 + \bar{d}_B\bar{\lambda} + \kappa = 0 \quad , \qquad (3.25)$$

where

$$\left.\begin{aligned} \bar{\lambda} &= \frac{\lambda}{\omega_R} && \text{related eigenvalue} \\ \omega_R &= \sqrt{\frac{k_R}{m}} && \text{natural frequency of rigidly supported rotor} \\ \bar{d}_B &= \frac{d_B}{\sqrt{k_R m}} && \text{related bearing damping} \\ \kappa &= \frac{k_B}{k_R} && \text{stiffness ratio.} \end{aligned}\right\} \quad (3.26)$$

Normally (3.25) has the roots (eigenvalues)

$$\bar{\lambda}_1, \bar{\lambda}_2 = \bar{\alpha}_1 \pm i\bar{\omega}_1', \qquad \bar{\lambda}_3 = \bar{\alpha}_2 \quad .$$

The third eigenvalue has no practical importance. It yields a monotonous decreasing natural motion which disappears very quickly. The first two eigenvalues stand for a damped natural vibration with related frequency $\bar{\omega}_1' = \omega_1'/\omega_R$. This frequency is plotted in figure 3.16 as a function of bearing stiffness and with bearing damping as parameter. The figure shows that at small bearing stiffnesses the influence of bearing damping on the natural frequency is considerable. With increasing bearing stiffness the curves approach to 1,0 ($\omega_1' \rightarrow \omega_R$) and the damping becomes unimportant.

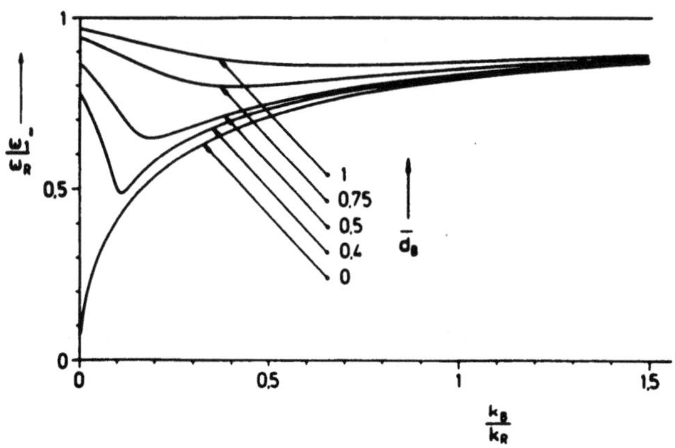

Fig. 3.16 Natural frequency of model figure 3.15

The negative real part of a complex eigenvalue λ_k indicates how much the corresponding natural vibration is damped. Therefore it may be called natural damping. Accordingly the expression

$$\Delta_k = \frac{-\alpha_k}{|\lambda_k|} = \frac{-\alpha_k}{\sqrt{\alpha_k^2+\omega_k'^2}} \tag{3.27}$$

is called related natural damping.

For our model the related natural damping is plotted in figure 3.17 as a function of bearing damping with parameter bearing stiffness.

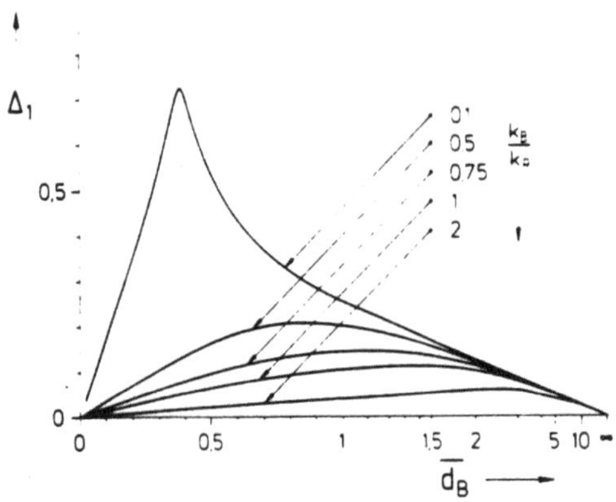

Fig. 3.17 Natural damping of model figure 3.15

The curves begin at zero, reach a maximum and end again towards zero. The maximum indicates optimum bearing damping. However the question is whether the optimum can be realized in practice. The influence of stiffness is important: large bearing stiffness compared to rotor stiffness lowers the natural damping.

For unbalance excitation

$$f(t) = \mathrm{Re}(me\Omega^2 e^{i\Omega t})$$

the steady state solution of (3.24) yields the harmonic vibration for the rotor

$$y(t) = \mathrm{Re}(\hat{y} e^{i\Omega t})$$

with the amplitude

$$\hat{y} = |\hat{\underline{y}}| = e\eta^2 \sqrt{\frac{(\kappa+0,5)^2+(\eta\bar{d}_B)^2}{[\kappa(1-\eta^2)-0,5\eta^2]^2+[\eta\bar{d}_B(1-\eta^2)]^2}} \quad , \tag{3.28}$$

where $\eta=\Omega/\omega_R$ is the related exciting frequency.

The amplitude-frequency plot is the well-known resonance curve with maximum amplitude in the neighbourhood of $\Omega=\omega_1'$. Figure 3.18 shows the related maximum $\hat{y}_{max}/e$ as a function of bearing damping with parameter bearing stiffness. For the assumed values and for small damping

$$\frac{\hat{y}_{max}}{e} \approx \frac{C}{\bar{d}_B} \quad , \tag{3.29}$$

where C depends very significantly on the related bearing stiffness (figure 3.19).

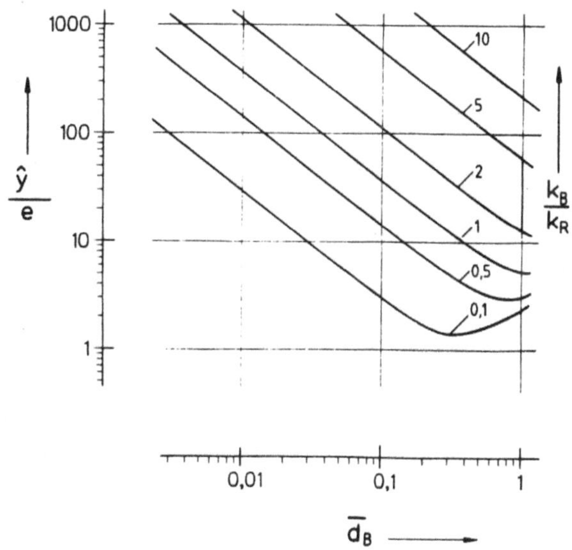

Fig. 3.18 Maximum amplitude of model 3.15

Remember that with a small bearing damping the natural damping decreases with increasing bearing stiffness. Therefore the maximum amplitude of the rotor may also be evaluated approximately by the natural damping.

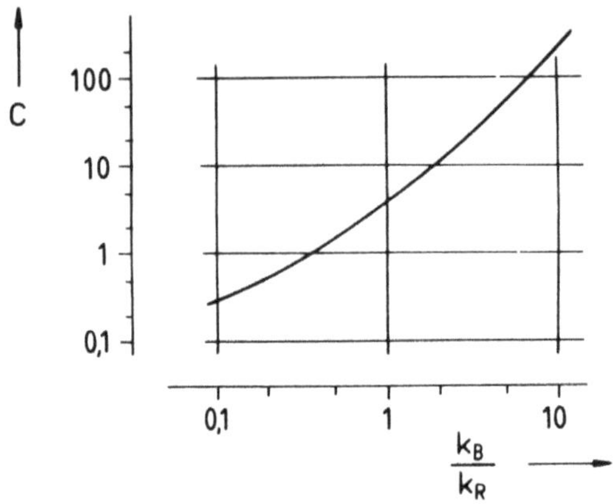

Fig. 3.19 Factor C [Eq. (3.29)] for small related bearing damping $\bar{d}_B$

3.3 Unbalance Response of Rotors in Bearings

In practice the most important problem in rotordynamics is unbalance response. In the previous chapters the principal behaviour of the Jeffcott rotor for this kind of excitation is described. As a further step the Jeffcott rotor with journal bearings may be studied. This has already been done in many published papers. The main results are shown in figure 3.20.

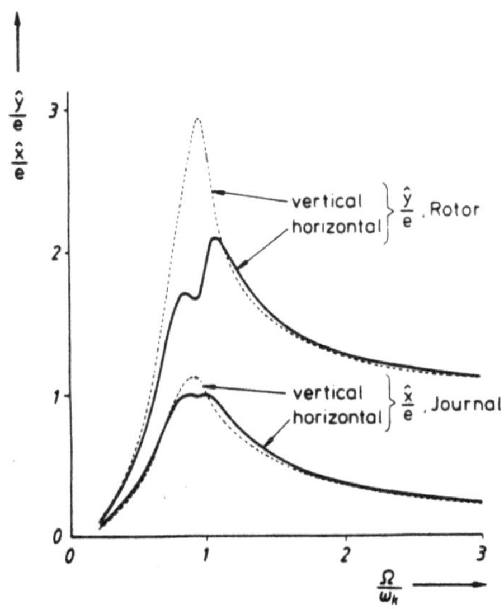

Fig. 3.20 Unbalance response of Jeffcott rotor with journal bearings (/6/)

Theoretically the rotor has two critical speeds, because the oil film behaves anisotropically. But usually the critical speeds differ so little that only one resonant peak appears. The example in figure 3.20 shows two peaks only for the horizontal rotor vibration. This problem is treated rather comprehensively by J. Glienicke in /14/.

Knowledge of unbalance behaviour of Jeffcott rotor is doubtless very important. For real rotors however more complicated models are necessary, which are characterized mainly by:

- different cross-sections of rotor along its axis,
- more than two bearings,
- consideration of additional effects, above all the gyroscopic effect,
- realistic models for the pedestals and the foundation.

The computation of such models is rather complicated and to date many efforts have been made to obtain a good and economical procedure. The development began in 1944 with the publication of the fundamental papers /15, 16/ of N.O. Myklestad and M.A. Prohl. Their method, called transfer method, was first applied to undamped, rigidly supported multi-bearing rotors. In the following years many researchers extended it for more complicated models. Even today most industrial computer programs are based on this method. For an introduction see the book of E.C. Pestel and F.A. Leckie /17/. The user must know, that numerical difficulties are possible, which are discussed in /18/.

By the transfer method displacements and forces of one end of the system are transfered by matrices to the other end. The solution is determined by the boundary conditions of the two ends. All intermediate conditions are considered during the transfer procedure. This method has the advantage that the number of unknowns is only the sum of the considered displacements and forces at both ends of the system.

Another way of computing was to extend the known procedures for the statics of large systems to dynamic problems. This leads to a large number of coupled equations, whose economical solution was possible only after fast computers were available. As unknown quantities either the displacements or the forces could be chosen. Both possibilities are investigated and compared by K. Marguerre and H. Wölfel in /19/. Nowadays the first possibility, called displacement method, is almost exclusively used in modern programs. The main advantage of this method is its flexibility. The system matrices are obtained in a simple manner from the matrices of the elements and of substructures. Therefore extensions or variations of systems may easily be introduced.

For the displacement method the equation of motion in matrix notation is

$$[M]\{\ddot{x}\} + [B]\{\dot{x}\} + [S]\{x\} = \{f(t)\} \quad , \tag{3.30}$$

where $\{x\}$ is the vector of the unknown displacements (n displacements and rotations: n-DOF system) and $\{f(t)\}$ is the vector of the known exciting forces or moments. The (n,n) system matrices [M], [B], [S] represent the masses and mass moments, damping and gyroscopic properties and conservative or nonconservative stiffnesses. The solution of (3.30) for most practical cases is well-known.

Let us look at some results concerning unbalance behaviour of multi-bearing systems, computed by the displacement method. Figure 3.21 shows a schematic picture of a multibearing rotor with its pedestals and foundation. For such a system usually models are assumed

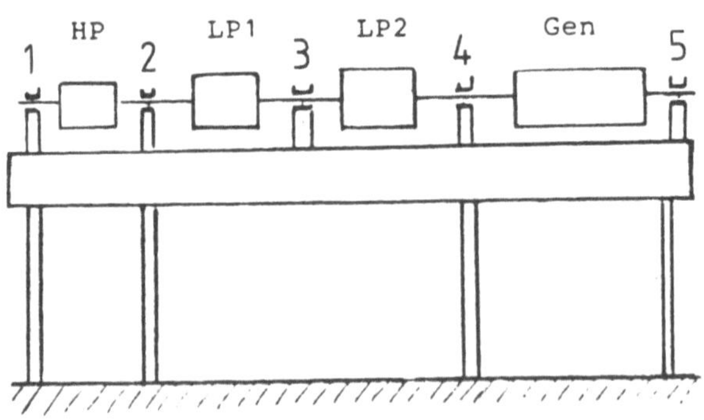

Fig. 3.21 Schematic picture of a turbo machine

with several hundred degrees of freedom with the same number of natural frequencies. Often 20 to 50 natural frequencies lie in the range of operating speed. For example figure 3.22 shows the natural frequencies of a five-bearing rotor with rigidly supported journal bearings. Aside from the twelve natural frequencies of the rotor the natural frequencies of foundation sometimes must be considered. The natural frequencies of the rotor depend on the rotational speed,

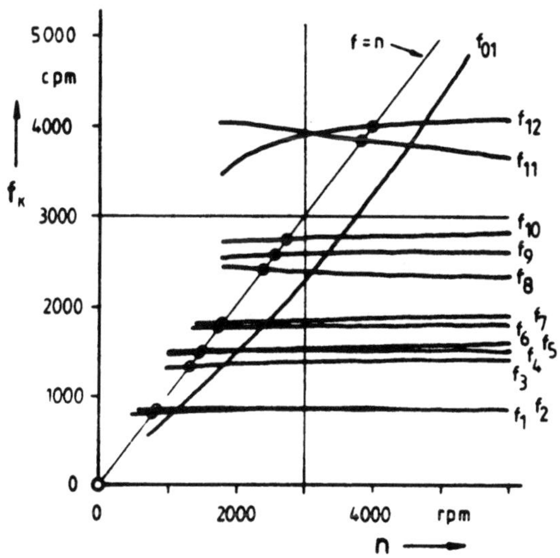

Fig. 3.22 Natural frequencies of a five-bearing rotor /20/

which is caused by the bearing properties. The influence is however mostly moderate. The curves marked by $f_1, f_3, \ldots$ belong to natural vibrations whirling backward and those marked by $f_2, f_4 \ldots$ belong to natural vibrations whirling forward. The critical speeds are determined by the intersection points of the line f=n.

The natural modes generally are not plane and they vary periodically with time. These special properties are however not very significant and it is sufficient to consider the mode shapes as plane and independent of time. Figure 3.23 shows the natural modes of the example, simplified as stated.

For computations of unbalance vibrations the distribution of unbalance along the rotor axis has to be assumed. In practice any kind of distribution is possible. For cases similar to that of our example

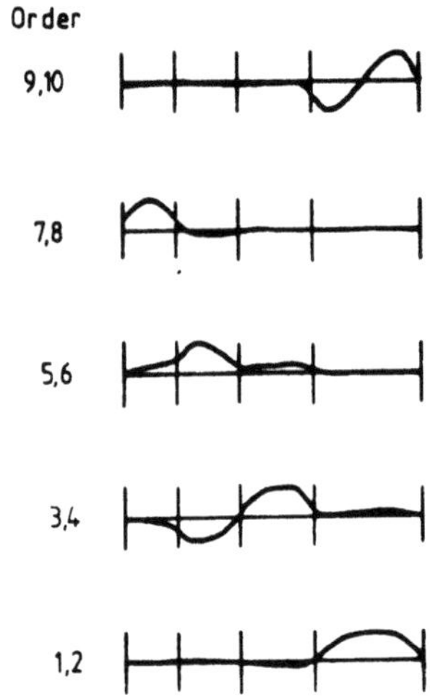

Fig. 3.23 Natural modes of a five-bearing rotor

it may be adequate to assume only two kinds of unbalance distributions: unidirected and opposite directed between two bearings. The unbalance may be constant distributed or concentrated to one or two single unbalances. Figure 3.24 shows for example the amplitudes of the shaft at bearing 5. In case ⓐ only one important peak appears at ≈660 rpm, while in case ⓑ two rather high peaks are produced in the neighbourhood of 2000 rpm.

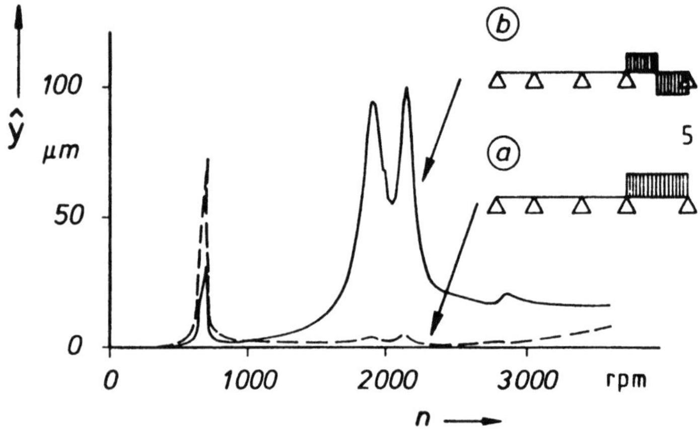

Fig. 3.24 Shaft amplitude at bearing 5 in case of unbalances in the generator rotor

Finally some remarks about the influence of foundation on unbalance response: Nowadays the rotor and the foundation are usually computed separately. Then the behaviour of the combined system is estimated either on the basis of experience with earlier machines or by computational approximations. The computation of the combined rotor-foundation system is time consuming, above all in the preparation of data. A result of combined computation is shown in figure 3.25. For comparison the behaviour of the rotor without foundation (that means with rigidly supported journal bearings) is also plotted. On the basis of such comparisons the influence of foundation may be studied, making it possible for future similar cases to avoid the time consuming combined computation.

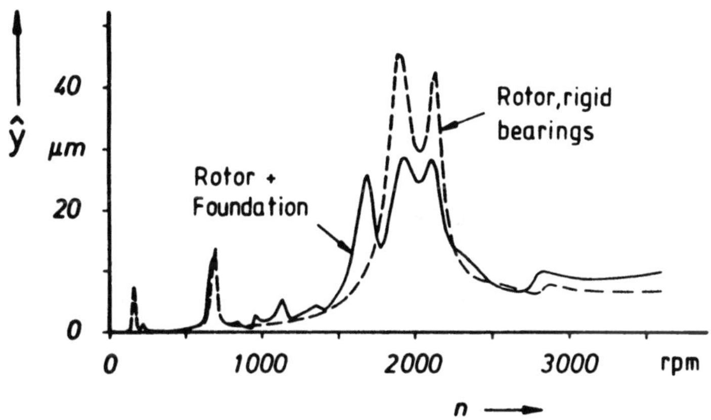

Fig. 3.25 The influence of foundation on the result of figure 3.24 ⓑ

References

/1/ Föppl, A.: Das Problem der Lavalschen Turbinenwelle, Der Civiling. 1895, S. 335.

/2/ Föppl, A.: Über den ruhigen Gang von schnell umlaufenden Hängespindeln, Der Civiling. 1895, S. 335.

/3/ Jeffcott, H.H.: The Lateral Vibration of Loaded Shaft in the Neighborhood of a Whirling Speed - the Effect of Want of Balance, Philosophical Magazine, Series 6, Vol. 37, 1919.

/4/ Biezeno, C.B., Grammel, R.: Technische Dynamik, 2. Band, Berlin, Göttingen, Heidelberg: Springer 1953.

/5/ Rao, J.S.: Rotordynamics, Wiley Eastern Ltd., 1983.

/6/ Krämer, E.: Maschinendynamik, Berlin, Heidelberg, New York: Springer 1984.

/7/ Pöschl, Th.: Das Anlaufen eines einfachen Schwingers. Ing.Arch. IV. Bd. (1933) S. 98-102.

/8/ Henning, G., Schmidt, B., Wedlich, Th.: Erzwungene Schwingungen beim Resonanzdurchgang. VDI-Berichte Nr. 113, 1967, S. 41-46.

/9/ Capello, A.: On the Acceleration of Rotors through their Critical Speed. Meccanica, Journal of the Italien Association of Theoretical and Applied Mechanics, Vol. 2, Nr. 3, 1967.

/10/ Markert, R.: Resonanzdurchfahrt unwuchtiger biegeelastischer Rotoren. VDI-Z Fortschrittsberichte Reihe 11, Nr. 34, 1980.

/11/ Hassenpflug, H.L., Flack, R.D., Gunter, E.J.: Influence of Acceleration on the Critical Speed of a Jeffcott Rotor. Journal of Engineering for Power, Vol. 103, S. 108-113, 1985.

/12/ Tsuchiya, K.: Passage of a Rotor through a Critical Speed. Journal of Mechanical Design, Trans. of ASME, 104 (2), S. 370-374, 1982.

/13/ Greb, J.: Berechnung des Hochlaufs und Ablaufs von Rotoren. Studienarbeit am FG Maschinendynamik, TH Darmstadt, 1985.

/14/ Glienicke, J.: Experimental Investigation of the Stiffness and Damping Coefficients of Turbine Bearings and Their Application to Instability Prediction. Symposium in Nottingham. London: The Institution of Mechanical Engineers, 1966.

/15/ Myklestad, N.O.: A new method of calculating natural modes of uncoupled bending vibrations. J. Aeronaut. Sci. 11 (1944) 153-163.

/16/ Prohl, M.A.: A general method for calculating critical speeds of flexible rotors. J. Appl. Mech., Trans. ASME, Series E 67 (1945) 142-146.

/17/ Pestel, E.C., Leckie, F.A.: Matrix methods in elastomechanics. New York: McGraw-Hill 1963.

/18/ Marguerre, K., Uhrig, R.: Das Übertragungsverfahren und seine Grenzen. Z. angew. Math. Mech. 44 (1964) 1-21.

/19/ Marguerre, K., H. Wölfel: Mechanics of structural systems - Mechanics of vibration. Sijthoff & Noordhoff, Alphen an den Rijn, The Netherlands, 1979.

/20/ Krämer, E.: Models for Computation of Turbomachine Vibrations. ASME 85-DET-138, 1985.

CHAPTER 4.1

PRINCIPLES OF BALANCING AND OF BALANCING MACHINES

N.F. Rieger
Stress Technology Incorporated, Rochester, New York, USA

ABSTRACT

The construction and operating principles of several types of modern balancing machines are reviewed. Soft support, hard support, and resonant machines are described. Plane separation and the need for Wattmeter filtering are discussed in relation to balancing requirements and signal conditioning. Larger rotor facilities and automated production facilities are described.

4.1.1 Introduction

The objective of rotor balancing is to minimize the effects of residual unbalance on the rotor system during normal operation. The main effects of excessive rotor unbalance are:

a) Undesirable vibratory forces applied at the rotor journals to the supporting structure and foundation.

b) Non-concentric rotor operation (rotor run-out).

c) Excessive noise level.

A perfectly balanced rotor will transmit no unbalance vibratory force or vibratory motion to its bearings or supports at any operating speed. Acceptable levels of residual unbalance are described in ISO balancing documents [1, 2]. A rotor is balanced when the c.g. of the rotor mass distribution in all normal modes of the system lies on the axis of rotation. The objective of the rotor balancing process therefore, is to achieve this condition. Typical rotor balancing involves:

a) Detection and measurement of the effect of unbalance on the rotor at selected locations.

b) Modification of the rotor mass distribution at the correction planes.

c) Repetition of (a) and (b) until the rotor residual unbalance is reduced to an acceptable level.

This procedure may be followed in a general purpose balancing machine, a special balancing machine, in a balancing facility, or on-site. Rigid rotors may be corrected in two planes in a general purpose balancing machine. Flexible rotors may be balanced in similar equipment with suitable high speed capability. Most rigid and flexible rotors are also trim balanced on-site. In most cases particular needs determine the type of balancing required.

4.1.2 Balancing Methods

Single Plane Rigid Rotor:

The simple single disk rotor consists of a thin uniform disk mounted eccentrically on a uniform shaft of circular cross section. The rotor unbalance then lies in the plane of the disk, and its effect may be removed by adding a suitable weight opposite the disk eccentricity. It is common practice to determine the angular location of the rotor unbalance by placing the shaft on two knife edges and allowing the rotor to roll until its c.g. finds its lowest position. A known trial weight is then added to the disk at a selected location, and the disk is again allowed to come to rest on the knife edges. The trail weight is then moved to another location, say 120° away from the first trial location, and the procedure is repeated. A third trial may be attempted with the weight an additional 120° from the previous location. The magnitude of the required correction weight may then be obtained by solving the resulting vector force problem: see Sommerville [3].

Two Plane Rigid Rotor:

Any rigid rotor may be balanced by the addition of correction weights in any two correction planes. The location of these correction planes is usually limited by available access to the rotor in its casing. Two plane balancing of stiff rotors of moderate length requires several measurements to be made with the rotor spinning at low speed, typically between 100 and 600 rpm. Modern rigid rotor balancing involves the following steps:

a) Spin the rotor at a suitable speed.

b) Measure the transmitted unbalance force and reference phase angle at the left bearing.

c) Measure the transmitted unbalance force and phase angle at the right bearing.

d) Use a suitable calculation to determine the angular location and correction weight required in each balance plane. This procedure must allow for force interaction between the correction planes, known as 'plane separation'.

e) Insert the weights at the required locations in both correction planes.

f) Measure the new transmitted forces at both bearings. Compare these forces with values permitted by balance criterion.

g) Repeat the above sequence if needed.

Multi-Plane Balancing:

Flexible rotors may require correction in more than two planes. Multi-plane balancing may be accomplished by a number of procedures, the best known of which are:

a) Modal balancing, in which the modal components of the unbalance are corrected mode by mode.

b) Influence coefficient balancing, in which the rotor correction masses are determined using a computer program to process the trial weight response data.

Multi plane balancing may be accomplished either in a balancing facility or in-situ. Techniques range from trial-and-error balancing to automated computer balancing. Multi plane balancing requires the following equipment:

a) Mechanical drive input for required balance speeds.

b) Vibration sensors for data acquisition (displacement probes, pedestal transducers, etc.).

c) Signal processing and data reduction equipment, e.g., tracking filters, wattmeter circuit, mini computer, etc.

d) Trial weights, and access to rotor correction planes.

Multi-plane balancing differs from two plane balancing in that it may require additional modal corrections for high speed operation. Appropriate two plane balancing of rigid rotors leads to smooth operation at all speeds.

4.3.3 Types of Balancing Machines

Types of Balancers:

General-purpose balancers. These machines are used to balance rotors of various types and sizes. They are usually designed for two plane balancing at low speeds although larger units have been designed for higher speeds. All recent general-purpose balancers use a mini computer to determine the correction weights.

Custom balancers. Special purpose balancers have been developed for such items as small fans, gyros, automotive wheels, satellites, etc.

Automated balancers. Special custom balancing facilities are part of the manufacturing process for crankshafts (with metal milling facilities), small electric motors, conveyor systems, and for production balancing of connecting rods: see Section 4.1.6.

Calibration Types:

Two important characteristics of a balancing machine are its calibration and readout capabilities. Balancing machines are classified with respect to calibration procedure, as follows:

a) Trial-and-error balancing machines.

b) Calibratable machines requiring a balanced prototype rotor.

c) Permanently calibrated hard bearing machines.

Support Type Classification:

Balancing machines are classified as soft support, hard support, or resonant machines. The influence of the rotor supports on the behavior of the balancing machine is shown in Figure 1. Figure 1 indicates the regions of operation of soft support, hard support, and resonant machines in relation to the dynamic support properties of the balancer. For a hard support machine, the natural frequency of the support is high, and the machine operates below the support natural frequency. The unbalance force and the support displacement are always in phase.

With a soft bearing machine, the balancing is always performed above the natural frequency of the rotor in its supports. Unbalance force and response are then 180° out of phase. Resonant balancing machines operate by passing down through the natural frequency of the support system, as rotor speed decreases. The associated resonant amplitude build-up is then used to amplify the unbalance signals.

Rotor Types:

Balancers are sometimes described as either rigid rotor, low speed, or flexible rotor, high speed types. These terms are imprecise. They omit other considerations in the balancing process. For example, a two bearing, low speed balancer may also be capable of multi-plane balancing. If such a balancer were capable of operating at higher speed, then a flexible multi-plane rotor balance may also be possible. It is therefore the operating speed of the balancer which defines its balancing capabilities.

4.1.4 Major Components of Balancing Machines

Rotor Supports:

The rotor support structure of a balancing machine includes:

a) Journal supports.

b) Journal pedestals.

c) Pedestal supports.

d) Foundation base.

There are two types of support in current use: soft and hard. The soft support principle is shown in Figure 2. It consists of a low stiffness horizontal spring support for which the period of free vibration may be one to two seconds. The vertical stiffness of this support is much higher. The advantage of soft supports is that, compared to hard supports, they give a larger signal strength for the same level of unbalance due to the larger displacements which these supports permit. Strong signals at rotational frequency require less sophisticated electronic signal-processing equipment. Soft support machines also tend to be simpler and less expensive than hard support machines. They are well suited to most rigid rotor balancing applications and are used for a variety of small to medium size universal balancers for the balancing of armatures, crankshafts, fan rotors, impellers, drive shafts, and so on.

A typical hard support balancing machine is shown in Figure 3. The hard support is moderately stiff in the horizontal direction, and very stiff in the vertical direction. Modern designs of such supports commonly make the journal pedestal, the hard springs, and the movable foundation block from a single piece of metal. Support motions may be sensed by displacement probes, strain gages, and other means. A typical arrangement is shown in Figure 4. Hard support machines may be more susceptible to extraneous vibrations.

The smaller unbalance signals of hard support machines are compensated for by the electronic equipment which is normally provided with such machines. This equipment may incorporate refined filter and amplifier circuits which condition the unbalance signals for amplitude and phase angle. Universal hard bearing balancers for both rigid and flexible rotor balancing are now available in a range of sizes. Many large special purpose balancers and facilities are now using hard support equipment. Rigidity of construction is generally desirable for high speeds. The electronic sophistication now available is also attractive because of its precision.

Rotor Drive:

A variety of techniques are used to impart rotary motion to the component being balanced; the belt drive and the end drive shaft are the most common. The drive selection is determined by rotor size, power involved, influence of bearing eccentricity, and system dynamics. Flat belt drives are common with small bench-type balancers. These drives allow for easy set-up, but additional vibration may be imparted to the pedestal by the belt motions.

The cantilever belt drive, shown in Figure 5 is used in resonant balancers. When lowered to contact the rotor upper surface it imparts rotation. The belt is moved away from the rotor when the desired speed is reached (somewhat above the rotor support critical speed). The rotor then drifts down in speed, passing through resonance. During measurement the belt is not in contact with the rotor. This eliminates any belt excitation effects.

For installations with large rotor inertia and drive power requirements, it is often desirable to use an end drive. The end drive may consist of a coupling shaft with a universal joint at each end. The shaft reaches from the balancer drive unit to the overhung end of the rotor as shown in Figure 6. Such drives are widely used in medium and large general purpose balancing machines, and in many custom machines because they transmit the higher power necessary for acceleration and regenerative braking. In very large installations, a specially designed drive shaft coupling may be required to supply adequate drive power.

Foundations:

A massive foundation is needed to support the rotor being balanced and to attenuate external vibrations. The foundation must permit axial adjustment of the balancer pedestals. For small balancing machines, e.g., Figure 7 the foundation may support only the pedestals, with the electronic equipment mounted elsewhere. Medium sized balancing installations, Figure 5, are often built as a unit with the electronic equipment mounted for convenience at one end, on the foundation. In large special purpose units, the foundation may merge with the protective equipment of the spin pit.

Electronics:

Electronic equipment used to acquire the vibration signals and to process the signals into unbalance information is an essential aspect of all modern balancing machines. The following electronic equipment is used:

a) Vibration sensors: inductance probes, accelerometers, strain gages.

b) Filtering circuits: wattmeter circuit, tracking filters.

c) Operational amplifiers.

d) Plane separation circuit.

Depending on their application, vibration sensors detect velocity, displacement, or acceleration. Though the sensor output may be weak or contain high background noise, this can usually be amplified and filtered out. Vibration sensors for a hard support machine are shown in Figure 3 and for a soft support machine in Figure 5.

Signal filtering can be accomplished by a number of special circuits. The most widely used is the so-called Wattmeter circuit, which acts as a filter to exclude all non-synchronous a.c. components from the balance signal. The resulting output signal is a clean harmonic waveform which may then be used to define the rotor unbalance. Other circuits use tracking filters with an analog/digital converter coupled to a micro-processor to achieve the same operation.

A balancing machine Wattmeter circuit is shown in Figure 8. Alternating current from an unbalance sensor is supplied to the Wattmeter field coil. Alternating voltage from an a.c. generator coupled to the drive shaft is supplied to the Wattmeter moving coil. The deflection of the moving coil is then proportional to the Wattmeter power $W = EI \cos \theta$, where E is the generator voltage, I is the unbalance sensor current, and θ is the phase angle between the voltage and current signals. Figure 9 shows how a wattmeter can combine waves of identical frequency, and how waves of differing frequency fail to produce a voltage and current when these components are in phase. Where the unbalance current signal leads the a.c. voltage, the unbalance power signal is reduced by $\cos \theta$. The unbalance Wattmeter also requires careful measurement of sensor signal phasing to avoid incorrect balance readings. Only voltages and currents having the same frequency can be combined in this instrument.

Accurate balancing depends on obtaining accurate signals which relate in a consistent known manner to the unbalance force being imparted to the pedestal supports. The Wattmeter method is a simple procedure for preventing unwanted signal components from appearing in the unbalance signal, and also for excluding unwanted signals arising from rotation such as non-circular journal harmonics drive stick-slip effects, small impacts, misalignment, and excitation from the external environment.

Plane Separation:

Figure 10 shows a rigid rotor to be corrected for unbalance in the two planes indicated. The rotor is supported by two end bearings. It has two correction planes inboard of the bearings, and the residual unbalance force is represented by the two applied forces shown. If the left bearing support is restrained so that no lateral motion is possible at that location, and if the right bearing support is free to move, then it is possible to select a balance weight for the right correction plane such that the right end of the rotor will run smoothly. The same rotor could also be corrected by restraining the right bearing and inserting a suitable correction weight in the left correction plane. The criterion for acceptable balance is that the rotor shall operate without transmitting any dynamic force to the bearings. Having corrected the rotor in the left and right planes independently, it might be thought that the rotor will run smoothly if both bearing restraints are released. This is not the case. The rotor will run roughly because the rotor force balances will not have been achieved independently of the support forces.

This problem can be overcome using the principle of plane separation, as follows. First, in Figure 11 instead of restraining the rotor motion at the left bearing, assume that lateral motion can be restrained at the left balance plane, Moment and Force equilibrium then give:

$$\sum M_1 = 0: \quad U(b-a) - C_2(c-a) = 0: \; C_2 = U\frac{b-a}{c-a}$$

$$\sum F = 0: \quad -F_1 + U - C_2 = 0: \quad F_1 = U\frac{c-b}{c-a}$$

Suitable weights are then added in the right correction plane until the rotor runs smoothly. The second step is to restrain the rotor at the right correction plane, and then balance it in the left correction plane until it again runs smoothly, i.e.,

$$\sum M_2 = 0: \quad U(c-b) - C_1(c-a) = 0: \quad C_1 = U\frac{c-b}{c-a}$$

$$\sum F = 0: \quad -C_1 + U - F_2 = 0: \quad F_2 = U\frac{b-a}{c-a}$$

Note from the above that $C_1 = F_1$ and $C_2 = F_2$. This indicates that taken together the balance corrections will automatically cancel the bearing forces, and both moment and force equilibrium for the rotor are thereby satisfied. If the rotor is then operated with corrections C_1 and C_2 installed, the transmitted forces at the bearings will be zero and the rotor will operate in a smooth, balanced manner.

4.1.5 Modern General Purpose Balancing Machines

Soft Bearing Machines:

The Gisholt Model 31S shown in Figure 12 is a typical soft bearing general purpose balancing machine. It is designed to operate well above the resonant speed of the supported test rotor at speeds between 500 rpm and 2000 rpm. It is able to accommodate rotors weighing between 1 lb. and 500 lbs. Drive power is supplied through a belt which wraps around the rotor, in a tension pulley arrangement. Unbalance data is obtained as velocity transducer readout from movements of the supporting pedestals. The electronic equipment includes plane separation circuits and Wattmeter filtering. The readout is calibrated in terms of the required balance correction units. The 'heavy' spot may be located by either stroboscope phase measurement or by contrasting surface photocell readings against numbered strips placed at the correction planes. The readout equipment is specified as being able to read down into the ten micro-inch range. The machine is not permanently calibrated, and pre-calibration against a rotor of known unbalance is required periodically. The soft pedestal construction is achieved with a pair of vertical flat springs. One end of each spring is attached to the machine frame while the other end is secured to the adjacent pedestal support.

Resonant Machines:

The Stewart-Warner 2380S machine, Figure 13, has a unique support which permits tuning so that the natural frequency of the rotor will occur at the desired balancing speed of the machine. In practice, the rotor to be balanced is loaded into the pedestal supports, whose dashpot damping is manually set to zero. The rotor is then bumped gently, causing it to vibrate in its lowest mode on the supports. A vibration meter indicates the frequency of this mode at the pedestal. A suitable value of damping is then chosen.

Both dashpots are then adjusted to the selected damping value. The required rotor speed is pre-set on the handle of the cantilever drive arm and an overhead cantilever belt drive brings the rotor up to speed. A common balancing speed for this machine is 450 rpm. The natural frequency of the rotor in its supports is adjusted to occur somewhat below this speed, so that large mechanical amplitude magnification occurs as the rotor passes through resonance when the drive is removed. Rotor unbalance is read on a graduated scale (no units) for the left and right correction planes independently. Unbalance angular orientation is determined stroboscopically, as with soft bearing machines. The balancing operations proceed by trial-and-error.

This machine has found widespread use in the automotive industry.

Hard Support Machines:

The Schenck H30V, Figure 14, is a typical hard bearing machine with plane separation circuits, Wattmeter filtering and measurement, push button set-up of the balancing operations and options for either direct or belt drive. Unbalance signals are detected at the pedestals using inductive force transducers. The balancing speed is usually within the range 275 rpm through 1500 rpm - considerably below the resonant frequency of the rotor on its support system. It accommodates a maximum rotor weight of 880 lbs., carried symmetrically on the pedestals, as well as asymmetric rotors of smaller weight.

Advantages are: easy set-up and operation, permanent calibrations with no trial weight runs, simple dial-in balance with vectormeters, precise balancing, and the ability to handle a wide range of rotors. Disadvantages are: higher installation cost and relative delicacy of the machine.

This machine is said to be capable of balancing down to 0.0035 oz. in. Such accuracy may be possible with light rotors, e.g., 30 lbs., though this corresponds to a c.g. eccentricity of 3.5 micro-inches. For 880 lbs. rotors, such balance precision would mean 0.25 micro-inches c.g. run-out in the balanced condition.

General Purpose Hard Support Balancing Machines:

A typical industrial hard support balancing machine is shown in Figure 15. The hard support general purpose machine provides an efficient procedure for two plane balancing. Strain gages or displacement sensors are attached to the hard support frame, as shown in Figure 4. These sensors transmit pedestal motions as electrical signals to the measurement console. Both pedestals are permanently factory calibrated prior to shipping. Special calibration is needed prior to balancing.

A synchronous phase reference voltage signal is taken from the drive, which may be either a universal coupling or quick-attach belt. Inductance transducer signals are then processed using a Wattmeter circuit in conjunction with a plane separation circuit. The vectormeter screen used in the model of Figure 16 shows either the residual unbalance and phase angle in both balancing planes or the required balance weights and their orientation.

With such equipment, an immediate, precise two plane balance is possible for a wide variety of ISO Class I and Class II rotors. The basic proportions of the rotor to be balanced (dimensions a, b, c, r_1, r_2 shown in Figure 16) are first dialed into the balancer console. The rotor is then run at the desired balancing speed, and readings of the magnitude and location of the required balance weights and their phase angles are displayed. The rotor is then stopped, the required correction weights are inserted in the two correction planes, and the rotor is re-run at the balancing speed to check the balance. The simplicity of this process permits a relatively unskilled operator to balance rotors rapidly and effectively.

4.1.6 Balancing Facilities

Turbine and Generator Balancing:

Turbine and generator balancing facilities are designed so that high speed balancing operations and overspeed testing of assembled rotors are both possible. Such facilities may incorporate:

a) Concrete overspeed burst-proof tunnel.

b) Vacuum spin test chamber.

c) Variable speed drive with 20 percent overspeed capability.

d) Provision for various fully bladed rotor assemblies.

e) Bearing pedestals designed to simulate machine support properties.

f) Transporter.

g) Control room.

h) Overhead crane.

i) Rigid attachment of transporter to tunnel foundation.

Figure 17 shows details of a transporter loaded with a rotor assembly and internal details of the tunnels. Figure 18 is a section schematic of a turbine-generator balance facility.

ISO Class I and Class II rotors will operate satisfactorily after having been balanced in a low speed balancing machine. Though such rotors may require load testing and overspeed testing, these additional operations may be conducted in a less fully instrumented facility, and probably without a vacuum chamber.

Class III flexible rotors require high speed balancing involving trial runs near one (or more) resonant speeds within the operating range. It is convenient to have a common high speed facility in which both balancing and overspeed testing can be undertaken. Class III rotors are usually large and long and may carry thousands of blades.

Modern spin pit balance facilities are complex installations. Costs may range from $5 million to $10 million or more for a complete, installed unit. Figure 19 (burst pit) is a general view of a concrete balance spin pit facility showing the rotor on its transporter, with special bearings and pedestals and other details. The bearings are bolted to the foundation during testing. A section through a balance facility, with a generator rotor installed, is shown in Figure 18. Details of the drive, drive coupling, lubrication system, and other features of such facilities are evident.

A special bearing pedestal support for use in turbine generator balance pits is shown in Figure 20. Such units have been designed to provide a tuned pedestal support in which the specific bearings of the rotor being balanced may be installed. This permits the rotor to be balanced while operating in its own bearings and it incorporates simulated operating pedestal effects.

Rotor support stiffness properties may exert a significant influence on the critical speeds and dynamic properties of a rotor bearing system. If a rotor is balanced in hard bearings, and then operated in softer bearing supports, the rotor system modes will be different, and the balance achieved in the balancing stand will not be fully realized during operation. The rotor may then run 'rough' unless corrected by further in-situ trim balancing. The support design shown in Figure 20 allows the rotor to be balanced in the bearings in which it will operate. Matched dynamic properties are especially important for large Class II rotors in which unbalance effects through the fourth flexural mode may be critical. Stiffness properties of such supports may be adjusted within the range 30,000 lbs./in. through 3.0×10^6 lbs./in., according to whether soft or hard support balancing is required. This preserves the balance quality achieved for operating conditions, and leads to less field balancing. Figure 20 shows details of such adjustable support bearings, including the lubrication inlet (foreground) and the tangential force transducers from which the transmitted unbalance force signals are read. Fluid film supports avoid the problem of 'brinelling' which may occur in rolling element bearing supports, due to rotor weight which may range up to 250,000 lbs. per pedestal.

Automated Balancing Facilities:

High volume production industries depend on automated equipment for large numbers of repeated operations. Automated and semi-automated balancing is widely used in the automotive industry for crankshafts, propeller shafts, tire-wheel assemblies, etc., and in the electric motor manufacturing industry for vacuum cleaner motors, blender motors, fan motors, and so on.

Figure 21 shows an example of an automated balance facility for crankshaft-clutch housing assemblies. This facility incorporates a production line for:

a) Initial measurement of residual unbalance in two planes.

b) Correction of the unbalance by drilling at the required locations.

c) Checking of the corrected assembly to meet required balance specifications.

The operation of this crankshaft balance facility is as follows. Incoming batches of crankshafts are loaded in an ordered fashion on to a conveyor with a gravity feed roller conveyor. The process is automatic from this point. Two displacement sensors record the unbalance as the crankshaft is rotated at low speed (400 rpm). The signals are stored and the crankshaft is automatically indexed to the required angular position for each plane in turn, using the data obtained during the previous measuring operation. Two separate drills then remove the required depth of metal at the desired orientation. The drilled assemblies are then spun again in the balance checking unit for a second measurement. Once the crankshaft is positioned, the two pneumatically controlled halves clamp it for drilling. The linkage and pneumatic cylinder of this equipment may be seen in Figure 21.

After checking, crankshafts are classified 'within tolerance' or 'outside tolerance' automatically. Accepted crankshafts pass out on the conveyor, while rejected crankshafts are removed for re-processing.

4.1.7 References

1) ISO Draft Document, 'Balance Quality of Rotating Rigid Bodies,' ISO/TC 108 DR 1940, 1967.

2) ISO Draft Document, 'The Mechanical Balancing of Flexible Rotors,' ISO/TC 108/SC1/WG2 (Secretariat-7) Document 12, 1976.

3) Sommerville, I. J., 'Balancing a Rotating Disc, Simple Graphical Construction,' Engineering, February 19, 1954.

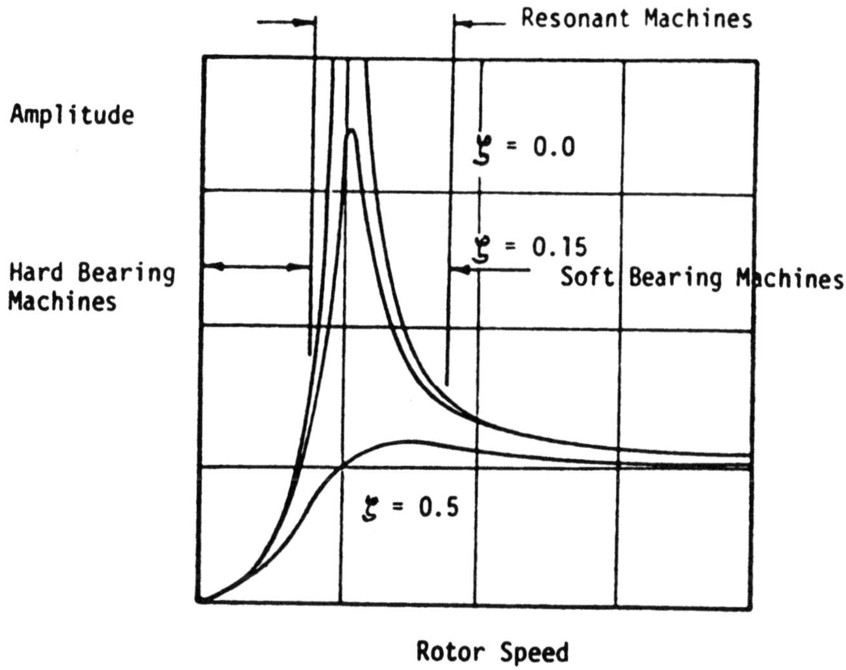

Figure 1. Relation Between Rotor Support Stiffness and Balancing Machine Type

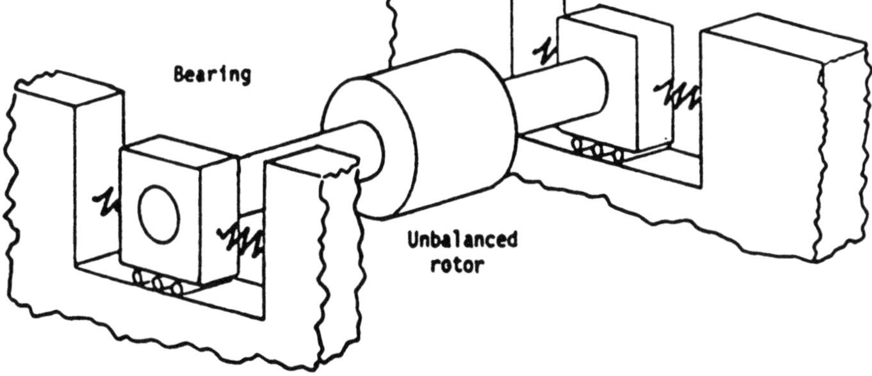

Figure 2. Soft Support Balancing Machine Principle

Figure 3. Hard Support Balancing Machine (Courtesy: Schenck-Trebel Corporation)

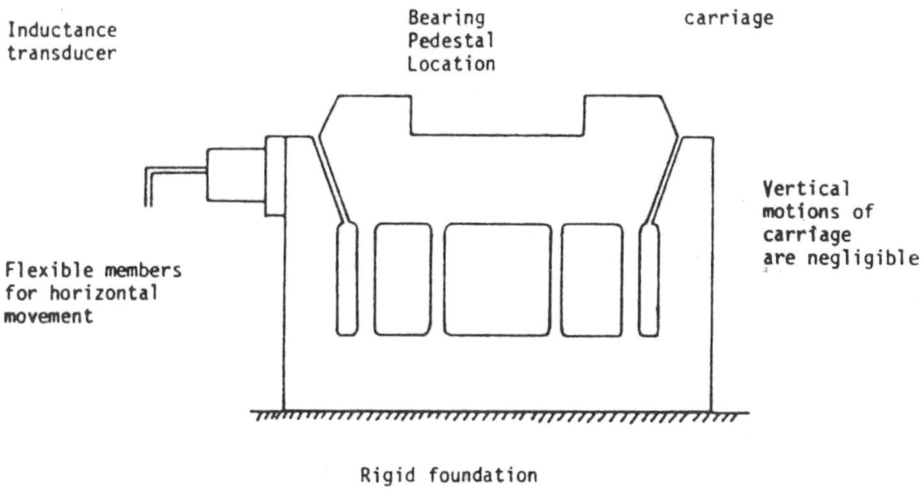

Figure 4. Hard Support Balancing Machine Principle

Figure 5. Cantilever Belt Drive
(Courtesy: Schenck-Trebel Corporation)

Figure 6. End Drive
(Courtesy: Schenck-Trebel Corp.)

Figure 7. Massive Foundation, External Electronics
(Courtesy: Schenck-Trebel Corporation)

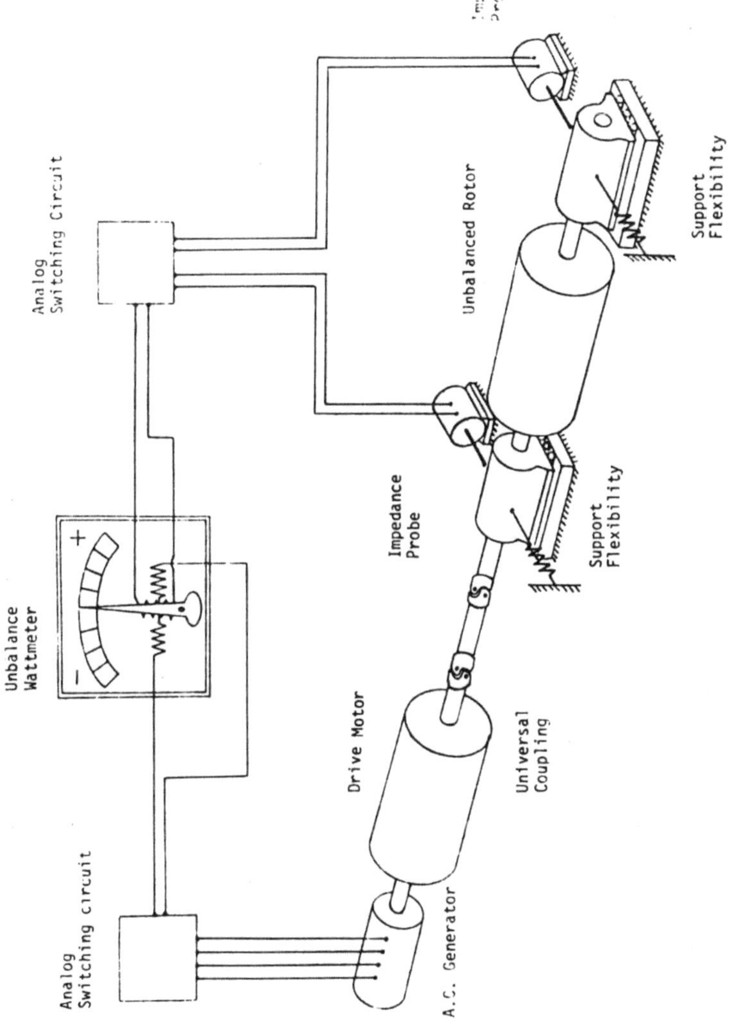

Figure 8. Wattmeter Filtering Principle

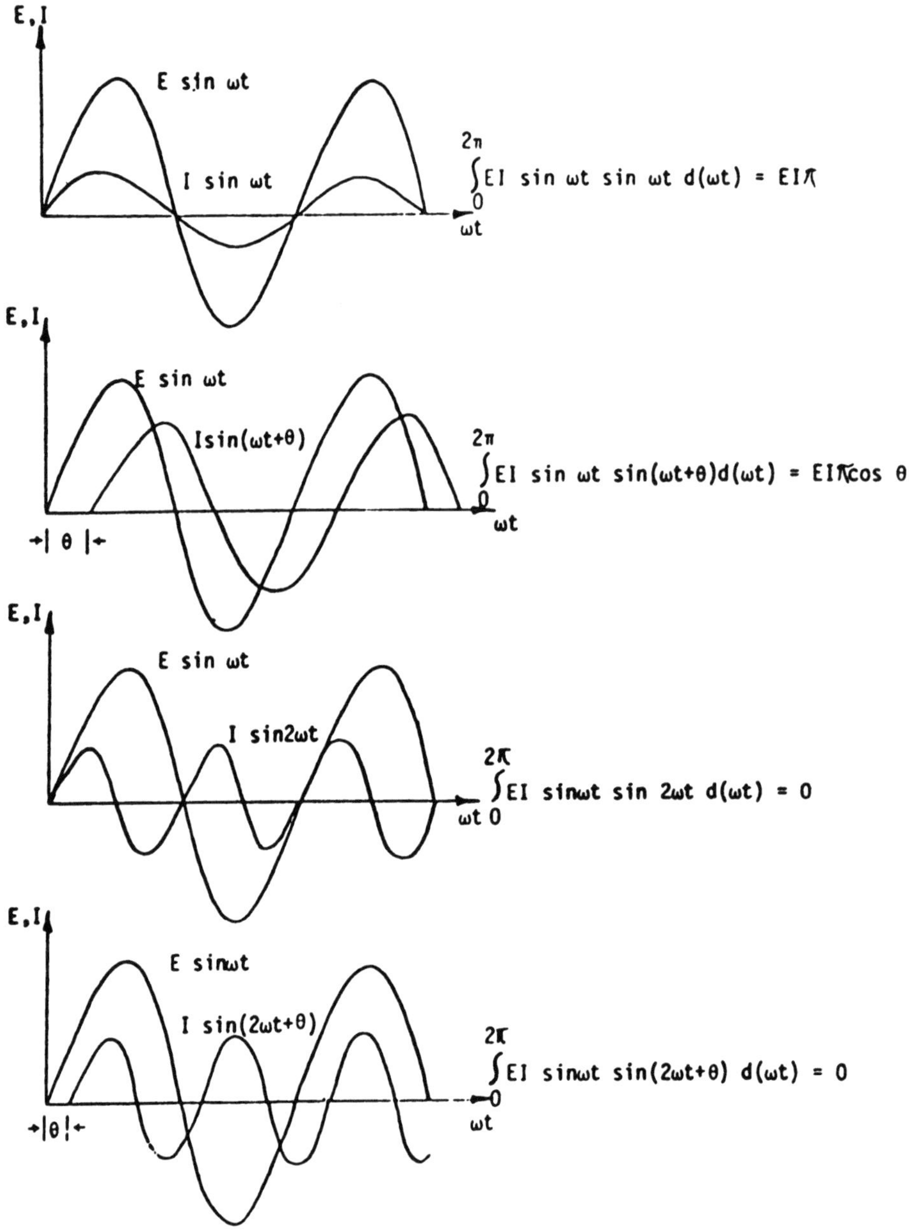

Figure 9. Wattmeter Elimination of Differing Frequencies

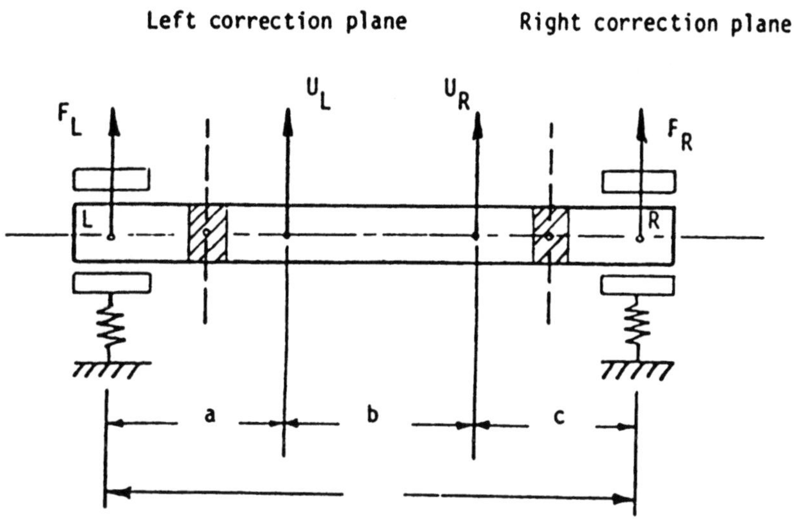

Figure 10. Rigid Rotor With Two Unbalance Forces

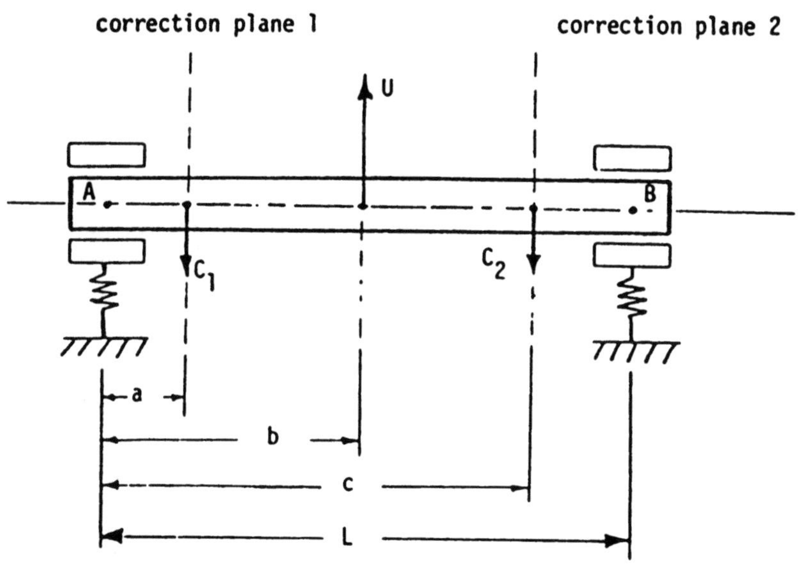

Figure 11. Correction Planes and Resultant Unbalance Force

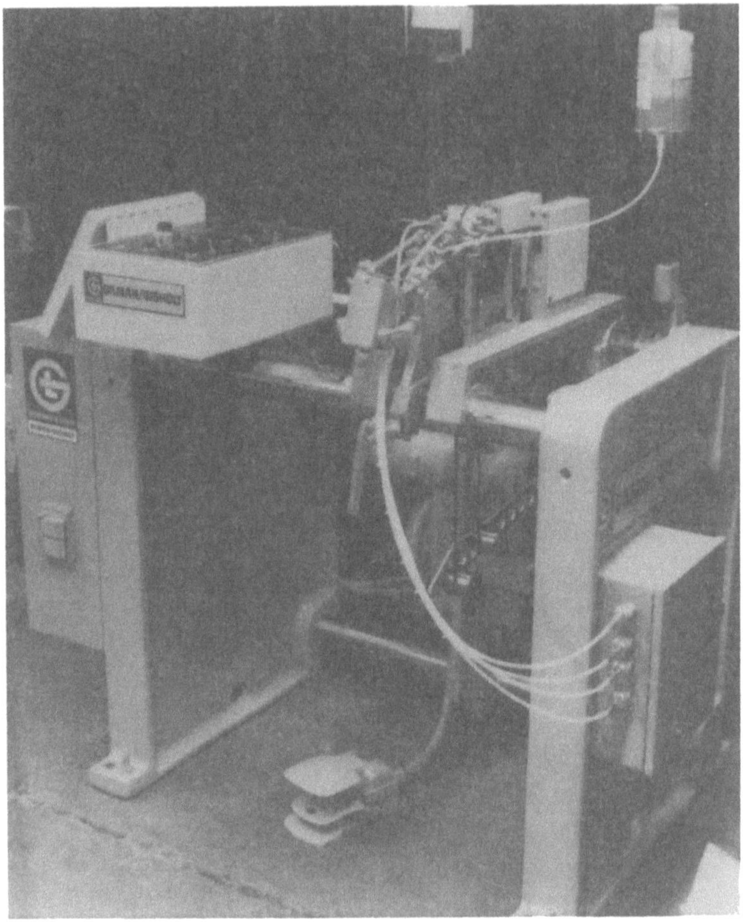

Figure 12. Gisholt Model 31S Soft Bearing Machine (Courtesy: Schenck-Trebel Corporation)

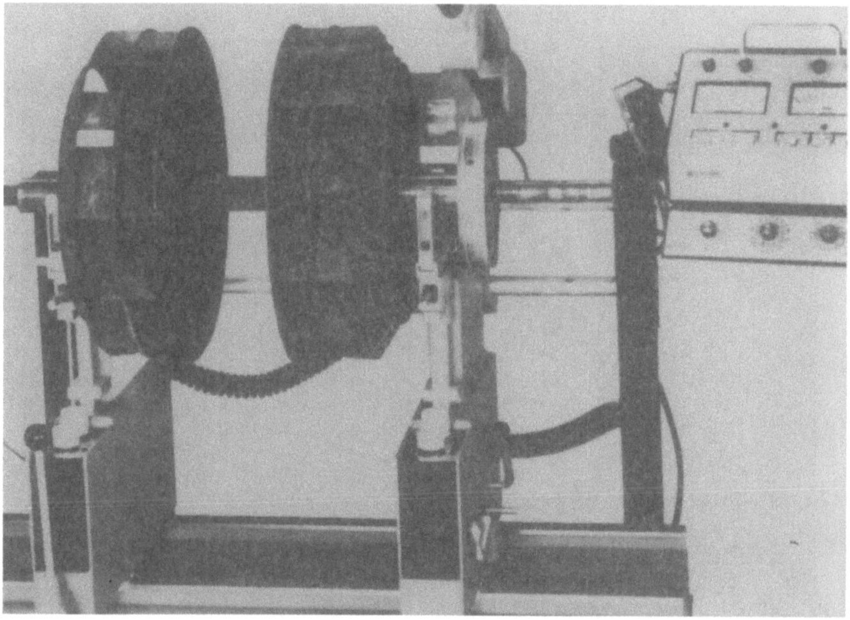

Figure 13. Stewart-Warner 2380S Resonant Support Machine
(Courtesy: Schenck-Trebel Corporation)

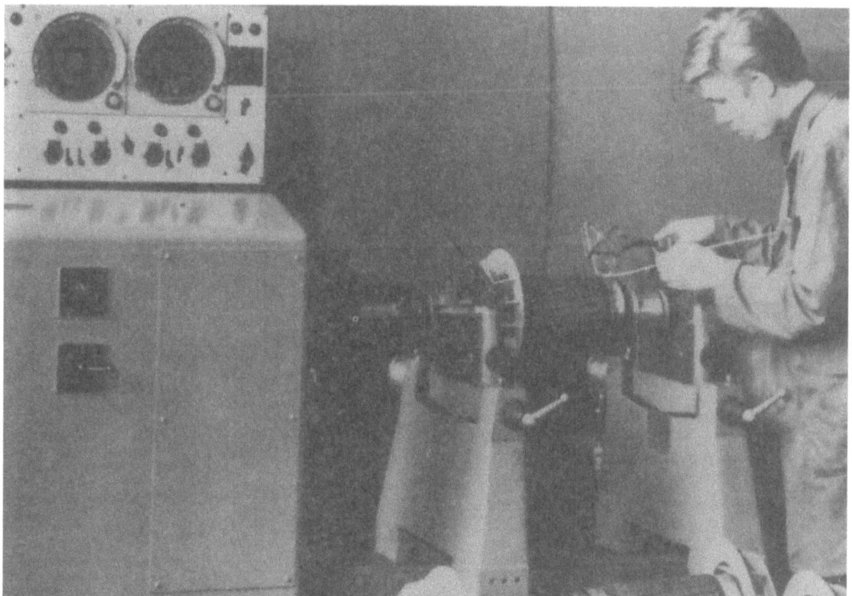

Figure 14. Schenck H30V Hard Bearing Support Machine
(Courtesy: Schenck-Trebel Corporation)

Figure 15. General Purpose Hard Support Machine (Courtesy: Schenck-Trebel Corporation)

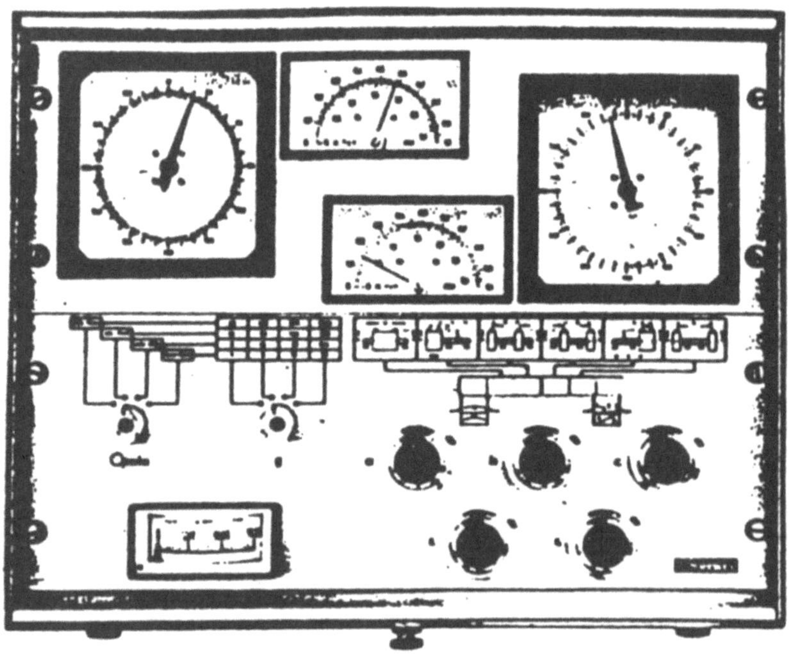

Figure 16. Vectormeter Two Plane Readout Console (Courtesy: Schenck-Trebel Corporation)

Figure 17. Transporter Supports With Generator Rotor
(Courtesy: Schenck-Trebel Corporation)

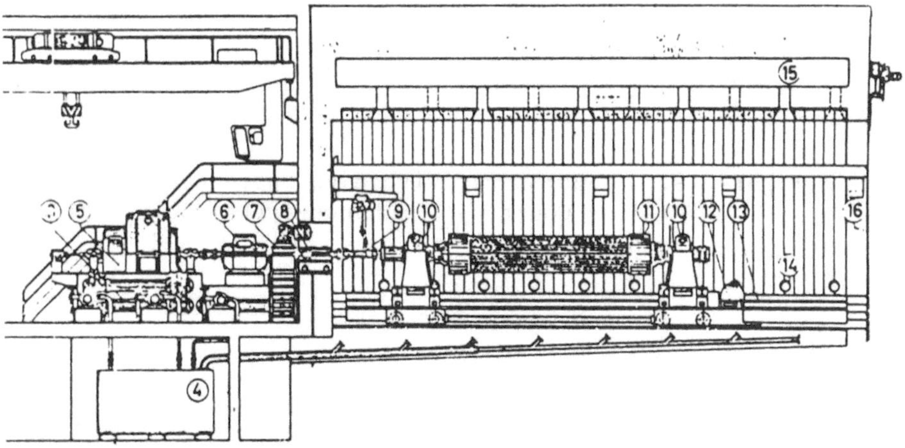

Figure 18. Schematic of High Speed Balancer Spin Tunnel

Figure 19. Burst Pit With High Speed Rotor
(Courtesy: Schenck-Trebel Corporation)

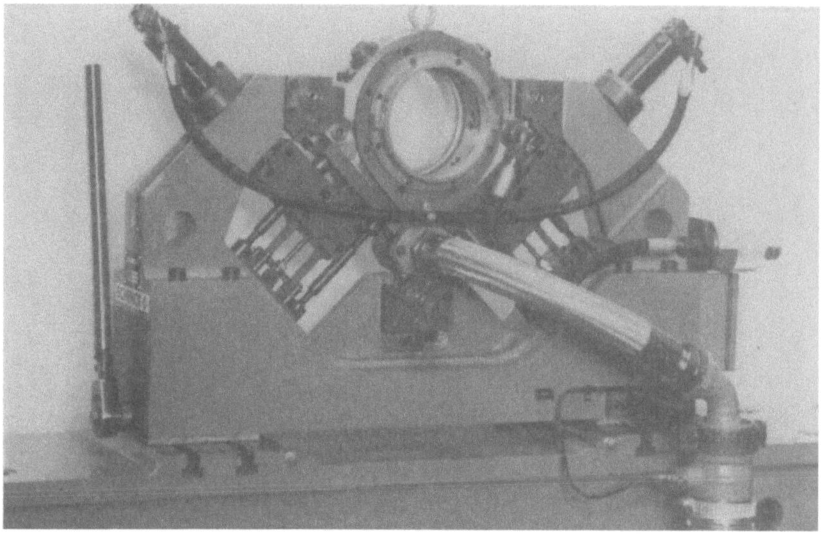

Figure 20. Tuned Dynamic Support Pedestal for Actual Bearings (Courtesy: Schenck-Trebel Corporation)

Figure 21. Automated Crankshaft Balance Facility (Courtesy: Schenck-Trebel Corporation)

CHAPTER 4.2

FLEXIBLE ROTOR BALANCING

N.F. Rieger
Stress Technology Incorporated, Rochester, New York, USA

ABSTRACT

Developments which led to modern flexible rotor balancing techniques are described. The theory of flexible rotor balancing is outlined, and the theoretical basis for Modal balancing, and for Influence Coefficient balancing, is presented. In both instances, the publications from which the source material for this section was obtained are identified, with discussion. Certain other balancing techniques are also discussed. Recent developments toward the establishment of criteria for appropriate levels of residual unbalance in flexible rotors following balancing operations are included. The state-of-art for flexible rotor balancing is summarized, and the work which remains is identified.

4.2.1 Development of Flexible Rotor Balancing

The earliest reference to flexible rotor balancing appears in the patent awarded to Linn in 1928, reference [1], in which a sequential process for correcting the whirl modes of flexible rotors in bearings is described. Figure 1 shows the title sheet of the Linn patent, which refers to a steam turbine rotor in end bearings, to demonstrate the proposed process. The next development appeared in a paper by Thearle [2] concerning a vector method for two-plane balancing of the rotors of a three-bearing turbine-generator set. The application of a proposed method is described in detail, but without mention of whether or not the operating speed is beyond any bending critical speed of the unit. The method proposed is in effect a trim balancing procedure, intended to obtain an improved final balance condition for the three-bearing shaft assembly, over the two-plane balance results achieved separately for the individual rotors. Two other early papers, by Kroon [3] deal with the theory of rigid rotor balancing. These papers also include certain practical details for field balancing of turbine-generators.

Grobel [4] is a discussion of practical flexible rotor balancing for large turbines and generators. The importance of accurate manufacturing methods to minimize inherent unbalance is stressed. Special problems arising from the generator windings, from thermal instability of the turbine rotor, and from the buckets around the turbine circumference are mentioned. An empirical trial-weight balancing method is described for the first three bending modes of the rotor, indicating that since Thearle's earlier paper improved techniques had become necessary to account for flexible rotor effects in larger machines. Details of residual vibration in installed units are given. The paper contains an interesting description of the practical aspects of turbine-generator balancing at that time (1953).

Moore and Dodd [5] is a detailed practical discussion of the Modal balancing technique applied to turbine-generator rotors. Only single-span rotors are considered, not entire units. The vibrations arising from each mode are first described, and then the Modal balancing technique and its application is described in detail, in relation to the suppression of these vibrations. An interesting example of a pump rotor which responded strongly in its second and third modes simultaneously at operating speed is described. The problem of separating and balancing mixed modes is discussed, with detailed numerical values. Recognition of the cause of this problem and its solution resolved persistent difficulties with this unit.

Another recent practical contribution is due to Lindsey [6] who described an empirical 'one-shot' method for the balancing of large turbine-generator rotors. The method is most effective when the interaction from adjacent spans is minimal, and where the whirl ellipse is circular. In essence, the method is a procedure for balancing single-span rotors in their first three modes without regard to adjoining spans. It relies on collected previous experiences with similar units. The method could probably be adapted to more general balancing of other types of supercritical rotors. This paper presents an outline of the method, but gives no theory nor supporting details of its performance.

4.2.2 Theory of Flexible Rotor Balancing

The theory of flexible rotor balancing was begun by Meldahl [7] who outlined the principles of modal balancing and their application to the three-plane balancing of an end-bearing flexible rotor. Meldahl recognized the inadequacy of two-plane balancing for flexible rotors, and observed that it was possible to balance a rotor mode-by-mode on the condition that the placement of corrections for one mode does not influence the response in any other mode. The justification for this requirement lies in the orthogonality relation[1] which shows that the interaction between any two modes will be minimized when the condition:

$$(\omega_i^2 - \omega_j^2) \int_0^{\ell} y_i \cdot y_j \cdot i \cdot dz = 0$$

between the modal displacements Y_i and Y_j is satisfied. This means that for non-trivial conditions,

$$\int_0^{\ell} y_i \cdot y_j \cdot j \cdot dz = 0 \begin{cases} > 0 & i = j \\ = 0 & i = j \end{cases}$$

where ω_i, ω_j are natural frequencies for the ith and j^{th} rotor modes respectively, and y_i and y_j are the modal amplitudes. It follows that a flexible rotor may be balanced by eliminating the effect of residual unbalance mode-by-mode. As the deflection of the rotor at any speed may be represented as the sum of various modal deflections multiplied by speed-dependant constants, a rotor which has been balanced for smooth operation at each of its critical speeds in accordance with the above condition will remain balanced at any other speed.

The remainder of Meldahl's paper outlines the procedure for balancing a rotor in its first three modes. Measurement of whirl amplitudes away from the bearing locations is implied by the statement that flexible rotor balancing in rigid bearings is not possible. This statement can evidently be relaxed if measurements are made at other locations along the rotor length. For overhung rotors, or with sufficient measurement and balance planes located along the rotor, rotors in rigid bearings may be readily balanced. The procedure described in this paper neglects damping effects, but this causes no problem, as damping merely renders the modes complex, and may become of minor significance where the system response is dominated by rotor flexibility.

---

[1]For a more complete discussion see, for example, Timoshenko [8] or Bishop and Gladwell [11].

4.2.3 Modal Balancing

The Modal balancing method may be applied to an axisymmetric elastic rotor of arbitrary stiffness and mass distribution in rigid bearings without damping such as that shown in Figure 2 as follows: first assume that all unbalances act in the same y-z plane. Let the rotor deflection shape y (z) at any speed be described by the series:

$$y(z) = \phi_1 Y_1(z) + \phi_2 Y_2(z) + \ldots + \phi_i Y_i(z)$$

$$+ \ldots = \sum_{i=1}^{\infty} \phi_i Y_i(z)$$

where the ϕ_i are coefficients, and the Y_i (z) are modal displacement expressions for each of the normal modes. The distribution of mass eccentricity throughout the rotor may also be expressed in series form as:

$$e(z) = e_1 Y_1(z) + e_2 Y_2(z) + \ldots + e_i Y_i(z)$$

$$+ \ldots = \sum_{i=1}^{\infty} e_i Y_i(z)$$

At any speed, the amplitude response resulting from the rotor mass eccentricity and the elastic deflections may also be expressed as an infinite series in the normal modes by the expression:

$$U(z) = \lambda_1 Y_1(z) + \lambda_2 Y_2(z) + \ldots + \lambda_i Y_i(z)$$

$$+ \ldots = \sum_{i=1}^{\infty} \lambda_i Y_i(z).$$

The orthogonality relations between the modes may now be used to determine the unknown coefficients in the above expressions:

$$\int_0^{\ell} m(z)\; Y_i(z) Y_u(z) dz \begin{cases} = 0 & i \neq j \\ = M_i & i = j \end{cases}$$

$$\int_0^{\ell} EI(z)\; Y_i''(z) Y_j''(z) dz \begin{cases} = 0 & i \neq j \\ = K_i & i = j \end{cases}$$

First let the rotor experience a virtual displacement y in the y-z plane such that:

$$\delta y = \delta\phi_i \; Y_i(z)$$

The work done by the centrifugal force during the virtual displacement is:

$$W_{cent} = \int_0^{\ell} m(z)\omega^2 [y(z) + e(z)](\delta\phi_i) Y_i(z) dz$$

Now utilize the first orthogonality expression by multiplying both sides of the rotor deflection expression by $m(z)\; Y_i\; (z)$ and integrating over the length of the rotor:

$$\int_0^{\ell} m(z) Y_i(z) y(z) dz = \phi_i \int_0^{\ell} m(z) Y_i^2(z) dz = \phi_i M_i$$

Hence:

$$W_{cent} = \omega^2 (\delta\phi_i)(\phi_i + \epsilon_i) M_i$$

The work done in the virtual displacement is stored as elastic energy in the rotor. For simplicity, let the rotor cross section be uniform along its length. Then the stored energy is:

$$V = \int_0^{\ell} \frac{1}{2} EI(z) \left[\frac{d^2y}{dz^2} \right]^2 dz = \int_0^{\ell} \frac{1}{2} EI(z) \left[\phi_i Y_i''(z) \right]^2 dz$$

$$= \frac{1}{2} \sum_{i=1}^{\infty} \phi_i^2 K_i$$

in keeping with the second orthogonality relation given above. The virtual work of the elastic forces may be expressed as:

$$W_{el} = - \frac{\partial V}{\partial \phi_i} \delta\phi_i = - (\delta\phi_i) \phi_i K_i$$

By the Principle of Virtual Work $W_{cent} + W_{el} = 0$. Performing this equation leads to the conclusion that:

$$\phi_i = \frac{\omega^2 M_i}{K_i - \omega^2 M_i}$$

and near the ith critical speed

$$K_i - \omega^2 M_i = 0, \text{ i.e., } \omega_{in}^2 = K_i / M_i .$$

Thus,

$$\phi_i = \frac{(\omega/\omega_{in})^2}{1 - (\omega/\omega_{in})^2} e_i \quad \text{or} \quad e_i = \frac{1 - (\omega/\omega_{in})^2}{(\omega/\omega_{in})^2} \phi_i$$

The unbalance coefficients λ_1 may also be evaluated by using the first orthogonality relationship. Multiply both sides of the unbalance equation by $m(z)Y_i(z)dz$ and get:

$$\int_0^{\ell} m(z)U(z)Y_i(z)dz = \int_0^{\ell} \lambda_i Y_i(z)m(z)Y_i(z)dz = \lambda_i M_i$$

$$\lambda_i = \frac{1}{M_i} \int_0^{\ell} m(z)U(z)Y_i(z)dz .$$

In order to balance the rotor with discrete masses at specific locations it is convenient to determine the effective unbalance at these locations. The deflection from a concentrated force P or from a uniformly distributed load q acting over a small length ξ distant c from z = 0 may be expressed as a modal series, viz.,

$$P = q\xi = \sum_{i=1}^{\infty} \alpha_i Y_i(z).$$

Determine the coefficients α_i using the first orthogonality relationship:

$$\int_0^{\ell} P m(z) Y_i(z) dz = \int_0^{\ell} \alpha_i Y_i(z) m(z) Y_i(z) dz = \alpha_i M_i$$

$$= \int_{c-\xi/2}^{c+\xi/2} q\, m(z) Y_i(z) dz$$

The coefficients α_i may be found for q = constant as ξ becomes vanishingly small as:

$$\alpha_i = \frac{q\, m(c)}{M_i} Y_i(c)\xi = \frac{P\, m(c) Y_i(c)}{M_i}$$

The condition for balance is that the ith modal component must vanish. To achieve this a correction weight must be inserted at a location z = c on the rotor such that U(z) = P or:

$$\lambda_i Y_i(z) + \alpha_i Y_i(z) = 0 \qquad \text{or} \qquad \lambda_i = -\alpha_i$$

Thus,

$$P\, m(c) Y_i(c) = -\int_0^{\ell} m^2(c) e(z) Y_i(z) dz$$

if

$$U(z) \approx m(z) e(z).$$

To balance a rotor in its ith mode it is therefore necessary to know the mass distribution $m(z)$, the natural frequency ω_i of the mode to be balanced, the modal amplitude $Y_i(z)$ and the measured mode shape $y(z)$. From these the ith deflection coefficient $\emptyset_i$ and the ith eccentricity coefficient e_i are calculated. The required unbalance correction P at location $z = c$ for the ith mode may then be obtained using the above formula.

4.2.4 Papers by Bishop and co-workers on Modal Balancing

Between 1959 and 1967 an important series of papers on flexible rotor balancing was published by Bishop and his co-workers at University College, London. These papers greatly extended the theory and application of the modal balancing method, and drew attention to the need for flexible rotor balancing in general.

Bishop [9] discusses the vibrations of a rotating circular shaft having distributed mass and elastic properties, in which the displacement amplitudes are represented as a power series in terms of the modes. The analysis describes the modal concept, with examples. Gladwell and Bishop [10] applies the analysis of [9] to an axisymmetric shaft of non-uniform section in flexible bearings. It is shown how natural frequencies and characteristic modal functions may be found. A discussion of free and of forced motions is given. The receptance method is generalized to achieve this, and the modal equations are again obtained. Bishop and Gladwell [11] presents the underlying theory of a general method for the balancing of flexible rotors, mode by mode. The Jeffcott whirl theory is presented for a distributed mass-elastic rotor, and the series modal whirl amplitude solutions are given. The implications of low speed balancing are first examined analytically, and then the balancing of flexible rotors is considered. An example involving a uniform shaft balanced in two modes is discussed. The effect of a bent shaft and of shaft weight are also examined. Bishop and Parkinson [12] discusses a procedure whereby the whirl modes may be isolated for balancing. This procedure requires the rotor to be run near each critical speed in turn, to magnify the modal distortions at the balancing speeds. Two disadvantages of modal balancing are discussed, together with an adaptation of the Kennedy-Pancu [13] method of resonance testing to overcome these shortcomings. The practical aspects of applying modal balancing methods are discussed in great detail, and the results of balancing tests on three experimental rotors are described.

Lindley and Bishop [14] discusses the balancing of large turbine rotors. After reviewing the overall balancing problem, certain laboratory experiments are mentioned, followed by a discussion of industrial turbine rotor problems. A 200 mw generator rotor with 500 oz.in. unbalance was measured for modal balancing at 90 percent of its lowest critical speed. It was then trim balanced at full speed (3000 rpm) when cold, and after hot overspeeding. Improvements over the initial unbalanced condition of between 10 and 20 times are reported. This paper contains an informative contributed discussion by several experienced balancing engineers, and a closure by the authors.

Parkinson and Bishop [15] discussed the problem of residual vibration in a rotor after the modal balancing method has been applied. The suitability of measurements made with seismic pedestal-mounted transducers is considered, following from a comment by Morton in the discussion of [14]. Residual vibration in rotors running below their (first) critical speed may be corrected by the addition of a single mass correction, and such is demonstrated for a boiler feed pump rotor. Modal balancing is proposed for all modes through which the shaft runs, and an averaging technique is recommended for the remaining higher modes.

Bishop and Parkinson [16] is a study of second-order whirl in flexible rotors caused by gravity sag, which remains after the modal synchronous unbalance has been removed. This problem is prominent in two pole generators, due to dissimilar lateral stiffness of the shaft. This paper contains a modal analysis of the motion of such a rotor, in rigid bearings for a uniform dissimilar-stiffness shaft, and a study of the stability of such a rotor in free vibration and forced vibration. Isolation of the second-order modes by the Kennedy-Pancu method is examined, and a brief discussion of industrial rotor problems is given which includes both dissimilar shaft stiffness and the effect of coupling stiffness fluctuation.

Bishop and Mahalingam [17] undertook experiments on a simple rotor with dissimilar lateral stiffness and gravity sag, considering both synchronous whirl and second-order whirl effects. First, an alternate method to proximity probes is proposed for measuring the whirl motions, utilizing a slotted disk shutter. This apparatus consists of a heavy disk on a heavy shaft, operating in rolling element bearings. The elastic and inelastic properties of the rotor-bearing system are discussed, and also the balancing of the shaft. During balancing, a vibration level was reached beyond which further improvement was erratic, due to changes in shaft initial bend and from the relaxation of strains incurred near the critical speed during balancing. Balancing of an asymmetric shaft at its critical speed is discussed, including the observed strain relaxation effect. The second order gravity vibrations are then examined, including the variation of gravity sag with speed.

Parkinson [18] describes a technique for balancing shafts which are not axially symmetric. In this, the modal balancing method is applied to find the single plane of unbalance corresponding to each pair of (close together) modes. This condition results from the orientation of the unbalance to the principal inertia axes. A procedure for locating the unbalance angle is described, using the Kennedy-Pancu polar plot.

Parkinson [19] examines the effect of asymmetric bearing properties in the x- and y- directions on the vibration and balancing of a uniform shaft, again using the modal technique to generalize the interpretation of the distributed mass-elastic result. The equations of motion from earlier works are utilized and the modal functions are developed in terms of the receptances of the system. Two series solutions for the modal displacements in the principal stiffness directions are used to obtain general expressions for any displacement and speed. The specific example of an undamped simple shaft in asymmetric bearings is considered, for the cases of free whirl stability and for unbalanced forced whirling. Bearings with large stiffness asymmetry and with small stiffness asymmetry are studied. Finally an extension of the modal technique to balancing with asymmetric bearing stiffness is presented.

Parkinson [20] summarizes the contributions made by this group of workers to the unbalance response of shafts, dissimilar stiffness whirl, and stability of rotors in bearings, and to the vibration and balancing of shafts. The basic principles of each aspect are presented and discussed in terms of the modal approach.

4.2.5 Other Modal Balancing Contributions

A number of modal balancing problems were examined analytically by Kushul' and Shlyakhtin [21]. The procedures described apply to complex rotor shapes with both concentrated and distributed mass-elastic properties, without gyroscopic or rotatory inertia effects, in two or more radially-rigid bearings. A number of theorems for balancing flexible rotors are given, and determination of the most effective distribution of balancing planes is discussed. Three numerical examples are given to demonstrate use of these theorems and procedures. Although complicated, the mathematical procedures are expressed in sufficient detail to make this a valuable paper. Generally speaking, the omission of gyroscopic and rotatory inertia effects detracts from the prediction accuracy attainable, but should not represent a serious problem except for single-span rotors with overhung end disks. The restriction of the method to rigid bearing rotors for which the critical speeds and mode shapes must be calculated as eigenvalues in each instance appears to be a serious shortcoming, especially as the authors' experience indicates slow convergence of the modal series, even for the simple examples chosen. The influence of damping is neglected, and so the modal forms are plane curves rather than space curves. Although the analysis is thereby made tractable, the results achieved are more hypothetical and less related to practical rotors. No correlation with experiment results is included, and so the effectiveness of the method cannot be assessed from the results given.

Kellenburger [22] made a systematic analysis of the balancing requirements for a flexible rotor in two flexible end bearings. Distributed, variable, mass-elastic rotor properties are assumed. The unbalance is continuously distributed. The required correction masses are determined for a specified number of balancing planes, and the errors caused by neglecting the effect of higher modes are determined. Rules to guide the location of balance planes, and for selection of the required number of correction planes, are specified. The author's intention is a precise formulation of the requirements for mass balancing of flexible shafts with an infinite number of degrees of freedom. Damping effects are not considered, foundation mass-elastic properties are assumed negligible, and bearing stiffnesses are assumed identical in both coordinate directions. The amount of residual bearing force is taken as the criterion of the balance obtained. It is claimed that a minimum of (r + 2) balancing planes suitably distributed are required to achieve an effective balance, where r is the number of critical speeds within the speed range of the machine. Instances where this formula may fail due to unfavorable positioning of the balance planes are discussed. Residual errors arising from insufficient planes, various distributions of planes, and from unbalanced whirl modes are studied using a sample problem. No experimental or test confirmation of the procedure is included, and no commentary on application of the method is provided.

4.2.6 Influence Coefficient Method

The practice of inserting a trial weight to determine the effect of unbalance on a given whirl mode is fundamental to all balancing methods. The influence coefficient method is a formalization of this procedure whereby: (a) trial weights are inserted at selected locations along a rotor in a specified sequence, (b) rotor amplitudes and phase angles are read at convenient locations along the rotor, and (c) the required correction masses are computed from the amplitude and phase data, and installed. This method requires no preknowledge of the system dynamic response characteristics, although such knowledge is helpful in selecting the most effective balance planes and readout locations. The earliest theoretical procedure to explain this method for flexible rotors appears to have been due to Goodman [23] around 1961.

The Influence Coefficient method may be applied to balance the elastic rotor in damped flexible bearings shown in Figure 3 as follows. The rotor system is axisymmetric, and may have any variation in its axial geometry. At speed it experiences circular synchronous whirl under the influence of residual unbalance and modal whirl displacements. The analysis requires only that the whirl amplitudes and phase angles should be available at the selected measurement locations, in accordance with the procedure given below for determining the speed-dependent influence coefficients a_{ij} (ω). These coefficients relate whirl amplitudes w_i to unknown rotor unbalances U_i, i.e.,

or

$$\begin{Bmatrix} w_A \\ w_B \\ \cdot \\ \cdot \\ \cdot \\ w_p \end{Bmatrix} = \begin{bmatrix} a_{A1} & a_{A2} & \cdots & \cdots & a_{An} \\ a_{B1} & a_{B2} & \cdots & \cdots & a_{Bn} \\ \cdot & \cdot & & & \cdot \\ \cdot & \cdot & & & \cdot \\ \cdot & \cdot & & & \cdot \\ a_{p1} & a_{p2} & \cdots & \cdots & a_{pn} \end{bmatrix} \begin{Bmatrix} U_1 \\ U_2 \\ \cdot \\ \cdot \\ \cdot \\ U_3 \end{Bmatrix}$$

$$\{w\} = [a]\{U\}$$

Assuming the unbalance to be concentrated in n correction planes, the balancing procedure is to first run the unbalanced rotor at speed 'ω_1 ' and to measure the whirl amplitudes w and phase angles ϕ at p locations along the rotor. This gives:

$$\{w_o^1\} = [a_{ij}]_{\omega 1}\{U_o\}$$

where $\omega = \omega\,(\cos\omega t + i\sin\omega t) = \omega_c + i\,\omega_s$ as shown in Figure 3. Now insert a trial weight T in correction plane 1, and re-run the rotor at ω_1. Again read w and ∅ which correspond to:

$$\begin{Bmatrix} w^1_{A1} \\ w^1_{B1} \\ \cdot \\ \cdot \\ \cdot \\ w^1_{p1} \end{Bmatrix} = \begin{bmatrix} a^1_{A1} & a^1_{A2} & \cdots & a^1_{An} \\ a^1_{B1} & a^1_{B2} & \cdots & a^1_{Bn} \\ \cdot & \cdot & & \cdot \\ \cdot & \cdot & & \cdot \\ \cdot & \cdot & & \cdot \\ a^1_{p1} & a^1_{p2} & \cdots & a^1_{pn} \end{bmatrix} \begin{Bmatrix} U_1 + T \\ U_2 \\ \cdot \\ \cdot \\ \cdot \\ U_n \end{Bmatrix}$$

Now

$$\begin{Bmatrix} w^1_{A1} - w^1_{A0} \\ w^1_{B1} - w^1_{B0} \\ \cdot \\ \cdot \\ \cdot \\ w^1_{p1} - w^1_{p0} \end{Bmatrix} = \begin{bmatrix} a^1_{A1} & a^1_{A2} & \cdots & a^1_{An} \\ a^1_{B1} & a^1_{B2} & \cdots & a^1_{Bn} \\ \cdot & \cdot & & \cdot \\ \cdot & \cdot & & \cdot \\ \cdot & \cdot & & \cdot \\ a^1_{p1} & a^1_{p2} & \cdots & a^1_{pn} \end{bmatrix} \begin{Bmatrix} U_1 + T - U_1 \\ U_2 \quad - U_2 \\ \cdot \quad\quad \cdot \\ \cdot \quad\quad \cdot \\ \cdot \quad\quad \cdot \\ U_n \quad - U_n \end{Bmatrix}$$

Thus

$$a^1_{A1} = \frac{w^1_{A1} - w^1_{A0}}{T}$$

$$a^1_{B1} = \frac{w^1_{B1} - w^1_{B0}}{T}$$

$$\vdots \qquad \vdots$$

$$a^1_{p1} = \frac{w^1_{p1} - w^1_{p0}}{T}$$

The balance weight is then removed from correction plane 1, and is then successively inserted in the remaining (n − 1) correction planes. Successively re-running the rotor at ω_1 and measuring w and ϕ for each trial balance gives, in general:

$$a^1_{iq} = \frac{w^1_{iq} - w^1_{io}}{T} \qquad \begin{array}{l} 1 < i < p \\ 1 < q < n \end{array}$$

This procedure must be repeated q = (n/2) (n even) and q = (n+1/2) (n odd), where n is the number of correction planes. The resulting simultaneous equations define the rotor residual unbalance, referred to the correction planes:

$$
\begin{Bmatrix} w_{A1} \\ w_{B1} \\ \cdot \\ \cdot \\ \cdot \\ w_{p1} \\ w_{A2} \\ w_{B2} \\ \cdot \\ \cdot \\ \cdot \\ w_{p2} \\ w_{A3} \\ w_{B3} \\ \cdot \\ \cdot \\ \cdot \\ w_{pn} \end{Bmatrix}
=
\begin{bmatrix}
a^1_{A1} & a^1_{A2} & a^1_{A3} & \cdots & \cdots & a^1_{An} \\
a^1_{B1} & a^1_{B2} & a^1_{B3} & \cdots & \cdots & a^1_{Bn} \\
\cdot & \cdot & \cdot & & & \cdot \\
\cdot & \cdot & \cdot & & & \cdot \\
\cdot & \cdot & \cdot & & & \cdot \\
a^1_{p1} & a^1_{p2} & a^1_{p3} & \cdots & \cdots & a^1_{pn} \\
a^2_{A1} & a^2_{A2} & a^2_{A3} & \cdots & \cdots & a^2_{An} \\
a^2_{B1} & a^2_{B2} & a^2_{B3} & \cdots & \cdots & a^2_{Bn} \\
\cdot & \cdot & \cdot & & & \cdot \\
\cdot & \cdot & \cdot & & & \cdot \\
\cdot & \cdot & \cdot & & & \cdot \\
a^2_{p1} & a^2_{p2} & a^2_{p} & \cdots & \cdots & a^2_{pn} \\
a^3_{A1} & {}^3_{A2} & a^3_{A3} & \cdots & \cdots & a^3_{An} \\
a^3_{B1} & a^3_{B2} & a^3_{B3} & \cdots & \cdots & a^3_{Bn} \\
\cdot & \cdot & \cdot & & & \cdot \\
\cdot & \cdot & \cdot & & & \cdot \\
\cdot & \cdot & \cdot & & & \cdot \\
a^q_{p1} & a^q_{p2} & a^q_{p3} & \cdots & \cdots & a^q_{pn}
\end{bmatrix}
\begin{Bmatrix} U_1 \\ U_2 \\ \cdot \\ \cdot \\ \cdot \\ U_p \\ \cdot \\ \cdot \\ \cdot \\ \cdot \\ \cdot \\ \cdot \\ \cdot \\ \cdot \\ \cdot \\ \cdot \\ \cdot \\ U_n \end{Bmatrix}
$$

from which $\{U\} = [a]^{-1}\{w\}$. The insertion of n correction masses, each equal in magnitude but opposite in sense to the corresponding resultant unbalance at these locations has the effect of cancelling the original unbalance of the rotor.

The above procedure has been described by Rieger [24] who made an analytical study of the effectiveness of the influence coefficient method. Three practical rotor-bearing systems were examined, (a) rigid rotor in gas bearings, (b) supercritical flexible three-disk overhung rotor in fluid-film bearings, and (c) supercritical three bearing rotor with one disk overhung, in fluid-film bearings. The influence of measurement errors and correlation weight installation errors on the resulting balance was studied, along with balance improvement with two, three and four balance planes. The number of bearing supports involved was shown to exert no influence on the quality of balance attainable. Bearing misalignment may affect the location of the system critical speeds and shaft bending stresses, but exerts little effect on the quality of balance attained unless the whirl ellipses are excessively elongated, or unless the ellipse axes are oriented at different angles by the misalignment and bearing damping.

This method has been verified experimentally for a flexible three-disk rotor operating through its bending critical speed by Tessarzik [25] who performed the balance weight calculations with a computerized procedure. The rotor-bearing system used was designed to contribute negligible damping to restrain the rotor whirl amplitudes at the bending critical speed. After balancing, maximum whirl amplitudes of 1.6×10^{-3} inches peak to peak were observed at this critical speed, thus demonstrating the effectiveness of the balancing technique used. Tessarzik, Badgley, and Anderson [26] discuss these results in detail. A least-squares influence coefficient procedure also has been proposed by Goodman [27] for obtaining a best-fit balance for a rotor which operates over a speed range which may contain several critical speeds.

Experimental results using a least-squares balancing approach have been obtained by Lund and Tonneson [28], which verify the effectiveness of this method. A further development has been given by Little [29] in which linear programming was used to optimize the balance of rotors operating through several bending critical speeds. Baier and Mack [30] have described the balancing of long helicopter drive shafts through six critical speeds to beyond 7000 rpm. These authors achieved smooth shaft operation using influence coefficient balancing after having previously tried several other flexible rotor balancing methods.

The influence coefficient method has the advantage of simplicity in application, which makes it suitable for a wide range of complex turbomachinery applications (helicopter shafts, multispool aircraft engines, ultra-centrifuges, etc.), plus the convenience that the correction mass calculations may be computerized. Effectiveness of this method is not influenced by the presence of damping in the system, nor by vibratory motions of the locations at which readings are taken. Initially-bent rotors may be balanced as readily as straight rotors, and no assumptions concerning perfect balancing conditions are involved to detract from the quality of balance attained. It shares the shortcomings of certain other methods, i.e., the number of readings required to acquire the input data, and the accuracy with which these amplitude and phase readings must be made. Existing versions of the Influence Coefficient method assume circular whirl orbits. The influence of elliptical orbits from certain fluid-film bearing conditions, or from dissimilar pedestal stiffness, is presently not precisely known, nor are effects drive torque changes (causing shaft wind-up), changes in bearing operating eccentricity, stiffness, etc., and system non-linearities.

Goodman [27] also proposed a balancing procedure which uses a least-squares technique to minimize the rms residual amplitudes at selected locations on the rotor. A second application of the technique then uses weighted least-squares to reduce the maximum residual vibration. For a rotor which has n balance planes, and on which m vibration readings ($m > n$) have been obtained for k different conditions, at p different locations. Then $m = k \times p$. The least-squares balancing procedure finds the optimum size and arrangement of the required balance correction masses and angular orientations in the n balance planes, by minimizing the sum of the squares of the m vibration readings. Initially, the uncorrected unbalance data w_o and the trial weight unbalance data w_i at the m measuring locations are obtained. Influence coefficients a_{ij} are then calculated as described previously. However, the previous case is for $m = n$, and the required values of the correction weights were computed directly. This reduces the whirl amplitudes to zero at p locations at the selected speeds, and generally reduces the amplitude of rotor whirl throughout the speed range: see Figure 4. At speeds other than the balancing speeds, a small residual unbalance e_i remains such that at the nth location and speed:

$$w_{em} = w^r_{em} + i\, w^i_{em} = w_{mo} + a_{m1}U_1 + \ldots + a_{mN}U_N$$

$$= w_{mo} + \sum_{n=1}^{\infty} a_{mn}U_n$$

where the response amplitudes w_{em}, w_{mo}, the influence coefficients $a_{mn}(\omega)$, and the unbalances U_n are complex quantities. Multiplying out and dividing this expression into its real and imaginary components gives:

$$w^r_{em} = w^r_{mo} + a^r_{m1} U^r_1 - a^i_{m1} U^i_1 + \ldots = w^r_{mo}$$

$$+ \sum_{n=1}^{\infty} (a^r_{mn} U^r_n - a^i_{mn} U^i_n)$$

$$w^i_{em} = w^i_{mo} + a^i_{m1} U^i_1 + a^r_{m1} U^r_1 + \ldots = w^i_{mo}$$

$$+ \sum_{n=1}^{\infty} (a^i_{mn} U^i_n + a^r_{mn} U^r_n).$$

Let:

$$S = \sum_{i=m}^{M} [w_e]^2 = \sum_{i=m}^{M} [(w_e^r)^2 + (w_e^i)^2]$$

The objective is now to select the balance weights for which S is a minimum. To do this requires that:

$$\frac{\partial S}{\partial w_1^r} = \frac{\partial S}{\partial w_1^i} = \ldots = \frac{\partial S}{\partial W_n^r} = \frac{\partial S}{\partial W_n^i} = 0$$

This leads to 2n linear equations of the form:

$$\sum_m \left\{ a^r_{mn} \left[w^r_{on} + \sum_n \left(a^r_{mn} U^r_n - a^i_{mn} U^i_n \right) \right] + a^i_{mn} \left[w^i_{on} \right. \right.$$

$$\left. \left. + \sum_n \left(a^i_{mn} U^r_n + a^r_{mn} U^i_n \right) \right] \right\} = 0$$

$$\sum_m \left\{ - a^i_{mn} \left[w^i_{on} + \sum \left(a^r_{mn} U^r_n - a^i_{mn} U^i_n \right) \right] + a^r_{mn} \left[w^i_{on} \right. \right.$$

$$\left. \left. + \sum_n \left(a^i_{mn} U^r_n + a^r_{mn} U^i_n \right) \right] \right\} = 0$$

The unknowns in these equations are the components of the balance weights U_{nr} and U_{ni} required to minimize the rotor response. These terms may be found using standard equation-solving routines when the above procedure is programmed.

Goodman described a first iteration using the above procedure, and then a weighted least-squares procedure which may be applied in several successive iterations to minimize the residual unbalance following each iteration, until a satisfactory final balance is achieved. This final set of balance corrections and orientations may be obtained automatically by continuing the iterations until a pre-specified balance criterion is achieved. Further discussion of computerized balancing is given in Goodman [24], Rieger [21], Lund and Tonneson [25], and Tessarzik, Badgley, and Anderson [23].

4.2.7 Other Analytical Balancing Procedures

Several other flexible rotor balancing techniques have been proposed, a number of which are variations of the modal balancing procedure. In most instances the shortcomings of these techniques lie not in the theoretical approach, but in the practical acquisition of suitable measurement data required to implement the method.

Church and Plunkett [31] applied the mobility method to the balancing of a simple uniform flexible rotor. They gave an analysis for determining the magnitude and orientation of correction weights. Shaft deflection is represented in terms of normal modes for a single span shaft in rigid bearings, without damping. The proposed method requires that a set of modal response curves should first be obtained, preferably by analysis. The static deflection and the dynamic deflections of the shaft are then measured. The vector difference between these measurements allows the modal equations to be solved. The required balance corrections are then computed from the forces calculated to be causing the dynamic deflections. For the test example given, practical difficulties caused the authors to resort to shake-test measurements to obtain the desired mobility response. Recent experiences with helicopter drive shaft balancing suggest that this was probably due to the flexibility of the test model used, and to the measuring technique employed. As presented, the method was restricted to a single span rotor, but in principle the method should apply equally well multi-span rotors in flexible bearings.

Den Hartog [32] presented an interesting discussion of flexible shaft balancing in which the requirements for perfect balance were reduced to a theorem. Assuming that perfect rotor balance is attainable, the bearings will then function as simple rigid supports, as no transverse whirl motions will then occur to generate dynamic displacements (and hence forces) at these locations. This assumption allows the related conditions for perfect balance to be deduced directly. Equations for balance equilibrium were then established, first for a two-bearing rotor, and then for multi-bearing rotors. Several numerical examples are included. For clarity, the theory is developed with all unbalance occurring in the same plane. The effects of damping are ignored (no bearing velocity) though the author recognizes that practical balancing may involve three-dimensional deflections of the rotor. The theory also assumes axial symmetry throughout the rotor-bearing system and small initial unbalance. The validity of the balance condition depends to some extent on the degree to which these conditions can be met in a given situation. No means (analytical, or through test examples) for assessing the effect of practical departures from the stated requirements of the method, are included.

More recently, Hundal and Harker [33] gave a modal balancing analysis of a flexible rotor with non-uniform stiffness and mass distribution, in rigid end bearings. This method also requires that the critical speeds and normal modes should first be calculated. The whirl amplitudes along the shaft are then measured at a selected speed, to calibrate the modal amplitudes. The rotor unbalance is then determined, and so balancing corrections may be selected in the desired planes. This method may also be extended to multi-bearing rotors in flexible bearings, though this does not appear in the theory presented. The theory has been developed by considering all unbalance as located in the same plane, for simplicity. However, it is noted that practical balancing involves finding independent solutions in two mutually perpendicular planes, and subsequently combining these solutions to obtain the required balance corrections.

A recent review paper by Findlay [34] contains comments on the Modal balancing method, and on the analysis and experiments of Hundal and Harker [33]. The fundamentals of the Modal approach are first discussed, followed by a discussion of the practical limitations of the Modal method. It is noted that the normal modes and natural frequencies must be determined accurately, and that the test measurements must be taken mode-by-mode to achieve the required accuracy for the balancing calculations. Questions unanswered by Hundal and Harker are: How to select balance planes and correction masses in combination to optimize balance: special problems where two critical speeds exist close together: and details of numbers of balance runs and measurements needed. Findlay has not commented on effects arising from an asymmetrical rotor, nor a multi-bearing rotor system (though this is implied by the Hundal-Harker analysis), but several other balancing procedures are compared in general terms with the Modal balancing method: in particular, the Influence Coefficient method, and the least-squares procedure. Findlay's conclusion is that modal balancing is not a generally applicable balancing technique, because of serious practical difficulties involved in acquiring useful response data, and because of its reliance on calculated modal inputs.

Gusarov and Dimentberg [35] studied the dynamic effect of distributed and concentrated unbalance forces on the balancing of rotors. The balancing problem for flexible rotors with a limited number of available correction planes is studied. The absence of dynamic reactions at the supports and of optimum reduction of rotor bending stress is the objective of the balancing procedure. Damping effects on balance quality are considered.

Gusarov [36] further investigated the problem of eliminating the first two modes of vibration for an unbalanced rotor operating at speeds below its second critical speed. Balancing is to be achieved using two correction planes. It is supposed that the influence of higher modes at low speed can be neglected. The question of optimal placing of the correction planes along the shaft length is also investigated. A shaft of constant cross-section without disks is considered. In cases where it is only necessary to remove rigid rotor modes, two correction plans are evidently sufficient to significantly improve the rotor balance quality. Similarly, a flexible rotor in rigid bearings may be balanced using two correction planes for two modes. The more general case of a flexible rotor in flexible bearings is noted to require more planes.

4.2.8 Criteria for Flexible Rotor Balancing

An International Standards organization document is now available to guide the balancing of flexible rotors. Flexible rotor balancing concepts are presented in the ISO Document ISO/TC108/SCI: 'Mechanical Balancing of Flexible Rotors' (1976)[37]. This document proposes a general classification of rotor types into specific Classes I through V, as shown in Table 4.2.1 from this ISO document. Specific rotor types are classified in accordance with their operating requirements, and guidance is provided on the selection of the most suitable method of balancing the rotors in each Class. Of the five recommended Classes, Class II rotors are quasi-flexible rotors, which may be corrected for smooth operation by the addition of masses in two balance planes. Only Class III rotors are truly flexible rotors in the sense that they require the use of some established flexible rotor balancing method to provide smooth operation. Flexible rotors are observed to require more than two correction planes to achieve smooth operation throughout their speed range. All other Classes of rotors may be adequately balanced by either two-plane techniques, or by simple three-plane balancing. Document ISO/TC108/SI1 (1976) is concerned with the adaptation of rigid rotor balancing concepts, equipment, and balancing machines to the balancing of high speed, flexible, two-bearing rotors. Brief comments are given to coupled rotor balancing. No detailed classification of flexible rotor types is presented (multi-bearing, multi-spool, etc.). Both the Modal method and the Influence Coefficient method are described in detail.

Criteria for flexible rotor balancing are presented in ISO Document ISO/TC108/SCI (1976), with a tabulation of typical numerical values for bearing velocity for the various rotor classes and measurement procedures used, to guide the selection of acceptable support vibration response. These criteria are reproduced in Table 4.2.2. This table accounts for rotor class, machine type, measurement location, and balance quality details. These criteria (readout velocity mm/s) should be regarded at typical, and are consistent with other machinery vibration criteria values for each rotor class.

4.2.9 Flexible Rotor Balancing: State-of-the-Art

Modal Balancing:

The Modal balancing method is the most developed and proven flexible rotor balancing method available in the open literature. Modal balancing has been discussed by many authors, but the most extensive contributions have been made by Bishop and his co-workers. These investigators have successfully applied Modal balancing to turbine-generator rotors, and their techniques are used in current practice. Modal balancing methods have also been applied for problems of dissimilar lateral stiffness, hysteretic whirl, and to complex shaft-bearing problems, such as a shaft with distributed dissimilar stiffness and mass properties operating in bearings with dissimilar x- and y- stiffnesses. Effective balancing has been achieved under laboratory conditions, in these instances. Such developments are in advance of the commercial applications of modal balancing at present. They are confined primarily to symmetrical two-bearing rotors.

Effects arising from the bearing fluid film have received only minor consideration in the modal balancing literature. The fluid film is the major source of system damping, and it also governs the rotor stability problem. The damping contributed by the fluid film strongly influences the phase angle and its rate of variation near the system critical speeds. This further contributes to the practical difficulties of taking effective balancing measurements. It should also be recognized that rolling-element bearing effects can be as important as shaft effects in lightly-damped systems, such as aircraft gas turbines.

Concerning the modal balancing technique itself, the problems which appear to confront its wider application to more general i.e. complex, damped, multi-shaft, or multi-bearing rotor systems are as follows:

o Calculation of modes and frequencies

To use modal methods, information is required on the mode shapes and natural frequencies of the rotor-bearing system to be balanced.[3] The accuracy of the computed results depends on the capabilities of the computer program used, and on the input data (dimensions, coefficients, effectiveness of system model) used in these calculations. In complex turbomachinery, this may create problems, especially where system damping is significant, such as with fluid-film bearings.

3Den Hartog, Kellenberger, Kushul', and Shlyaktin, and others have required that rotor modes only need to be obtained, using the criterion that rotor amplitudes at the bearings are to be balanced to zero. In complex systems this is frequently impossible, e.g., for an aircraft turbine damper bearing, and in any case the effects of non-zero bearing amplitudes on this approach remains unexplored. The alternative of accepting the existence of bearing amplitudes, as proposed by Bishop appears to be more realistic and moe general, judging from practical experience to date.

The mode shapes and resonant frequencies of heavily damped systems often bear little resemblance to undamped mode shapes and frequencies. The use of a damped response program is therefore essential in such cases to obtain accurate resonant frequencies and mode shapes. At present, the available circular orbit response programs are usually adequate for single-level systems, but few two-level response programs are available. The reliance of modal balancing on predicted modes and frequencies is, at least, an inconvenience. It could constitute a significant disadvantage for the application of modal balancing to complex machine structures involving multi-spool, or shaft-structure system unless adequate response programs are available to obtain the analytical balancing inputs. It should further be noted that damped rotor response also depends on unbalance distribution. The influence of an unrepresentative distribution of unbalance on the accuracy of modal balancing should be further examined. Evidently a procedure which accurately calibrates the unbalance distribution model would be a useful contribution to modal balancing.

o Measurement of Modal Amplitudes and Frequencies

Experience has shown the difficulty of trying to extract amplitude and phase data from broad-band measurements taken where several modes are present simultaneously. It is well known that the resonant amplification associated with any given mode may be utilized to obtain the required modal data (shape, frequency, phase) by running the rotor at or close to resonance in that mode. Where sharp or severely resonant modes are involved, approximate (near-resonant) data may first be used to obtain an improved initial balance, before a final balance is attempted. A further measurement problem results from the rapid variation in phase angle which occurs in a lightly damped system in the vicinity of resonance. Even when practical modal balancing is performed on a mode-by-mode basis, it is evident that a certain residual whirl amplitude will still result at operating speed, due to the effect of the higher modes. Recognizing this situation, Parkinson and Bishop [15] proposed a correction for such modes which uses data acquired at the rotor operating speed.

Multi-Mode Balancing:

The Modal theory suggests that several modes may be balanced simultaneously, but this does not appear to have been achieved in practice. Bishop and Gladwell [11] recommend balancing mode-by-mode. Church and Plunkett [31] resorted to shaking the rotor to ascertain the mode shapes. Hundal and Harker [33] were unable to achieve low amplitudes in the vicinity of the first critical speed, and appear to have been unable to pass through the second critical speed.

Influence Coefficient Method:

Successful applications of the Influence coefficient technique appear to include a wider range of rotating machinery types than those which have been reported for other flexible rotor balancing methods. Known applications include: a high speed (24000 rpm) pump simulator, a long helicopter drive shaft (20 ft. unsupported span, six critical speeds), certain ultracentrifuge applications, several small steam turbines, and small aircraft gas turbine applications. Notable features reported in these applications were:

o Convenience of Application

The method is widely recognized as being simple to apply, but it usually requires the acquisition and careful analysis of a large amount of readout data. 2n sets of amplitude and phase angle data are required for the Exact-Point-Speed method, and preferably several more than 2n sets for the least-squares method. This data acquisition is fairly straightforward for operation through one or two critical speeds, but the amount of data required in cases where the speed range involves many criticals, i.e., helicopter drive shafts, may become very large. Some automated form of data-taking and recording on magnetic tape is evidently advantageous, and preferably arranged in a form suitable for direct use as input for the balance mass and angle calculations which follow. With such an arrangement, the Influence Coefficient method can provide a rapid and efficient balancing procedure.

o Accuracy of Balance Attainable

The factors which limit the present form of the Influence Coefficient method are (a) the precision to which the measurements of amplitude and phase may be acquired, (b) the assumption of a circular rotor whirl orbit in the present procedures, and (c) the constancy of the test set-up, when readings are being taken. These shortcomings are, of course, inherent in every flexible rotor balancing method yet proposed. To remove them will require: for (a) the development of a precision electronic data-sampling system which is capable of reading and storing amplitude and phase data at all rotor locations simultaneously, (b) development of advanced analytical procedures which include elliptical orbit effects in the rotor balancing procedure, and for (c) a programmed statistical technique for evaluation of the sampled amplitudes, phase and speed data. Experience shows that the acquired input data may vary substantially, even over a very short period of time. One reason for this variation is rotor speed fluctuations, which may be especially serious in the vicinity of a critical speed. Each data reading could be electronically sampled many times over a fairly short time period (1000 revs. = 1000 readings). The statistically-analyzed results could then be used to compute the required balance weights and angles to give a refined statistical balance. An indication of the balance accuracies attainable with present methods is given in the results of Bishop et. al., Hundal and Harker, and Church and Plunkett. The effect of errors of measurement and of balance weight installation on the quality of balance obtained has been studied by Rieger [24].

Other Methods:

Most of the other methods proposed have been variations of the Modal balancing method. Where experimental verification has been sought, tests have usually been made on long slender shafts. The success achieved in most cases has been marginal. This has usually been due either to inherent limitations in the techniques themselves or to practical difficulties of measuring the modes, including the skill of the operators. Other practical problems of amplitude and phase measurement, speed fluctuation, data variation, and of shaft crookedness, etc., have also detracted from quality of correlation achieved.

Other techniques such as the empirical one-shot method are restricted to the applications for which they were developed, and seem to offer little promise of further development. Still other methods require that amplitudes should be reduced to zero at the bearings, which may not always be possible due to system restrictions in cases involving flexible bearings, e.g., with an overhung rotor, for multi-bearing rotors, and with variation the operating drive loads, etc. The effect of practical factors such as these do not yet appear to have been investigated. Finally, suggestions such as the zero-amplitude method which depends on accurately calculated whirl amplitudes for selecting the balance weights and angles, do not appear to offer any practical advantage.

4.2.10 Comment

It should be noted that the flexible rotor balancing techniques discussed in this section are effective only against synchronous rotor residual unbalance. In each instance, some procedure for filtering out and measuring the synchronous unbalance component is an essential part of the balancing technique. Where a machine has becomes unstable in its bearings, no amount of balancing will restore its stability. Rotor instability is an entirely separate phenomenon for which procedures other than balancing are needed.

4.2.11 References

1. Linn, F. C., 'Method of and Means for Balancing Flexible Rotors', U.S. Patent 1,776,125, Filed September 17, 1928.

2. Thearle, E. L., 'Dynamic Balancing of Rotating Machinery in the Field,' Trans. ASME, Vol. 56, pp. 745-753, 1934.

3. Kroon, R. P., 'Balancing of Rotating Apparatus,' Part I, Journal of Applied Mechanics, Vol. 10, pp. A-225, 1943. Part II, Journal of Applied Mechanics, Vol. 11, pp. A-47, 1944.

4. Grobel, L. P., 'Balancing Turbine-Generator Rotors,' GE Review, Vol. 56, No. 4, pp. 22, 1953.

5. Moore, L. S., Dodd, E. G., 'Mass Balancing of Large Flexible Rotors,' GEC Journal, Vol. 31, No. 2, 1964.

6. Lindsey, J. R., 'Significant Developments in Methods for Balancing High Speed Rotors,' ASME 69-Vibr-53, Vibrations Conference, Philadelphia, PA, March 1969.

7. Meldahl, A., 'Auswuchten Elastischer Rotoren,' Z angew, Math und Mech. 34, 1954.

8. Timoshenko, S. P., Vibration Problems in Engineerings, D van Nostrand Company, Inc., Princeton, NJ, 3rd Edition, 1955.

9. Bishop, R. E. D., 'The Vibration of Rotating Shafts,' Journal of Mechanics and Engineering Sciences, 1, 50, 1959.

10. Gladwell, G. M. L., Bishop, R. E. D., 'The Vibration of Rotating Shafts Supported in Flexible Bearings,' Journal of Mechanical Science, 3, 195, 1959.

11. Bishop, R. E. D., Gladwell, G. M. L., 'The Vibration and Balancing of an Unbalanced Flexible Rotor,' Journal of Mechanics and Engineering Sciences, 1, 66, 1959.

12. Bishop, R. E. D., Parkinson, A. G., 'On the Isolation of Modes in the Balancing of Flexible Shafts,' Proceedings IMechE, 177, 16, 407, 1963.

13. Kennedy, C. C., Pancu, D. D. P., 'Use of Vectors in Vibration Measurements and Analysis,' Journal of Aeronautical Science, 14, 1947.

14. Lindley, A. G., Bishop, R. E. D., 'Some Recent Research of the Balancing of Large Flexible Rotors,' Proceedings IMechE, 177, 30, 811, 1963.

15. Parkinson, A. G., Bishop, R. E. D., 'Residual Vibration in Modal Balancing,' Journal of Mechanics and Engineering Science, 1, 33, 1965.

16. Bishop, R. E. D., Parkinson, A. G., 'Second-Order Vibration of Flexible Shafts,' Phil. Trans. Royal Society of London, 259, Series A, 1, 1965.

17. Bishop, R. E. D., Mahalingam, S., 'Some Experiments in the Vibration of a Rotating Shaft,' Proceedings Royal Society of London, 292, Series A, 537, 1966.

18. Parkinson, A. G., 'On the Balancing of Shafts with Axial Asymmetry,' Proceedings Royal Society of London, 292, Series A, 66 1966.

19. Parkinson, A. G., 'The Vibration and Balancing of Shafts Rotating in Asymmetric Bearings' Journal of Sound and Vibration, 4, 477, 1965.

20. Parkinson, A. G., 'An Introduction to the Vibration of Rotating Flexible Shafts,' Bulletin Mechanical Engineering Education, 47, 1967.

21. Kushul, M. Ya., Shlyaktin, A. V., 'Modal Approach to Balancing with Additional Constraints,' Isvestiya Akademii Nauk SSSR, Mekhanikai Mashinostroyeniye, No. 2, 1966.

22. Kellenburger, W., 'Balancing Flexible Rotors on Two Generally Flexible Bearings,' Brown Boveri Review 54, No. 9. 1967, pp. 603-617.

23. Goodman, T. P., 'A Least-Squares Method for Computing Balance Corrections,' ASME 63-WA-295, September 1964.

24. Rieger, N. F., 'Computer Program for Balancing of Flexible Rotors,' MTI Report 67TR68, September 1967.

25. Tessarzik, J. M. 'Flexible Rotor Balancing by the Exact Point-Speed Influence Coefficient Method,' NASA Technical Report CR-72774, Lewis Research Center, October 1970.

26. Tessarzik, J., Badgley, R. H., and Anderson, W. J. 'Flexible Rotor Balancing by the Exact Point-Speed Influence Coefficient Method,' Paper Submitted to ASME Third Vibrations Conference, Toronto, Canada, September 1971.

27. Goodman, T. P., 'Correction of Unbalance by Force-Canceling Bearing Pedestals,' GE Company Report No. 61GL110, May 15, 1961.

28. Lund, J. W., Tonneson, J. 'Analysis and Experiments on Multi-Plane Balancing of a Flexible Rotor,' ASME Third Vibrations Cconference, Toronto, 1971.

29. Little, R. M., 'Current State of the Art of Flexible Rotor Balancing Technology,' Ph.D. Thesis Bibliography, University of Virginia, Charlottesville, 1971.

30. Baier, R., Mack, J., 'Design and Test Evaluation of a Supercritical Speed Shaft,' The Boeing Company, Vertol Division, Morton, PA, USAAVLABS Technical Report 66-49/R458, June 1966.

31. Church, A. H., Plunkett, R., 'Balancing Flexible Rotors,' Trans. ASME, Journal of Engineering for Industry, Vol. 83, Series B, No. 4, pp. 383-389, November 1961.

32. Den Hartog, J. P., <u>The Balancing of Flexible Rotors</u>, Air, Space and Instruments, Stark Draper Commemoration Volume, McGraw-Hill, 1963, pp. 165.

33. Hundal, M. S., Harker, R. J., 'Balancing of Flexible Rotors Having Arbitrary Mass and Stiffness Distribution,' Journal of Basic Engineering, Trans. ASME, Paper No. 65-MD-B, June 1965.

34. Findlay, J. A., Review of paper: 'Balancing of Flexible Rotors,' by Hundal and Harker, ASME Paper 65-MD-8.

35. Gusarov, A. A., Dimentberg, F. M., 'Balancing of Flexible Rotors with Distributed and Concentrated Mass,' Proceedings, Procnosti v Mashinistroenii (Problems of Elasticity in Machinery) Publication House Academy of Sciences, 1960, Moscow No. 6, pp. 5-37.

36. Gusarov, A. A., 'On Placing of Balancing Planes on Flexible Rotor,' Kolebania i Prochnost pri Peremennyh Naprjeniah (Vibrations and Elasticity for Variable Stresses), Proceedings, National Research Institute of Machinery, Soviet Academy of Sciences, Publication House, Nauka Moscow, 1965, pp. 112-124.

37. ISO Draft Document, 'The Mechanical Balancing of Flexible Rotors', ISO/TC 108/SC1/WG2 (Secretariat-7) Document 12, 1976.

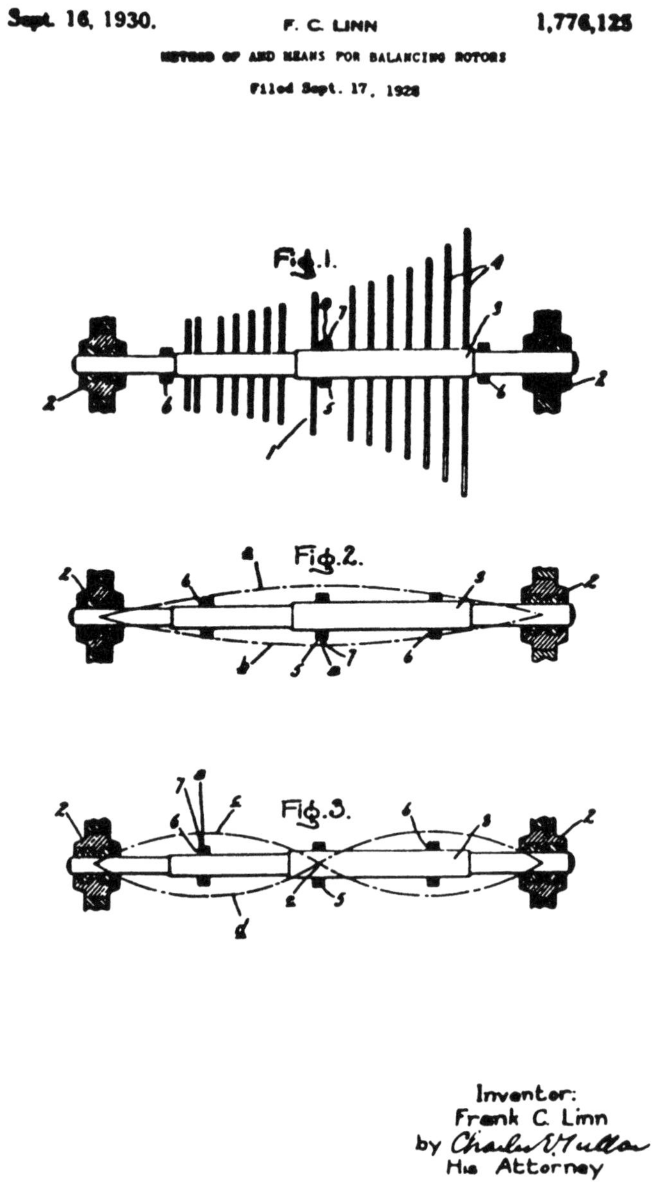

Figure 1 Title Sheet of Linn Patent for Flexible Rotor Balancing (1928)

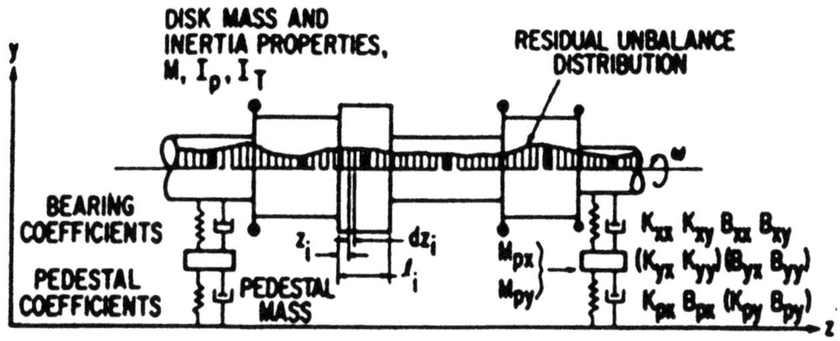

Figure 2 Y-Z Distribution of Mass, Elasticity, and Residual Unbalance Within a Generalized Rotor (X-Z Distribution is Similar)

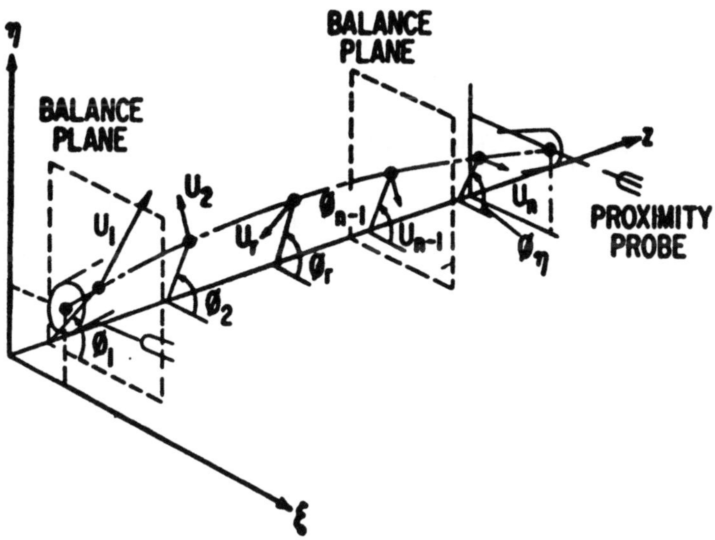

Figure 3 Representation of Axially-Distributed Rotor Unbalance by n Discrete Unbalance Moments

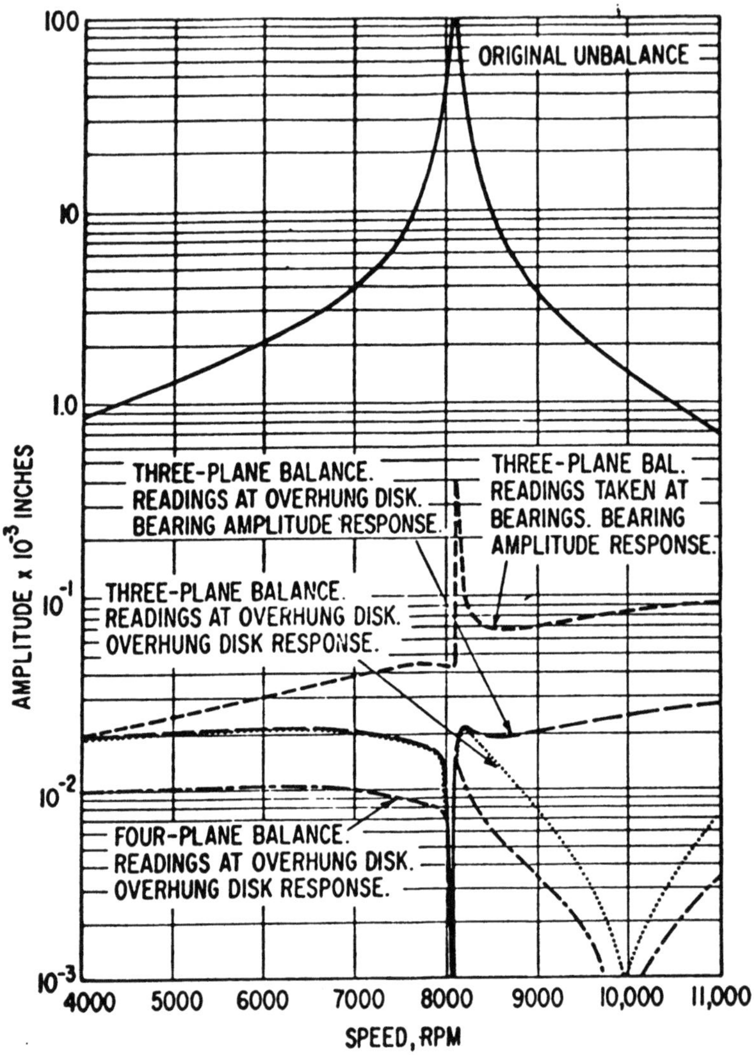

Figure 4 Effect of Number of Planes on Balancing Response at Bearings and Overhung Disk

Table 1 Flexible Rotor Balance Criteria From ISO Draft International Standard Document 5343 (1978)

Rotor Category	Ranges of effective pedestal vibration velocity at once-per-rev frequency, V_1 (mm/sec) rms: 0.28, 0.45, 0.71, 1.12, 1.8, 2.8, 4.5, 7.1, 11.2, 18, 28, 45, 71	Correction Factor C_1	C_2	C_3
I	Quality Bands: A, B, C, D			
	Small electric motors up to 20 HP.	0.63		
	Superchargers.	0.63	2	
II	A, B, C, D			
	Paper making machines	0.63		
	Medium size electric motors & generators, 20-100 HP on normal foundations.	0.63	4	
	Electric motors and generators up to 400 HP on special foundations.	0.63	4	20
	Pumps and compressors.	0.63	8	15
	Small turbines.	0.63	4	8
III	A, B, C, D			
	Large electric motors.	0.63	5	
	Turbines and generators on rigid and heavy foundations.	0.63	5	20
IV	A, B, C, D			
	Large electric motors, turbines, and generators on lightweight foundations.	0.63	3	10
	Small jet engines.	0.63		
V	A, B, C, D			
	Jet engines larger than category IV.	0.63	2	10

CHAPTER 4.3

CASE HISTORIES IN BALANCING OF HIGH SPEED ROTORS

N.F. Rieger
Stress Technology Incorporated, Rochester, New York, USA

ABSTRACT

Three case histories concerning rotor balancing problems from practice are described in detail. These case histories refer to problems in which the cause was rotor unbalance and/or unsuitable rotor dynamic conditions. In each instance, a diagnostic phase preceded the eventual corrections made to the system. These cases demonstrate both the diagnostic techniques used, and the successful corrections made.

4.3.1 Introduction

The case histories described in this chapter demonstrate the application of diagnostic techniques to balancing problems and the successful application of balance correction procedures to practical machine situations. In all instances it is important that a correct diagnosis of the problem should precede any attempt to apply a solution. The following examples are drawn from actual experience. They demonstrate both the procedure used for identification of the problem cause, and the balancing procedure which was subsequently used.

4.3.2 Case History 1 - Thermal Unbalance Vibrations of a Generator Rotor

Severe vibrations of the generator casing structure shown in Figure 1 were observed. End plate amplitudes up to 11.0 mils were felt adjacent to the generator bearings. It was considered that the fabricated end plates would experience fatigue failures of the welds if operation in this condition was continued.

The generator had a maximum output of 4000 kw, and it was operated at 3600 rpm to produce 60 Hz power for a paper mill. The generator casing was of standard welded steel fabrication, and the generator rotor was direct-driven by a steam turbine unit. Both the generator rotor and the turbine rotor were supported in two elliptical oil film journal bearings. The generator rotor was driven through a Falk gear tooth type coupling. The entire unit was located on the third floor of a concrete frame building, mounted on a steel bedplate, and supported upon massive reinforced concrete beams and columns.

A chart of axial vibration amplitude response vs. frequency at the generator number 4 bearing is shown in Figure 2, after the first stiffening fix had been applied to the generator end plates. A frequency spectrum for the number 4 bearing in the vertical direction for operation at 3600 rpm is shown in Figure 3.

Initial operation had indicated that strong vibrations were occurring at the generator bearing supports, in the horizontal, axial, and vertical directions. The end plates were subsequently stiffened. This reduced the vibration to the levels shown in Figure 2, but did not eliminate the vibration.

It was observed that the vibrations typically developed some time after unit startup. Initially, the unit ran smoothly. Both the turbine rotor and the generator rotor had recently been balanced in a local shop. Once started, the vibrations persisted, and the unit eventually had to be shut down.

Characteristics of the Problem:

a) Axial vibrations of the generator end plates was the worst aspect.

b) Vibrations were worst under high generator load conditions. Low load vibration was typically low amplitude.

c) Vibrations were worst when structure temperature was high.

d) End plate axial mode natural frequency was around 68-71 Hz.

e) Rotor system natural frequency was around 34 Hz.

f) Measured vibration frequency was mostly synchronous, at 60 Hz.

g) Other harmonics were typically smaller than the 60 Hz component in the measured vibration frequency spectra.

Possible Causes: The following possible causes were each considered in detail during the problem diagnosis phase.

a) Residual rotor unbalance arising from:
   Uneven mass distribution
   Bent shaft
   Thermal bow
   Elliptical journals

b) Bearing fluid film instability due to:
   Half frequency whirl
   Resonant whip

c) Magnetic structure interaction:
   Waveform harmonics
   Structure shell modes
   Eccentric armature

d) Misalignment:
   Hot vs. cold alignment
   Coupling forces
   Thrust bearing force
   Foundation distortion
   Multi-bearing support alignment

e) 'Flat' Shaft effect:
   1 /rev, 2/rev, possible 3/rev
   4/rev

Observations on Possible Causes:

a) Mechanical residual unbalance can be discounted as a serious cause because the startup vibration at no load was small and within acceptable limits. Bent shaft unbalance effect was likewise rejected as a possible cause for the same reason.

b) Sequential testing (removal of excitation, increased generator cooling, decreased oil inlet temperature) of the system indicated that:

- Magnetic effects were not a major cause of excitation, or their removal would have minimized the vibration immediately.

- Thermal effects could be likely cause. Under cooler operating conditions the observed vibration decreased in each instance.

c) Fluid film instability is discounted because:

- No sustained half-speed frequency component was consistently observed in the test frequency results. Therefore there was no half-frequency whirl occuring in the system.

- No strong 34 Hz component was observed in the bearing frequency spectrum, when operating at 3600 rpm.

- The rotor operates below twice its bending critical speed: therefore the cause is not resonant whip.

d) No evidence of any strong foundation vibration was found. Measurements showed that vibrations were strongest at the generator bearing caps, and were small elsewhere in the concrete foundation.

e) Flat Shaft effect: Exactly 2x, 3x, 4x, etc., harmonics observed in the vibration spectra suggest that some flat shaft effect may be present in the generator, most likely due to the two-pole rotor construction.

f) Hot vs. cold rotor alignment checks were made to identify possible casing distortion. No significant misalignment was found.

g) Coupling alignment was checked by sling test. Both the drive coupling and the exciter coupling were found to be well aligned.

h) Output current waveforms could also be checked. Casing 'shell' modes are typically high, but they have significant structural damping due to coupled winding movements. No evidence of an eccentric armature was found other than a strong 60 Hz component, as expected.

i) Elliptical journal may cause 2/rev radial vibration but not 1/rev.

Natural Frequencies of End Plates:

Vibration measurements were made on the generator pedestals and on both sets of end plates. Finite element natural frequency calculations were made for (a) the original unstiffened end plates, (b) for the first stiffening of the end plates, and (c) for a proposed second stiffening. The originally stiffened construction is shown in Figure 1. Details of the finite element model are shown in Figure 4. The first mode of this structure is shown in Figure 5. The first mode is a diaphragm type vibration mainly involving motions of the upper end shield. The second mode has a vertical diameter with anti-symmetrical motions.

The original end plates had a measured natural frequency of 63.5 Hz. This agrees with the calculated natural frequency of 64 Hz. Test results showed that the lowest natural frequency of the initially stiffened end plate structure was at 68 Hz at the turbine bearing end, and approximately 71 Hz at the generator bearing end. The finite element result for the initially stiffened case was 69 Hz, with reinforcing members added to the original F.E. model. It also is noted that the measured second natural frequency of the end plates occurred at 86 Hz for the turbine end bearing, and at 87 Hz for the generator end bearing. The calculated second natural frequency of the end plates occurs at 84 Hz.

The correlation obtained between the test results and the calculated results for natural frequencies is close enough to validate the finite element models of the original structure and of the existing stiffened structure.

Positive Indications:

a) Increased generator output power leads to increased operating temperature. Some time lag occurs until thermal equilibrium is reached. Vibration intensity is directly related to higher temperatures, and vibration intensity follows temperature growth.

b) Mechanical unbalance and bent shaft are discounted as causes of the problem because the shaft runs smoothly during no-load startup, and when it cools down.

c) The strongest response component is always at 60 Hz under synchronized operating conditions.

d) Electrical excitation per se appears to have a minor effect on the vibration, as indicated by the sequential test results.

e) System bending critical speed occurs at 34 Hz. This was observed in the run-down test data. This means shaft resonance is not aggravating the problem.

f) Calculated and measured end plate first-mode natural frequencies occur at 68 Hz and 71 Hz. Increased end plate natural frequency values resulted in decreased end plate axial vibration amplitudes. This suggests that plate resonance is contributing to the problem. It also suggests that further detuning, i.e., further stiffening, should lead to smaller end plate vibrations.

g) Thermal effects could cause the shaft to bow. The relatively slow development of thermal bowing in the generator rotor due to the slotted generator construction is consistent with the observed slow worsening vibration problem.

h) Shaft thermal bowing would also explain the correlation between high vibration and high generator power output (when steady state conditions exist), and the strong synchronous component observed in the results.

i) Slow cooling of the rotor and a residual bow would also explain the persistence of the vibration when generator load is reduced, and the lack of correlation between generator load and power level.

j) The 2x, 3x, etc. components may be due to either flat shaft effect (most probably), or to uneven electrical airgap effects. Flat shaft effect seems likely because the multiple harmonics remain after excitation is removed. Airgap effect could be checked in detail by examining the electrical waveform.

k) Bearing (or casing) misalignment (mechanical, thermal) could cause some 1x effect.

l) Coupling may be occurring between radially transmitted forces from shaft thermal unbalance excitation at 60 Hz, and the axial end plate mode at 68-71 Hz. This coupling could occur because:

- The end plate mode is a zero diameter (umbrella) mode. This is strongly evident from measurements, and is confirmed by the finite element calculations.

- The thermally bowed shaft could be applying a transverse rotating couple to the bearing support as well as a radial rotating force. This would tend to excite the end plate in the observed phase relation of this mode.

- The observed anti-phase relations between the end plate vibration displacements agrees with the thermal bow/bearing support excitation hypothesis.

4.3.3 Conclusions

- o The observed end plate axial vibrations were caused by synchronous vibrations transmitted from the generator rotor.

- o The transmitted vibrations were caused by mechanical unbalance due to thermal bowing of the generator rotor.

- o Rotor thermal bowing most likely results from uneven expansion of the slotted armature, possibly due to coil binding, local hot spots, etc.

- o The end plates respond strongly to this excitation because the bearings are overhung, and the lowest end plate mode is close to the synchronous frequency of 60 Hz.

- o The higher harmonic components in the response waveform are due to dissimilar shaft moments of inertia, the 'flat' shaft effect.

- o Damping is low throughout the system. This causes sharply defined high amplitudes responses to occur at 34 Hz and at 68-71 Hz, as observed in the run-down test results.

4.3.4 Corrections Made to System

- o The shaft was mechanically balanced to compensate for the thermal bow at full power and 3600 rpm. This caused some increase in the transmitted cold vibration amplitudes. The thermal balance correction was adjusted to reduce the hot vibration amplitudes to minimal values consistent with very smooth operation, and low axial vibration of the end plates.

- The end plates were further stiffened with members which oppose the first mode shape. This further detuned the end plate modes. This backup fix was recommended in case the shaft becomes otherwise thermally unbalanced sometime in the future.

4.3.5 Case History 2 - Unbalance Vibrations of a Boiler Feed Pump Motor

A schematic of a boiler feed pump rotor is shown in Figure 6. An unslotted forging of this rotor was successfully balanced in a test stand. The balanced rotor forging ran smoothly, close to its observed first and second critical speeds, Figure 7, and at the intended operating speed of 2300 rpm. Problems appeared when a final balance of the completed rotor was attempted. The rotor was successfully balanced for operation near its first critical speed, but severe vibrations were encountered as the operating speed was approached. It was assumed that these vibrations were caused by the second mode of the rotor, which lay beyond the rotor overspeed condition. An attempt was then made to correct the second mode by inserting correction masses on opposite sides of the rotor, near the points of maximum second mode deflection, Figure 8. It was found that one arrangement of correction masses could be selected to make one or the other bearing run smoothly, but the remaining bearing would then experience rough vibrations, as shown in Figure 9. With these corrections, none of the operating speed vibrations could be brought within acceptable limits. It was therefore important to locate the root cause of these vibrations, to balance the rotor.

No calculations were available for locations of the rotor critical speeds. It was observed that near 2300 rpm the vibrations were in-phase at the ends of the rotor. It was therefore concluded that there was an error in the first mode correction, due to mis-proportioning of the masses added for the attempted second mode correction. It was eventually deduced that the rotor was responding to unbalances in its second and third modes simultaneously, even though it was running well below it second critical speed. The schematic in Figure 10 indicates how this 'mixed mode' problem could arise.

Comments on Mixed Mode Correction Procedure:

When a rotor, running at full speed, is under the influence of unbalance in both second and third modes, the vibration of the bearing pedestals will normally contain an out-of-phase component from the second mode and an in-phase component from the third mode. These two components may be at any angle with respect to each other, and the resultant vibrations of the bearings will likely appear as two unequal vectors, not in the same plane.

The solution lies in resolving the resultant vectors back into their respective components, as shown in Figure 9. In addition, because either mode may have unequal effects on the two bearings from rotor to rotor, the constant of proportionality may vary widely. This information cannot easily be deduced with sufficient accuracy from the calculated modal shapes, since many relevant factors such as foundation stiffness and bearing stiffnesses enter these calculations. However, even if the unbalance could be positively resolved into definite amounts in the second and the third modes at specific angular positions, the information would not yet have practical value because it is virtually impossible to add correction masses which will influence only one mode at a time on a real rotor, given a limited number of available balance correction planes.

Balancing of Mixed Modes:

First assume that the rotor has been balanced in its lowest mode around 2300 rpm. At some higher speed, well below its second critical speed, it now is deflecting in its second and third modes simultaneously. The mode shapes are those shown in Figure 6. The most suitable locations for balancing planes are at 25 inches and 70 inches from bearing A for the second mode, and 10 inches, 44 inches, and 90 inches from bearing A for the third mode. Also assume that all correction masses are to be added at the same radius in these planes.

The first requirement is that there shall be no disturbance of the condition of balance already achieved near the first critical speed. Considering the relative deflection curve shown in Figure 7, if a calibrating mass were to be moved along the length of the rotor, its effect on the rotor when running near its first critical speed would be proportional to the relative modal deflection at the location concerned. Taking measurements from Figure 7, it can be deduced that, for no disturbance to occur in the first mode:

a) If p ounces are added on one side of the rotor 25 inches from bearing A, then $p \times 5.4/3.9$ ounces must be added on the opposite side of the rotor 70 inches from bearing A.

b) If q ounces are added on one side of the rotor 10 inches from bearing A, and also at 90 inches from bearing A, then $q \times (3.4/1.6)/5.9$ ounces must be added on the opposite side of the rotor 44 inches from bearing A.

These proportions must be maintained for both calibrating and correction masses, in determining the true correction mass proportions for the desired modal balance.

The calibrating masses added for either second or third mode will influence the deflection of the rotor in both modes simultaneously. The best understanding is achieved by viewing the two effects separately and adding the results.

Figure 7a shows the change of vibration level at the bearing pedestals, caused by deflection of the rotor in the second mode, as a calibrating weight traverses the length of the rotor at a given radius, in a given radial plane. It can be seen that this pedestal response is similar to the second mode shape, and takes account of the fact that the bearings vibrate in antiphase, and that the amplitude of vibration at pedestal A is roughly 1.75 times that of B. Figure 7b is the equivalent diagram for third mode, using the same weight, at the same radius, in the same radial plane.

Assuming that the rotor is running well below its second critical speed, the response of the rotor will be in the same plane for both diagrams, i.e., if positive readings occurred at an angle of 90°, negative readings would occur at 270°, for both modes. It is therefore permissible to consider only positive or negative responses rather than vectors.

When adding corrections mostly for the second mode, the total mass added at any location must be divided between the two planes. The proportions into which the mass is best divided were decided by the reasoning given above. The overall effect of these correction masses on the vibration of the bearing pedestals is determined by summing their individual effects in both modes. These sums are, in turn, deduced from the corresponding magnitude of the response curves given in Figure 7. The values involved are given in Table 1. A further computation is then required to determine the overall effect of corrections added mostly for the third mode: (see Table 2).

The effects on both modes were found as follows. When mass was added mostly for second mode, the resultant change in vibration at the bearings was:

At A: $(8.64x - 1.41x) = 7.23x$ units

At B: $(-4.95x - 2.35x) = -7.3x$ units

where x is the desired correction in the second mode.

The ratio [Effect at end A/Effect at end B] = 0.99, numerically, instead of the 1.75 value noted above. This latter is the ratio for deflection only in second mode.

Similarly, when correction masses are added mostly for the third mode, the resultant change in vibration at the bearings was:

At A: (2.31y + 4.52y) = 6.83y units

At B: (-1.32y + 7.53y) = 6.21y units

where y is the desired correction in the third mode.

The ratio [Effect at end A/Effect at end B] = 1.1 numerically, instead of the 0.6 value noted previously. This ratio applies for deflection only in the third mode.

Although the original readings of vibration may have involved an error in second mode and another in third, for balancing purposes the vectors must be resolved into components of modified proportions, to compensate for the influence of the added masses on both modes. A suitable graphical construction for achieving this resolution is given in Figure 11. If during application this construction gives an unattainable answer it means that the unbalance condition cannot be corrected by the addition of masses in the required proportions, in the proposed correction planes. Other planes must then be selected, and the procedure repeated.

Figure 12 gives the complete worked example from original readings and calibrating runs, for the boiler feed pump rotor. The original readings are resolved into their appropriate components,and the calibrating weights are adjusted to nullify these components.

4.3.6 Comments

The construction shown in Figure 11 was used to identify the required correct masses in the second and third modes, simultaneously.

These corrections were then added to the rotor. With some further minor adjustments, the rotor was then able to run smoothly.

4.3.7 References

1. Moore, L. S., Dodd, E. G., 'Mass Balancing of Large Flexible Rotors,' G.E.C. Journal (England) Vol. 31, No. 2, 1964.

4.3.8 Case History No. 3 - Coffee Roaster Rotor Vibrations

A hot air fan for roasting coffee beans was experiencing troublesome vibrations and excessive bearing wear. A schematic of the fan installation is shown in Figure 13. The fan impeller operated at 250°F when blowing hot air. Both bearings were single row spherical roller bearings, drip fed with lubricant from sight glasses, installed in 1947. The bearings were mounted in cast iron pedestal supports bolted to a concrete machine foundation, on the second floor of the coffee plant. The fan was driven by an AC induction motor through a gear tooth type coupling.

Details of Problem:

a) The fan was observed to be vibrating while operating at its normal running speed of 5,000 rpm.

b) Strong vibrations were observed in the impeller end bearing, and in the fan hot air duct surrounding the impeller.

c) Local noise level was high, but not excessive. No noise readings had been taken.

d) Shaft operating temperature at the impeller bearing was about 200°F.

e) No spectral analysis of the vibrations had been made.

f) The drive end bearing of the unit experienced only low vibration levels and ran at 140°F.

g) Drive motor temperature was not excessive. The motor casing was warm (100°F) but not hot.

h) Foundation vibrations at the impeller were noticeable but not excessive.

Investigation:

a) The fan rotor had a calculated critical speed of 7000 rpm in its bearings. This was based on the assumption that the impeller was a point mass located at the overhung end. No computer critical speed calculations had been made.

b) Impeller bearing wear rates had been high. The impeller bearing had been replaced three times during the previous 12 months.

c) The fan operating environment was dusty. It was observed that the bearing seals were not effective in preventing dust from entering the bearings and contaminating the lubricating oil. Bearing failure was frequent and appeared to involve lubrication failure. There was evidence of cage wear and brass race smearing over the bearing outer ring contact surface.

d) Some relative axial sliding movement had apparently taken place between the shaft and the impeller bearing, probably due to thermal expansion. The thrust bearing was located in the drive end race which was an angular spherical contact bearing.

Calculations:

o The calculated critical speed value was confirmed to be around 7000 rpm, assuming a uniform shaft, and point mass for the disk.

o The critical speed for the same uniform shaft with a massive overhung inertia disk was at 5060 rpm.

4.3.9 Conclusions

o The original critical speed calculation was found to be incorrect because gyroscopic-inertia effects of the disk had not been included in the calculation. Inclusion of these factors showed that the rotor critical speed was actually occurring around 5000 rpm.

o The true calculated critical speed of the rotor system was found to lie within 60 rpm of the 5,000 rpm running speed. During operation the rotor system was almost resonant with its unbalance excitation.

o Residual unbalance and thermal bow could, for this case, evidently cause high synchronous resonant vibrations at the fan operating speed.

o It is concluded that rotor resonant operation at 5000 rpm caused the impeller end bearings to vibrate as observed, and that this vibration together with lubricant contamination caused the bearing wear and failures.

4.3.10 Recommendations

- The critical speed of the rotor system was raised by welding a steel stiffening sleeve on to the shaft between its bearings. This construction is shown in Figure 14. The calculated rise in critical speed for this condition is from 5,000 rpm to 6300 rpm.

- Improved bearing seals were introduced to exclude coffee dust from entering the bearing lubricant.

- Bearing operating temperature could also be reduced by increasing the oil flow through the bearing.

4.3.11 Effect of Corrective Action

As a result of the shaft stiffening action, and the improved dust sealing, the bearing failures were eliminated.

(a) Generator and Turbine Coupling

(b) Stiffened End Plate

Figure 1. Paper Mill Generator Casing Structure

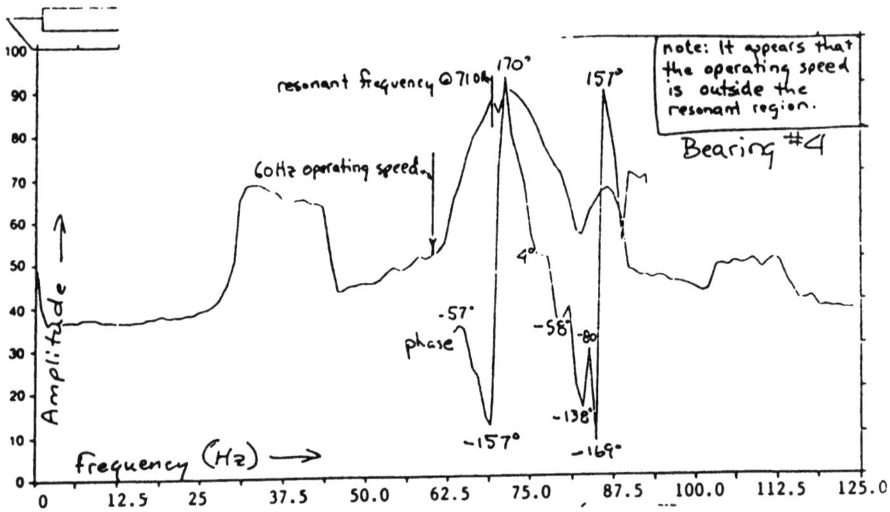

Figure 2. Axial Acceleration at No. 4 Bearing - Resonant Peak at 71.0 Hz

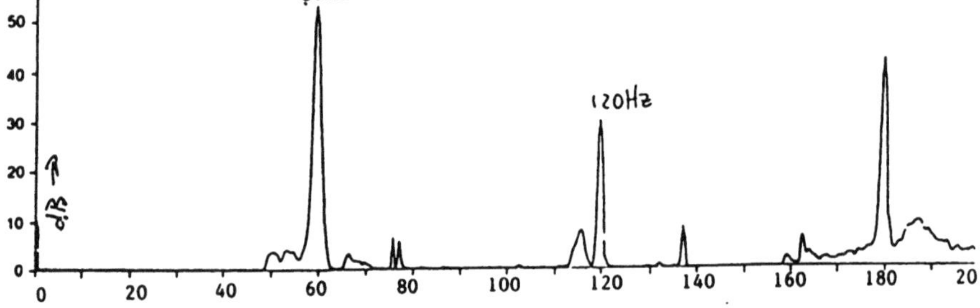

Figure 3. Vertical Acceleration at No. 3 Bearing - 2.8 Mw Load

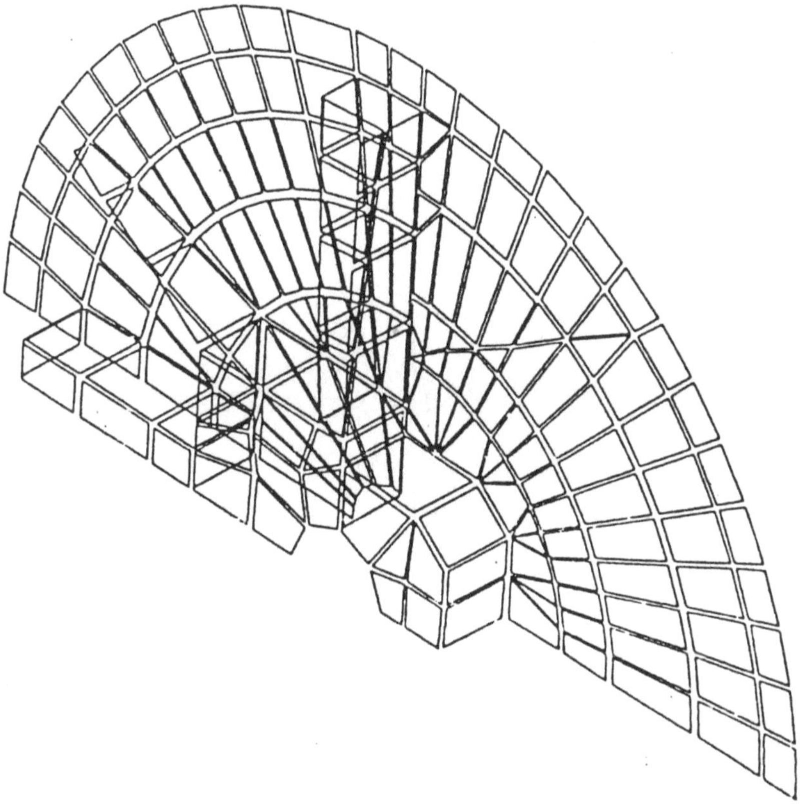

Figure 4. Finite Element Representation of Stiffened Generator End Plate

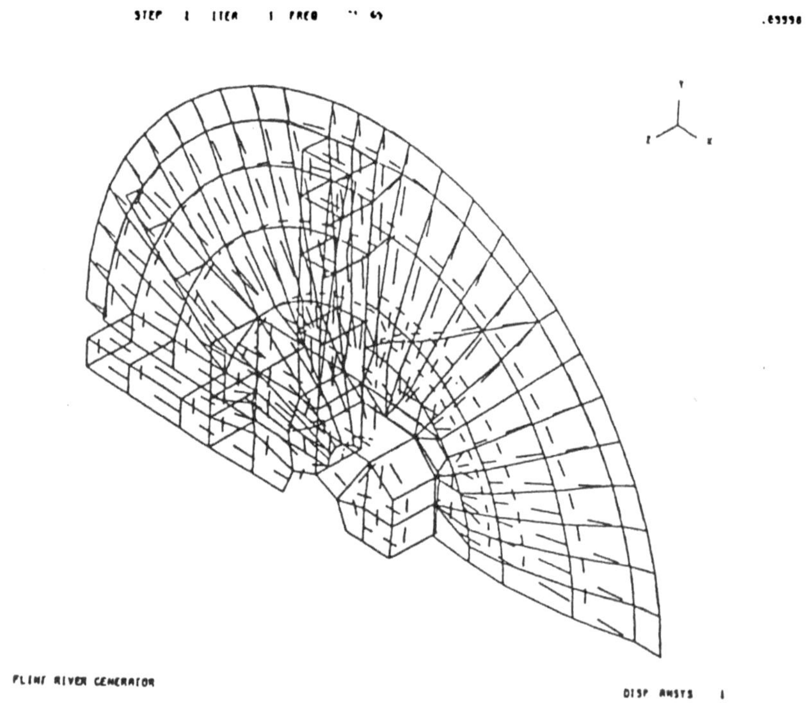

Figure 5. Lowest Mode of Generator End Plate - 69 Hz

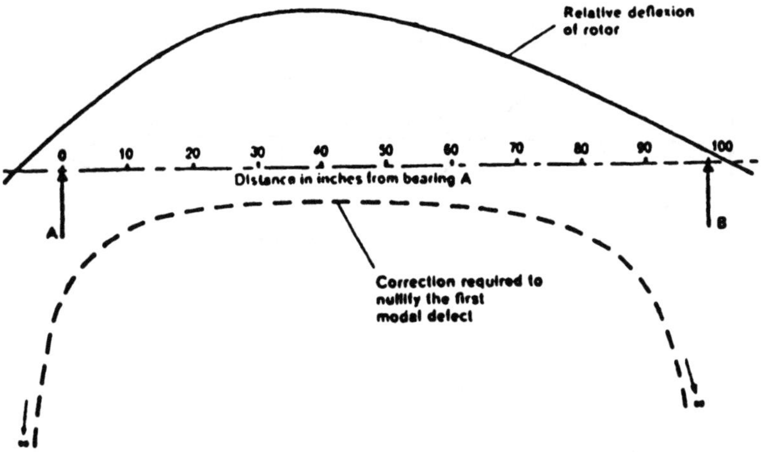

(a) Unbalance Response in First Mode

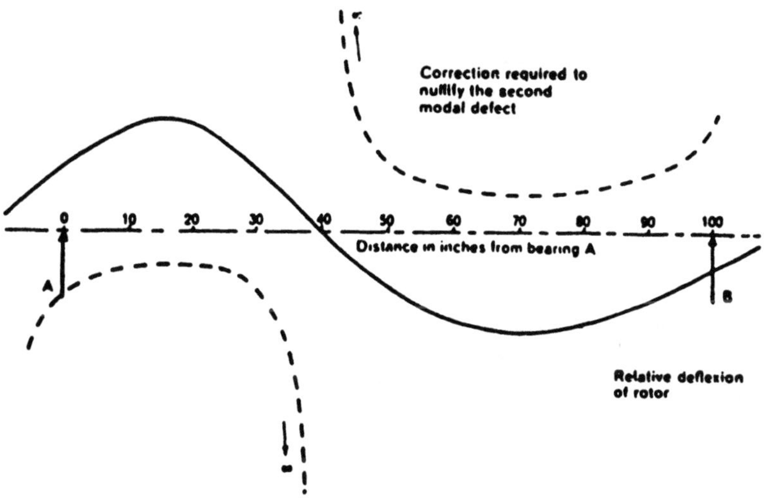

(b) Unbalance Response in Second Mode

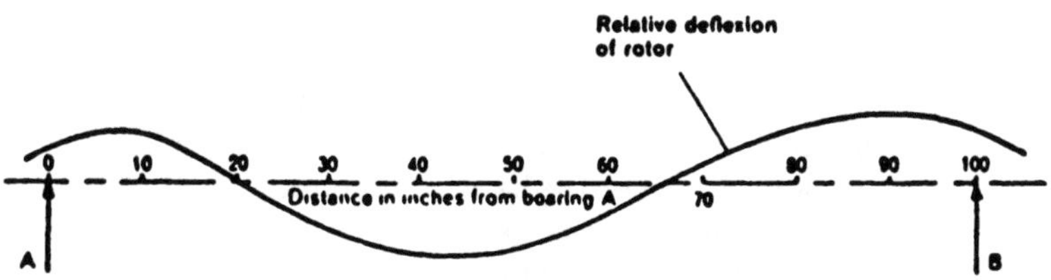

(c) Unbalance Response in Third Mode

Figure 6. Undamped Mode Shapes of Rotors in Bearings Near Critical Speeds

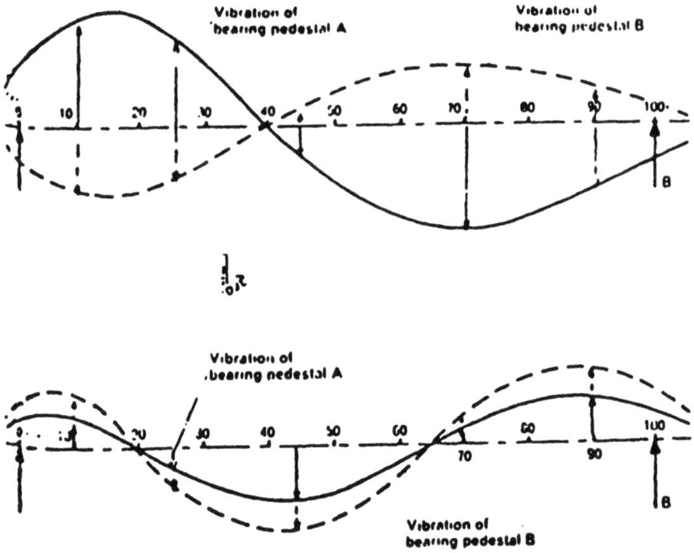

Figure 7. Influence Curves for Response to Correction Weights Along Rotor Length

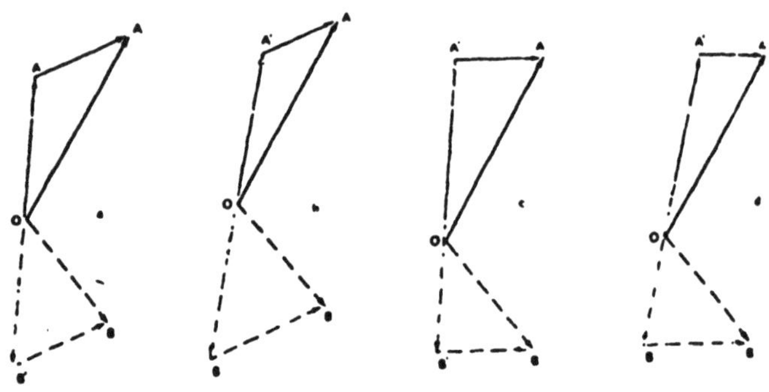

Figure 8. Explanations for Readings of Vibration Vectors OA and OB:

a) Equal out-of-phase component, equal in-phase component.
b) Equal out-of-phase component, unequal in-phase component.
c) Unequal out-of-phase component, equal in-phase component.
d) Unequal out-of-phase component, unequal in-phase component.

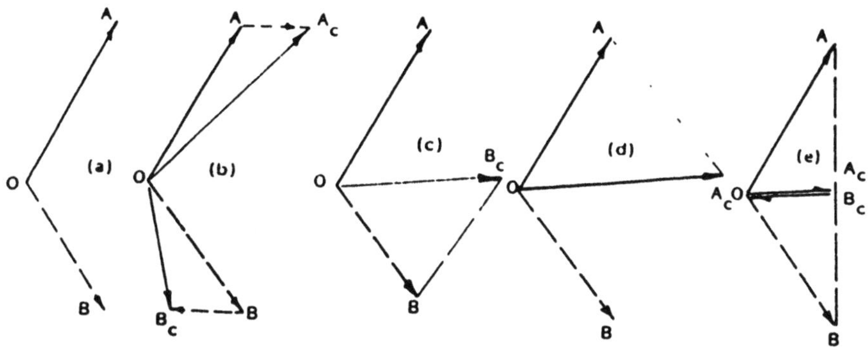

Figure 9. Vibration Response of Vectors at Operating Speed
a) Original readings of vibration of rotor pedestals.
b) Effect of out-of-phase pair of calibrating weights at ends of rotor.
c) Effect of adjusting the calibrating weights to eliminate vibration of Pedestal A.
d) Effect of adjusting the calibrating weights to eliminate vibration of Pedestal B.
e) Effect of adjusting calibrating weights for best compromise.

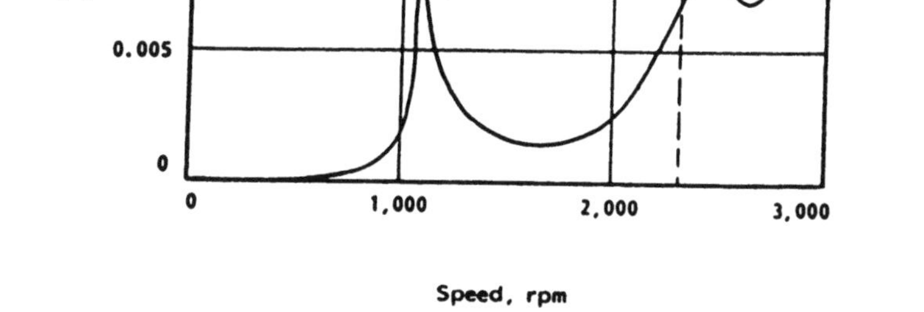

Figure 10. Rotor Response Due to Second and Third Modes

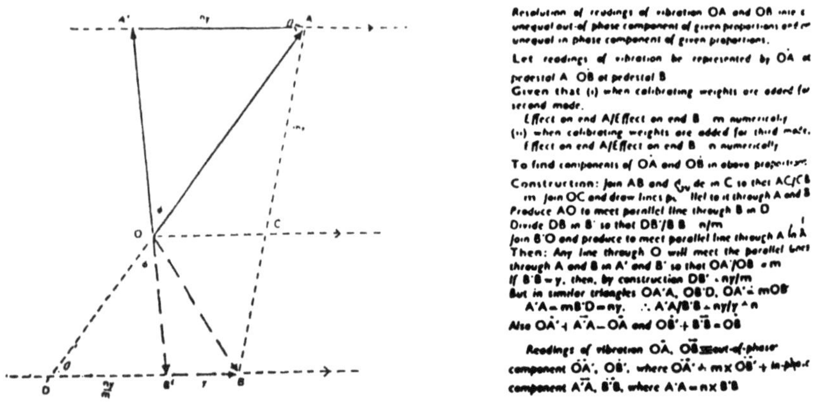

Figure 11. Graphical Construction for Identifying Correction Weights for Mixed Modes

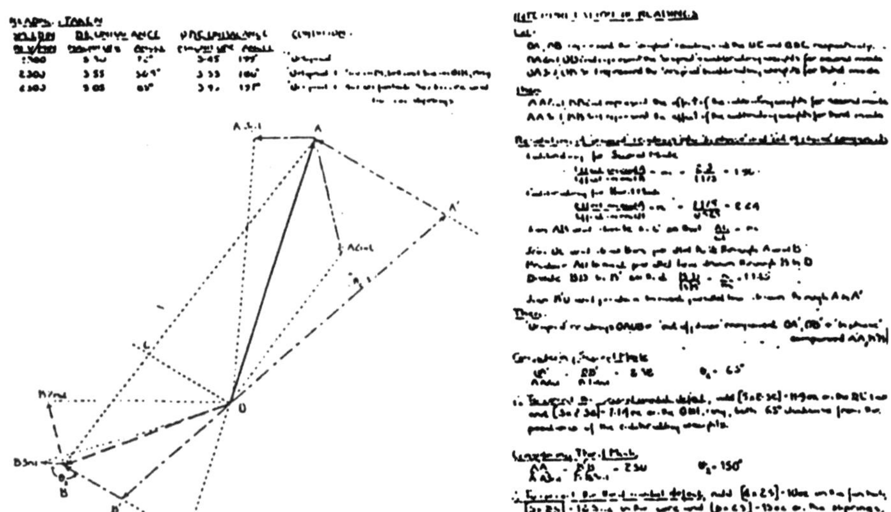

Figure 12. Application of Graphical Construction to Balance Boiler Feed Pump Unit

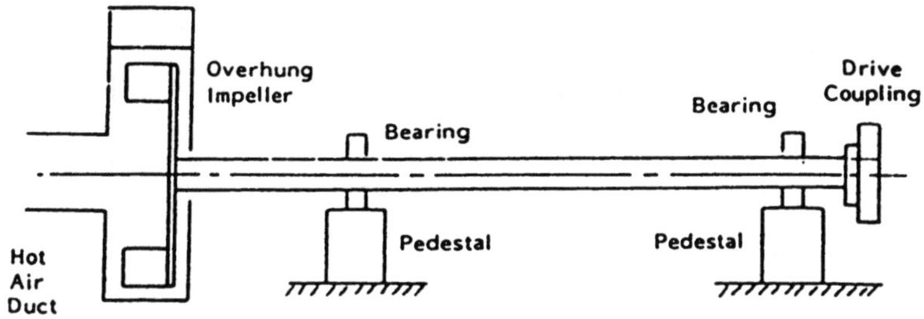

Figure 13. Overhung Fan Impeller in Bearings

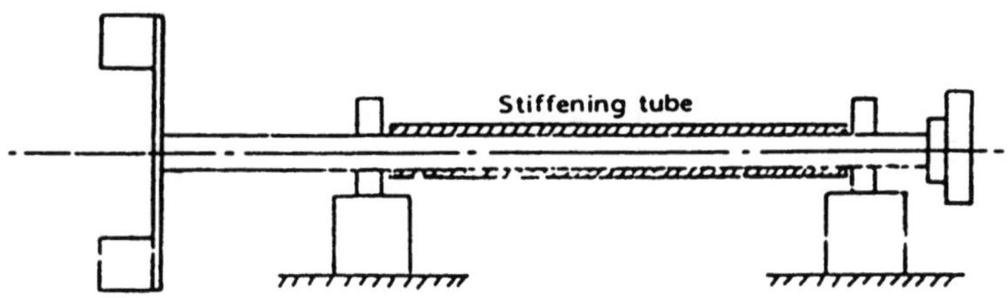

Figure 14. Stiffening Tube Over Shaft Between Bearings

Table 1. Computation of Overall Effect of Weights Added Mostly for Second Mode

Weight added, expressed as a multiple of the calibrating weight used to obtain fig. 7.	Distance in inches from bearing A to plane of added weight	Units of unbalance effect in second mode		Units of unbalance effect in third mode	
		Bearing A	Bearing B	Bearing A	Bearing B
x	25	$+3{\cdot}3x$	$-1{\cdot}9x$	$-0{\cdot}65x$	$-1{\cdot}1x$
$\frac{5{\cdot}4}{3{\cdot}9}x$	70 (opposite side of rotor)	$+3{\cdot}85 \times \frac{5{\cdot}4}{3{\cdot}9}x$	$-2{\cdot}2 \times \frac{5{\cdot}4}{3{\cdot}9}x$	$-0{\cdot}55 \times \frac{5{\cdot}4}{3{\cdot}9}x$	$-0{\cdot}9 \times \frac{5{\cdot}4}{3{\cdot}9}x$
TOTAL FOR EACH MODE		$+8{\cdot}64x$	$-4{\cdot}95x$	$-1{\cdot}41x$	$-2{\cdot}35x$
Therefore, actual vibration of bearings would be $+7{\cdot}23x$ units at A and $-7{\cdot}3x$ at B					

Table 2. Computation of Overall Effect of Weights Added Mostly for Third Mode

Weight added, expressed as a multiple of the calibrating weight used to obtain fig. 7	Distance in inches from bearing A to plane of added weight	Units of unbalance effect in second mode		Units of unbalance effect in third mode	
		Bearing A	Bearing B	Bearing A	Bearing B
y	10	$+3{\cdot}95y$	$-2{\cdot}25y$	$+1{\cdot}2y$	$2{\cdot}0y$
$\frac{5{\cdot}0}{5{\cdot}9}y$	44 (opposite side of rotor)	$+0{\cdot}9 \times \frac{5{\cdot}0}{5{\cdot}9}y$	$-0{\cdot}5 \times \frac{5{\cdot}0}{5{\cdot}9}y$	$+1{\cdot}85 \times \frac{5{\cdot}0}{5{\cdot}9}y$	$-3{\cdot}1 \times \frac{5{\cdot}0}{5{\cdot}9}y$
y	90	$-2{\cdot}4y$	$+1{\cdot}35y$	$+1{\cdot}75y$	$+2{\cdot}4y$
TOTAL FOR EACH MODE		$+2{\cdot}31y$	$-1{\cdot}32y$	$+4{\cdot}52y$	$7{\cdot}53y$
Therefore, actual vibration of bearings would be $+6{\cdot}83y$ units at A and $6{\cdot}21y$ at B					

CHAPTER 5

SEAL PROPERTIES

R. Nordmann
University of Kaiserslautern, Kaiserslautern, FRG

ABSTRACT

Hydraulic forces in annular seals may have a large influence on the vibrations of high speed centrifugal pumps. To have a better knowledge about such seal effects, theoretical as well as experimental studies are important in the actual research of rotordynamics. In this chapter a theoretical model and an identification procedure are presented to determine the dynamic coefficients of seals by calculation and measurements as well. The identified parameters confirm the assumptions in modelling and point out, that the stiffness and damping characteristics of seals are significant for the stability behavior of pumps.

5.1 IDENTIFICATION OF STIFFNESS, DAMPING AND INERTIA COEFFICIENTS OF ANNULAR TURBULENT SEALS

An important assumption for the reliability of high speed centrifugal pumps is a good rotordynamic behavior. Connected to this problem hydraulic forces acting on the rotor are of major importance. It is well known that neck ring seals as well as interstage seals (Fig. 1) may have a large influence on the bending vibrations of a pump rotor. Besides their designed function of reducing the leakage flow between the impeller outlet and inlet or between two adjacent pump stages, the contactless seals have the potential to develop significant forces. This type of forces, created by lateral rotor vibrations can be described by a linear

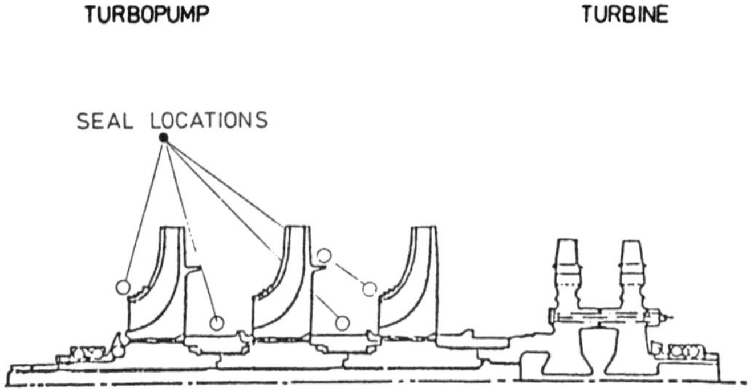

Fig. 1 Seals of a High-Pressure-Turbopump Rotor in Aerospace Engineering

model with stiffness, damping and inertia coefficients.

If contactless seal elements are used in a turbopump the fluidmechanical interactions have to be considered when predicting the vibration behavior of the pump rotor in the design process. However, there is often a uncertainty, concerning the data for the dynamic coefficients. Up to now the stiffness and damping characteristics of seals are not very well known and there is a need for additional research in this area. This is particularly the case for grooved seals, which are very common in practice. Different research projects have been started to investigate the dynamics of seals by theoretical models as well as experimental procedures. The following chapter presents a possible model, based on a bulk flow theory and describes an experimental procedure to identify the stiffness, damping and inertia coefficients.

Modelling of Annular Seals with Turbulent Flow Conditions

Seal Model. To explain the seal model, we consider a very simple geometrical form, consisting of a cylindrical shaft with circular cross section, surrounded by a cylindrical housing (Fig. 2). This annular seal seperates the two chambers with pressure p_1

and pressure p_2, respectively. The pressure difference is $\Delta p = p_1 - p_2$. Caused by this pressure difference there is a leakage flow in axial direction, which is always almost a turbulent flow

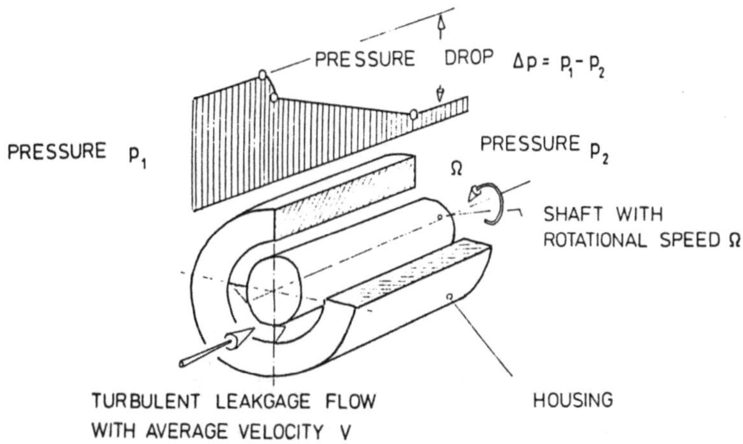

Fig. 2 Modelling of an Annular Seal with Turbulent Flow

with average velocity V. A velocity in circumferential direction is superimposed due to the rotation of the shaft with angular of velocity Ω. In order to obtain the governing equations for the presented seal we assume pure translational movements of the shaft in radial direction.

To derive the pressure around the shaft and then the force-motion-relationships for the vibrating rotor we are using a bulk flow model, which was originally derived by Hirs /1/. Childs /2/ introduced this bulk flow theory for seal elements. The first basic idea of this theory is, that the fluid velocity distributions in radial direction are substituted by average velocities. For a fluid element between the rotor and stator surface (Fig. 3), located at the axial coordinate Z and the circumferential coordinate Θ, the average axial velocity is $U_z(Z,\Theta,t)$ and the average circumferential velocity is $U_\theta(Z,\Theta,t)$. The corresponding pressure for this locations is $p(Z,\Theta,t)$ and the seal radial clearance $H(\Theta,t)$. The shaft circumferential velocity is $U^* = R\,\Omega$.

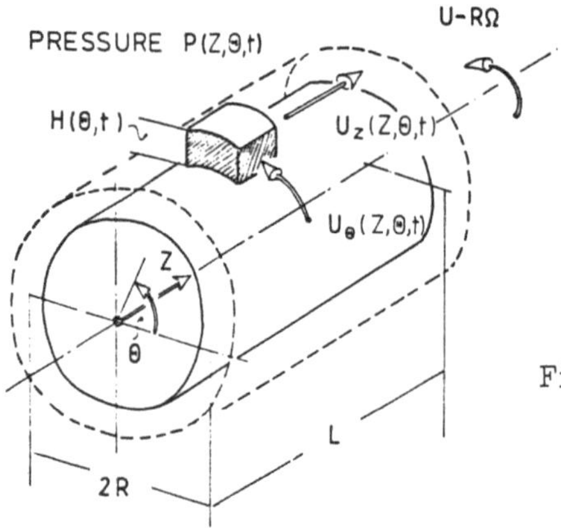

Fig. 3 Velocities, Pressure and Radial Clearance for a Fluid Element

The second basic assumption in Hirs theory is his empirical finding, that the relationship between the shear stress at the wall and the mean velocity of the bulk flow - relative to the wall - can be expressed by well known formulas. With the bulk flow velocity $V_R = \{(U_\theta - R\Omega)^2 + U_z^2\}$ relative to the rotor surface we obtain the wall shear stress at the rotor

$$\tau_R = n_o \left(\frac{2H\, V_R}{\nu}\right)^{m_o} \frac{\rho}{2} V_R^2 \tag{1}$$

n_o, m_o are empirical turbulence coefficients, ρ is the fluid density and ν the kinematic viscosity of the fluid. In a similar way $V_s = \{U_\theta^2 + U_z^2\}$ is the bulk flow velocity relative to the stator surface and the corresponding stator shear stress is

$$\tau_s = n_o \left(\frac{2H\, V_s}{\nu}\right)^{m_o} \frac{\rho}{2} V_s^2 \tag{2}$$

Formulas (1) and (2) correlate the shear stresses τ_R, τ_s with the Reynolds numbers, which are defined in parantheses.

We can now derive the two momentum equations, expressing the "equilibrium" for the fluid element in axial and circumferential direction (Fig. 4). If we introduce the shear forces at the walls, the pressure forces and the inertia forces we end up with the two

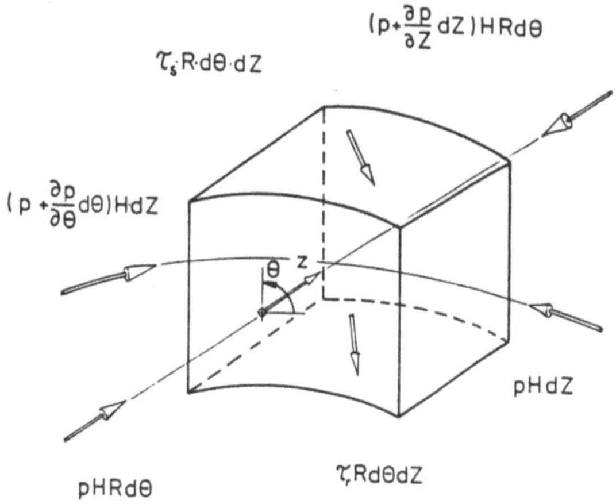

Fig. 4 Pressure and Shear Forces acting on a Fluid Element

equilibrium equations (3) and (4), which are shown in Fig. 5 together with the continuity equation (5)

Axial Momentum Equation

$$-\frac{H^2}{\eta U}\frac{\partial P}{\partial Z} = \frac{n_o}{2} R_\theta^{\;1+m_o} \{U_Z (U_\theta^2+U_Z^2)^{\frac{1+m_o}{2}} + U_Z((U_\theta-1)^2+U_Z^2)^{\frac{1+m_o}{2}}\}$$

$$+ R_\theta \{\frac{H}{U}\frac{\partial U_Z}{\partial t} + \frac{HU_\theta}{R}\frac{\partial U_Z}{\partial \theta} + H\, U_Z \frac{\partial U_Z}{\partial Z}\} \qquad (3)$$

Circumferential Momentum Equation

$$-\frac{H^2}{\eta U}\frac{1}{R}\frac{\partial P}{\partial \theta} = \frac{n_o}{2} R_\theta^{\;1+m_o} \{U_\theta (U_\theta^2+U_Z^2)^{\frac{1+m_o}{2}} + (U_\theta-1)((U_\theta-1)^2+U_Z^2)^{\frac{1+m_o}{2}}\}$$

$$+ R_\theta \{\frac{H}{U}\frac{\partial U_\theta}{\partial t} + \frac{HU_\theta}{R}\frac{\partial U_\theta}{\partial \theta} + H\, U_Z \frac{\partial U_\theta}{\partial Z}\} \qquad (4)$$

Continuity Equation

$$H\frac{\partial U_Z}{\partial Z} + \frac{1}{R}\frac{\partial}{\partial \theta}(H\, U_\theta) + \frac{1}{R\Omega}\frac{\partial H}{\partial t} = 0 \qquad (5)$$

Reynolds number in circumferential direction

$$R_\theta = \frac{HU}{\nu} = \frac{\rho\, H\, U}{\eta}$$

Fig. 5 Governing Equations for a Fluid Element

Perturbation Analysis. With the assumption of small motions of the shaft about a centered position we can expand the equations (3), (4), (5) by means of a perturbation analysis /2/

$$\begin{aligned} H\,(\theta,t) &= H_0 + \varepsilon\, H_1 \quad ; \quad H_0 = \Delta R \\ p\,(Z,\theta,t) &= p_0 + \varepsilon\, p_1 \\ U_Z(Z,\theta,t) &= U_{Z0} + \varepsilon\, U_{Z1} \quad ; \quad U_{Z0} = V \\ U_\theta(Z,\theta,t) &= U_{\theta 0} + \varepsilon\, U_{\theta 1} \end{aligned} \tag{6}$$

ε is a perturbation parameter. H_0, p_0, U_{Z0}, $U_{\theta 0}$ are the quantities for the zero eccentricity flow condition (centered position of the shaft). H_1, p_1, U_{Z1}, $U_{\theta 1}$ correspond to the flow conditions for small shaft motions.

If we introduce (6) into the governing equations (3), (4), (5) we obtain zeroth-order and first order perturbation equations. The substitution and the solution procedures of the remaining equations are very extensive. They are described in more detail in /2/ and /3/. In this presentation we discuss only the essential results.

The solution of the zeroth-order equations ($\varepsilon = 0$, shaft without radial motion) defines the steady state leakage or the pressure drop Δp in axial direction (Fig. 6).

$$\Delta p = p_1 - p_2 = \{1 + \xi + 2\,\sigma\}\, \frac{\rho}{2}\, V^2 \tag{7}$$

The pressure drop Δp is proportional to ρ and the squared velocity V and consists of three parts. The first one shows the change of pressure energy to kinetic energy. The second part points out the pressure loss at the seal entrance and the third part expresses the pressure drop along the seal, caused by friction.

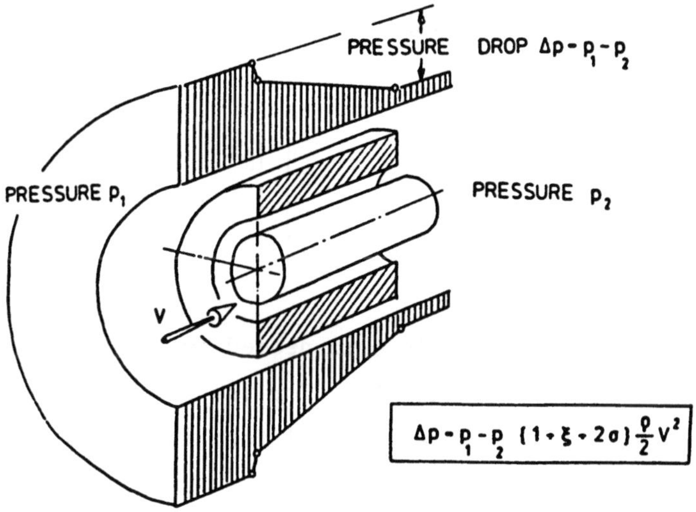

Fig. 6 Pressure Drop in Axial Direction

The seal behaves like a hydrostatic bearing. A static displacement of the shaft in radial direction causes a restoring force and the fluid acts like a spring. The second important result of the zeroth-order equation is the development of the circumferential velocity of a fluid element proceeding axially along the seal. This velocity influences the cross coupled stiffness coefficients.

The first order equations describe the pressure and flow quantities due to a small shaft motion $H(\theta,t)$ about the centered position. These equations can be solved numerically. When we introduce a circular harmonic orbit as a special motion, we can express $H(\theta,t)$ in terms of this motion. By the further assumption of a harmonic pressure and velocity distribution in circumferential direction, the first order equations can be reduced to a system of three coupled complex ordinary differential equations for the unknowns U_{Z1}, $U_{\theta 1}$, p_1 /2/, /3/. From the pressure field solution the reaction forces acting on the rotor due to the circular shaft motion have to be determined by integration of the pressure along the seal and in circumferential direction. Finally the force motion relationship is established. The dynamic system of a seal can be modeled by a linear system with stiffness, damping and inertia terms, if small movements about the seal centre are assumed (Fig. 7).

$$\begin{vmatrix} F_y \\ F_z \end{vmatrix} = \begin{vmatrix} m_{yy} & \\ & m_{zz} \end{vmatrix} \begin{vmatrix} \ddot{y} \\ \ddot{z} \end{vmatrix} + \begin{vmatrix} c_{yy} & c_{yz} \\ c_{zy} & c_{zz} \end{vmatrix} \begin{vmatrix} \dot{y} \\ \dot{z} \end{vmatrix} + \begin{vmatrix} k_{yy} & k_{yz} \\ k_{zy} & k_{zz} \end{vmatrix} \begin{vmatrix} y \\ z \end{vmatrix} \qquad (8)$$

In general a numerical procedure is needed to calculate the dynamic coefficients of equation (8). For the special case of a short seal a solution in an analytical form is possible /4/, /5/.

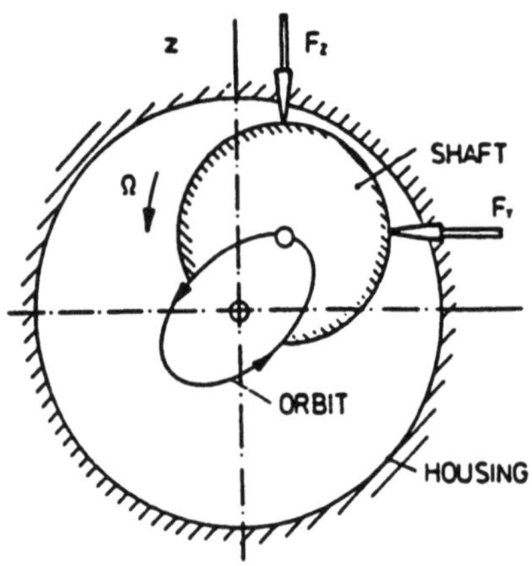

Fig. 7 Dynamic Seal Forces caused by Small Shaft Motions about a Centered Position

The main diagonal elements in each of the matrices are equal and the cross-coupled terms are opposite in sign. The coefficients are mainly dependent on the pressure drop, the average axial velocity V, the rotational speed Ω of the shaft the seal geometry (seal length L and Radius R) and on some quantities characterizing the friction in a seal. It is important to note, that the cross coupled stiffnesses $k_{yz} = - k_{zy}$ are strongly influenced by the rotational speed Ω and the fluid entry swirl, which is the circumferential velocity of the fluid at the seal entrance. This effect may cause serious instability problems in high speed rotating machinery, when the cross coupled stiffness terms become dominant.

In one of his first publications about seals, Black /4/ has derived stiffness, damping and inertia coefficients for short seals. Fig. 7 presents this coefficients in dependence of the most important influence parameters Δp, V, Ω, $T = L/V$ and some friction coefficients μ_0, μ_1, μ_2. In Black's derivatives the entry

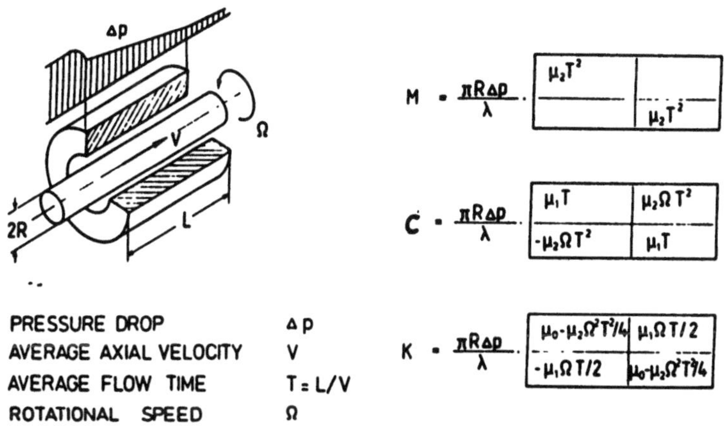

Fig. 8 Dynamic Coefficients of a Short Seal /4/

swirl was assumed to be half of the circumferential velocity of the shaft: $R\Omega/2$.

With the à priori knowledge about the seal dynamic coefficients we have valuable informations about the structure of the model (linearity, order of the model, skewsymmetric etc.), which will be used in the following parameter identification procedure.

Identification of the Dynamic Seal Coefficients

Most of the experimental methods to determine dynamic coefficients of bearings or seals are working with test forces (input signals) and are measuring the relative displacement between shaft and housing (output signals). The unknown seal parameters can then be calculated by means of input-output relations in the time domain or in the frequency domain (Fig. 9).

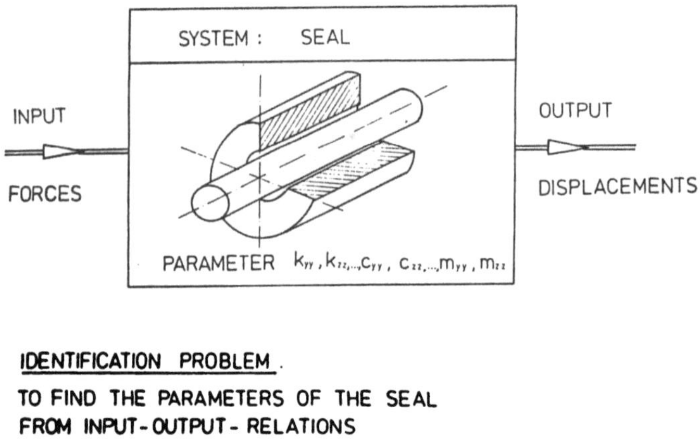

Fig. 9 Identification of the Parameters of a Seal

The basic steps of the applied identification procedure are pointed out in Fig. 10. Measurements are carried out at a test rig, that consists of a very stiff rotating shaft and an elastically mounted rigid housing with two symmetric seals between shaft and housing. Water flows axially across the two seals in opposite directions, while the shaft is running with a fixed rotational speed Ω. The housing is excited by a test force and the system response is measured as a radial motion between the seal surfaces. From measured input and output time signals frequency response functions can be determined by signal processing.

Corresponding to the seal test rig a linear mechanical model ex ists, consisting of the rigid mass (housing) and the stiffness and damping coefficients of the seals. The analytical frequency

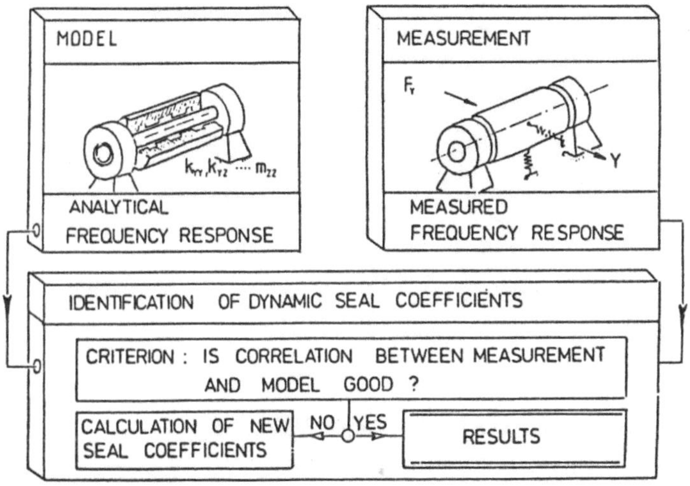

Fig. 10 Basic Steps of the Identification Procedure

response functions of the model depend on the unknown seal parameters, which have to be determined.

The dynamic seal coefficients are estimated by a linear procedure and uses the measured mobility frequency response function and the analytical functions of the model. The main steps of the identification are following in more detail.

Mechanical model. Fig. 11 shows the mechanical model with a rigidly supported very stiff shaft, the rigid mass m of the housing and the stiffness and damping coefficient elements corresponding to the seals and the flexible springs supporting the casing. If test forces are acting in the center of the housing, the system responds only with translatory motions in the two directions y and z. The equations of motion for the model

$$\begin{vmatrix} m+2m_{yy} & \\ & m+2m_{zz} \end{vmatrix} \begin{vmatrix} \ddot{y} \\ \ddot{z} \end{vmatrix} +2 \begin{vmatrix} c_{yy} & c_{yz} \\ c_{zy} & c_{zz} \end{vmatrix} \begin{vmatrix} \dot{y} \\ \dot{z} \end{vmatrix} +2 \begin{vmatrix} k_{yy} & k_{yz} \\ k_{zy} & k_{zz} \end{vmatrix} \begin{vmatrix} y \\ z \end{vmatrix} = \begin{vmatrix} F_y \\ F_z \end{vmatrix} \quad (9)$$

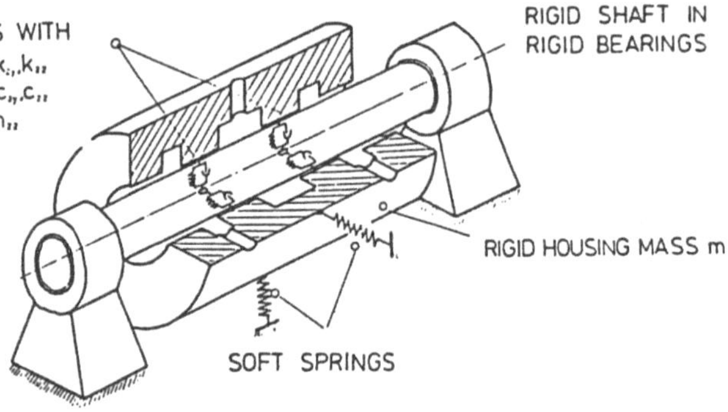

Fig. 11 Mechanical Model of the Seal Test Rig

describe the equilibrium of the inertia forces (housing), the seal forces and exciter forces.

For the considered two degree of freedom test rig system a total of four stiffness frequency functions as well as four flexibility frequency functions can be derived. They depend on the seal coefficients and the exciter frequency ω, as well. The exciter frequency ω is usually different from the frequency of rotation Ω. Fig. 12 points out the mathematical expressions of the two types of frequency response functions $\overset{*}{\underline{H}}(\omega)$ and $\overset{*}{\underline{K}}(\omega)$. By inversion of the frequency response $\overset{*}{\underline{H}}(\omega)$ we obtain the stiffness response $\overset{*}{\underline{K}}(\omega)$. Both functions are used in the identification procedure.

Measurements of the Frequency Response Functions $\overset{*}{H}(\omega)$. Up to now we have considered frequency response functions of the model. For the determination of the frequency response functions from measured input and output time data, we take advantage of the fact that the ratio of the Fourier transformed signals is equal to the frequency response. Due to this possibility, excitation signals with broadband character in the frequency domain (impact, random etc.) can also be applied. The force and response signals are measured in the time domain, transformed to the fre-

Complex Flexibility Frequency Response

$$\underline{H}^{*} = \begin{vmatrix} H_{yy}(\omega) & H_{yz}(\omega) \\ H_{zy}(\omega) & H_{zz}(\omega) \end{vmatrix} = \frac{1}{\Delta} \begin{vmatrix} k_{zz}-\omega^2(\frac{m}{2}+m_{zz})+i\omega c_{zz} & -(k_{yz}+i\omega c_{yz}) \\ -(k_{zy}+i\omega c_{zy}) & k_{yy}-\omega^2(\frac{m}{2}+m_{yy})+i\omega c_{yy} \end{vmatrix}$$

$$\Delta = (k_{yy}-\omega^2(\frac{m}{2}+m_{yy})+i\omega c_{yy})(k_{zz}-\omega^2(\frac{m}{2}+m_{zz})+i\omega c_{zz})$$
$$- (k_{yz}+i\omega c_{yz})(k_{zy}+i\omega c_{zy})$$

Complex Stiffness Frequency Response

$$\underline{K}^{*} = \begin{vmatrix} K_{yy}(\omega) & K_{yz}(\omega) \\ K_{zy}(\omega) & K_{zz}(\omega) \end{vmatrix} = 2 \begin{vmatrix} k_{yy}-\omega^2(\frac{m}{2}+m_{yy})+i\omega c_{yy} & k_{yz}+i\omega c_{yz} \\ k_{zy}+i\omega c_{zy} & k_{zz}-\omega^2(\frac{m}{2}+m_{zz})+i\omega c_{zz} \end{vmatrix}$$

Fig. 12 Frequency Response Functions $\underline{H}^{*}(\omega)$ and $\underline{K}^{*}(\omega)$ of the Mechanical Model

quency domain by means of Fast Fourier Transformation and the quotient is calculated (Fig. 13). This procedure is executed by efficient two channel Fourier Analyzers. Fig. 14 shows in principal the measurement equipments. A hammer was used in this case to excite the housing by an impulse force (see also Fig. 13). By this excitation the signal contains energy in a desired frequency range, which can be influenced by the hammer mass, the flexibility of the impact cap and the impact velocity. The relative displacements between housing and shaft are measured with displacement pick ups. The time signals are amplified, digitized by the

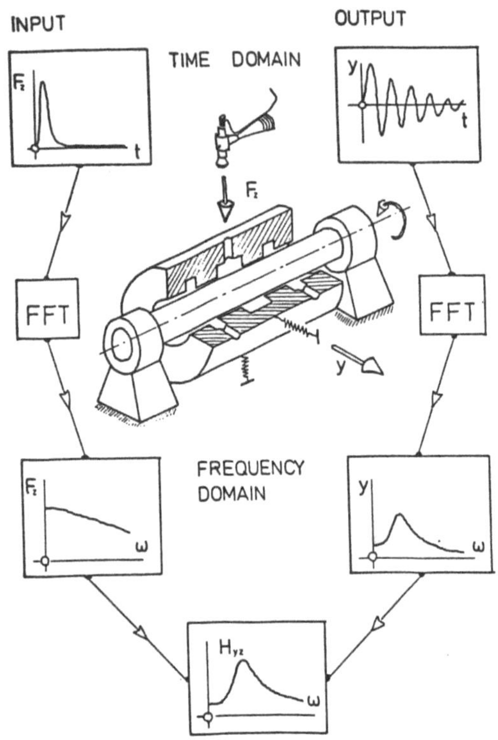

Fig. 13 Measurement of Frequency Response Functions by Means of Fast Fourier Transformation

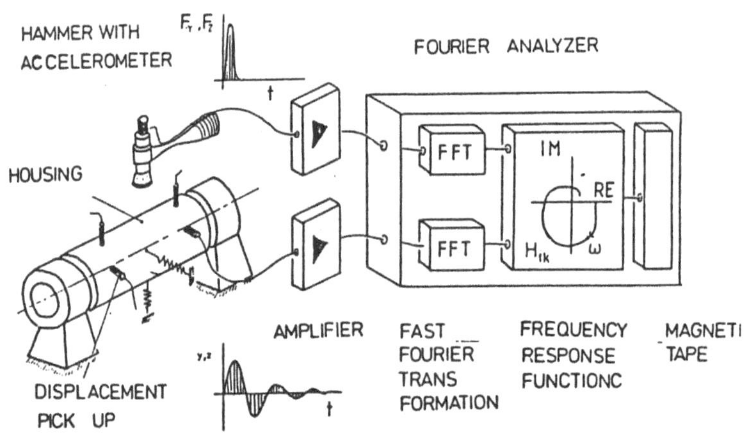

Fig. 14 Measurement Equipments

Analyzer and the frequency response functions are calculated.

Estimation of the Dynamic Seal Coefficients. Different possibilities exist to determine the seal coefficients from measured frequency response curves. One idea is to fit analytical flexibility functions to the measured ones. When working with the output error, this leads to nonlinear equations for the unknown parameters. Another method will be presented here with the definition of an error for the input signals.

From a theoretical point of view the product of the complex mobility matrix $\underline{H}^*$ and the complex stiffness matrix $\underline{K}^*$ should be the unity matrix $\underline{E}$. By combining the measured matrix $\underline{H}^*$ with the analytical matrix $\underline{K}^*$ the result will be $\underline{E}$ plus an additional error matrix $\underline{S}$, caused by measurement noise. Fig. 15 points out this fact and shows either the complex equation or the two real equa-

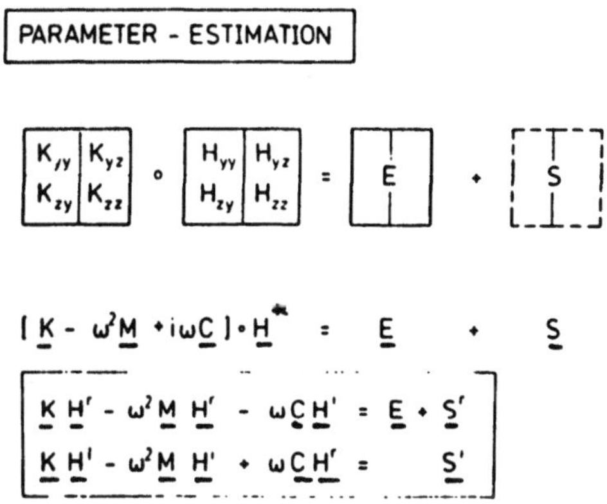

Fig. 15 Parameter Estimation

tions for the unknown parameters, concentrated in $\underline{M}$, $\underline{C}$, $\underline{K}$. The two real equations belong to one exciter frequency. In the case of broadband excitation (impulsive force) we have as much equations as frequency lines, generally much more, than unknown parameters. The overdetermined equation system is presented in Fig. 16. The rectangular matrix A contains all information about the measured frequency response functions ($\underline{H}^r$, $\underline{H}^i$ real and imaginary part of frequency response) and the related exciter frequencies ω. X contains the unknown matrices $\underline{M}$, $\underline{C}$, $\underline{K}$ and $\underline{E}'$ is a modified unity matrix. Applying the criterion to this equation, that $\underline{S}'$ shall be-

EQUATIONS FOR THE DETERMINATION
OF UNKNOWN SEAL PARAMETER $\underline{X}$

$$\underline{A} \cdot \underline{X} = \underline{E}' + \underline{S}'$$

$\underline{A}$ CONTAINS THE MEASURED FREQUENCY RESPONSE DATA $\underline{H}^{*}(\omega)$

$\underline{X}$ CONTAINS THE UNKNOWN PARAMETERS $\underline{M}, \underline{D}, \underline{K}$

CRITERION : $\underline{S}' \longrightarrow$ MINIMAL

$$\underline{A}' \underline{A} \cdot \underline{X} = \underline{A}' \underline{E}'$$ NORMAL EQUATIONS

Fig. 16 Normal Equations for Unknown Parameters

come a minimum, we find the so called normal equations. This is a determined system of equations for the twelve unknown seal parameters in the matrices $\underline{M}$, $\underline{C}$, $\underline{K}$. Due to the definition of an input error the solution procedure for the linear system may fail. This can be avoided by introducing instrumental variables /6/. In this case the matrix $\underline{A}^T$ is substituted by a matrix $\underline{W}^T$, containing the instrumental variables. In our case they have the meaning of the frequency response data, determined with the estimated dynamic coefficients from a previous step. Then the solution is found by an iterative procedure.

Test Rig and Some Measurement Results

Seal Test Rig. The mechanical part of the test facility is shown in Fig. 17. The main components are a very stiff rotating shaft, driven by a speed controled motor and the stiff housing, supported in flexible springs. The shaft is rigidly supported in roller bearings. The fluid enters the housing in the center, flows across the two seals in axial direction and is ex iting the housing at the two ends. With removable stator parts of the seal different geometries and roughnesses can be realized. The static position (zero eccentricity) is adjusted by a special mechanism and measured by eddy current pick ups, which are also used to

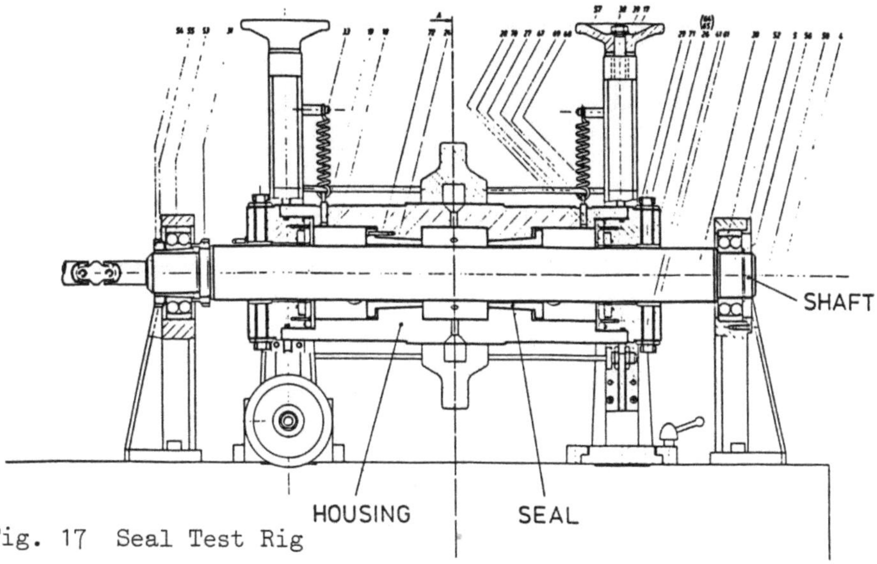

Fig. 17 Seal Test Rig

measure the shaft motions.

The range for the rotational speed is from 0 to 6000 rpm and for the axial fluid velocity from 0 to 14 m/sec. With this we achieve Reynolds numbers up to 15000, when the fluid temperature is about 30° Celsius.

Dynamic measurements. In case of a dynamic measurement four frequency response functions have to be determined for one measurement set, defined by a working condition with constant axial fluid velocity, rotational speed and fluid temperature. A computer takes over the measured data and calculates the dynamic coefficients by means of the described estimation procedure. Both sets of frequency response functions, the measured and the fitted one, can then be displayed and plotted.

Some Measurement Results. The process of the dynamic measurement is demonstrated for one working condition with V=12 m/sec, n = 3450 rpm and a temperature of 30° Celsius. Fig. 18 shows the corresponding measured and fitted response functions as magnitude and phase characteristics in the frequency range 0 to 100 Hz. The

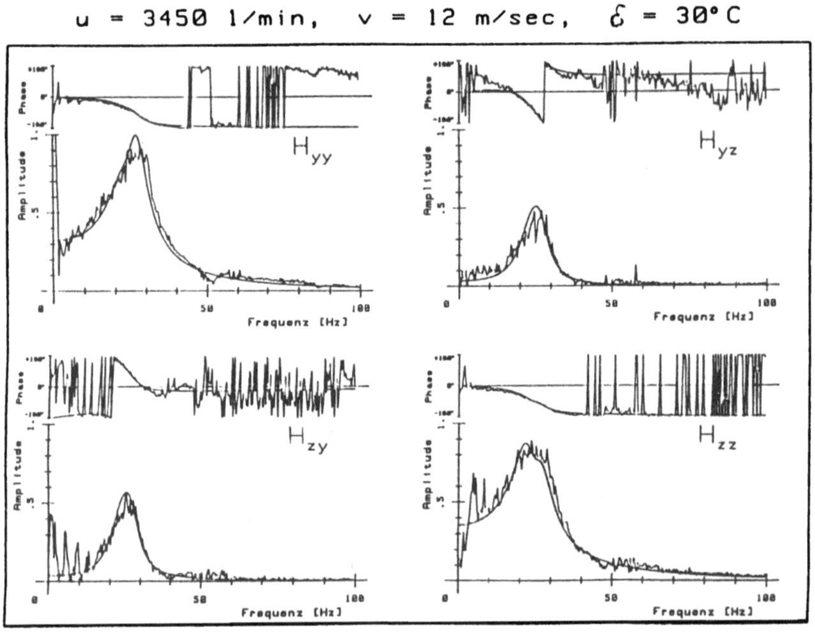

Fig. 18 Measured and Fitted Frequency Response Functions

two direct functions H_{yy} and H_{zz} should be equal and the cross coupled functions H_{yz} and H_{zy} should be equal in magnitude but opposite in phase. The correlation between fitted and measured functions is more or less good.

Several measurements were carried out for different rotational speeds with constant temperature and axial velocity. For each set of functions the inertia, damping and stiffness coefficients were calculated. They are shown in Figures 19, 20 and 21 versus the rotational speed. The values for the complete system with two seals and all known additional terms (mass of the housing, soft springs etc.) are presented. It was found, that the direct coefficients are not equal, furthermore the expected skew-symmetry could not be found exactly in the measured results. The coefficients, which should be equal in magnitude are shown in one diagram and treated as two values for the same operating condition. Besides the measured coefficients the corresponding values of the above mentioned theoretical model are also shown.

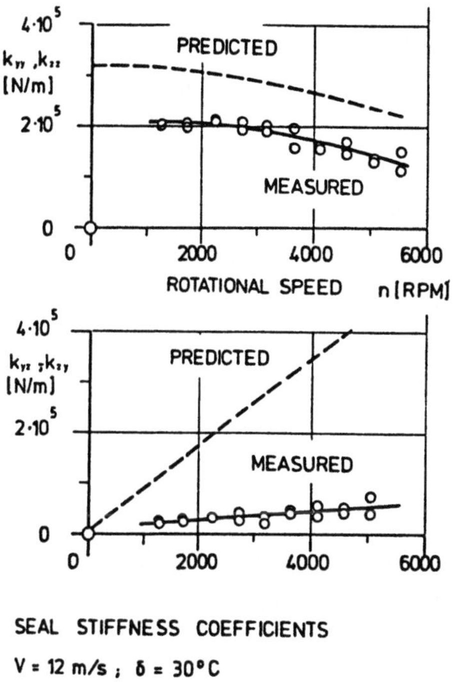

Fig. 19 Stiffness Coefficients

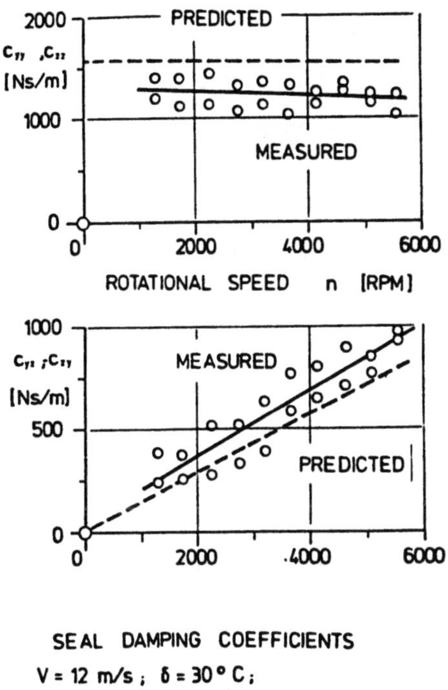

Fig. 20 Damping Coefficients

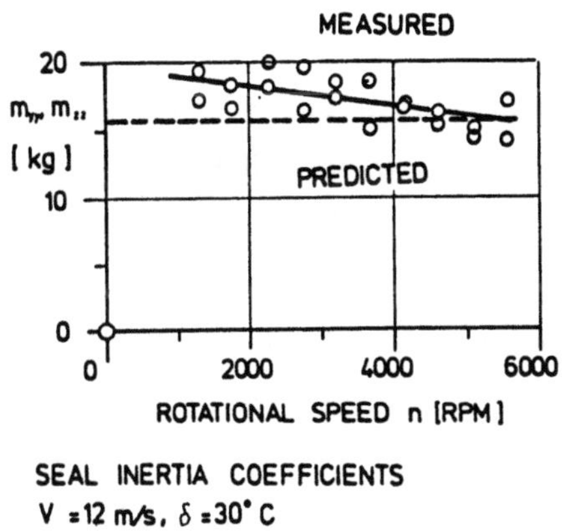

Fig. 21 Inertia Coefficients

The correlation is not that good for the stiffness terms, which are presented in Fig. 19. The dependence on the rotational speed is reasonable for both quantities, a parabolic decrease for the main stiffness and a linear increase for the cross coupled terms. However, the main stiffnesses are found out 40 to 50 % to small and the cross coupled terms much more than this. Meanwhile it is known, that the reason for the small measured cross coupled terms is the low entry swirl in the test rig, which has a large influence on k_{yz} and k_{zy}.

Fig. 20 shows the damping values. The theoretical model predicts nearly constant values for the main damping and an increase with rotational speed for the cross coupled terms. The correlation between measurements and predictions are good to fair.

Finally the total inertia terms are presented in Fig. 21. The theory predicts constant inertia terms. The correlation of measured and calculated values looks good. However, it has to be noted, that the mass of the housing is approximately 15 kg. This mass is included in the results. Therefore the relative error of the seal inertia coefficients, related to the model prediction, is much higher.

References

/1/ Hirs, G.G.: "Fundamentals of a bulk-flow theory for turbulent lubricant films"; Diss. TH Delft, Niederlande, 1970.

/2/ Childs, D.W.: "Finite-Length solution for rotordynamic coefficients of turbulent annular seals"; ASME 82-Lub-42.

/3/ Nordmann, R. et. al: "Rotordynamic Coefficients and Leakage Flow for Smooth and Grooved Seals in Turbopumps"; IFTOMM-Conference Proceedings, Tokyo 1986.

/4/ Black, H.F.: "Effects of hydraulic forces in annular pressure seals on the vibrations of centrifugal pump rotors"; J. Mech. Eng. Sci., Vol 11, No 2, 1969, S. 206-213.

/5/ Childs, D.W.: "Dynamic analysis of turbulent annular seals based on Hirs' lubrication equation"; ASME 82-Lub-41.

/6/ Maßmann, H.: "Ermittlung der dynamischen Parameter turbulent durchströmter Ringspalte bei inkompressiblen Medien"; Diss. Universität Kaiserslautern 1986.

CHAPTER 6

STABILITY OF ROTORS

R. Nordmann
University of Kaiserslautern, Kaiserslautern FRG

ABSTRACT

Instability in rotating machinery may be caused by different effects like oil film forces in journal bearings or forces in contactless seals. A machine designer needs to know, whether a rotor will run stable during operation and what size the stability threshold speed will have. Furthermore he needs information about the parameters influencing the stability. This chapter treats the special system of a rigid rotor, running in two oil film bearings. The equations of motion are derived and the stability behavior is evaluated by means of the complex eigenvalues. Analysis results show a good correlation with measurement results from a test rig.

6.1 RIGID ROTOR STABILITY

Why do we investigate the stability behavior of a rigid rotor? Is this useful from a practical point of view or is such an investigation of more academic character? Every rotor is an elastic system, of course. However, in special operating conditions, a rotor can be regarded as a rigid shaft, approximately. To explain this, let us consider a 40 MW gasturbine-compressor shaft, running in two lobe journal bearings with an operating speed of 4485 rpm (Fig. 1).

The shaft has a mass of 16000 kg and a length of 4,4 m. If we consider the dynamic characteristic of this rotor in the oper-

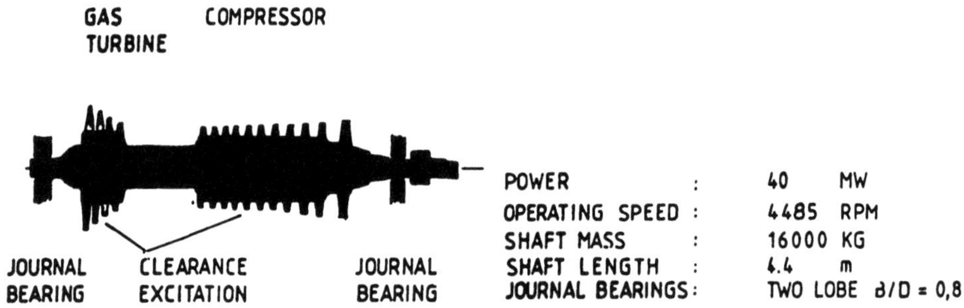

Fig. 1 Gasturbine Compressor Shaft

ating range, its behavior is dominated by the first eigenfrequency combined with a very low damping constant. The change in sign for the damping value points out the onset of instability at a speed near the operating speed. The corresponding mode shape is characterized by nearly same displacements for all points along the shaft, showing that the bearings are relatively weak compared to the shaft (Fig. 2). We recognize, that this Gasturbine Compressor Shaft can be modeled by a rigid rotor and two elastic journal bearings with stiffness and damping coefficients, to study the stability behavior in the interesting speed range.

The rigid rotor- oil film bearing model has only a few degrees of freedom and is relatively easy to handle. However, it is a good approximation for many practical cases. It is also useful to study the dynamic behavior of different bearing types.

Mechanical Model

The mechanical model in our investigation consists of a symmetrical rigid rotor with mass m, equatorial moment of inertia Θ_E and polar moment of inertia Θ_p (Fig. 3). The rotor is running in two equal oil film bearings, characterized by stiffness coefficients k_{ik} and damping coefficients c_{ik} (i,k = 1, .

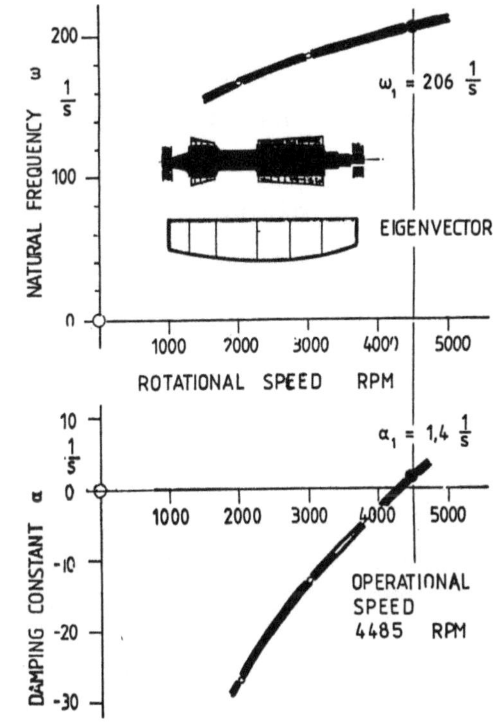

Fig. 2 First Natural Frequency and Damping Constant of the Gasturbine Compressor Shaft

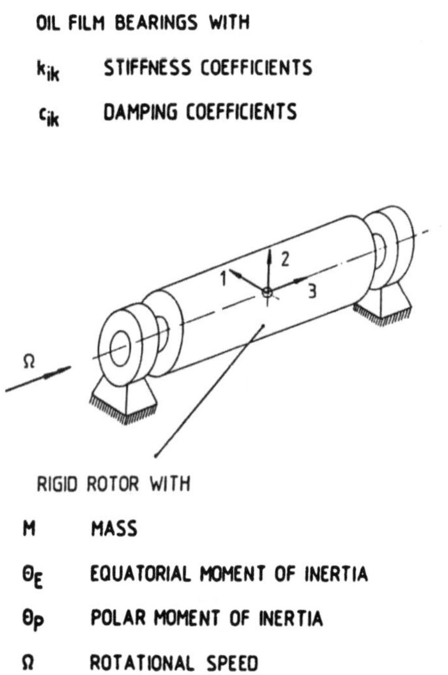

Fig. 3 Rigid Rotor with Oil Film Bearings

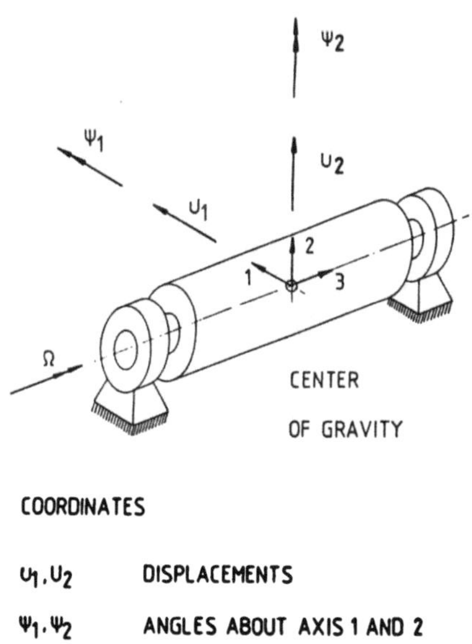

Fig. 4 Coordinates for the Rigid Rotor

To describe the motion we introduce coordinates (Fig. 4). A rigid body has a total of six degrees of freedom, three translations and three rotations. For the rotation about the x_3-axis we suppose, that the angular velocity Ω is constant. Therefore this is no real degree of freedom. Furthermore we do not consider movements in axial directions. The remaining four degrees of freedom are the displacements u_1, u_2 of the center of gravity and the angles ψ_1, ψ_2 about the axis 1 and 2.

From lubrication theory it is well known, that the pressure distribution in the oil film of a cylindrical bearing with a time dependent journal motion can be described by the Reynolds equation (Fig. 5). The solution of this differential equation leads to the pressure in dependence of the journal displacement and velocity. By integrating the oil pressure in circumferential

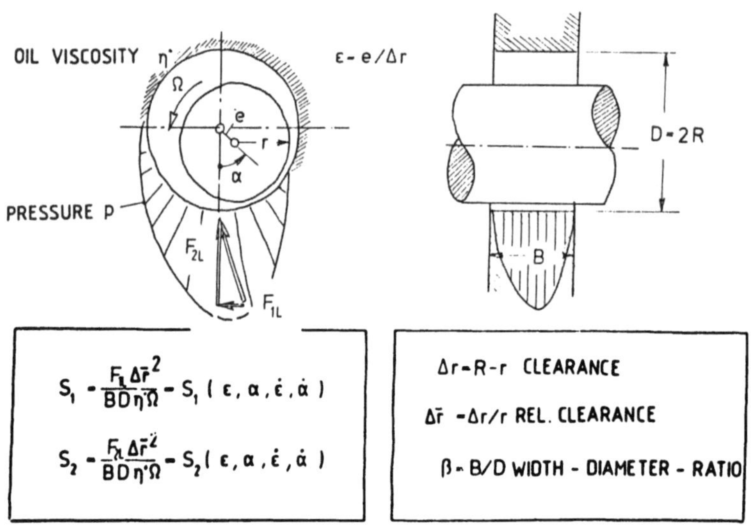

Fig. 5 Pressure Distribution and Forces in a Journal Bearing

as well as in axial direction, we obtain the resultant forces F_{1L} in horizontal and F_{2L} in vertical direction, respectively. These forces can be written in dimensionless form: S_1, S_2 with the relative clearance $\Delta\bar{r}$, the width B, diameter D, oil viscosity η^* and the rotational frequency Ω. They are nonlinear functions of the relative eccentricity $\varepsilon = e/\Delta r$, the angle α and the corresponding velocities $\dot{\varepsilon},\dot{\alpha}$ (Fig. 5).

$$S_1 = \frac{F_{1L}\ \Delta\bar{r}^2}{B\ D\ \eta^*\ \Omega} = S_1(\varepsilon,\ \alpha,\ \dot{\varepsilon},\ \dot{\alpha})$$

$$S_2 = \frac{F_{2L}\ \Delta\bar{r}^2}{B\ D\ \eta^*\ \Omega} = S_2(\varepsilon,\ \alpha,\ \dot{\varepsilon},\ \dot{\alpha}) \tag{1}$$

If only a static load F_{stat} is acting from the vertical weight (no force in horizontal direction), the corresponding equilibrium position of the journal is characterized by

$$\dot{\varepsilon} = 0\ , \quad \dot{\alpha} = 0\ , \quad \varepsilon \neq 0\ , \quad \alpha \neq 0$$

$$S_1 = 0$$

The remaining vertical dimensionless static force

$$S_2(\varepsilon,\ \alpha,\ 0,\ 0) = \frac{F_{stat}\ \Delta\bar{r}^2}{B\ D\ \eta^*\ \Omega} = So \tag{2}$$

is the wellknown Sommerfeld number So.
We recognize, that the static equilibrium position of a rotating journal in a bearing is determined by the So-number, or by the average bearing pressure $\bar{p}$, the relative clearance $\Delta\bar{r}$, the oil viscosity η^* and the rotational speed Ω. Different So-numbers correspond to different equilibrium positions, shown in the static equilibrium curve (Fig. 6).
For $\Omega=0$ (So=∞) the journal center is at a low position and for $\Omega=\infty$ (So=0) the journal center moves up to the bearing center.

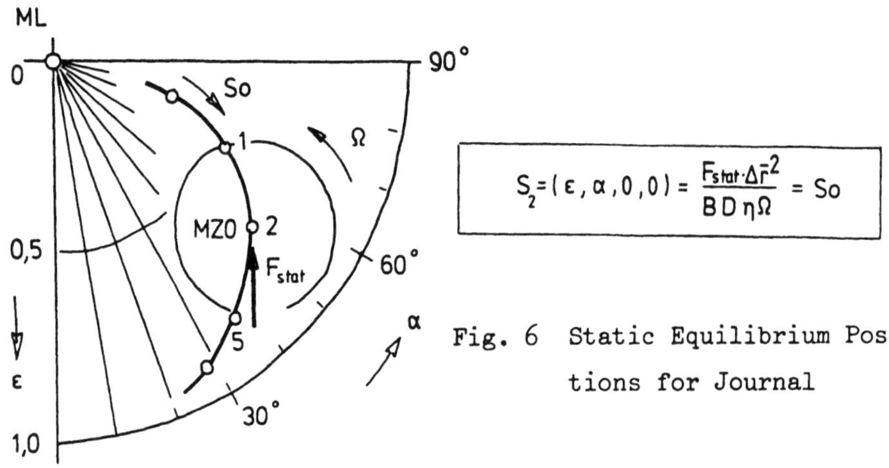

Fig. 6 Static Equilibrium Positions for Journal

If small vibrations about a static position occur, additional dynamic forces are acting from the oil film to the journal. They depend on the displacements u_{1L}, u_{2L} and the corresponding velocities $\dot{u}_{1L}$, $\dot{u}_{2L}$, represented in an cartesian coordinate system (Fig. 7).

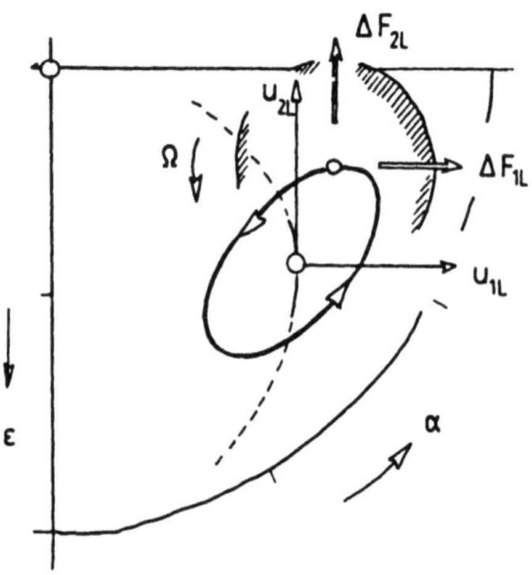

Fig. 7 Dynamic Forces due to Journal Vibrations

$$\Delta F_{1L} = \frac{\partial F_{1L}}{\partial u_{1L}} u_{1L} + \frac{\partial F_{1L}}{\partial u_{2L}} u_{2L} + \frac{\partial F_{1L}}{\partial \dot{u}_{1L}} \dot{u}_{1L} + \frac{\partial F_{1L}}{\partial \dot{u}_{2L}} \dot{u}_{2L}$$

$$\Delta F_{2L} = \frac{\partial F_{2L}}{\partial u_{1L}} u_{1L} + \frac{\partial F_{2L}}{\partial u_{2L}} u_{2L} + \frac{\partial F_{2L}}{\partial \dot{u}_{1L}} \dot{u}_{1L} + \frac{\partial F_{2L}}{\partial \dot{u}_{2L}} \dot{u}_{2L} \qquad (3)$$

ΔF_{1L}, ΔF_{2L} are first order expansions of the oil film reaction forces (Taylor expansion). Equation (3) points out, that the dynamic system oil film bearing can be modeled by means of stiffness and damping coefficients

$$k_{ik} = -\frac{\partial F_{iL}}{\partial u_{kL}} \quad \text{Stiffness coefficient of oil film } (i,k = 1,2)$$

$$c_{ik} = -\frac{\partial F_{iL}}{\partial \dot{u}_{kL}} \quad \text{damping coefficient of oil film } (i,k = 1,2) \tag{4}$$

for small motions about the equilibrium position.

The four stiffness and damping coefficients can be assembled to the 2x2-matrices $\underline{K}$ and $\underline{C}$, respectively. They show, that the oil film behavior is anisotropic ($k_{11} \neq k_{22}$, $c_{11} \neq c_{22}$) and that the coupling coefficients k_{12}, k_{21} generally are not equal. The stiffness and damping coefficients are very often defined as nondimensional quantities, for example as

$$\gamma_{ik} = k_{ik} \frac{\Delta r}{F_{stat}} \quad , \quad \beta_{ik} = c_{ik} \frac{\Delta r \, \Omega}{F_{stat}} \tag{5}$$

For a given bearing type γ_{ik}, β_{ik} are dependent only on the Sommerfeld number or on the static equilibrium position of the journal, respectively. Fig. 8 points out the dependence of this dimensionless coefficients on the Sommerfeld number for a circular cross section bearing. The values are measured quantities, determined by Glienicke /1/. We recognize the anisotropic and nonsymmetric behavior.

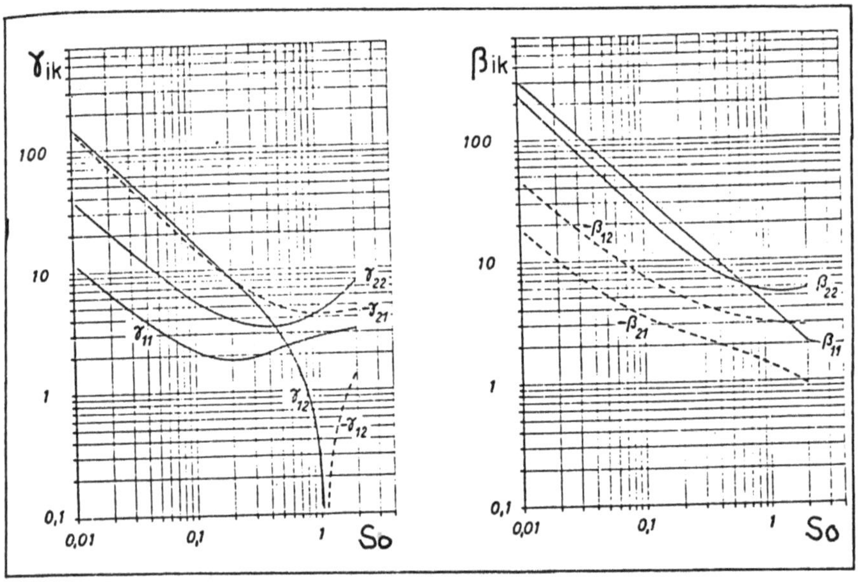

Fig. 8 Dimensionless Stiffness and Damping Coefficients /1/

Equations of Motion

Based on the force-motion relationships for the components of our mechanical system, we are now able to derive the equations of motion. This can be achieved by application of Newton's law for the four coordinates u_1, u_2, ψ_1, ψ_2. Figure 9 points out the bearing forces in two planes acting on the rotor, when it is displaced (u_1, u_2) and twisted (ψ_1, ψ_2) about the center of gravity.

The upper diagram in Figure 9 represents bearing forces in the x_1-x_3 plane. With a displacement u_1 and an angle ψ_2 the left bearing displacements is (u_1- a ψ_2) and the corresponding force component k_{11} (u_1- a ψ_2). Caused by the cross-coupled coefficient k_{12}, another force k_{12} (u_2 + a ψ_1) has to be added from movements in the other plane. Furthermore the damping forces have to be completed.

By taking into consideration all bearing forces the four differential equations of motion are found, describing the dynamic equilibrium of the rotor bearing system. Because of the special

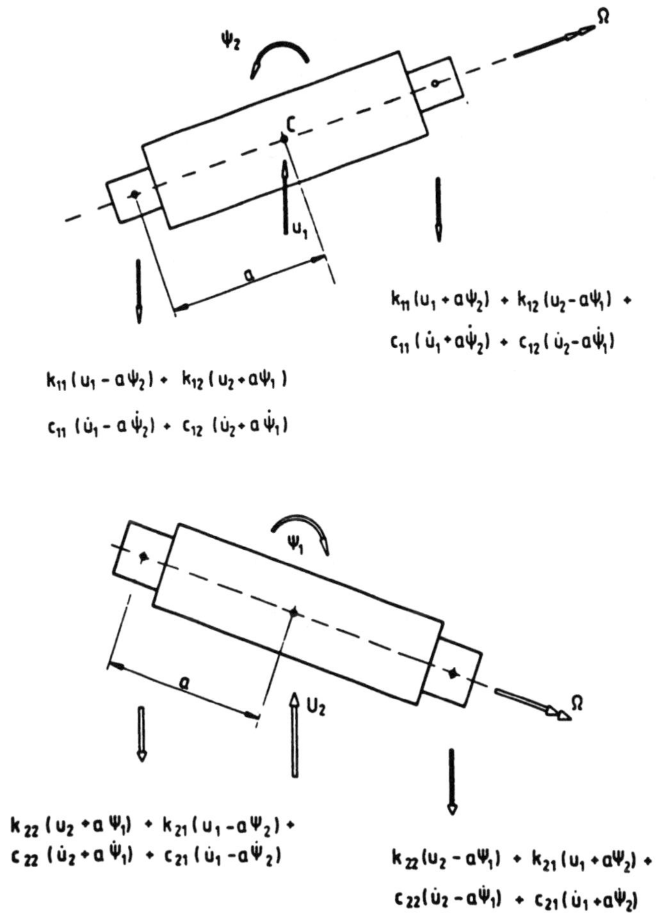

Fig. 9 Dynamic Bearing Forces acting on the Rotor

nature of the system (symmetrical rotor, same bearings on both sides), the equations are partly independent. There are two coupled equations for the translatory motion and two coupled equations for the rotatory motion (Figures 10 and 11). The first one contain inertia terms, bearing forces and in addition external forces (Fig. 10). Concerning to the rotatory motion, we have again inertia terms from the equatorial moment of inertia, additional gyroscopic effects with the polar moment of inertia (damping matrix), forces (moments) from the bearings and external moments (Fig. 11).

$$\begin{bmatrix} m & \\ & m \end{bmatrix} \cdot \begin{bmatrix} \ddot{u}_1 \\ \ddot{u}_2 \end{bmatrix} + 2 \cdot \begin{bmatrix} c_{11} & c_{12} \\ c_{21} & c_{22} \end{bmatrix} \cdot \begin{bmatrix} \dot{u}_1 \\ \dot{u}_2 \end{bmatrix} + 2 \cdot \begin{bmatrix} k_{11} & k_{12} \\ k_{21} & k_{22} \end{bmatrix} \cdot \begin{bmatrix} u_1 \\ u_2 \end{bmatrix} = \begin{bmatrix} F_1 \\ F_2 \end{bmatrix} \quad (6)$$

INERTIA FORCES OF RIGID ROTOR — OILFILM FORCES (BEARING) — EXCITER FORCES

Fig. 10 Equations for Translatory Motion

$$\begin{bmatrix} \Theta_E & \\ & \Theta_E \end{bmatrix} \cdot \begin{bmatrix} \ddot{\psi}_2 \\ \ddot{\psi}_1 \end{bmatrix} + 2a^2 \begin{bmatrix} c_{11} & -c_{12} - \frac{\Theta_P \Omega}{2a^2} \\ -c_{21} + \frac{\Theta_P \Omega}{2a^2} & c_{22} \end{bmatrix} \cdot \begin{bmatrix} \dot{\psi}_2 \\ \dot{\psi}_1 \end{bmatrix} + 2a^2 \begin{bmatrix} k_{11} & -k_{12} \\ -k_{21} & k_{22} \end{bmatrix} \cdot \begin{bmatrix} \psi_2 \\ \psi_1 \end{bmatrix} = \begin{bmatrix} M_2 \\ M_1 \end{bmatrix} \quad (7)$$

INERTIA MOMENTS — MOMENTS, CAUSED BY BEARING FORCES, GYROSCOPIC EFFECTS — EXCITER MOMENTS

Fig. 11 Equations for Rotatory Motion

Stability Analysis

To analyze the stability behavior of the rigid rotor with oil film bearings we neglect any external forces and moments, respectively. Now the two sets of homogeneous equations without the right hand side terms have to be solved. The solution procedure for the translatory motion is explained in Fig. 12. The equations of motion can be written in short form with mass matrix $\underline{M}$, damping matrix $\underline{C}$ and stiffness matrix $\underline{K}$. It has to be noticed once more, that $\underline{C}$ and $\underline{K}$ are generally nonsymmetric and depend on the operating conditions, particularly on the rotational speed Ω.

The natural motion of the rotor has the general form $\underline{u}(t) = \hat{\underline{u}}\, e^{\lambda t}$. Substituting this into the above equations yields a quadratic eigenvalue problem. Corresponding to the 2x2 matrices the

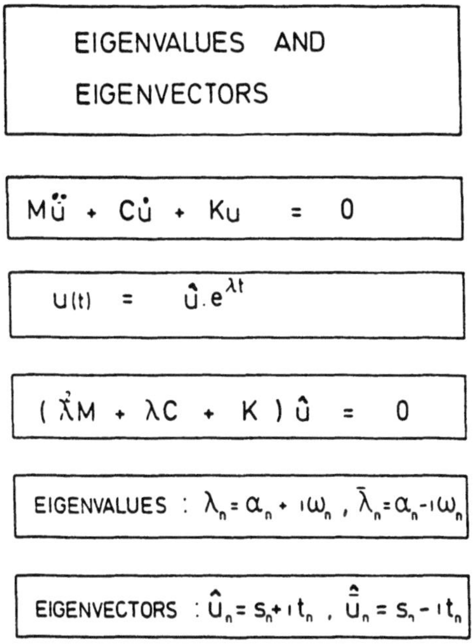

Fig. 12 Solution Procedure for Eigenvalues and Eigenvectors

system has four eigenvalues and eigenvectors. They normally occur in two conjugate complex pairs (Fig. 12). The part of the solution, which belongs to such a pair, can be written as

$$u_n(t) = B_n\, e^{\alpha_n t} (S_n \sin(\omega_n t + \gamma_n) + t_n \cos(\omega_n t + \gamma_n) \qquad (8)$$

It consists of an exponential function and of harmonic functions. ω_n is the circular natural frequency, α_n is the damping constant. By means of this damping constant, which is the eigenvalue real part, one can evaluate whether the corresponding natural motion $u_n(t)$ increases ($\alpha_n > 0$) or decreases ($\alpha_n < 0$) (Fig. 13). If all eigenvalue real parts are negative, the system is stable. If only one real part is positve, we have an unstable system.

Fig. 14 shows the circular natural frequency and the damping constant of the dominant first eigenvalue versus the running speed. These results correspond to the translatory motion and are presented in dimensionless form. ω_o is a reference frequency, defined by $\omega_o^2 = g/\Delta r$ (g acceleration due to gravity). So_o is the Sommerfeld number defined with $\Omega = \omega_o$, which is constant in the presentation of Figure 14.

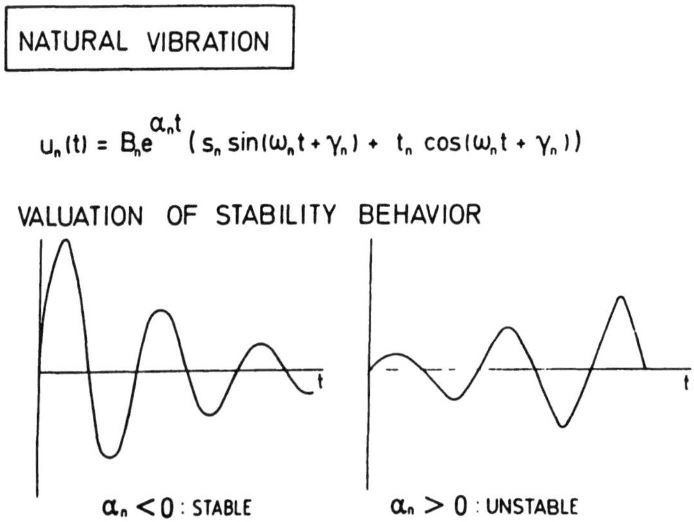

Fig. 13 Evaluation of Stability Behavior by Means of the Eigenvalue Real Part

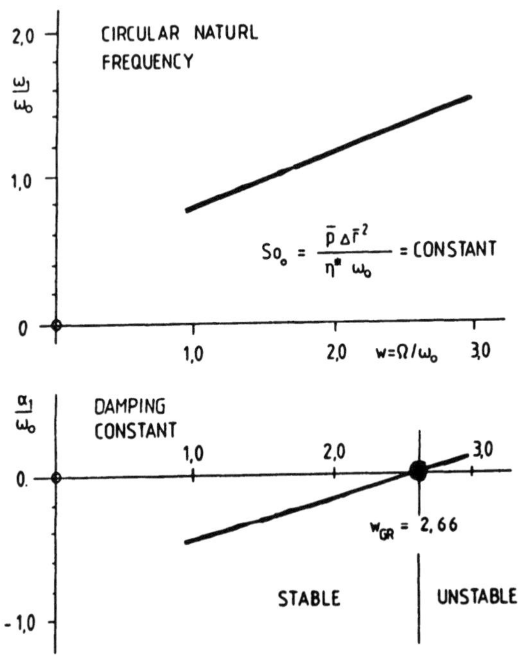

Fig. 14 Natural Frequency and Damping Constant of the Dominant Eigenvalue versus Running Speed

Real and imaginary part of the eigenvalue are changing with the running speed because of the changing dynamic coefficients of the bearings. The real part becomes zero at the dimensionless speed of rotation $w_{GR} = \Omega_{GR}/\omega_o = 2{,}66$, pointing out the instability threshold speed. The corresponding natural frequency is about half of the instability onset speed. Therefore the whirl motion of the journal in the bearing is often called "half frequency whirl".

In Fig. 15 the previously calculated eigenvalue is shown in the complex plane (imaginary part versus real part). The dominant first eigenvalue for the rotatory motion is also included in this

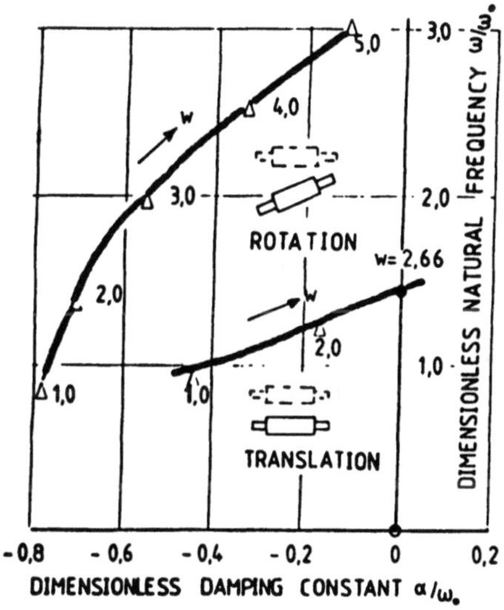

Fig. 15 Eigenvalues presented in the Complex Plane

diagram. Parameter in both curves is the dimensionless rotational speed $w = \Omega/\omega_o$. In both cases it can be seen again, that the dynamic behavior of the system changes with speed, caused by the different bearing characteristics. Both natural frequencies increase with increasing speed w, both damping constants decrease. Once again the instability of the translatory motion occurs, when the dimensionless speed reaches the value $w_{GR} = 2{,}66$. At this speed the eigenvalue for the rotatory motion is still well damped and not dangerous at all. The remaining eigenvalues of the different motions are also very highly damped and not presented in diagram 15.

The stability threshold speed W_{GR} = 2,66 was found for the translatory motion of the rigid rotor, characterized by a special system parameter So_o. If the type of bearing is fixed, So_o is the only parameter defining the stability onset speed. For other parameters So_o other limit speeds W_{GR} have to be calculated. If another bearing type is selected, the threshold speeds will also change. Fig. 16 shows a stability diagram for the rigid rotor with oil film bearings, pointing out the rotational speed limit W_{GR} in dependence of the Sommerfeld number So_o. The different curves belong to different bearing types. Such diagrams are very useful in the design process for rotating machinery.

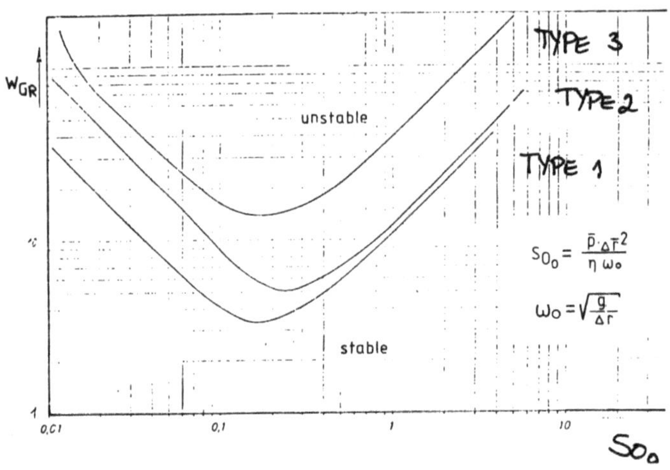

Fig. 16 Stability Diagram for Rigid Rotor in Journal Bearings

Comparison with Test Results

Figure 17 shows a rotor test rig, which was originally designed to measure stiffness and damping coefficients of oil film bearings /2/. The system consists mainly of a relative stiff rotor and two cylindrical journal bearings with circular cross sections. A d.c. electric motor with speed control drives the shaft in a speed range up to 6000 rpm.

This test rig is also well suited to measure damping constants of a rigid rotor-bearing system. To achieve this, the rotating shaft can be excited by a special impulse hammer. The displacements of the shaft, caused by the impulse are measured with contactless inductive pick ups near the bearings.

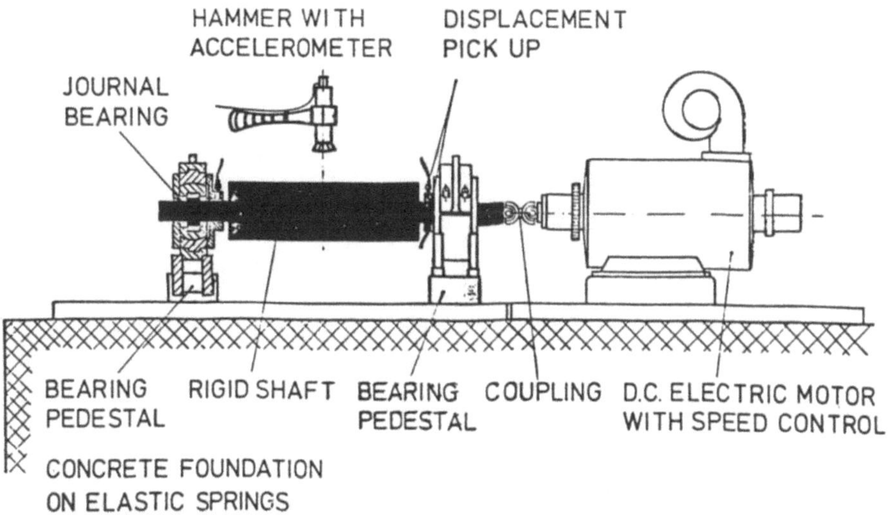

Fig. 17 Test Rig with Rigid Rotor and Oil Film Bearings

The measured system response signals mainly contain one eigenfrequency with the corresponding damping constant. The other eigensolutions decay very rapidly, pointing out the high damping of this eigenvalues. From the decay rate of the displacement signal the damping constant can be determined. There is another possibility to find the damping from frequency response functions, calculated with the force input and the displacement output.

In Figure 18 measured damping constants α/ω_o are presented for different rotational speeds $w = \Omega/\omega_o$. Exept the speed all other bearing quantities are held constant during the tests. There is a good correlation between measured results and the calculated values from the above presented model. In both cases the

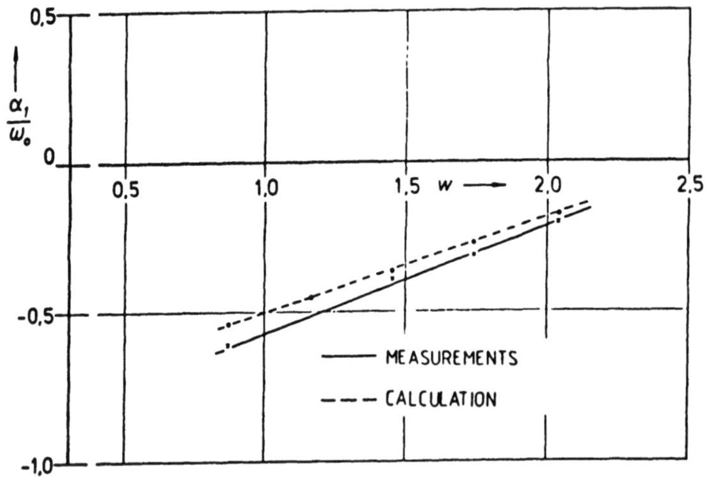

Fig. 18 Damping Constant of a Rigid Rotor with Oil Film Bearings in Dependence of the Rotational Speed

damping constant decreases with the rotational speed. Measurements were carried out up to $w \approx 2.0$. The extended curve shows, that a zero crossing of the damping constant is expected at $W_{GR} \approx 2{,}6$, as predicted by the model.

References

/1/ Glienicke, J.: Experimental Investigation of the Stiffness and Damping Coefficients of Turbine Bearings and Their Application to Instability Prediction. Symposium in Nottingham. The Institution of Mechanical Engineers, 1966.

/2/ Nordmann, R.: Identification of Stiffness and Damping Coefficients of Journal Bearings by Means of the Impact Method. Dynamics of Rotors, CISM Course. Springer-Verlag Wien-New York, 1984.

CHAPTER 7

COMPUTER ANALYSIS OF ROTOR BEARING SYSTEMS

P.A.L.L.A.: A PACKAGE TO ANALYZE THE DYNAMIC BEHAVIOR OF A ROTOR-SUPPORTING STRUCTURE SYSTEM

G. Diana, A. Curami, B. Pizzigoni
Dipartimento di Meccanica, Politecnico di Milano, Milan, Italy

ABSTRACT

A package of programs which can be used to study the static and dynamic behaviour of a multi-supported shaft line is illustrated. The analysis is carried out by keeping account of the interaction with a supporting structure of same (foundation, casings, etc.). Together with the structure of the package, this paper is aimed at illustrating the mathematical model used for static and dynamic analysis.

7.1.1 Foreword

"P.A.L.L.A." is the name of a package of programs developed by a research group working at the Sezione di Meccanica dei Sistemi of the Dipartimento di Meccanica of the Politecnico of Milan. This package is the result of a series of research work, carried out over the last ten years, in collaboration with ANSALDO, ENEL-CRTN, ENEL- DPT.SAS, TECNOMASIO ITALIANO BROWN BOVERI and WORTHINGTON NORD ITALIA. The package can be used to study the different problems that arise in rotor dynamics: the analysis is carried out by keeping account of the iteraction between the shaft line and the supporting structure of same (foundation, case, etc.).

7.1.2 Description of the Mathematical Model

Schematization of the system

The system analysed consists of a shaft-line composed of several rotors, bearings, possible seals and a carrying structure: each single element is schematized in turn as follows :

- ROTOR - The rotor is schematized with finite elements of the "beam" type with four d.o.f. per node, as shown in fig. A.1. The choice of the number of elements is conditioned both by the rotor geometry as well as by the frequency range which one wants to analyse. In fig. A.2. the schematization of a rotor is shown as an example.
- BEARINGS and SEALS - Bearings and seals are schematized with equivalent springs and dampers. As will be explained in more detail further on, this equivalent elements are calculated by evaluating, in the neighbourhood of the rotor's equilibrium position, the force field due to the bearings and the seals fluid film. The system composed of rotor and bearings and/or seals is shown in fig. A.3.
- SUPPORTING STRUCTURE - By "supporting structure" we mean the entire structure surrounding the rotor except the bearings and the seals. The dynamic behaviour of the supporting structure is defined by means of a modal approach. The modal parameters i.e. generalized mass, stiffness and damping relative to the different eigenmodes of the supporting structure are calculated by means of modal identification techniques with one or more degrees of freedom ([1], [2]) starting from the transfer function of the foundation itself. The transfer function (or the frequency response) can be defined both by means of a mathematical model via F.E.M. ("beam" finite elements) of the structure itself, either starting from the experimental measurements carried out on a real structure or on a physical reduced scale model. The modal parameters are used to define the mechanical impedances of the carrying structure through which, in the rotor-fluid film- carrying structure overall model, the dynamic behaviour of the latter is kept account of.

The complete model and the fundamental equations

An example of the complete model rotor-oil film-carrying structure is shown in fig. A.4., where, as can be seen, as far as the carrying structure is concerned, the single nodes connecting the rotor-structure, together with the correspondent transmitted forces are considered. If $\underline{X}$ indicates the d.o.f. vector relative both to the rotor $\underline{X}_R$ as well as the carrying structure $\underline{X}_F$, the equations of motion of the entire system are:

$$[M^*]\ \ddot{\underline{X}} + [R^*]\ \dot{\underline{X}} + [K^*]\ \underline{X} = \underline{F}^* \tag{A.1}$$

where

$$\underline{X} = \begin{Bmatrix} \underline{X}_R \\ \underline{X}_F \end{Bmatrix} \qquad \underline{F}^* = \begin{Bmatrix} \underline{F}_R \\ \underline{F}_F^* \end{Bmatrix} = \underline{F} + \tilde{\underline{R}}_F \tag{A.2}$$

where $\underline{X}_R$ and $\underline{F}_R$ are the subvectors relative to the degrees of freedom of the single rotor, while $\underline{X}_F$ and $\underline{F}_F^*$ are the subvectors relative to the d.o.f. of the connection nodes. In eq. (A.1) $[M^*]$, $[R^*]$ and $[K^*]$ are respectively the mass, stiffness and damping matrices of the system obtained by assembling the rotor matrices and where the equivalent stiffness and damping coefficients of the bearings and seals are kept account of. $\underline{F}^*$ is the excitation forces vector composed, in turn, of forces $\underline{F}$ effectively external, applied directly to the line rotors and/or to the connection points and forces $\tilde{\underline{R}}_F$ transmitted to the connection nodes by the carrying structure. These vectors can be subdivided into subvectors:

$$\underline{F} = \begin{Bmatrix} \underline{F}_R \\ \underline{F}_F \end{Bmatrix} \qquad \tilde{\underline{R}}_F = \begin{Bmatrix} 0 \\ \underline{R}_F \end{Bmatrix} \tag{A.3}$$

By using the matrix partitioning technique eq. (A.1) becomes:

$$\begin{bmatrix} [M_{RR}] & [M_{RF}] \\ [M_{FR}] & [M_{FF}] \end{bmatrix} \begin{Bmatrix} \ddot{\underline{X}}_R \\ \ddot{\underline{X}}_F \end{Bmatrix} + \begin{bmatrix} [R_{RR}] & [R_{RF}] \\ [R_{FR}] & [R_{FF}] \end{bmatrix} \begin{Bmatrix} \dot{\underline{X}}_R \\ \dot{\underline{X}}_F \end{Bmatrix} + \begin{bmatrix} [K_{RR}] & [K_{RF}] \\ [K_{FR}] & [K_{FF}] \end{bmatrix} \begin{Bmatrix} \underline{X}_R \\ \underline{X}_F \end{Bmatrix} = \begin{Bmatrix} \underline{F}_R \\ \underline{F}_F \end{Bmatrix} + \begin{Bmatrix} 0 \\ \underline{R}_F \end{Bmatrix} \tag{A.4}$$

In eq. (A.4) the forces $\underline{R}_F$ transmitted by the foundation are unknown functions. In order to define these functions, as has been said, a modal approach was used. The vector $\underline{q}$ of the modal variables of the foundation can be correlated to the displacements of the connection nodes $\underline{X}_F$ of the foundation itself by means of the following equation :

$$\underline{X}_F = [\Phi]\ \underline{q} \tag{A.5}$$

where $[\Phi]$ is a matrix containing the eigenvectors of the foundation examined. $[\Phi]$ is generally a rectangular matrix with as many columns as modes considered and as many lines as degrees of freedom of the connection nodes. It is possible to define the vector of the connection forces between the rotor and the carrying structure $\underline{R}_F$ as a function of the variables $\underline{X}_F$ using equation (see e.g. [3], [4]):

$$[m_F]\,\ddot{\underline{q}} + [r_F]\,\dot{\underline{q}} + [K_F]\,\underline{q} = [\Phi]^T\,\underline{R}_F \tag{A.6}$$

where $[m_F]$, $[r_F]$ and $[K_F]$ are diagonal matrices containing the modal parameters of the supporting structure. If as many modes as degrees of freedom associated to the connection nodes are assumed, $[\Phi]$ is a square matrix and can be inverted thus allowing the transformation

$$\underline{q} = [\Phi]^{-1}\,\underline{X}_F \tag{A.7}$$

which substituted in eq.(A.6) gives:

$$[\Phi']^{-1}[m_F]\,[\Phi]^{-1}\ddot{\underline{X}}_F + [\Phi']^{-1}[r_F]\,[\Phi]^{-1}\dot{\underline{X}}_F + [\Phi']^{-1}[K_F]\,[\Phi]^{-1}\underline{X}_F = \underline{R}_F \tag{A.8}$$

where prime indicates the transposed matrix. Equation (A.8) (*) correlates the displacements of the connection nodes $\underline{X}_F$ to the forces $\underline{R}_F$ transmitted to the rotor through these nodes. By now substituting the expression of the forces $\underline{R}_F$ of (A.8) in equation (A.4) and by moving the terms containing $\ddot{\underline{X}}_F$, $\dot{\underline{X}}_F$ and $\underline{X}_F$ to the left hand side, the fundamental equation (A.1) can be rewritten as:

$$[M]\,\ddot{\underline{X}} + [R]\,\dot{\underline{X}} + [K]\,\underline{X} = \underline{F} \tag{A.9}$$

where $[M]$, $[R]$ and $[K]$ represent the mass, damping and stiffness matrices of the overall system that keep account not only of the

---

(*) If the eigenvectors are orthogonal in the restrict sense, eq. A.7) can also be written as $\underline{q} = [\Phi']\,\underline{X}_F$.

inertia, damping and stiffness of the rotor but also of the bearing/seals fluid films and the effects of the carrying structure. Vector $\underline{F}$ in eq. (A.9) therefore represents the vector of the single external forces applied to the nodes of the rotor and to the connection nodes. Equation (A.9) therefore permits, once $\underline{F}$ is known and different from zero, any problem of forced motion to be solved. Always with eq. (A.9) on the other hand, by assuming $\underline{F} = 0$ it is possible to study the free motion and to analyse instability condition by solving the associated homogenous problem.

Methods of solution

As will be explained further on, the different problems that are encountered when studying rotor dynamics (forced vibrations and instability) are cast, by the P.A.L.L.A. package programs, into the form of the evaluation of the frequency response to a harmonic excitation force of the entire rotor-fluid films-foundation system, thus adopting a more stable algorithm of solution.

As has been said, all the problems are studied by analysing the frequency response to a harmonic excitation force of the type:

$$\underline{F} = \underline{F}_0 \exp(i\Omega_E t) \tag{A.10}$$

Both here and further on, we will indicate the excitation frequency by Ω_E and the rotating speed of the rotor by Ω_R, both expressed in r/s (radians per second). With this excitation force, the forces at the connection nodes are also of the type :

$$\underline{R}_F = \underline{R}_{Fo} \exp\ (i\Omega_E t) \tag{A.11}$$

In this case eq. (A.8) simply becomes :

$$[I(\Omega_E)]\ \underline{X}_{Fo} = \underline{R}_{Fo} \tag{A.12}$$

Since $\underline{X}_{Fo}$ is the vector of the (complex) vibration amplitudes of the connection nodes :

$$\underline{X}_F = X_{Fo} \exp(i\Omega_E t) \tag{A.13}$$

In the case of the harmonic excitation force considered, by using the modal approach it is easy to define the matrix of the mechanical impedances $[I(\Omega_E)]$. In fact if we assume that a unitary harmonic excitation force is applied to the carrying structure at one of the connection nodes "along" a d.o.f. (fig. A.5), the vibrations at all the other nodal points can be obtained by using eq. (A.6) and by substituting vector $\underline{R}_F$ with vector $\underline{F}_1$ whose components are all null except for the d.o.f. where the excitation force was applied. Having obtained the components q_i of vector $\underline{q}$ from eq. (A.6), composed of all uncoupled equations, $\underline{X}_F$ is obtained from eq. (A.5); beeing $\underline{F}_1 = \underline{F}_{10}\exp(i\Omega_E t)$ it is also $\underline{X}_F = \underline{X}_{F0}\exp(i\Omega_E t)$. Since the flexibility matrix is defined by equation:

$$\underline{X}_{F10} = [H(\Omega_E)]\ \underline{F}_{10} \tag{A.14}$$

the solution obtained above represents the column of $[H(\Omega_E)]$ corresponding to the d.o.f. where the unitary excitation force was applied. By changing the application node of the excitation force it is possible to define the entire deformation matrix $H[(\Omega_E)]$ and from its inversion

$$[I(\Omega_E)] = [H(\Omega_E)]^{-1} \tag{A.15}$$

it is possible to obtain the matrix of the mechanical impedances.

In this calculation it is only necessary to check that the number of natural modes is equal or higher than the number of d.o.f. relative to the connection nodes. This inversion is not possible otherwise. Let us now go back to eq. A.1); since $\underline{F}^* = \underline{F}_0^*\exp(i\Omega_E t)$ and therefore $\underline{X} = \underline{X}_0\exp(i\Omega_E t)$, it can be rewritten as:

$$(-\Omega_E^2[M^*] + i\Omega_E[R^*] + [K^*])\ \underline{X}_0 = \underline{F}_0^* \tag{A.16}$$

where $\underline{F}_0^*$ contains both the external forces as well as the forces transmitted from the foundation through the connection nodes, i.e.:

$$\underline{F}_0^* = \begin{Bmatrix} \underline{F}_{Ro} \\ \underline{F}_{Fo} \end{Bmatrix} + \begin{Bmatrix} 0 \\ \underline{R}_{Fo} \end{Bmatrix} = \underline{F}_0 + \underline{\tilde{R}}_{Fo} \tag{A.17}$$

By substituting $\underline{\tilde{R}}_{Fo}$ with the expression (A.12) and by bringing the term in $\underline{X}_{Fo}$ of (A.12) to the left hand side, in the same way as done for equation (A.9), eq. (A.16) can be rewritten as:

$$[E(\Omega_E)]\ \underline{X}_0 = \underline{F}_0 \tag{A.18}$$

where $[E(\Omega_E)]$ is the elastodynamic matrix of the overall system and $\underline{F}_0$ the vector of the single external forces applied to the rotor and to the rotor-carrying structure connection nodes. All the problems relative to rotor dynamics can be solved by using the frequency response of the system given by (A.18), which represents a complex non-homogenous algebraic expression. By assuming that Ω_E=0, eq. (A.18) represents the response of the system to a static excitation force. The program offers the user the simplified possibility of schematizing the foundation by means of uncoupled mass-spring-damper systems for each bearing or seal as in fig. A.6. In this case, as regards the foundation, it is necessary to simply assign the parameters relative to mass, spring and viscous damper.

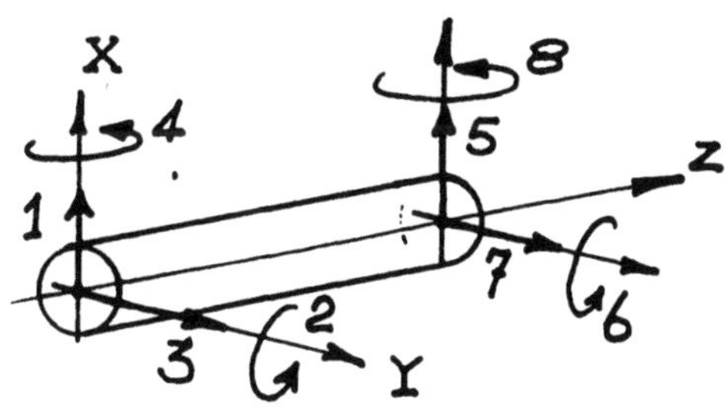

Fig. A.1 - 8 d.o.f. "beam" finite element

Fig. A.2 - Shaft line schematization

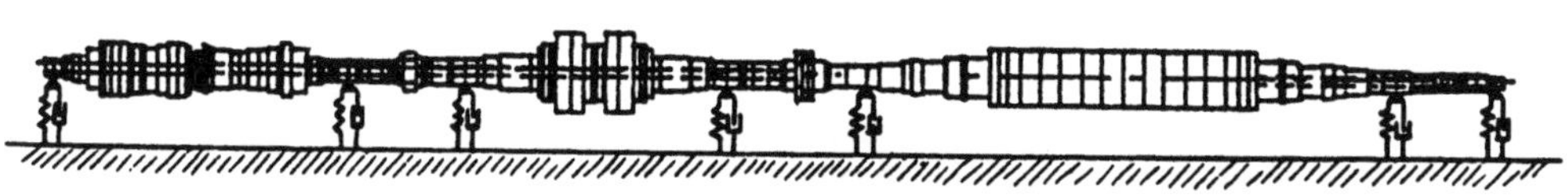

Fig. A.3- Shaft line + bearings system

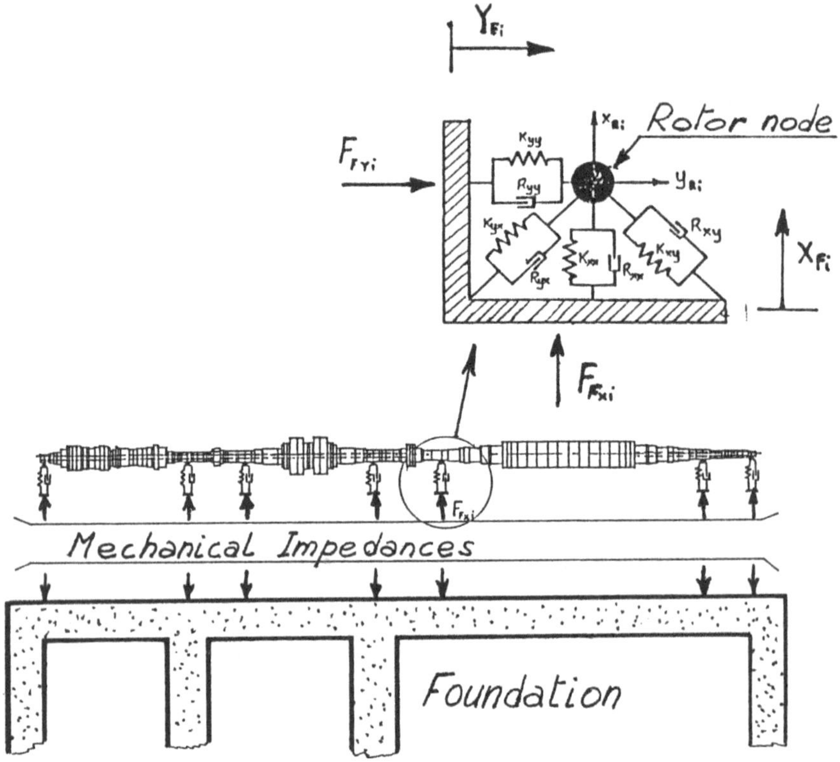

Fig. A.4 - Shaft line-fluid film-carrying structure overall system

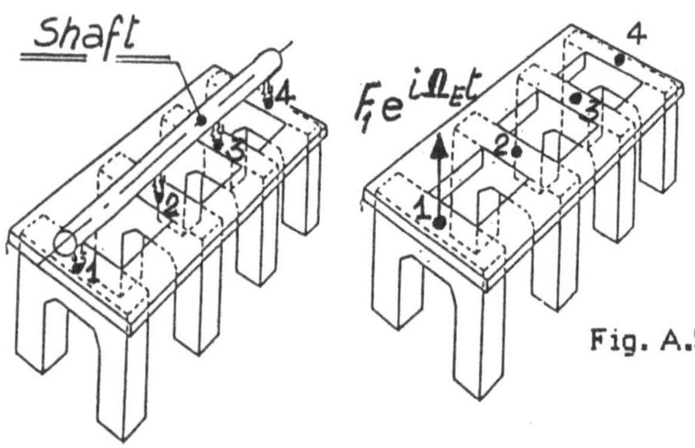

Fig. A.5 - Carrying structure connection nodes

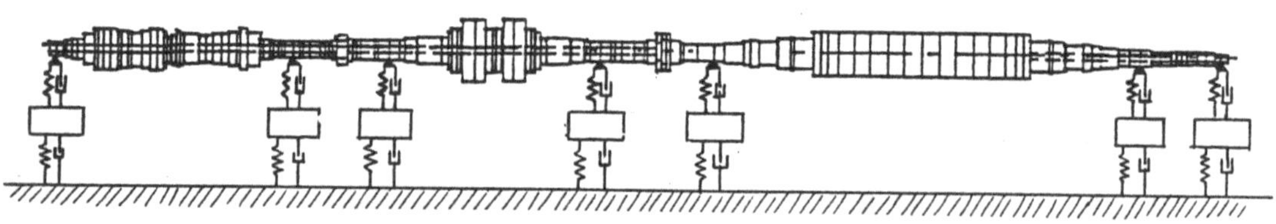

Fig. A.6 - A mass-spring-dampers simplified representation of the carrying structure

7.1.3 Package Configuration

The P.A.L.L.A. package is basically interactive both as regards the data input as well as the choice of the type of processing to be used. Calculation can be carried out either using the "demand" and/or "batch" modes. The handling of the set of the different programs is carried out by means of a "Package General Menu" which carries out the various functions by transferring control to different main programs which in turn recall specific subroutines. Data transfer takes place by means of interconnection logical units which support the different input and/or output files of the various programs. As specified further on, some of these files are created interactively by the user, others automatically as the output of subprograms. The preparation and use of the above mentioned files allows the calculation to be carried out even in the "batch" mode. The logic chart of the package that evidences the different functions is shown in fig. B.1. with the numbers of the relative "Functional Blocks". Block 0 is the "Package General Menu" for the handling of the various programs: this guides the user in the choice of the different functions carried out in the other blocks. Blocks 1,2,3 and 4 are recalled for the introduction, in colloquial form, of the data relative to the shaft line, bearings, seals and the carrying structure, which can be schematized by means of the mechanical impedances or by means of uncoupled mass-spring-damping systems. These same blocks supply the plots of the shaft, bearings and carrying structure in order to check the accuracy of the data introduced: eventual correction are carried out interactively. Once the accuracy of the data introduced has been verified, data files corresponding to the different components previously mentioned are created automatically from Blocks 1,2, 3 and 4. Blocks 5 to 13 carry out intermediate calculation functions and are activated by means of sequences piloted by the "General Menu" in function of the final calculation which the user intends to carry out. Block 5, which calculates the overall mass and stiffness matrices of the shaft line (without fluid films), is carried out automatically, in that it is an obligatory step, whatever the final type of calculation expected. The same goes for Block 6 which carries out the static analysis of the shaft line, analysis that always have to be done since it supplies the loads on supports once

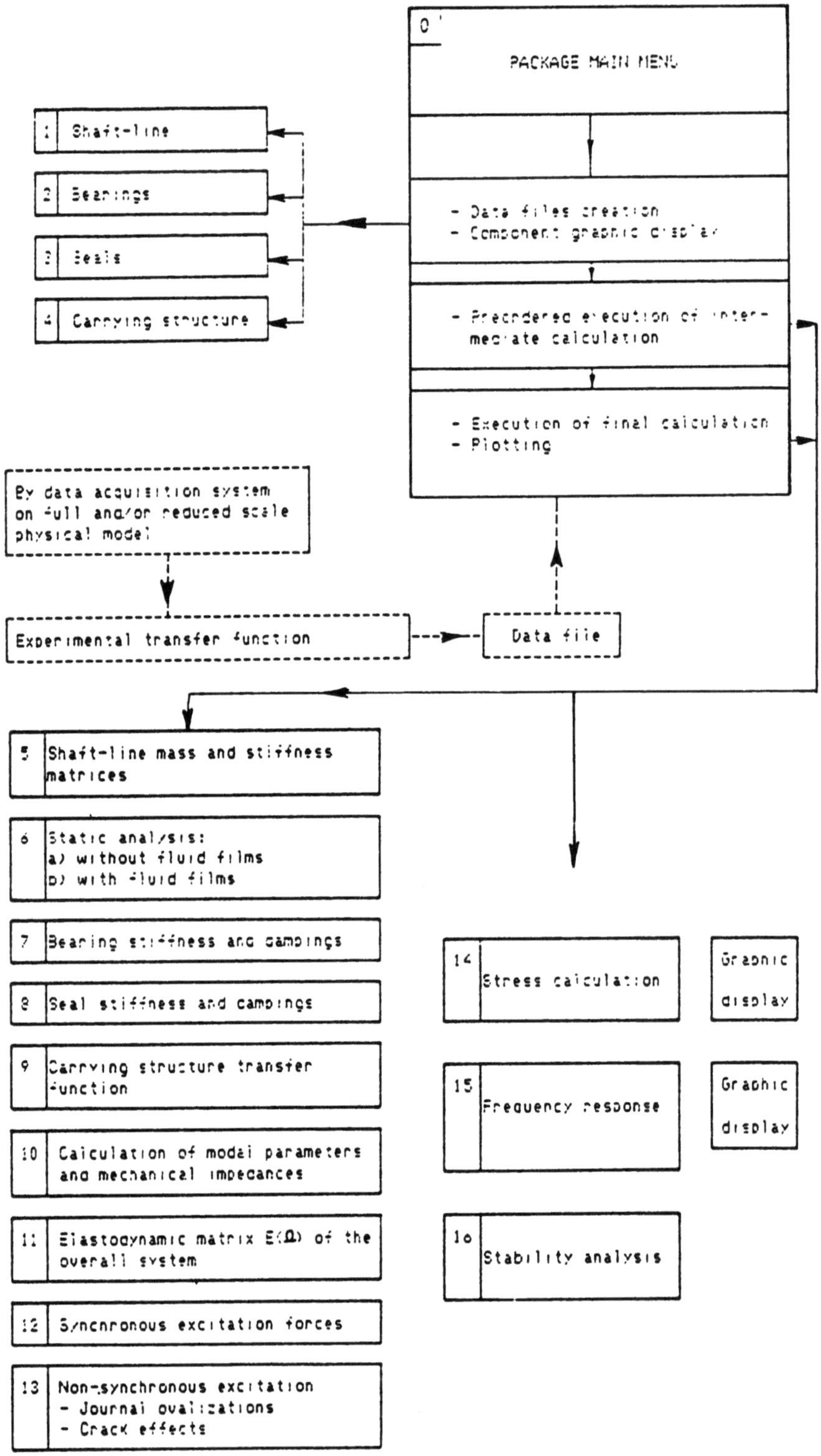

Figure B.1 - Flow chart of program showing function blocks

alignment has been assigned, or the alignment, once the loads and the position of some of the supports have been assigned. In this same Block it is also possible to keep account of the stiffness of the lubricating film to define the loads on supports once the alignment has been assigned or vice-versa. For this reason it is necessary to know the fluid film stiffness of bearings or seals, characteristics which are calculated at a given angular speed of the shaft line Ω_R in Blocks 7 and 8, by assigning the first attempt loads defined in Block 6 as input loads. To improve the calculation an iterative procedure between Block 6 and Blocks 7 and 8 is followed. Block 6, therefore, calculates the alignment conditions while the shaft is stationary (in the absence, therefore, of an oil film, or by keeping account of the oil film stiffness corresponding to an assigned angular speed). Once the alignment conditions have been defined, Block 14, on request, gives the stress calculation along the entire shaft line due both to the alignment conditions as well as, for the horizontal shafts, to its own weight. Block 9 is activated should the user wish to keep account of the carrying structure through the mechanical impedances. In this Block, the dynamic behaviour of the carrying structure itself whose transfer functions are defined in correspondence to the connection nodes with the shaft is simulated through a Finite Element schematization. These transfer functions can also be assigned by the user, as in the case where they are defined starting from the experimental data. Block 10 utilizes the transfer functions derived from the calculation (Block 9), or assigned by the user to define the modal parameters of the carrying structure. The matrix of the mechanical impedances is also calculated from the modal parameters for an assigned frequency Ω_E. The data processed in Blocks 5, 6, 7, 8, and 10 are used in Block 11 to create the elastodynamic matrix $[E(\Omega_E)]$ of the whole system, matrix defined by relation:

$$[E(\Omega_E)]\ \underline{X}_o = \underline{F}_o \tag{B.1}$$

This matrix, besides the mass and stiffness matrix of the shaft, contains the matrix of the mechanical impedances of the foundation and the stiffness and damping matrices of the oil film which are a function of the angular speed of the rotor Ω_R which coincides with the frequency of the excitation force Ω_E in the event of synchronous excitation which is the case of a rotating unbalance.

If the frequency Ω_E of the excitation force is different from the angular speed Ω_R of the rotor, matrix [E] also becomes a function of Ω_R, since the stiffness and damping constants of the bearings and the seals are dependent on the rotating speed of the rotor. Block 15, by using the elastodynamic matrix assembled in Block 11, solves the forced problem with sinusoidal excitation i.e. solves eq. (B.1). Through Blocks 12 and 13 the user can introduce different excitation forces to be applied to the system for the calculation of the frequency response (Block 15). Possible different excitation forms are:

Block		Type of function
12	Synchronous excitation	- Unbalance response - "Crank" effect
13	Non-synchronous excitation	- 2 per rev. due to shaft variable stiffness - 2 per rev. due to journal ovalization - transverse crack

Block 16, using matrix $[E(\Omega_E)]$ defined in Block 11 and the stiffness and damping defined in Blocks 7 and 8 carries out, at the user's request, the stability analysis of the shaft line. Graphic outputs are also available for Blocks 14 and 15. Block 3 relative to the introduction and creation of data files for the seals is still in the implementation stage. So is Block 8 relative to the calculation of the equivalent stiffness and damping coefficients of same.

7.1.4 Block 1: Definition of the Input Data of the Shaft Line

The introduction of data regarding the shaft line is carried out in this Block. As has already been said, the rotor is represented with beam finite elements. The subdivision into elements (see fig. A.2) must be done in such a way as to respect the usual schematization techniques for the use of this approach. The elements are thickened in correspondence to sudden diameter variations. Length/diameter ratios of the various elements higher than a limit value l_s and lower than a minimum value l_i are avoided. The value l_s is limited by the fact that the single element should have a natural frequency sufficiently higher than the frequencies at which we are interested in calculating the behaviour of the shaft line being examined. On the other hand, value l_i is linked to the requirement of not having an element with too small a length-to-diameter ratio with respect to the surrounding elements: the stiffness of this small element would negatively influence the overall stiffness matrix of the system. For each element the external diameter, internal diameter and length must be given, besides the value of a fourth quantity, the added mass, which will be specified further on. The shaft material is assumed to be steel (specific weight 7860 Kg/m^3, Young modulus 21000 Kg/mm^2). The automatic calculation program starting from this data (element external and internal diameter, length), the weight of the generic element (considered as uniformly distributed along the element itself) and its bending stiffness are calculated. If the area of the transverse section which is effectively resistent (from the stress point of view) does not coincide with the real one (as in the case of bladed disks, etc. being present) it is necessary to input the values of the external and internal diameter of the effectively resistent section and an extra weight (added mass) given by the difference between the effective weight of the element and the one calculated by the program starting from the assigned geometric data. The added mass is also considered by the program as uniformly distributed along the element. The introduction of shaft line data occurs element by element, added to which there is the possibility of controlling the elements already introduced using graphic outputs (fig.1.1.) and, consequently, to correct the input data.

7.1.5 Block 2: Definition of the Bearing Geometry

The part of the program described in this Block deals with the introduction of the geometric data of the bearings of the shaft line being examined. The three types of bearings presently considered by the program and which can therefore be chosen by the user are:

- multi-lobe

- tilting pad (20 pads max.)
- roller bearings

The lobe of the bearing is the circumferential portion of the bearing bounded by lubricant inlet and outlet pockets, in correspondence to which known (supply and discharge) pressures are assigned. The fundamental parameters to be assigned are the journal radius R and the radial clearance δ, which represents the fluid film thickness when the journal axis coincides with the center of the lobe (or generic pad). For this reason it is assumed that the lobe (or the generic pad) has a circular profile. Non circular profile are assigned through the method described further on. In the event of the different lobes and/or pads having different clearances, a reference clearance is assumed: the other values of the radial clearance will be specified through a parameter e_{oi}, which defines the ratios between the clearances of the different lobes and the reference one.

Multi-lobe bearings

With reference to the generic i-th lobe (see fig. 2.1), the geometric parameters to be assigned are :

- x_{ci}, y_{ci}: coordinates of the center of the lobe i-nth in a reference system with origin in the theoric center of the bearing and oriented as in fig. 2.1;
- e_{oi} : ratio between the clearance of the lobe i-nth (assumed circular) and the reference clearance;
- ϑ_{oi} : start angle, measured as in fig. 2.1, of the i-th lobe;
- ϑ_{2i} : angular extension of the i-th lobe;
- JD(i) : number of cylindrical parts into wich the i-th lobe is axially divided (max. 3).

Assuming that the journal axis is displaced by x_{ci}, y_{ci} with respect to the center of the i-th lobe, the film thickness is defined by a function $h(\vartheta)$ (see [5]) which is dependent on the bearing characteristics. When the journal center assumes the coordinates x,y in the reference system of fig. 2.1, the film thickness $h(\vartheta)$ assumes the value given by:

$$h(\vartheta) = h_{oi}(\vartheta) + (x - x_{ci})\sin\vartheta + (y - y_{ci})\cos\vartheta \qquad (2.1)$$

In the case of the circular lobe considered up till now we have equation

$$h_{0i} = \text{const.} = \delta_i \tag{2.2}$$

which defines clearance δ_i of the i-th lobe. This clearance is linked to clearance δ, assumed as a reference, by equation:

$$\delta_i = e_{0i}\delta \tag{2.3}$$

which specifies the meaning of parameter e_{0i}. For the cases in which the bearing profile is not circular, function $h(\Theta)$ assumes the expression:

$$h_{0i}(\Theta) = \delta_i + a\Theta^3 + b\Theta^2 + c\Theta + d \tag{2.4}$$

within intervals of amplitude $\Delta\Theta$ (in a maximum number specified in the User Instructions Manual) such as eq. 2.4) approximates the effective profile with the desired accuracy (see fig. 2.2). Eq. (2.4) is used, with a = b= 0, for bearings equipped with "grooves" with a profile that varies linearly with the angle (fig. 2.3). In its most general form eq. (2.4) it is used for cases when the effects of wear of the bearing has to be considered (fig. 2.2). In the case in which, in the i-th lobe, there are circumferential pockets with a large clearance (see fig. 2.1), the single cylindrical portions (which the program considers equal one with the other and in a maximum number of 3) are considered active, as far as carrying capacity is concerned. The width of each of these portions is represented by parameter b_i previously specified. The so called "elliptic (lemon) standard" bearing is a two-lobe bearing with the following geometric characteristics (fig. 2.4.) already fixed by the program :

- radial clearance δ = 2.66‰ of journal radius R
- $x_{c1} = .5\,\delta$; $x_{c2} = -.5\,\delta$
- $y_{c1} = y_{c2} = 0$
- $e_{01} = e_{02} = 1$
- ϑ_{01} = 25 deg. ; ϑ_{02} = 205 deg.
- ϑ_{2i} = 130 deg.
- JD(1) = 1 ; JD(2) = 3

Therefore, only the widths of the single cylindrical portion of the lower lobe (b_1) and that of the upper one (b_2) (fig. 2.4) have to be assigned. The cylindrical bearing without pockets or grooves should be treated as a one-lobe bearing with a 360 deg. extension.

Tilting pad bearings

For this type of bearing the film thickness of the generic pad assumes the expression (see fig. 2.5)

$$h(\vartheta) = h_{oi}(\vartheta) + (X-X_{ci})\sin\vartheta + (y-y_{ci})\cos\vartheta - \psi r_0 \sin(\vartheta-\varphi_0) \qquad (2.5)$$

in which r_o and φ_o are the polar coordinates of the hinge and ψ is the generic small rotation of the pad from the position assumed to evaluate $h_{oi}(\vartheta)$. The quantities x_{ci} and y_{ci} determine what in literature is normally indicated by preload, which in the program is assumed to be a radial displacement of the tilting pad center with respect to the theoretical center of the bearing (generally the center of the circle passing through pad hinges). The geometric definition of a tilting pad bearing besides the quantities r_o and φ_o which appear in eq. (2.5) and the radial clearance already defined at the beginning, (it is important to note that different radial clearances are not expected for the various pads and is therefore always $e_{oi} = 1$) requires, for each of the tilting pads, the following data :

- the preload of the tilting pad, defined as has already been said. Non-dimensional ratio preload/radial clearance is used;
- the moment of inertia J_p of the tilting pad with respect to the hinge, assigned by means of an equivalent mass $m_e = J_p/r_o^2$.
- ϑ_{oi} start angle of the i-th tilting pad measured as in fig. 2.5;
- ϑ_{2i} angular amplitude of the i-th tilting pad.

Moreover, the axial length b of the tilting pads must be assigned.

Roller bearings

For these type of bearings the stiffness matrix, i.e. the coefficients K_{xx}, K_{xy}, K_{yx}, K_{yy} in the previously defined reference system, must be directly assigned. The introduction of the geometric data of the bearings takes place bearing by bearing, with the possibility of controlling the data introduced by means of graphic outputs.

7.1.6 Block 3: Seals Geometry

This Block has not yet been implemented in the package.

7.1.7 Block 4: Carrying Structure

In this phase of the program all the data required to define the parameters of the mathematical model of the carrying structure are prepared. The type of mathematical model used is described in order to justify the necessary input data.

Mathematical Model

The carrying structure is schematized via F.E.M. with 12 d.o.f. "beam" elements ([6], [7],). These elements (fig. 4.1) represent homogenous beams with a constant section, undergoing axial action, shear, twisting moment and bending moment along the two axis of inertia of the transverse section. In Tab. 4.I the corresponding stiffness $[k_t]$ and mass $[m_t]$ matrices evaluated in the local reference system are shown.

The corresponding structural matrices $[K_t]$ and $[M_t]$ calculated in the absolute reference assumed for the overall system are calculated by means of the well known relation:

$$
\begin{aligned}
[K_t] &= [\Lambda]^T [k_t] [\Lambda] \\
[M_t] &= [\Lambda]^T [m_t] [\Lambda]
\end{aligned}
\qquad (4.1)
$$

where $[\Lambda]$ represents the coordinate transformation matrix.The mathematical model that reproduces the carrying structure in the P.A.L.L.A. package program can also simulate the presence of rigid bodies, linked to the carrying structure itself, whose inertia contribution is not negligible. The equivalent corresponding mass matrix, in the absolute reference system, is shown in Tab. 4.II, where R_x, R_y and R_z are the coordinates of the center of gravity G of the body in the local reference system $\bar{x}$, $\bar{y}$, $\bar{z}$, (see fig. 4.2), m^* is the total mass of the rigid body and T_{ij} (i,j = 1,2,3) represent the components of a matrix [T] defined as:

$$
[T] = [\Lambda_\varphi]^T [I_G] [\Lambda_\varphi] \qquad 4.2)
$$

where $[I_G]$ is the matrix of the polar inertia moments of the rigid body with respect to the three principal axis:

$$
[I_G] = \begin{bmatrix} I_{\bar{x}G} & 0 & 0 \\ 0 & I_{\bar{y}G} & 0 \\ 0 & 0 & I_{\bar{z}G} \end{bmatrix} \qquad (4.3)
$$

and $[\Lambda_\vartheta]$ is the coordinate transformation matrix which correlate rotations in the local reference system to those in the absolute reference system:

$$\begin{Bmatrix} \bar{\vartheta}_x \\ \bar{\vartheta}_y \\ \bar{\vartheta}_z \end{Bmatrix} = [\Lambda_\vartheta] \begin{Bmatrix} \vartheta_x \\ \vartheta_y \\ \vartheta_z \end{Bmatrix} \tag{4.4}$$

Once all the structural matrices of the elements that compose the complete model of the carrying structure have been defined, it is possible to assemble the total stiffness matrix $[K_s]$ and the mass matrix $[M_s]$ using the usual techniques. Having assigned the constraint conditions to the structure, after partitioning the equations of motion relative to the constrained and unconstrained ([6], [7]) d.o.f., the equations of motion of the entire carrying structure therefore becomes:

$$[M_s]\ddot{\underline{X}}_s + [R_s]\dot{\underline{X}}_s + [K_s]\underline{X}_s = \underline{F}_s(t) \tag{4.5}$$

where $[R_s]$ represents the damping matrix calculated as:

$$[R_s] = \alpha[M_s] + \beta[K_s] \tag{4.6}$$

By assuming $\underline{F}_s = 0$ in eq. (4.5) and $[R_s] = [0]$, the equation is reduced to the form:

$$[M_s]\ddot{\underline{X}}_s + [K_s]\underline{X}_s = 0 \tag{4.7}$$

whose solution:

$$\underline{X}_s = \underline{X}_s e^{i\omega t} \tag{4.8}$$

enables us to obtain the natural pulsations of the system and relative principal vibrating modes(*). On the other hand by assuming

---

(*) Frequency response calculation and modal identification techniques are used instead of the eigenvalues and eigenvectors calculation. Reasons are given further on.

$$\underline{F}_s(t)=\underline{F}_{so}e^{i\Omega t} \qquad (4.9)$$

in 4.5) it is possible to calculate the frequency response of the carrying structure for the assigned harmonic excitation forces. This possibility is used to calculate the transfer functions of the carrying structure (Block 9), functions evaluated to calculate the modal parameters of the same in order to define the corresponding mechanical impedances matrix (Block 10).

7.1.8 Block 5: Shaft Line Mass and Stiffness Matrices

The rotors of the shaft line analysed by the P.A.L.L.A. package are schematized, as already mentioned in paragraph A, by means of homogenous beam elements, with a constant section, that reproduce the different mass and stiffness characteristics of the same rotors ([8]. Each beam element has 8 d.o.f. (see fig. A.1) to reproduce the deformation shape assumed by the rotor in space: the vertical X and horizontal Y transverse displacements and the rotations of the node section are therefore represented. The axial and torsional displacements of the shaft are negligible in that the associated vibrations (axial and torsional) are generally uncoupled from the bending motions considered herein. The corresponding stiffness [K] and mass [m] matrices of the generic "beam" element which reproduce the elastic and inertia characteristics of an element of the rotor are shown in Tab. 5.I. For more details see [4],[8],[9] and [10].

7.1.9 Block 6: Statis Analysis

This phase of the program enables the user to carry out the static analysis of the shaft line by also keeping account of the bearing lubricating films. In the hyperstatic system which is the one constituted by a shaft line on several supports, the loads acting on the latter are also functions of the bearings alignment, i.e. of the position that these hold in the reference planes orthogonal to the undeformed shaft line axis. The knowledge of these loads is important in order to establish the working conditions of the machine. This determination is obtained, in the static field, using a P.A.L.L.A. package program.The aims of this program are:

a) determination of the alignment of the supports to obtain pre-established loads on bearings.This is the method used to calculate the supports alignment on starting up the machine. For the rotors with horizontal axis an alignment is normally imposed so that the loads on the bearings correspond to the so called "single beam reactions", i.e. reactions exerted by the uncoupled rotors. Each rotor is normally placed on two bearings: the constraint reactions determined in this way are also maintained when, by means of the end joints, the aforementioned rotor is constrained, flexurally and torsionally, to the other rotors composing the shaft line.

b) determination of the loads once the alignment has been assigned. This method is useful when we wish to evaluate the effects of the support displacements: e.g. thermal distortions of the foundation structure can cause, on one or more supports, load variations with respect to the nominal with consequent phenomena such as oil film instability.

As previously indicated, the shaft line is schematized with 8 d.o.f. "beam" finite elements. To obtain the stiffness and mass matrices of the rotor-oil film system it is necessary to suitably assemble the stiffness matrices $[k]$ (already described in Block 5) of the single elements and the equivalent oil film stiffness matrices (see Block 7). The d.o.f. number of the shaft nodes is generally very large so that condensation becomes necessary (fig. 6.1). In the static analysis carried out whether the lubricating film is kept account of or not, the condensation adopted, of the static type, proves to be rigorous. The use of this approach decomposes the generic shaft line into a set of substructures (called "superelements") each composed of several beam elements with connection nodes between superelement and superelement as extremity end nodes (called "supernodes"). As a substructure the program assumes the portions of the rotor connecting two supernodes. Extremity nodes and bearing nodes are always assumed to be supernodes (see fig. 6.2). For the generic superelement the equation of static equilibrium can be expressed as:

$$[K_e]\underline{X}_e=\underline{P}_e \tag{6.1}$$

where $\underline{X}_e$ is the vector of the displacements relative to the generic superelement, $[K_e]$ is the relative stiffness matrix assembled as shown in fig. 6.3. Finally, in eq.(6.1), $\underline{P}_e$ represents the vector of the generalized forces due to the weight of the rotor portion schematized by means of the superelement. Having called $\underline{X}_{ee}$ the displacement vector of the nodes external to the condensation (nodes 1 and 10 of fig. 6.3) and $\underline{X}_{ei}$ the vector of the remaining degrees of freedom, it is possible to express equation (6.1) in partitioned form, by reordering the vector Xe as:

$$\underline{X}_e = \begin{Bmatrix} \underline{X}_{ee} \\ \underline{X}_{ei} \end{Bmatrix} \qquad (6.2)$$

thus obtaining:

$$\left[\begin{array}{c|c} K_{ee} & K_{ei} \\ \hline K_{ie} & K_{ii} \end{array}\right] \begin{Bmatrix} \underline{X}_{ee} \\ \hline \underline{X}_{ei} \end{Bmatrix} = \begin{Bmatrix} \underline{P}_{ee} \\ \hline \underline{P}_{ei} \end{Bmatrix} \qquad (6.3)$$

or rather:

$$\begin{aligned} &[K_{ee}]\underline{X}_{ee}+[K_{ei}]\underline{X}_{ei}=\underline{P}_{ee} \qquad a) \\ &[K_{ie}]\underline{X}_{ee}+[K_{ii}]\underline{X}_{ei}=\underline{P}_{ei} \qquad b) \end{aligned} \qquad (6.4)$$

Obtaining from (6.4):

$$\underline{X}_{ei}=[K_{ii}]^{-1}\underline{P}_{ei}-[K_{ii}]^{-1}[K_{ie}]\underline{X}_{ee} \qquad (6.5)$$

and substituting (6.5) in (6.4a), we obtain:

$$([K_{ee}]-[K_{ei}][K_{ii}]^{-1}[K_{ie}])\underline{X}_{ee}=\underline{P}_{ee}-[K_{ei}][K_{ii}]^{-1}\underline{P}_{ei} \qquad (6.6)$$

By defining the stiffness matrix of the statically condensed generic superelement with

$$[K_c]=([K_{ee}]-[K_{ei}][K_{ii}]^{-1}[K_{ie}]) \qquad (6.7)$$

and with

$$\underline{P}_c=\underline{P}_{ee}-[K_{ei}][K_{ii}]^{-1}\underline{P}_{ei} \qquad (6.8)$$

the vector of the generalized condensed forces, it is possible to rewrite equation (6.6) as:

$$[K_c]\underline{X}_{ee}=\underline{P}_c \tag{6.9}$$

In this way it is possible, to solve the static equation in rigorous form, by using as degrees of freedom those relative to the single supernodes. Having calculated the matrices of the single superelements and the matrices of the equivalent stiffness of the oil film (Blocks 7 and 8) in this way, it is possible to proceed with the assembly of the complete rotor + oil film system as shown in figure 6.4. As regards the only shaft line, the equations system for the static situation considered is therefore given by:

$$[K]\underline{X}=\underline{F} \tag{6.10}$$

in which $\underline{F}$ is the vector of the nodal forces equivalent to the external forces (natural weights) and $\underline{X}$ is the vector containing all the shaft line d.o.f. (4 d.o.f. per node times the number of shaft nodes). Matrix [K] which, for now, is assumed not to contain the stiffness matrices of bearings or seals, is singular. Elimination of [K]'s singularity (static equilibrium condition) is obtained by imposing at least fo· displacements (isostatic case) equally partitioned in the horizontal and vertical directions. By partitioning matrix [K], the displacement vector $\underline{X}$ and the nodal forces vector $\underline{F}$ we obtain :

$$\begin{bmatrix} K_{LL} & K_{LV} \\ K_{VL} & K_{VV} \end{bmatrix} \begin{Bmatrix} \underline{X}_L \\ \underline{X}_V \end{Bmatrix} = \begin{Bmatrix} \underline{F}_L \\ \underline{F}_V \end{Bmatrix} \tag{6.11}$$

From system 6.11) we can formally obtain the displacements of the free nodes $\underline{X}_L$ and the constraint reactions $\underline{F}_V$ according to the system:

$$\underline{X}_L=[K_{LL}]^{-1}\underline{F}_L-[K_{LL}]^{-1}[K_{Lv}]\underline{X}_v \tag{6.12}$$

$$\underline{F}_v=[K_{vL}]\underline{X}_L+[K_{vv}]\underline{X}_v \tag{6.13}$$

by solving system 6.12) first and by substituting the values found thus for $\underline{X}_L$ in eq. 6.13). Coordinates $\underline{X}_V$ are always, as is obvious, bearing node coordinates. They are as has been said, in a minimum number of two per plane, horizontal and vertical, and in a maximum number equal to twice the number of bearings (at the bearing nodes rotations are never constrained). The problem, discussed in point b) of the previous paragraph, of determining the support forces when the alignment is assigned can be solved by means of equations 6.12) and 6.13). Obviously, the same system enables us to determine $\underline{X}_V$ once forces $\underline{F}_V$ have been assigned . This situation is described in point a) of the previous paragraph. However, some caution is necessary. Vector $\underline{F}_V$ cannot be completely arbitrarily assigned, since it has to respect the equilibrium conditions between the external forces and the constraint reactions. To avoid possible mistakes in this sense, it is preferable to impose, in this case too, the minimum number of two displacements at any two supports, thus obtaining the isostaticity of the structure; on the two supports thus constrained it will not be possible to impose constraint reactions. It is now possible to assign any set of constraint reactions to the n-2 remaining supports and to obtain their alignment always by means of eqs. 6.12) and 6.13), cast for this purpose in another form. At the end of this calculation phase the loads acting on the supports are known. It is therefore possible to keep account of the presence of the film of the bearings and/or seals, once an angular reference speed has been assigned. For this reason and for the two situations previously described the program operates in the following way:

1) the loads on the bearings in correspondence to a certain rotation speed of the shaft line are imposed.

   The program, once the displacements have been calculated in absence of the lubricating film as previously described, determines the relative journal-bearing displacements for the assigned rotation speed. The relative coordinates thus obtained and added to the ones indicating the positions of the bearing centers determine the alignment to be assigned to the latter during the assembly phase.

2) the alignment is assigned.

   Once the loads on supports have been calculated as previously described, the relative displacements which are added to the assigned alignment are therefore calculated. The deformation of the shaft line thus obtained does not however correspond to the loads calculated on the basis of the initial alignment due to the relative displacements determined by the lubricating film. This accounts for the need to use an iterative process which, for the sake of calculation economy, is limited to only one iteration.

The calculation method followed to define the characteristics of the lubricating film is illustrated in the description of Block 7. A more detailed explanation of the program is given in [8] where the calculation procedures followed to avoid matrix partitions and inversions which appear in the equations previously shown are also described.

7.1.10 Block 7: Determination of the Bearing Static and Dynamic Characteristics

The part of the program relative to this Block defines the force field that is set up between the journal and the bearing due to the lubricating film, as a function of the relative journal-bearing position. This field is linearized in the neighbourhood of the static equilibrium position, thus defining the equivalent stiffness and damping coefficients. The calculation of the equilibrium position of the journal inside the relative bearing and the corresponding values of the stiffness and damping constants of the oil film is obtained by integrating Reynolds' equation for both a laminar and turbulent regime, keeping account of the temperature variability along the film. Reynolds' equation is written, in cylindrical coordinates (fig. 7.1) in the form:

$$\frac{\partial}{\partial\vartheta}\left(\frac{h^3}{\mu_\vartheta}\cdot\frac{\partial p}{\partial\vartheta}\right)+\left(\frac{2R}{b}\right)^2\cdot\left(\frac{h^3}{\mu_\eta}\cdot\frac{\partial^2 p}{\partial\eta^2}\right) = 6\Omega_R R^2\cdot\left(\frac{dh}{d\vartheta}+\frac{2V}{\Omega_R}\right) \qquad 7.1)$$

having indicated as

b	axial width of the bearing
h	film thickness
p	pressure
R	journal radius
ϑ	non-dimensional cylindrical coordinates
μ	turbulent viscosity of the lubricant for axial and circumferential flow
Ω	angular speed of the journal
V	velocity of the journal center.

The program automatically verifies the type (laminar or turbulent) of the fluid film flow.

<u>Calculation of the force acting on the journal</u>

Both for the laminar as well as the turbulent regimes the program integrates equation (7.1) following an approximate method shown in [11], whose applications are shown in [5]. After integration of eq. (7.1), the values of the horizontal Fy and vertical Fx components of the fluid film force acting on the journal (fig. 7.2) are calculated:

$$F_x = F_x(x, y, \dot{x}, \dot{y}, \Omega_R)$$
$$F_y = F_y(x, y, \dot{x}, \dot{y}, \Omega_R) \qquad (7.2)$$

as functions of the assumed values of the coordinates x,y and velocities $\dot{x}, \dot{y}$ of the journal center and of the rotation speed Ω_R. To carry out this calculation it is necessary to also assign, besides the geometric quantities that enable oil film thickness $h(\vartheta)$ to be calculated, the values of density ρ of the specific heat C_s and the viscosity μ of the lubricant considered, together with the temperature (in °C) of same at the entrance. The values of C_s and ρ are considered constant while the viscosity μ is assumed variable according to the equation:

$$\mu = 1/(AT^2+BT+C) \qquad (7.3)$$

where A,B and C are dependent on the lubricant type considered. As regards the temperature, the program assumes, as described in [11], that the walls of the meatus are adiabatic, and, as a first approximation calculation, considers the pressure effect negligible thus obtaining expression:

$$C\ dT/d\Theta = 2\ \nu(T)UR/h^2 \qquad (7.4)$$

in which U represents the peripheral speed of the journal ($\Omega_R R$) and ν the kinematic viscosity. From equation (7.4), having assigned a temperature T_o at the beginning of the meatus, solution $T(\Theta)$ and the local viscosity value $\mu(\Theta)$ from eq. (7.3) are obtained. In the case of laminar flow regime the viscosity is defined as:

$$\mu = \mu(\vartheta) \qquad (7.5)$$

On the other hand, in the case of turbulent flow regime, the program extends the validity of Reynolds' equation by introducing the expressions of a turbulent viscosity derived in [12] on the basis of papers [13] and [14]. According to the findings shown, it is assumed that there are two different expressions of the turbulent viscosity, i.e.:

$$\mu_\theta = \mu\,(1 + C_\theta) \tag{7.6}$$

for flow in the circumferential direction and:

$$\mu_\eta = \mu\,(1 + C_\eta) \tag{7.7}$$

in the axial direction. In [14] the values of C_ϑ and C_η as a function of the local Reynolds' number R_h are not directly given. For use inside the program interpolating functions with expressions

$$\begin{aligned} C_\theta &= 0.00327\ R_h^{4/5} - 0.36 \\ C_\eta &= 0.00178\ R_h^{4/5} - 0.24 \end{aligned} \tag{7.8}$$

were determined which give results that are fairly close to those of the paper quoted above for values of Rh ranging from 900 up to 200000. The experimental results obtained in a test campaign, carried out on elliptic bearings normally used in real turbosets showed the necessity to modify these expressions of the turbulent viscosity. For this purpose (see [15]) other modifications were also introduced to analytically reproduce the distribution of the oil film temperature measured experimentally. As shown in [16],[17],[18] an excellent agreement between the experimental locus of journal center and the analytical one was obtained by using, instead of eq. 7.8), expressions:

$$\begin{aligned} C_\theta &= (0.00327\ R_h^{4/5} - 0.755)\beta + 0.395 \\ C_\eta &= (0.00178\ R_h^{4/5} - 0.411)\beta + 0.171 \end{aligned} \tag{7.9}$$

with $\beta = 2$. The systematic comparisons carried out using the experimental results of those test campaigns did however show that this correction alone is not enough. Finally, for the turbulent viscosity expressions:

$$\mu_\vartheta = \mu\,(\alpha\, C_\vartheta + 1)$$
$$\mu_\eta = \mu\,(\alpha\, C_\eta + 1) \tag{7.10}$$

were used to obtain a good agreement between the experimental values in which C_ϑ, C_η are given by eq. 7.9) and α has the expression:

$$\alpha = a\left(\frac{dh}{d\vartheta}\right)^2 + b\,\frac{dh}{d\vartheta} + c \tag{7.11}$$

The mathematical model used to calculate the thermal field is modified in the second iteration and in the case of turbulent flow regime, in order to keep account of the influence of the pressure flow. The expression of the fluid rate of flow in a circumferential direction was assumed to be:

$$Q_\vartheta = U\,\frac{h}{2} - h^3\,\frac{\partial p}{\partial \vartheta}\,\frac{1}{12\mu R} \tag{7.12}$$

Moreover, in the expression of the thermal gradient, account was kept of the power W_d dissipated as result of the harmonic journal oscillations with amplitude U_o and pulsation Ω_E. When calculating the temperature, it is in fact necessary to evaluate the power dissipated locally (in every small fluid element) due to journal vibration. The calculation was carried out in an approximate manner by evaluating the overall power dissipated by the journal with the simplified expression:

$$W_d = \frac{\gamma}{2}\,\Omega_E^2 U_o{}^2 R_{xx}\,\sin\varphi \tag{7.13}$$

where R_{xx} represents the direct dissipation term of the oil film and γ is a corrective term. By dividing this expression by the carrying surface of bearing $\pi R\, b$, we obtain an averaged power per unit area. Moreover, this power was "weighed" with the local pressure $p(\vartheta)$ compared with the average pressure p_m, thus obtaining expression:

$$W_u = \frac{W_d}{b\,R\,\pi}\,\frac{p(\vartheta)}{p_m} = \frac{W_d}{Q_x}\,p(\vartheta) \qquad 7.14)$$

since Q_x is the bearing load. Finally, the temperature gradient, keeping account of the pressure flow and the dissipation associated with the vibration, becomes:

$$C_s\,dT/d\vartheta = \frac{UT}{Q_\vartheta} - \frac{\partial p}{\partial \vartheta} + \left(\frac{\omega^2}{2}\,\frac{p_m}{F_{st}}\,R_{xx}U_o{}^2\sin\varphi\right)\frac{R}{Q_\vartheta} \qquad 7.15)$$

where Q_ϑ should be evaluated through equation (7.12) and R_{xx} can be given the value r_{xx} calculated by the program.

<u>Determination of the static equilibrium position</u>

The procedure to determine the static equilibrium position of the journal inside the single bearing consists of an iterative procedure enabling the determination of coordinates x and y of the journal center so that equations:

$$\begin{aligned} F_y(x,y,0,0,\Omega_R) &= Q_y \\ F_x(x,y,0,0,\Omega_R) &= Q_x \end{aligned} \qquad 7.16)$$

are satisfied and where Q_x and Q_y are the load components on the journal, according to the vertical x and horizontal y directions, evaluated as previously described in Block 9.

<u>Determination of the equivalent stiffness and damping coefficients</u>

The determination of the stiffness and damping constants, to obtain the equilibrium position calculated in this way, is made by developing equations 7.2) in Taylor series, thus obtaining expressions:

$$\begin{aligned} F_x(x+\delta x, y+\delta y, \delta\dot{x}, \delta\dot{y}, \Omega_R) &= Q_x + K_{xx}x + K_{xy}y + R_{xx}\dot{x} + R_{xy}\dot{y} \\ F_y(x+\delta x, y+\delta y, \delta\dot{x}, \delta\dot{y}, \Omega_R) &= Q_y + K_{yx}x + K_{yy}y + R_{yx}\dot{x} + R_{yy}\dot{y} \end{aligned} \qquad (7.17)$$

where

$$\begin{aligned} K_{xx} &= \partial F_x/\partial x \qquad & K_{xy} &= \partial F_x/\partial y \\ K_{yx} &= \partial F_y/\partial x \qquad & K_{yy} &= \partial F_y/\partial y \end{aligned} \qquad (7.18)$$

represent the stiffness constants and:

$$R_{xx} = \partial F_x / \partial \dot{x} \qquad R_{xy} = \partial F_x / \partial \dot{y}$$
$$R_{yx} = \partial F_y / \partial \dot{x} \qquad R_{yy} = \partial F_y / \partial \dot{y} \tag{7.19}$$

the damping constants of the lubricating film. The numerical values of the derivatives in expression 7.18) and 7.19) are calculated by assuming that the functions F_x and F_y have a parabolic trend in the neighbourhood of the equilibrium position and are defined by separately giving increments δx, δy, $\delta \dot{x}$ and $\delta \dot{y}$.

7.1.11 Block 8: Seal Stiffness and Damping Coefficients

This block has not yet been implemented in the "P.A.L.L.A." package program.

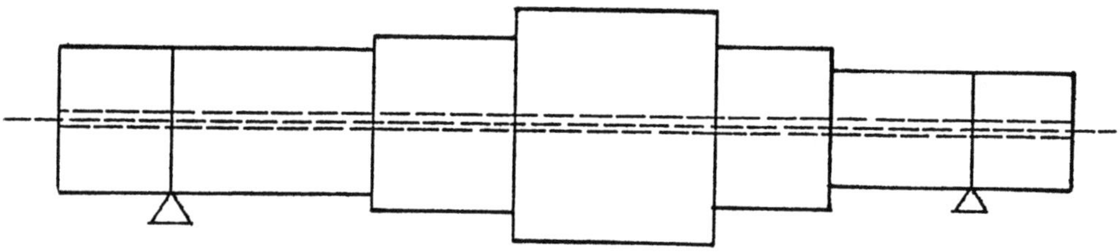

Fig. 1.1 - Graphic output example

Fig. 2.1 - Geometric characteristics of a general lobe

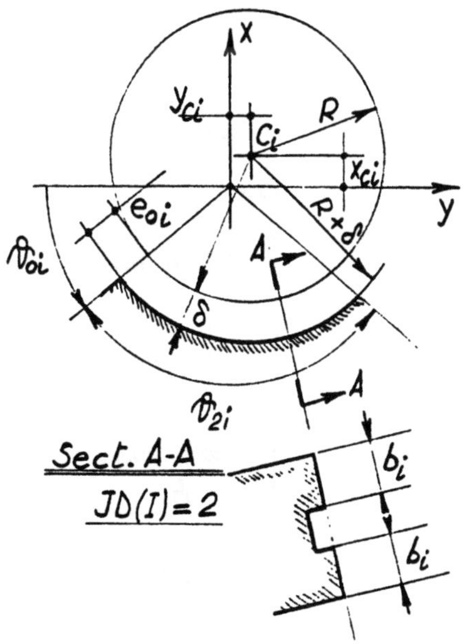

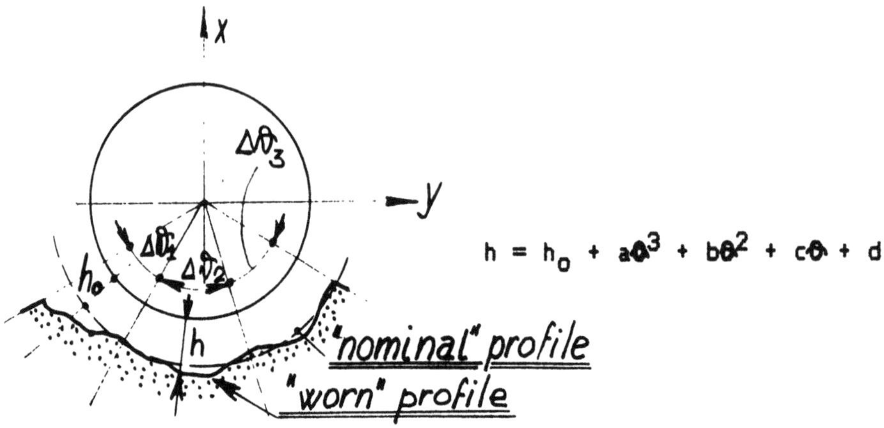

Fig. 2.2 – Stepwise representation of a worn bearing profile

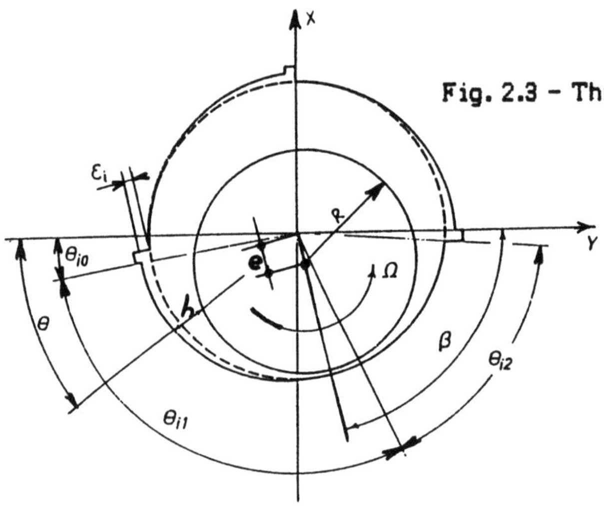

Fig. 2.3 – Three lobe grooved bearing

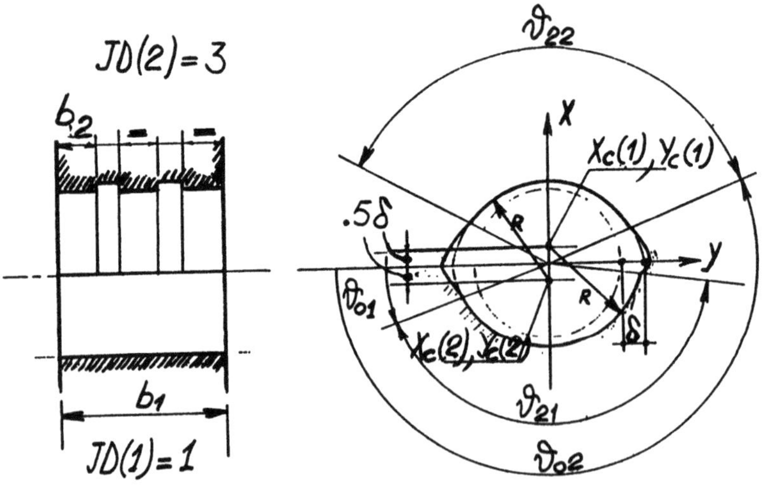

Fig. 2.4 – Elliptic bearing

$[K_t] =$

1	2	3	4	5	6	7	8	9	10	11	12
EA/l											
	$12EJ_z/l^3$										
		$12EJ_y/l^3$									
			GJ_x/l								
		$-6EJ_y/l^2$		$4EJ_y/l$							
	$6EJ_z/l^2$				$4EJ_z/l$						
$-EA/l$						EA/l					
	$-12EJ_z/l^3$				$-6EJ_z/l^2$		$12EJ_z/l^3$				
		$-12EJ_y/l^3$		$6EJ_y/l^2$				$12EJ_y/l^3$			
			$-GJ_x/l$						GJ_x/l		
		$-6EJ_y/l^2$		$2EJ_y/l$				$6EJ_y/l^2$		$4EJ_y/l$	
	$6EJ_z/l^2$				$2EJ\ /l$		$-6EJ_z/l^2$				$4EJ_z/l$

SYMMETRIC

(elements not shown are zero)

$[M_t] = \rho A l \cdot$

1	2	3	4	5	6	7	8	9	10	11	12
1/3											
	$13/35+6J_z/5Al^2$										
		$13/35+6J_y/5Al^2$									
			$J_x/3A$								
		$-11l/210-J_y/10Al$		$l^2/105+2J_y/15A$							
	$11l/210+J_z/10Al$				$l^2/105+2J_z/15A$						
1/6					$-(12,2)$	$(1,1)$					
	$9/70-6J_z/5Al^2$			$-(11,3)$			$(2,2)$				
		$9/70-6J_y/5Al^2$						$(3,3)$			
			$J_x/6A$						$(4,4)$		
		$13l/420-J_y/10Al$		$-l^2/140-J_y/30A$				$-(5,3)$		$(5,5)$	
	$-13l/420+J_z/10Al$				$-l^2/140-J_z/30A$		$-(6,2)$				$(6,6)$

SYMMETRIC

(elements not shown are zero)

TAB. 4.I - 12 d.o.f. beam element stiffness and mass matrices

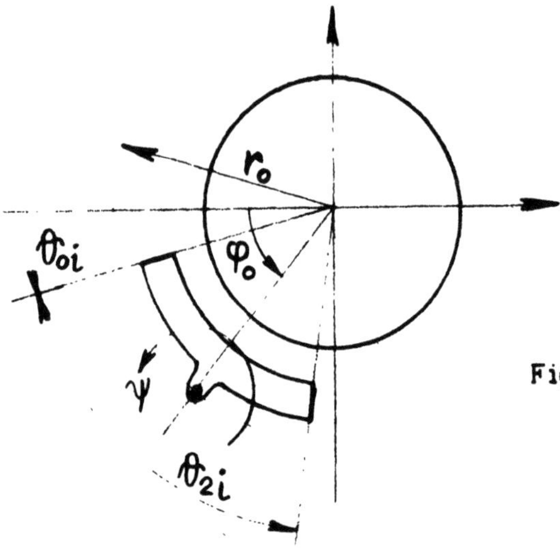

Fig. 2.5 - Tilting pad bearing

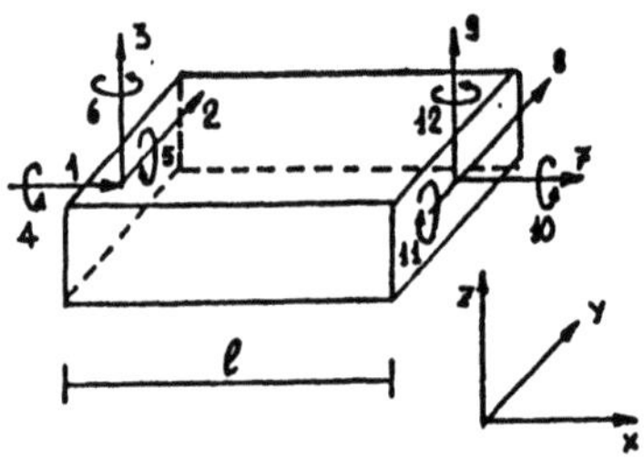

Fig. 4.1 - 12 d.o.f. "beam" element

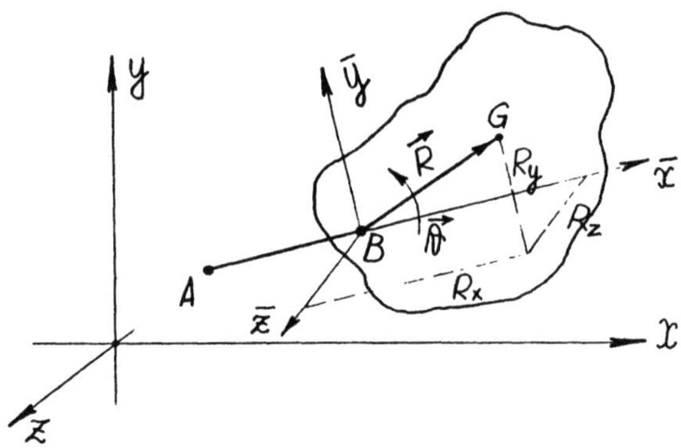

Fig. 4.2 - A beam element A B + rigid body system

$$
\begin{bmatrix}
1 & 0 & 0 & 0 & R_z & -R_y \\
0 & 1 & 0 & -R_z & 0 & R_x \\
0 & 0 & 1 & R_y & -R_x & 0 \\
0 & -R_z & R_y & T_{11}(R_y{}^2 + R_z{}^2)/m^* & T_{12}/m^* - R_x R_y & T_{13}/m^* - R_x R_z \\
R_z & 0 & -R_x & T_{21}/m^* - R_x R_y & T_{22}(R_x{}^2 + R_z{}^2)/m^* & T_{23}/m^* - R_y R_z \\
-R_y & R_x & 0 & T_{31}/m^* - R_x R_z & T_{32}/m^* - R_y R_z & T_{33}(R_x{}^2 + R_z{}^2)/m^*
\end{bmatrix} m^*
$$

TAB 4.II - Equivalent mass matrix of system shown in fig. 4.2

$[K] =$

$12EJ_z/1^2$							
	$12EJ_y/1^2$						
	$-6EJ_y/1^2$	$4EJ_y/1$					
$6EJ_z/1^2$			$4EJ_z/1$				
-(1,1)			-(8,1)	(1,1)			
	-(2,2)	-(7,2)			(2,2)		
	(3,2)	$2EJ_y/1$			-(3,2)	(3,3)	
(4,1)			$2EJ_z/1$	-(8,1)			(4,4)

SYMMETRIC

(elements not shown are zero)

$[m] =$

156							
221	41^2						
		156					
		-221	41^2				
54	131			156			
-131	-31^2			-221	41^2		
		54	-131			156	
		131	-31^2			221	41^2

SYMMETRIC

(elements not shown are zero)

$$\left(\frac{\rho \pi 1 (D_e^2 - D_i^2)}{4} + m \right) / 420$$

Tab. 5.I – 8 d.o.f. "beam" element stiffness and mass matrices

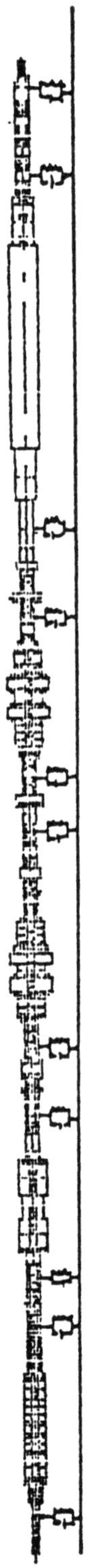

Fig. 6.1 – Shaft line + oil film bearings system

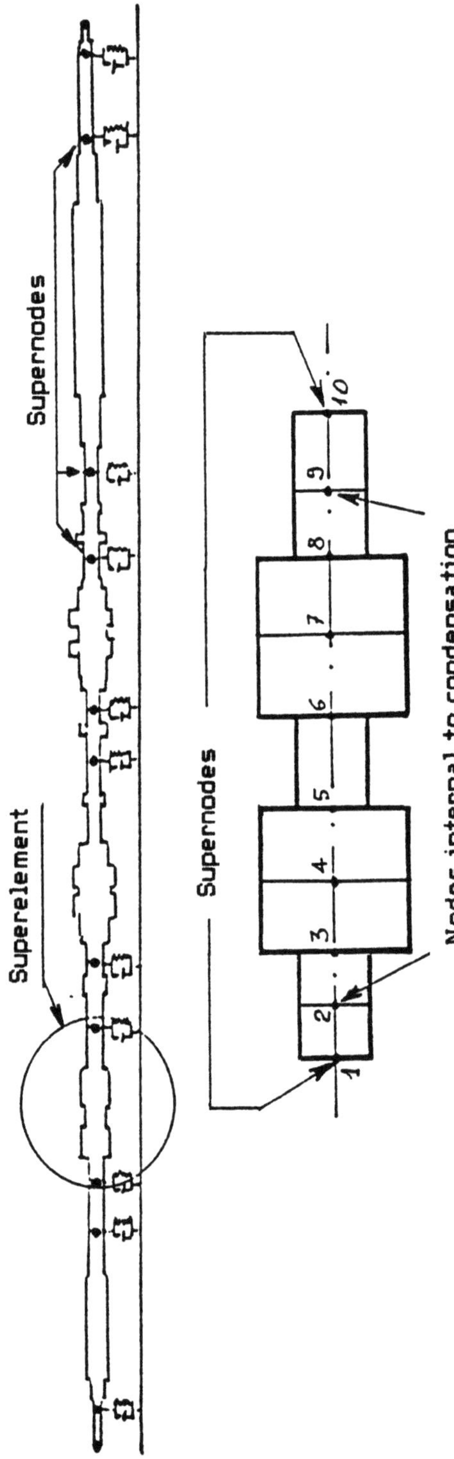

Fig. 6.2 – Subdivision into superelements

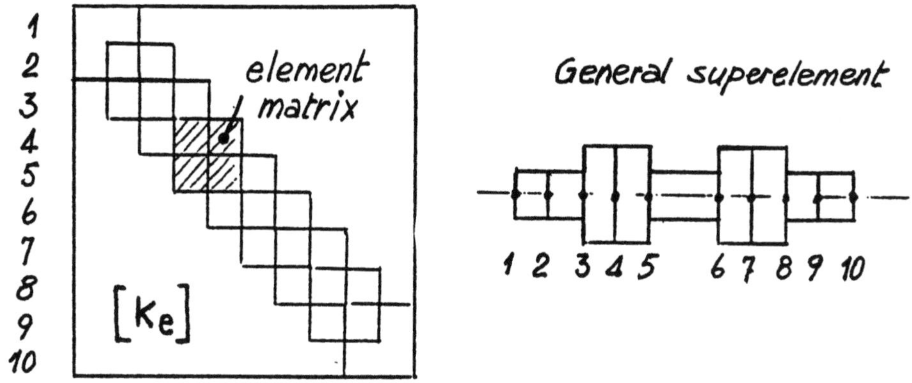

Fig. 6.3 - Superelement matrix

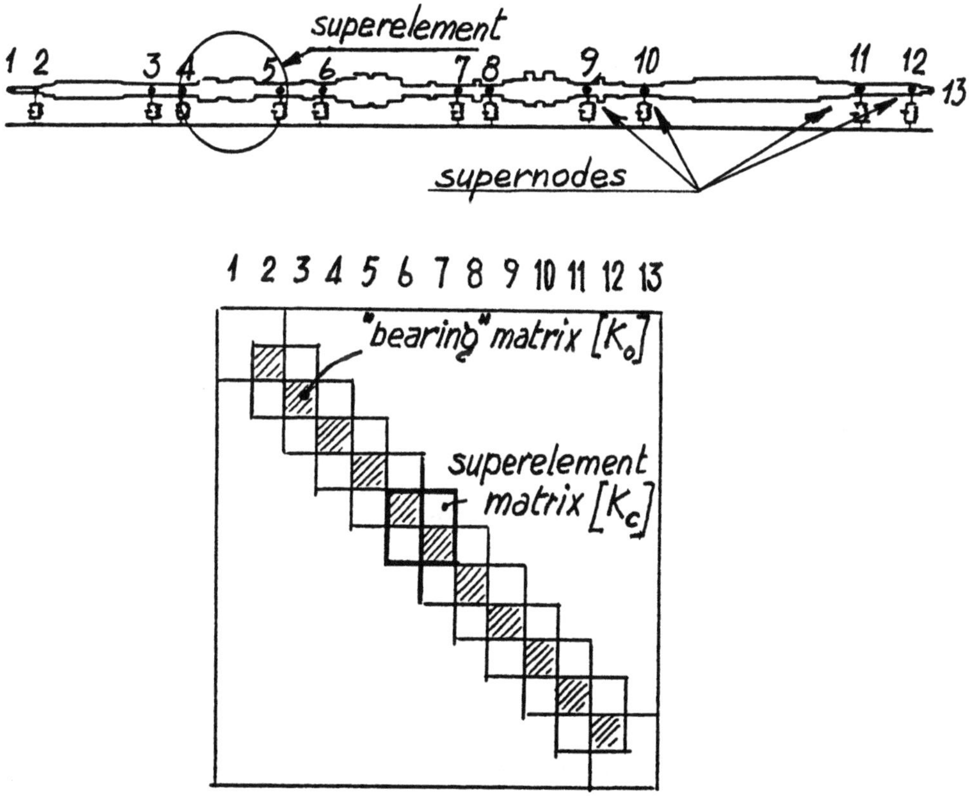

Fig. 6.4 - Overall stiffness matrix of the system (on rigid foundation)

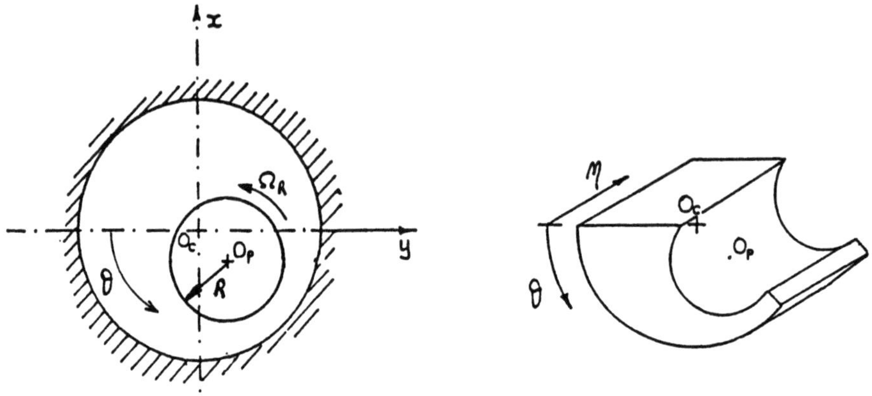

Fig. 7.1 - Fluid film geometry

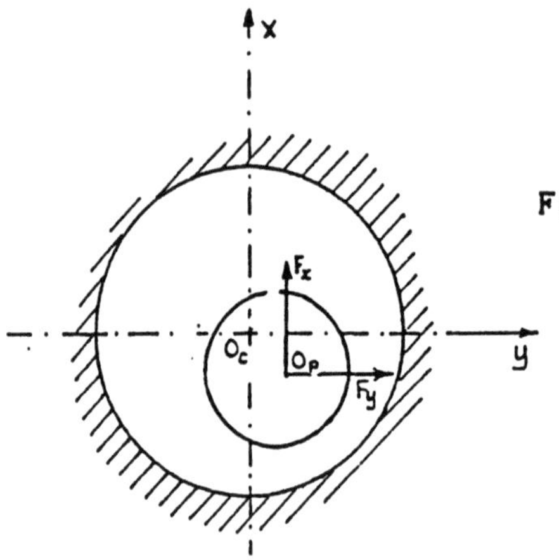

Fig. 7.2 - Fluid film force components

7.1.12 Block 9: Transfer Functions of the Carrying Structure

In this part of the program the frequency response of the carrying structure is calculated using the mathematical model already described in Block 4: the equations of motion of the support structure can be expressed as:

$$[M_S]\ddot{\underline{X}}_S + [R_S]\dot{\underline{X}}_S + [K_S]\underline{X}_S = \underline{F}_S(t) \qquad (9.1)$$

where $[M_S]$, $[R_S]$ and $[K_S]$ are respectively the matrices of mass, damping and stiffness of the foundation. In equation (9.1) $\underline{X}_S$ represents the node displacements of the same supporting structure, while $\underline{F}_S$ is the vector of the applied excitation forces.

In the modal approach used to define the mechanical impedances of the foundation, eq. (9.1) is cast in the frequency domain, thus assuming a harmonic excitation force of the type:

$$\underline{F}_s=\underline{F}_{so}\ e^{i\Omega_E t} \tag{9.2}$$

to which a response corresponds (also harmonic):

$$\underline{X}_s=\underline{X}_{so}\ e^{i\Omega_E t} \tag{9.3}$$

In the frequency domain eq. (9.1), keeping account of eqs. (9.2) and (9.3), can be rewritten as:

$$(-\Omega_E^2[M_s]+i\Omega_E[R_s]+[K_s])\underline{X}_{so}=\underline{F}_{so} \tag{9.4}$$

In order to calculate the matrix of the mechanical impedances of the foundation $[I(\Omega_E)]$ defined by relation:

$$\underline{F}_{soc}=[I(\Omega_E)]\underline{X}_{soc} \tag{9.5}$$

where $\underline{X}_{soc}$ is the displacement vector of the foundation relative to the only nodes that connect the rotor and $\underline{F}_{soc}$ the vector of the forces exchanged in the same places, it is necessary to evaluate the flexibility matrix $[H(\Omega_E)]$ defined by:

$$\underline{X}_{soc}=[H(\Omega_E)]\underline{F}_{soc} \tag{9.6}$$

This matrix is evaluated for a generic pulsation Ω_E, starting with the modal parameters of the foundation, parameters that can be evaluated if only one column of $[H(\Omega_E)]$ is known, evaluated for a certain number of pulsations Ω_E. The generic l-th column of the flexibility matrix $[H(\Omega_E)]$ represents the (complex) vibration amplitudes of the d.o.f. of all the connection nodes for a unitary harmonic excitation force applied to the l-th d.o.f., therefore corresponding to the index of the column considered. The generic term h_{kl} of the column l-th is, as has been said, complex and is called a transfer function between the k-th d.o.f. and the l-th d.o.f. of the rotor-foundation connection nodes. To obtain the l-th column of the matrix $[H(\Omega_E)]$ it is enough to therefore impose a unitary harmonic excitation force in correspondence to the d.o.f. relative to the column considered (*), thus solving equation (9.4). The frequency range in which it is necessary to evaluate the generic column $\underline{h}_1$ and for which it is therefore necessary to solve eq. (9.4) is defined by means of the following criteria:

1) it must start from a null frequency and extend itself at least above twice the rotor regime frequency;
2) it must extend itself above the working frequency so that a number of modes, respectively vertical and horizontal, equal or higher than the number of supports for each of the two excitations (vertical and horizontal) can be determined.

Equation (9.4) is solved by means of a Gauss algorithm with double total pivoting.

7.1.13 Block 10: Definition of Modal Parameters and Mechanical Impedances

This Block of the program enables one to carry out the modal analysis of the structure carrying the shaft line i.e. to identify the natural frequencies of the system, the damping factors and the generalized masses associated to them. Moreover, for each natural frequency, the corresponding vibration modes of the structure are determined. The modal parameters of the carrying structure will be used subsequently to determine the matrix of the mechanical impedances of the foundation, in relation to the d.o.f. corresponding to the support points of the axis line. The modal parameters are determined by analysing the transfer functions, evaluated at the shaft line-carrying structure connection nodes, and, if necessary, in other significant points to identify the vibration modes. These functions can be obtained both experimentally, by taking measurements on a real model, and analytically, by means of a mathematical model of the system. In this case the frequency response due to a unitary harmonic excitation force is calculated (see Block 9).

(*) In effect, the vertical and horizontal modes are uncoupled; it is therefore necessary to calculate two columns of the matrix [H], one relative to a vertical d.o.f. and one relative to a horizontal d.o.f., having in this way to solve equation (9.4) twice, once with a vertical and once with a horizontal unitary excitation force.

The definition of the modes depends, in the experimental case, on the measurement points used, which should be at least all the support points. In the following, with "foundation d.o.f." we will mean either the degrees of freedom of the mathematical model or those of the measurement points of a real structure used to define the transfer functions. The modal parameters should be determined by analysing separately the transfer functions obtained by applying one harmonic excitation force at a time in a suitable place of the carrying structure. The modal analysis program is of the iterative type and requires the user to have a basic knowledge of some of the modal analysis notions. The single modal parameters can be determined by carrying out in succession a series of logic and calculation operations described in a menu and organized according to a pre-established procedure. In the context of each calculation phase of the modal parameters it is possible to use different algorithms, characterized by different applicability fields. The applicability of the methods depends particularly on the level of independence of the natural vibration modes, on the damping factor entities that characterize the system and finally on the frequency resolution with which the transfer functions have been evaluated. The user has the possibility of intervening in the results given by the single methods, discarding the values considered unsatisfactory. The modal parameters must therefore be determined by using the results obtained with different algorithms.

<u>Identification of the modal parameters</u>

In matrix form, the equations of motion of any vibrating system can be written as:

$$[M]\underline{\ddot{X}}+[R]\underline{\dot{X}}+[K]\underline{X} = \underline{F} \qquad (10.1)$$

where $[M]$ $[R]$ and $[K]$ respectively represent mass, damping and stiffness matrices of the system, $\underline{X}$ is the vector relative to "n" d.o.f. and $\underline{F}$ is the vector of the excitation forces. The modal matrix $[\Phi]$, generally rectangular n x p (where p is the number of the vibration modes considered) is formed by p eigenvectors $\underline{\Phi}^{(j)}$, ordered in columns (these values are obtained as eigenvectors of the matrix $[M]^{-1}[K]$ and for a stable conservative system are real), i.e.:

$$[\Phi]=[\underline{\Phi}^{(1)};\underline{\Phi}^{(2)};\ldots\underline{\Phi}^{(P)}] \tag{10.2}$$

in which every eigenvector is formed by n terms $X_K^{(j)}$ (also known as "residues"), so that:

$$\underline{\Phi}^{(j)}=\begin{Bmatrix} X_1^{(j)} \\ X_2^{(j)} \\ \vdots \\ X_n^{(j)} \end{Bmatrix} \tag{10.3}$$

By means of the coordinate transformation:

$$\underline{X} = [\Phi]\underline{q} \tag{10.4}$$

it is possible to rewrite eq. (10.1) in principal coordinates:

$$[\ulcorner m \lrcorner]\ddot{\underline{q}}+[\ulcorner r \lrcorner]\dot{\underline{q}}+[\ulcorner k \lrcorner]\underline{q}=[\Phi]^T\underline{F} \tag{10.5}$$

where $[\ulcorner m \lrcorner]$, $[\ulcorner r \lrcorner]$ and $[\ulcorner k \lrcorner]$ are the mass, damping and stiffness matrices in principal coordinates (square and diagonal matrices pxp) and $\underline{q}$ are the principal coordinates themselves (the matrix $[\ulcorner r \lrcorner]$ is diagonal if $[R]$ is of the type: $[R]=a[M]+b[K]$). The generic diagonal element m_j of eq. (10.5) represents the generalized mass of the j-th vibration mode r_j is the associated damping and k_j the relative stiffness. If the vector of the excitation forces $\underline{F}$ of (10.1) is harmonic:

$$\underline{F}=\underline{F}_0 \, e^{i\Omega_E t} \tag{10.6}$$

the solution will also be harmonic:

$$\underline{X}=\underline{X}_o e^{i\Omega_E t} \tag{10.7}$$

and in principal coordinates:

$$\underline{q}=\underline{q}_0 \, e^{i\Omega_E t} \tag{10.8}$$

Equation 10.5) could then be rewritten as:

$$(-\Omega_E^2[\ulcorner m \lrcorner]+i\Omega_E[\ulcorner r \lrcorner]+[\ulcorner K \lrcorner])\underline{q}_o=[\Phi]^T\underline{F}_o \quad (10.9)$$

or rather

$$\underline{q}_o=(-\Omega_E^2[\ulcorner m \lrcorner]+i\Omega_E[\ulcorner r \lrcorner]+[\ulcorner K \lrcorner])^{-1}[\Phi]^T\underline{F}_o \quad (10.10)$$

Once the modal parameters m_j, r_j, and k_j and the corresponding vibration modes have been defined, with the algorithms that will be described hereinafter and having applied the transformation (10.4) we obtain:

$$\underline{X}_o=[\Phi](-\Omega^2_E[\ulcorner m \lrcorner]+i\Omega_E[\ulcorner r \lrcorner]+[\ulcorner K \lrcorner])^{-1}[\Phi]^T\underline{F}_o \quad (10.11)$$

i.e.:

$$\underline{X}_o=[H(\Omega_E)]\underline{F}_o \quad (10.12)$$

having indicated the flexibility matrix with $[H(\Omega_E)]$:

$$[H(\Omega_E)]=[\Phi](-\Omega^2_E[\ulcorner m \lrcorner]+i\Omega_E[\ulcorner r \lrcorner]+[\ulcorner k \lrcorner])^{-1}[\Phi]^T \quad (10.13)$$

As can be seen, once the modal parameters are known and once the pulsation Ω_E of the excitation force has been assigned, the matrix $[H(\Omega_E)]$ is numerically determined and therefore, from its inversion it is possible to obtain (see first chapter) the matrix of the mechanical impedances $[I(\Omega_E)]$. The generic term $h_{Kl}(\Omega_E)$ of the matrix $[H(\Omega_E)]$ is also named as, as has already been mentioned, transfer function and represents the complex frequency response of the system evaluated at point K, having applied a unitary harmonic excitation force of pulsation Ω_E at the point l. From (10.13) the generic transfer function can therefore be expressed in function of the only modal parameters as:

$$h_{Kl}(\Omega_E)= \sum_{j=1}^{m} \frac{X_K^{(j)}X_l^{(j)}}{m_j(\omega_j^2-\Omega_E^2+2i\sigma_j\Omega_E)} \quad (10.14)$$

where

$\omega_j^2 = k_j/m_j$ represents the j-th natural frequency of the system

$X_K^{(j)}$ represents the generic K-th component of the j-th vibration mode.

$X_l^{(j)}$ is the l-th mode component in the application point of the excitation force

σ_j is the associated damping factor, linked to the nondimensional damping $\xi_j = (r_j/r_{cj})$ through the following relation:

$$\sigma_j = \omega_j \xi_j = \omega_j \frac{r_j}{r_{cj}} = \omega_j \frac{r_j}{2m_j\omega_j} = \frac{r_j}{2m_j} \qquad (10.15)$$

<u>Determination of the natural frequencies</u>

The natural frequencies ω_j of the carrying structure can be determined by using two different algorithms. The first enables us to calculate the undamped natural frequencies and the second the damped natural frequencies of the system.

Method n.1

The first mathematical method consists in defining the "spectral power" function obtained by adding, for each sample frequency, the quadratic value of the imaginary component of each transfer function. The "spectral power" function thus defined will show very accentuated peaks in correspondence to the undamped natural frequencies of the structure. Furthermore, this method enables us to evidence all the resonances included in the frequency range analysed, even in the case where some transfer functions were evaluated in proximity to the vibration nodes; the lesser the damping the more accurate the results obtained with this method.

Method n.2

Once first approximation values are known (defined with the first method), the natural frequencies of the structure can be more rigourously determined by means of an algorithm which simultaneously allows the determination of the natural frequencies and the relative damping factors (see the method in following paragraph).

<u>Identification of the damping parameters</u>

In order to determine the damping factors σ_j three different methods were implemented. Some of these methods offer reliable results if the damping is low: it has, however, been experimentally shown that these conditions are satisfied in a large number of structures.

Method n.1

If the coupling between the vibration modes is negligible, the transfer function phase presents a variation near to 180° in a frequency range near to a resonance. A method which enables us to evaluate the damping factors is based on the analysis of the derivative of the phase evaluated in correspondence to the natural frequencies of the structure [19]. In fact, phase ψ of the transfer function in a system with one d.o.f. can be expressed as follows:

$$\psi = \tan^{-1} \frac{2\xi \frac{\Omega_E}{\omega}}{1 - \left(\frac{\Omega_E}{\omega}\right)^2} \tag{10.16}$$

where ω represents the natural frequency of the system and ξ the non-dimensional damping r/rc. By deriving equation (10.16) with respect to ω we have:

$$\frac{d\psi}{d\omega} = \frac{2}{\omega} \frac{1 + \left(\frac{\Omega_E}{\omega}\right)^2}{\left[\left[1 - \left(\frac{\Omega_E}{\omega}\right)^2\right]^2 + \left(\frac{2\xi\Omega_E}{\omega}\right)^2\right]^{1/2}} \tag{10.17}$$

In correspondence to the natural frequency ω this derivative assumes the following expression:

$$\left.\frac{d\psi}{d\omega}\right]_{\omega=\Omega_E} = \frac{2\xi}{\omega} \frac{1}{2\xi^2} = \frac{1}{\omega\xi} \tag{10.18}$$

Therefore, an extension to n d.o.f. systems, if the natural vibration modes of the system are uncoupled, for the j-th natural frequency ω_j gives (see eq. (10.15)):

$$\sigma_j = \xi_j \omega_j = \left.\frac{d\omega}{d\psi}\right]_{\omega=\omega_j} \tag{10.19}$$

Method n.2

This algorithm is based on the calculation of the frequencies ω_d and ω_b near to each resonance of the structure, in correspondence of which the amplitudes of the transfer function is $1/\sqrt{2}$ of the amplitude of the resonance peak. The calculation of the frequencies ω_d and ω_b is performed by determining the solutions of equation [20]:

$$\left(\frac{\Omega_E}{\omega_j}\right)^2 = (1-2\xi_j)^2 \pm 2\xi_j\sqrt{1+\xi_j} \tag{10.20}$$

assuming $\xi_j=(r_j/r_{cj}) \ll 1$ and neglecting higher order terms we obtain:

$$\left(\frac{\Omega_E}{\omega_j}\right)^2 = 1\pm 2\xi_j \tag{10.21}$$

Indicating the roots of this equation with ω_d and ω_b we have:

$$4\xi_j = \frac{\omega_b{}^2-\omega_d{}^2}{\omega_j} \tag{10.22}$$

or

$$\sigma_j = \frac{\omega_b{}^2-\omega_d{}^2}{2} \tag{10.23}$$

The accuracy of the results depends on the acquisition step of the frequencies.

Method n.3

If $h_{Kl}(\omega_j)$ is the transfer function evaluated in correspondence to the j-th natural frequency and $\Delta\omega$ the acquisition step (the step with which the transfer functions are known experimentally or analytically) we have:

$$\begin{aligned} \Delta h_{Kl}{}^{(j)} &= h_{Kl}(\omega_j) - h_{Kl}(\omega_j-\Delta w) \\ \Delta y_{Kl}{}^{(j)} &= h_{Kl}(\omega_j)\omega_j - h_{Kl}(\omega_j-\Delta\omega)(\omega_j-\Delta w) \end{aligned} \tag{10.24}$$

The damping factor associated to the j-th vibration mode can be defined by the imaginary part which from ratio $\Delta y^{(j)}{}_{Kl}/\Delta h^{(j)}{}_{Kl}$ [2] is:

$$\frac{\Delta y_{Kl}{}^{(j)}}{\Delta h_{Kl}{}^{(j)}} = \omega_{dj} + i\sigma_j \tag{10.25}$$

where with ω_{dj} the j-th pulsation of the damping system is indicated.

Identification of the vibration modes

Method n.1

The evaluation of the imaginary part of the transfer function in correspondence to the natural frequencies is the simplest identification method of the vibration modes; in this case too the accuracy of the results is strongly influenced by the value of the damping factors and the coupling of the modes themselves.

Method n.2

A more sophisticated method, known as the Nyquist method, consists in determining the circumference arc which approximates the transfer function in the neighbourhood of each resonance; the value of the diameter of the circumference is proportional to the modal amplitude, while the corresponding phase is determined by the coordinates of the center. If the coupling between the modes is not negligible but contribution given by modes near to the one considered can be approximated (in the neighbourhood of the frequency range where the j-th vibration mode is predominant) by a constant, we have ([1]):

$$\left(R_e(\Omega_E) + \frac{U_j}{2\sigma_j}\right)^2 + \left(I_m(\Omega_E) + \frac{V_j}{2\sigma_j}\right)^2 = \frac{U_j{}^2 + V_j{}^2}{4\sigma_j{}^2} \tag{10.26}$$

where with $U_j + iV_j$ the residue associated $X_K{}^{(j)}$ to the j-th vibration mode is indicated. The terms $Re(\Omega_E)$ and $Im\ (\Omega_E)$ represent respectively the real part and the imaginary part of the transfer function $h_{Kl}(\Omega_E)$ evaluated at frequency Ω_E. Equation (10.26) is the equation of a circumference whose center has the coordinates given by:

$$X_{cj} = -U_j/2\sigma_j$$
$$Y_{cj} = -V_j/2\sigma_j \tag{10.27}$$

while the value of the radius is:

$$R_j = \frac{\sqrt{U_j^2+V_j^2}}{2\sigma_j} \qquad (10.28)$$

from which the residue value relative to the generic j-th mode is proportional to the diameter of the circumference (eq. (10.26)). The accuracy of the results obtained with this method depends directly on the solving method used to determine the transfer function and on the number of points used to calculate the interpolating circumference.

Determination of the generalized mass

Method n.1

Let us now consider an equation of the system (10.9), relative to the generic j-th vibration mode: by assuming that we have only one harmonic excitation force applied to the 1-th degree of freedom we will have:

$$(-\Omega^2_E m_j + i\Omega_E r_j + K_j) q_{jo} = X_1^{(j)} F_{1o} \qquad (10.29)$$

where $X_1^{(j)}$ represents the amplitude of the j-th vibration mode evaluated at the point 1, F_{1o} is the 1-th term of vector $\underline{F}_o$ and q_{jo} the j-th term of vector $\underline{q}_o$ of eq. (10.9). By assuming resonance condition ($\Omega_E = \omega_j$) we will have:

$$i\Omega_E r_j q_{jo} = i\omega_j r_j q_{jo} = X_1^{(j)} F_{1o} \qquad (10.30)$$

or, with suitable simplifications:

$$\frac{2m_j\omega_j}{2m_j\omega_j} \; i\omega_j r_j q_{jo} = X_1^{(j)} F_{1o} \qquad \text{a)}$$

$$2im_j\xi_j\omega_j^2 q_{jo} = X_1^{(j)} F_{1o} \qquad \text{b)} \qquad (10.31)$$

from which:

$$m_j = \frac{X_1^{(j)} F_{1o}}{2i\xi_j\omega_j^2 q_{jo}} \qquad (10.32)$$

In eq. (10.32) F_{1o} is known, ω_j, ξ_j, $X_1^{(j)}$ can be determined by the methods described in the previous paragraphs. For the evaluation of q_{jo} the relation:

$$[\Phi]^{-1} = [\Phi]^T \tag{10.33}$$

is assumed, which is rigorously true only in the case of orthogonal eigenvectors in a restricted sense (mass matrix [M] of equation 10.1 diagonal). Using this hypothesis, equation (10.4) can be rewritten as:

$$\underline{q} = [\Phi]^T \underline{X} \tag{10.34}$$

or

$$q_{jo} = \sum_{k=1}^{n} X_k^{(j)} X_{ko} \tag{10.35}$$

where X_{ko} is the generic k-th element of the vector $\underline{X}_o$, evaluated experimentally by applying to the system an excitation force $F = F_o \exp(i\Omega_E t)$ or by analytically calculating the latter in the same way (see Block 9). By keeping account of equation (10.34) eq. (10.32) becomes:

$$m_j = \frac{X_1^{(j)} F_{1o}}{2i\xi_j \omega_j^2 \sum_{k=1}^{n} X_k^{(j)} X_{ko}} \tag{10.36}$$

From the definition of transfer function:

$$h_{k1} = \frac{X_{ko}}{F_{1o}} \tag{10.37}$$

and by keeping account of equation (10.19):

$$m_j = \frac{X_1^{(j)}}{2i\xi_j \omega_j} \sum_{k=1}^{n} \frac{1}{h_{k1}(\omega_j) X_k^{(j)}} \tag{10.38}$$

It is possible to reach the same result in another way, by assuming that the coupling between the different vibration modes is negligible.

Using this hypothesis, the error between the given transfer function $h_{kl}*(\omega_j)$ and the approximating function $h_{kl}(\omega_j)$, evaluated in the neighbourhood of the j-th natural frequency, is given by:

$$E(\omega_j) = h_{kl}^*(\omega_j) - h_{kl}(\omega_j) = h_{kl}^*(\omega_j) - \frac{X_k^{(j)} X_l^{(j)}}{2 i m_j \sigma_j \omega_j} \tag{10.39}$$

The sum of the quadratic errors evaluated for the transfer function determined in N measurement points is given by:

$$E_t(\omega_j) = \sum_{k=1}^{n} E^2\left(\omega_j, \frac{1}{m_j}\right) \tag{10.40}$$

It is possible to minimize this error by imposing the following condition:

$$\frac{\partial E_t}{\partial (1/m_j)} = 0 \tag{10.41}$$

from which we obtain:

$$m_j = \frac{X_l^{(j)}}{2 i \sigma_j \omega_j} \sum_{k=1}^{n} \frac{1}{h_{kl}(\omega_j) X_k^{(j)}} \tag{10.42}$$

Method n.2

The second method adopted tries, for each k-th d.o.f., to simultaneously calculate the generalized masses m_j of all the vibration modes: the user is thus able to accept, mediate or discard the generalized masses calculated for the generic j-th vibration mode in correspondence to the different degrees of freedom. To identify the generalized masses the method minimizes, for all the ω_j considered in correspondence to the generic k-th d.o.f., the function $h_{kl}(\omega_j)$ (j=1,2,.....p) defined by equation (10.14), with the same known functions (determined experimentally or analytically) $h_{kl}^*(\omega_j)$.

Analytically this can be expressed as:

$$\begin{Bmatrix} h_{kl}(\omega_1) \\ \cdots \\ \cdots \\ \cdots \\ h_{kl}(\omega_n) \end{Bmatrix} = \begin{bmatrix} \dfrac{X_k^{(1)} X_l^{(1)}}{i 2 \sigma_1 \omega_1} & \cdots & \dfrac{X_k^{(i)} X_l^{(i)}}{\omega_i^2 - \omega_1^2 + 2 i \sigma_i \omega_1} & \cdots & \dfrac{X_k^{(n)} X_l^{(n)}}{\omega_n^2 - \omega_1^2 + 2 i \sigma_n \omega_1} \\ \cdots & & \cdots & & \cdots \\ \dfrac{X_k^{(1)} X_l^{(1)}}{\omega_1^2 - \omega_n^2 + 2 i \sigma_1 \omega_n} & & \dfrac{X_k^{(i)} X_l^{(i)}}{\omega_i^2 - \omega_n^2 + 2 i \sigma_i \omega_n} & & \dfrac{X_k^{(n)} X_l^{(n)}}{i 2 \sigma_n \omega_n} \end{bmatrix} \begin{Bmatrix} \dfrac{1}{m_1} \\ \cdot\cdot \\ \dfrac{1}{m_n} \end{Bmatrix}$$

10.43)

or

$$\underline{h} = [T] \frac{1}{\underline{m}} \tag{10.44}$$

In the case in which the number of the sampling frequencies is higher than the generalized masses to be determined, it is possible to define the following error function:

$$\underline{E} = \underline{h}^{*}-\underline{h} = \underline{h}^{*}-[T] \frac{1}{\underline{m}} \tag{10.45}$$

To minimize the quadratic error:

$$\underline{E}^{T}\underline{E} = \left[H-[T] \frac{1}{\underline{m}}\right]^{T} \left[\underline{H}-[T] \frac{1}{\underline{m}}\right] \tag{10.46}$$

it is necessary to impose the condition:

$$\frac{\partial(\underline{E}^{T}\underline{E})}{\partial(1/\underline{m})} = 0 \tag{10.47}$$

thus reaching:

$$\frac{1}{\underline{m}} = ([T]^{T}[T])^{-1} ([T]^{T}[H]) \tag{10.48}$$

<u>Synthesis</u>

The frequency response in correspondence to a point K of a structure forced in a point l from a unitary harmonic excitation force depends on the excitation frequency Ω_E and on the single modal parameters of the structure (eq. 10.14)). Once the modal parameters are known, it is therefore possible to calculate the transfer function in correspondence to the frequencies included in the frequency range analysed. This operation is commonly called "synthesis". By indicating with Y_{Kl} and Z_{Kl} the contributions relative to the vibration modes having frequencies which are respectively lower and higher than the frequencies Ω_1 and Ω_2 delimiting the frequency range analysed, we have:

$$h_{kl}(\Omega_E) = -\frac{Y_{kl}}{\Omega_E^2} + \sum_{j=1}^{m} \frac{X_k^{(j)} X_l^{(j)}}{m_j(\omega_j^2 - \Omega_E^2 + 2i\sigma_j\Omega_E)} + z_{kl} \qquad (10.49)$$

The more accurate the evaluation of the single modal parameters, the more the function $h_{kl}(\omega_j)$ approximates the starting function $h_{kl}^*(\omega_j)$. The comparison between the starting transfer function and the one obtained by means of the synthesis operation therefore enables us to evaluate the reliability of the modal parameters calculated. The modal parameters ω_j, σ_j and m_j of the carrying structure (natural frequencies, damping factors and generalized masses) are memorized, together with the vibration modes $\underline{\Phi}^{(j)}$, on a file to subsequently be used to calculate the matrix of the mechanical impedances. As far as the vibration modes are concerned, only the mode components corresponding to the connection points (i.e. between the carrying structure and the shaft line) are memorized. In fact, these components are the only ones that are actively involved in the definition of the matrix of the mechanical impedances of the foundation. Nonetheless, as has been mentioned, to better define the vibration modes of the carrying structure it is convenient to also calculate the transfer function (Block 9) in correspondence to foundation points different from the connection nodes. It is important to underline that, in order to define the matrix of the mechanical impedances [I (Ω_E)] it is necessary, for each single modal analysis (i.e. for each direction along which the excitation force acts) to identify a number of natural frequencies at least equal to the number of supports. Moreover, the natural vibration modes identified must have significant components in correspondence to the connection nodes between the carrying structure and the shaft line. Therefore, all the vibration modes in which a considerable number of connection nodes coincide with a vibration "node" (i.e. a non vibrating point) have to be discarded. This fact might be verified e.g. for torsional vibration modes of a foundation slab. (fig. 10.1). The modal parameters are determined by analysing the transfer function $h_{kl}(\Omega_E)$ (with k=1,2,...n) evaluated in n points of the carrying structure whith a unitary harmonic excitation force applied at point l. As already been said, the transfer functions can be expressed as a function of the modal parameters only (see equation (10.14):

the transfer functions evaluated at node K, with an excitation applied at node r, can be determined by using the modal parameters of the foundation, without having to once again carry out the calculation of the forced response of the structural model (see Block 9). The set of functions $h_{Kr}(\Omega_E)$, evaluated in the vertical and horizontal planes, enables us to define a square matrix $[H(\Omega_E)]$ of order 2*n, known as flexibility matrix. As is known, the inverse of matrix [H] represents the matrix of the mechanical impedances $[I\ (\Omega_E)]$.The terms of the matrix $[I(\Omega_E)]$ will be added to the terms of the elastodynamic matrix $[E(\Omega_E)]$ (rotor + oil film) corresponding to the degrees of freedom associated with the shaft line connection points (Block 11).

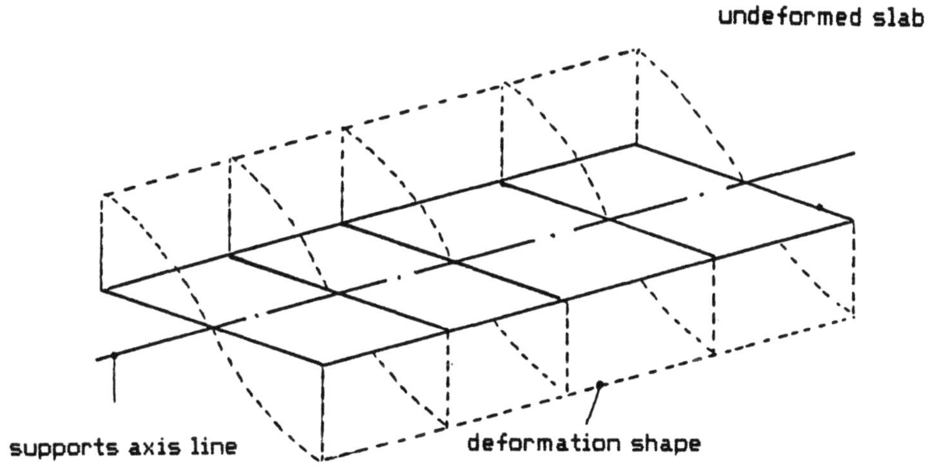

Fig. 10.1 - Torsional vibration mode of a foundation slab

7.1.14 Block 11: Creation of the Elastodynamic Matrix

To obtain the elastodynamic matrix of theoverall system rotor + oil film, it is necessary to suitably assemble the mass [m] and stiffness [k] matrices of the single beam element together with the mechanical impedances matrix of the foundation (already defined in Block 10) and the stiffness and damping matrices of the fluid film of bearings and seals (Blocks 7 and 8).

The d.o.f. number associated with the shaft nodes, as already stated for that which concerns the static analysis (Block 6), is generally large and a dynamic condensation becomes necessary. In the dynamic analysis carried out in the frequency domain used to evaluate the response of the shaft line to a harmonic excitation, this condensation is rigorous. Hereinafter we will firstly describe the method adopted to reduce the shaft d.o.f. by means of dynamic condensation and subsequently will show the assembly of the condensed matrices of the shaft with the matrices of the mechanical impedances of the foundation as well as the stiffness and damping ones of the oil film and/or seals.

Dynamic condensation

We will now analyse, for the sake of convenience, the shaft line shown in fig. 11.1. Let us consider a set of beam elements corresponding to a portion of the rotor which will be considered as a superelement. By indicating with $\underline{X}_e$ the relative d.o.f. vector, the equations of motion of this system can be written as:

$$[M_e]\,\ddot{\underline{X}}_e + [R_e]\,\dot{\underline{X}}_e + [K_e]\,\underline{X}_e = \underline{F}_e \tag{11.1}$$

where $[M_e]$, $[R_e]$ and $[K_e]$ are respectively the mass, damping and stiffness matrix of the generic superelement analysed. For the example proposed, these matrices are assembled as shown in fig. 11.1. In eq. (11.1), $\underline{F}_e$ represents the vector of the external excitation forces applied to the nodes. In the P.A.L.L.A. package the steady state response of the rotor to harmonic excitation forces is analysed, so that we have:

$$\underline{F}_e = \underline{F}_{eo}\, e^{i\Omega_E t} \tag{11.2}$$

In this case, the solution of (11.1) given by (11.2) is :

$$\underline{X}_e = \underline{X}_{eo}\, e^{i\Omega_E t} \tag{11.3}$$

By substituting equations (11.2) in (11.1) we obtain:

$$(-\Omega_E^2[M_e]+i\Omega_E[R_e]+[K_e])\underline{X}_{eo}=\underline{F}_{eo} \tag{11.4}$$

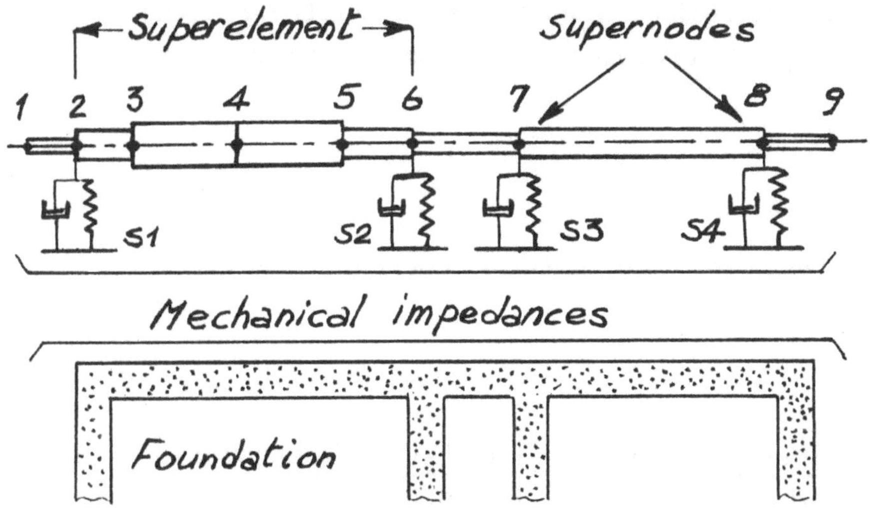

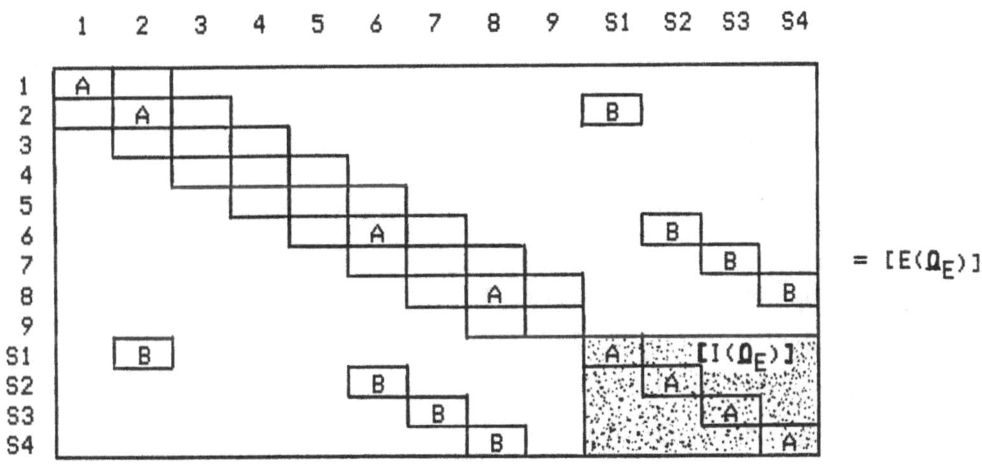

$$[A] = \begin{bmatrix} K_{xx}(\Omega_R) + i\Omega_E R_{xx}(\Omega_R) & 0 & K_{xy}(\Omega_R) + i\Omega_E R_{xy}(\Omega_R) & 0 \\ 0 & 0 & 0 & 0 \\ K_{yx}(\Omega_R) + i\Omega_E R_{yx}(\Omega_R) & 0 & K_{yy}(\Omega_R) + i\Omega_E R_{yy}(\Omega_R) & 0 \\ 0 & 0 & 0 & 0 \end{bmatrix}$$

$$[B] = -[A]$$

Fig. 11.1 - Overall assembly of shaft line - foundation system

Having called $\underline{X}_{ee}$ the vector of the "external" displacements, i.e. the vector containing the d.o.f. of extremity nodes of the portion analysed (nodes 1 and 10 of fig. 6.2) and $\underline{X}_{ei}$ the vector relative to the other nodes ("internal" nodes), it is possible to renumber the equations as:

$$\underline{X}_e = \left\{ \begin{array}{c} \underline{X}_{ee} \\ \hline \underline{X}_{ei} \end{array} \right\} \tag{11.5}$$

With this new order of the equations it is possible to rewrite eq. (11.4) in partitioned form as:

$$\left(-\Omega_E^2 \left[\begin{array}{c|c} M_{ee} & M_{ei} \\ \hline M_{ie} & M_{ii} \end{array} \right] + i\Omega_E \left[\begin{array}{c|c} R_{ee} & R_{ei} \\ \hline R_{ie} & R_{ii} \end{array} \right] + \left[\begin{array}{c|c} K_{ee} & K_{ei} \\ \hline K_{ie} & K_{ii} \end{array} \right] \right) \left\{ \begin{array}{c} \underline{X}_{eeo} \\ \hline \underline{X}_{eio} \end{array} \right\} = \left\{ \begin{array}{c} \underline{F}_{eeo} \\ \hline \underline{F}_{eio} \end{array} \right\} \tag{11.6}$$

or rather, by splitting eq. (11.6) into two subsystems:

$$\begin{array}{l} (-\Omega_E^2[M_{ee}]+i\Omega_E[R_{ee}]+[K_{ee}])\underline{X}_{eeo}+(-\Omega_E^2[M_{ei}]+i\Omega_E[R_{ei}]+[K_{ei}])\underline{X}_{eio}=\underline{F}_{eeo} \\ (-\Omega_E^2[M_{ie}]+i\Omega_E[R_{ie}]+[K_{ie}])\underline{X}_{eio}+(-\Omega_E^2[M_{ii}]+i\Omega_E[R_{ii}]+[K_{ii}])\underline{X}_{eio}=\underline{F}_{eio} \end{array} \tag{11.7}$$

For a better understanding, by defining the complex matrices:

$$\begin{array}{l} [A_{ee}(\Omega_E)]=-\Omega_E^2[M_{ee}]+i\Omega_E[R_{ee}]+[K_{ee}] \\ [A_{ei}(\Omega_E)]=-\Omega_E^2[M_{ei}]+i\Omega_E[R_{ei}]+[K_{ei}] \\ [A_{ie}(\Omega_E)]=-\Omega_E^2[M_{ie}]+i\Omega_E[R_{ie}]+[K_{ie}] \\ [A_{ii}(\Omega_E)]=-\Omega_E^2[M_{ii}]+i\Omega_E[R_{ii}]+[K_{ii}] \end{array} \tag{11.8}$$

it is possible to rewrite equation (11.7) as:

$$\begin{array}{ll} [A_{ee}(\Omega_E)]\underline{X}_{eeo}+[A_{ei}(\Omega_E)]\underline{X}_{eio}=\underline{F}_{eeo} & \text{a)} \\ [A_{ie}(\Omega_E)]\underline{X}_{eeo}+[A_{ii}(\Omega_E)]\underline{X}_{eio}=\underline{F}_{eio} & \text{b)} \end{array} \tag{11.9}$$

From (11.9b) it is possible to express $\underline{X}_{eio}$ as a function of $\underline{X}_{eeo}$:

$$\underline{X}_{eio}=[A_{ii}(\Omega_E)]^{-1}\underline{F}_{eio}-[A_{ii}(\Omega_E)]^{-1}[A_{ie}(\Omega_E)]\underline{X}_{eeo} \tag{11.10}$$

By substituting (11.10) in (11.9a) we obtain:

$$([A_{ee}(\Omega_E)]-[A_{ei}(\Omega_E)][A_{ii}(\Omega_E)]^{-1}[A_{ie}(\Omega_E)])\underline{X}_{eeo}= \\ = \underline{F}_{eeo}-[A_{ei}(\Omega_E)][A_{ii}(\Omega_E)]^{-1}\underline{F}_{eio} \tag{11.11}$$

By defining

$$[E(\Omega_E)]_e=[A_{ee}(\Omega_E)]-[A_{ei}(\Omega_E)][A_{ii}(\Omega_E)]^{-1}[A_{ie}(\Omega_E)] \tag{11.12}$$

as the condensed elastodynamic matrix of the generic superelement and

$$\underline{F}_{ec}(\Omega_E)=\underline{F}_{eeo}-[A_{ei}(\Omega_E)][A_{ii}(\Omega_E)]^{-1}\underline{F}_{eio} \tag{11.13}$$

as the vector of the condensed forces, it is possible to rewrite (11.11) as:

$$[E(\Omega_E)]_e\underline{X}_{eeo}=\underline{F}_{ec}(\Omega_E) \tag{11.14}$$

In other words, it is possible , without introducing any approximation, to solve the equations of motion of the complete system using as degrees of freedom only those ones relative to the supernodes (external nodes) of the single superelements, thus reducing the computer memory occupation. Should the user wish to know the displacements of the internal nodes, known from (11.14), it is possible from (11.10) to obtain the value of $\underline{X}_{eio}$.

Assembly

Having dynamically condensed the shaft elements to obtain the elastodynamic matrix of the complete system shaft + oil film (or seals) + foundation, we then proceed with the overall assembly as illustrated in the case given in fig. 11.1 as an example. As can be seen from the figure, the vector of the total d.o.f. number contains, in order, the d.o.f. relative to the shaft and, subsequently, the d.o.f. relative to the shaft-foundation connection nodes. The squared matrix $[I(\Omega_E)]$ relative to the degrees of freedom of the connection nodes (shaded area of fig. 11.1) is the matrix of the mechanical impedances, to which matrices [A] and [B], relative to the oil film or seals, are added.

7.1.15 Block 12: Synchronous Excitation

The synchronous excitation forces that can be assigned in this Block are basically the ones caused by unbalance, both of the concentrated and distributed type. In the first case we are dealing with concentrated mass, situated in a nodal point of the F.E. schematization of the shaft line at a distance R from the rotation axis and with a certain phase in the relative reference system (fig. 12.1). The distributed unbalances that can be assigned within the generic element simulate the situation in which the center of gravity locus of the generic section do not coincide with the rotation axis. The eccentricity distribution considered by the program is a function e(z) of the type:

$$e(z)=az^2+bz+c \tag{12.1}$$

lying in an assigned orientation plane (see fig. 12.2). For the definition of 12.1 the eccentricities $e_1(z_1)$, $e_2(z_2)$ and $e_3(z_3)$ must be assigned in correspondence to three distinct points z_1, z_2 and z_3 ofthe element axis. The origin of z axis is assumed in correspondence to the extreme left of the element itself.

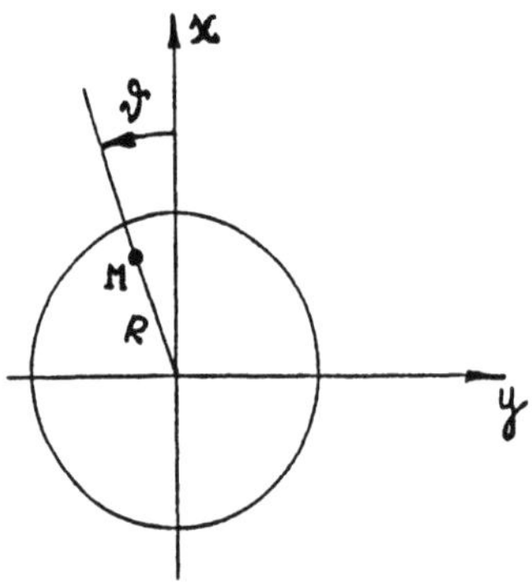

Fig. 12.1 - Concentrated unbalance

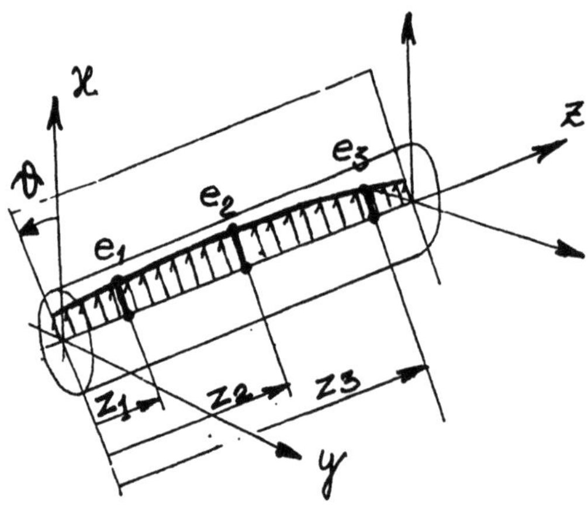

Fig. 12.2 - Distributed unbalance

7.1.16 Block 13: Non-Synchronous Excitation

In this paragraph, the mathematical model used to simulate the effects of an excitation frequency twice the rotation frequency of the shaft line is described. Double frequency excitation is considered here to be due to transverse cracks and to journal ovalization. Since the mathematical model simulating the crack effect is not at present implemented in the package, we will limit ourselves to describing the model relative to the ovalization effect. The normal section of a journal of a generic bearing can exhibit some deviations compared with the circular one, deviations indicated hereinafter as "surface irregularities". These irregularities are detected when the rotor is stationary. When the rotor is rotating the air gap in correspodence to a proximitor fixed on a bearing and facing the surface of the corresponding journal is determined not only by the journal-bearing relative vibrations but also by the surface irregularities and by the "run-out". The latter only appears when the journal is rotating and can be considered as the rotation of a circular section around an axis different from the geometric one. As far as the steps to be taken to determine the various quantities are concerned refer to [21]. With reference to fig. 13.1, we assume that there are two pick-ups placed in two "radial" directions parallel to the axis of an absolute reference system X,Y. While the journal is rotating at a certain angular speed Ω_R sufficiently low to consider the displacements x and y of same as negligible, the signals of the two pick-ups are only representative of a surface irregularity and eccentricity. These signals are periodic functions, with the fundamental period given by $2\pi/\Omega_R$, with a phase angle $\pi/2$ between them. By developing these function in a Fourier series up to the second order term we have:

$$
\begin{aligned}
X_i &\cong I_1\cos(\Omega_R t+\beta_1)+I_2\cos(2\,\Omega_R t+\beta 2)\\
Y_i &\cong I_1\sin(\Omega_R t+\beta_1)+I_2\sin(2\,\Omega_R t+\beta 2)
\end{aligned}
\tag{13.1}
$$

The 2 per rev. component is indicated as "ovalization". The deviation of the journal surface from circularity can be attributed to a harmonic motion of the journal center itself. This motion has the same frequency as that of the rotation or twice the same, depending on whether the first or second terms of (13.1) are considered. For a generic rotation speed Ω_R and assuming that the irregularity given by Xi and Yi of (13.1) is limited to the ovalization alone, the journal-bearing relative displacement in the X,Y directions is therfore given by:

$$
\begin{aligned}
X_r &= X_c - X_s + I_2 e^{2i\Omega_R t}\\
Y_r &= Y_c - Y_s + I_2 e^{i(2\Omega_R t-\pi/2)}
\end{aligned}
\tag{13.2}
$$

in which X_c and Y_c are the journal displacements and X_s and Y_s those of the bearing. For small relative journal bearing displacements, in the neighbourhood of the static equilibrium position, the components of the force due to the lubricating film are given by:

$$\begin{bmatrix} K_{xx} & K_{xy} \\ K_{yx} & K_{yy} \end{bmatrix} \begin{Bmatrix} X_r \\ Y_r \end{Bmatrix} + \begin{bmatrix} R_{xx} & R_{xy} \\ R_{yx} & R_{yy} \end{bmatrix} \begin{Bmatrix} \dot{X}_r \\ \dot{Y}_r \end{Bmatrix} = \begin{Bmatrix} F_x \\ F_y \end{Bmatrix} \qquad (13.3)$$

where $\dot{X}_r$ and $\dot{Y}_r$ are the relative speeds. If second order effects are neglected in a congruent manner with the linearization carried out by (13.3), one can, as has already been said, attribute the ovalization to a 2 per rev frequency motion of a journal with a circular section (fig. 13.2). Therefore, by substituting (13.2) and its first derivatives with respect to time in (13.3), we obtain for F_x and F_y expressions that are both a function of the independent variables X_c, Y_c, X_s and Y_s and the ovalization, which should be considered assigned: the first are already included in the stiffness and damping matrices of the system while the second are put in the right hand side member. The following expressions are found with the above mentioned substitution:

$$F_x = \sqrt{(K_{xx}+2\Omega_R R_{xy})^2+(2\Omega_R R_{xx}-K_{xy})^2}\; I_2 e^{i(2\Omega_R t+\psi_x)}$$
$$\psi_x = \mathrm{arctg}\,[(-2\Omega_R R_{xx}+K_{xy})/(K_{xx}+2\Omega_R R_{xy})] \qquad (13.4)$$

$$F_y = \sqrt{(2\Omega_R R_{yx}-K_{yy})^2+(K_{yx}+2\Omega_R R_{yy})^2}\; I_2 e^{i(2\Omega_R t+\psi_y)}$$
$$\psi_y = \mathrm{arctg}\,[(-2\Omega_R R_{yx}+K_{yy})/(K_{yx}+2\Omega_R R_{yy})] \qquad (13.5)$$

in which F_x and F_y represent the excitation forces to be assigned as a known term.

7.1.17 Block 14: Stress Calculation

In this paragraph we will describe the method adopted to calculate the maximum nominal stresses that are set up in a rotor on account of:

- static load (own weight) and the assigned alignment of the axis line
- dynamic load that can act on the same line due different vibration phenomena.

For this reason the P.A.L.L.A. package programs exploit the results obtained in terms of supernode displacements in Blocks 6 and 15, respectively to calculate the rotating bending stresses due both to the own weight and the alignment as well as due to the rotor dynamic loads.

Determination of rotating bending stresses

In Block 6 the P.A.L.L.A. package programs enable us, as previously seen, to calculate the static deformation assumed by the rotor with or without a lubricating film, determining in particular:

- the alignment to be assigned to the machine to obtain the desired reactions with supports;
- the loads acting on the supports for a given imposed alignment.

In this analysis the stiffness matrices of several elements of the shaft were statically condensed, thus decomposing the axis line in several superelements with supernodes for extremity nodes. In this way the final equations of the entire system are functions of the supernodes d.o.f. only. Finally, to calculate the stresses, the displacements of the shaft nodes (internal with respect to the condensation made) are computed ("retracking"), as herein shown. For the sake of convenience we hereby give the equations of static equilibrium (Block 6, eq. 6.4) relative to the single superelement partitioned into external (supernodes) and internal nodes, with respect to the condensed system:

$$[K_{ee}]\underline{X}_{ee}+[K_{ei}]\underline{X}_{ei}=\underline{P}_{ee} \qquad \text{a)}$$
$$[K_{ie}]\underline{X}_{ee}+[K_{ii}]\underline{X}_{ei}=\underline{P}_{ei} \qquad \text{b)} \qquad (14.1)$$

The relation between $\underline{X}_{ei}$ and $\underline{X}_{ee}$, as has already been seen in Block 6 (eq. 6.5) is given by:

$$\underline{X}_{ei}=[K_{ii}]^{-1}\underline{P}_{ei}-[K_{ii}]^{-1}[K_{ie}]\underline{X}_{ee} \qquad (14.2)$$

By substituting (14.2) in (14.1a) it is possible (Block 6) to obtain the static condensation of the superelement stiffness matrix. Since supernode displacements $\underline{X}_{ee}$ are known from the solution of the static problem, from (14.2) it is also possible to determine the displacements of the internal nodes $\underline{X}_{ei}$. By repeating this operation for all the superelements with which the line is schematized it is now possible to determine the static displacements of all the nodes of the rotor. For the generic finite "beam" element it is possible to evaluate the generalized forces at the extremity nodes by means of relation:

$$\underline{F}_j=[K_j]\underline{X}_j-\underline{P}_j \tag{14.3}$$

where $\underline{X}_j$ are the generalized displacements of the extremity nodes of the generic j-th element, $[K_j]$ is the stiffness matrix of same, $\underline{P}_j$ the "perfect clamp" reaction vector or of the generalized forces applied to the nodes themselves calculated to evaluate the static deformation of the overall system. The vector $\underline{F}_j$ of equation (14.3) thus represents the value assumed by shear and by bending moment M_f in the two deflection planes. Once the value assumed by the bending moment M_f is known in every node, bending stresses for the generic finite element are calculated by means of the known relation:

$$\sigma_{nfj} = \frac{M_{fj}}{J_j} \frac{D_j}{2} \tag{14.4}$$

where J_j represents the moment of inertia of the shaft section and D_j the external diameter.

Determination of the stresses due to dynamic effects

In Blocks 11 and 15 we described the methods adopted by the P.A.L.L.A. package programs to evaluate the frequency response of the rotor for different dynamic load situations. In this paragraph, we will describe how, starting from the knowledge of the displacements of the dynamic condensation supernodes, it is possible to evaluate the maximum nominal stresses due to dynamic effects. As in the static case, to calculate these quantities it is necessary first of all to determine, from the displacements of the supernodes, the displacements of all the nodes of the shaft line. As described in Block 11, the dynamic condensation of the elasto-dynamic matrices of several elements has been performed in order to have the displacements of the only supernodes as the only unknowns of the problem. We will now go on to describe the relationship that exists between the displacements of the internal nodes and those of the supernodes already expressed in Block 11 (eq. 11.10):

$$\underline{X}_{eio}=[A_{ii}(\Omega_E)]^{-1}\underline{F}_{eio}-[A_{ii}(\Omega_E)]^{-1}[A_{ie}(\Omega_E)]\underline{X}_{eeo} \qquad (14.5)$$

Known, from the solution of the forced problem, the response $\underline{X}_{eeo}$ of the rotor at the supernodes for different frequencies (Block 15), from eq.(14.5) it is possible to obtain the response in all the points of the shaft line. To obtain the maximum bending stresses we used the shape function of the finite "beam" element which permits, once the displacements and the rotations of the end nodes are known, to obtain the bending moment in the vertical M_x and horizontal M_y plane of the generic section (see fig. 14.1), by means of :

$$M_y = EJ\left\{\frac{6}{l^3}\,(-2X_d+2X_s+l\vartheta_{ys}+l\vartheta_{yd})\xi+\frac{2}{l}\left(\frac{3}{l}X_d-l\vartheta_{yd}-2\vartheta_{ys}-\frac{3}{l}X_s\right)\right\}$$

$$(14.6)$$

$$M_x = EJ\left\{\frac{6}{l^3}\,(-2Y_d+2Y_s+l\vartheta_{xs}+l\vartheta_{xd})\xi+\frac{2}{l}\left(\frac{3}{l}Y_d-l\vartheta_{xd}-2\vartheta_{xs}-\frac{3}{l}Y_s\right)\right\}$$

The program recognizes the bending neutral axis and calculates the maximum value assumed by the bending moment in each section of the rotor successively calculating the nominal stress using eq. (14.4). These stresses will give rise to fatigue phenomena should the rotation speed of the rotor differ from the pulsation of the dynamic forces acting on same or should the orbit of the shaft not be circular.

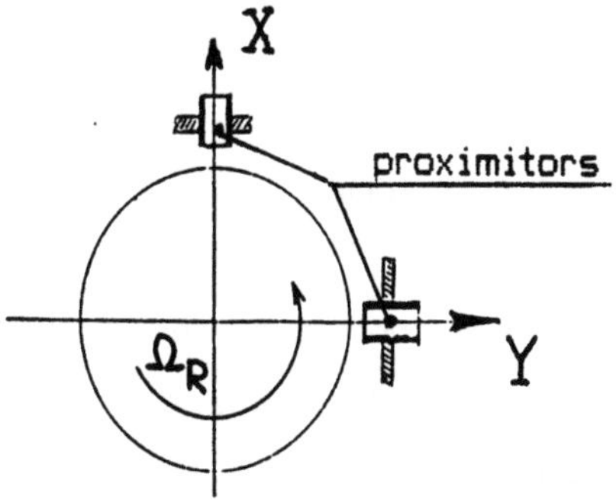

Fig. 13.1 - Detecting run-out and ovalization

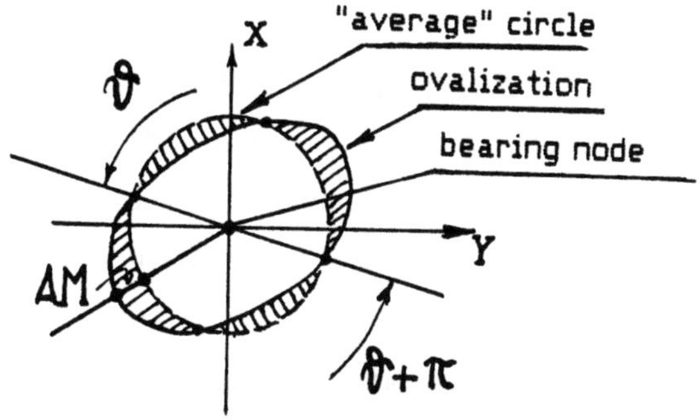

Fig. 13.2 - Ovalization

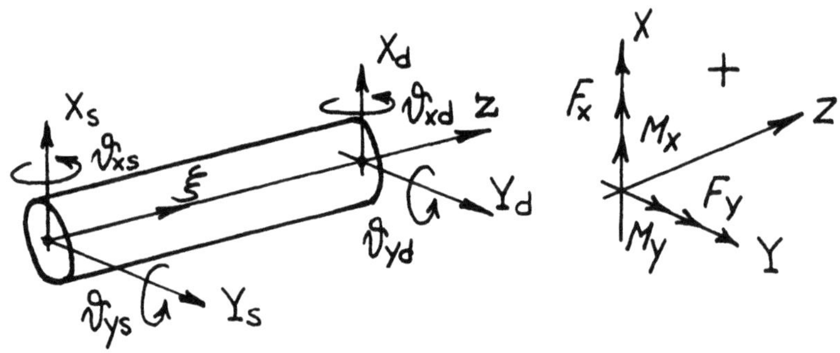

Fig. 14.1 - Reference system and sign conventions

7.1.18 Block 15: Frequency Response

Once the elasto-dynamic matrix of the entire system , rotor+ oil film + foundation $[E(\Omega_E)]$, has been calculated, as described in Block 11, and once the dynamic forces $F = F_0 \exp(i\Omega_E t)$ applied to the rotor (see Blocks 12 and 13) have been evaluated, it is possible to calculate (see eq.(A. 18)) the response of the complete system by solving equation:

$$[E(\Omega_E)]\underline{X}_0=\underline{F}_0 \qquad (15.1)$$

The displacements in terms of module and phase, both of the d.o.f. of the rotors of the line as well as of the connection nodes of the foundation, can be printed or visulized by means of graphic output that show the displacements along the generic d.o.f. as a function of the frequency or displacements of all the nodes of the line (spatial deformation), for a certain pulsation of the excitation force.

7.1.19 Block 16: Stability Analysis

The study of the dynamic behaviour of large turbogenerator groups includes, as a fundamental phase, the verification of the stability of the dynamic equilibrium of the different rotors that constitute the machine, commonly known as the shaft line. This paragraph describes the method adopted to evaluate, by means of the frequency response of the rotor-oil film-foundation system to a harmonic excitation force, the overall stability of the entire line and the of single bearings. The method used in the P.A.L.L.A. package, called "forced method" is described in paper [23]. This method is essentially based on the determination of the energy introduced by an external excitation force with an assigned pulsation into the overall system. As will be more fully described in the following paragraph, the relationship between the energy introduced by the excitation force and the total energy of the system enables us to calculate the ratio r/r_c between the damping of the entire system and the corresponding critical value. The sign of this relationship permits the evaluation of the instability threshold.

The "forced" method: one d.o.f. systems

To illustrate the method used it is convenient to start off by considering a one d.o.f. system. Let us consider the system of fig. 16.1 in which K_s and r_s represent the stiffness and damping of the system while r_i is an active viscous element capable of introducing energy into the system i.e. capable of reproducing the instability conditions into the one d.o.f. system ($r_i<0$). The equation of motion of the system is:

$$m\ddot{x}+(r_s+r_i)\dot{x}+K_s x = 0 \tag{16.1}$$

which, by putting $r = r_s+r_i$, becomes:

$$m\ddot{x}+r\dot{x}+K_s x = 0 \tag{16.2}$$

The solution of equation (16.2) is, as is known:

$$x = Xe^{\lambda t} \tag{16.3}$$

where

$$\lambda = \alpha \mp i\omega_d$$
$$\alpha = -r/2m \ ; \quad \omega_d^2 = \omega_n^2 - \alpha^2; \quad \omega_n^2 = K_s/m \tag{16.4}$$

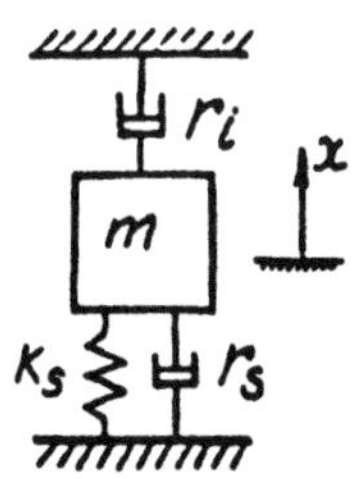

Fig. 16.1 - One d.o.f. system

If $r<0$, coefficient of eq. (16.4) becomes positive and solution $x(t)$ grows exponentially in time: the system is unstable and the instability threshold is defined by the sign of r. This sign can be determined by also analysing the steady state response of the system to a harmonic excitation force. In fact, if the system of fig. 16.1 is forced by an excitation of the type $F = F_0 \exp(i\Omega_E t)$ the response will be of the type $x = X_0 \exp(i\Omega_E t)$: the instability conditions can be verified by the relationship between the energy E_{ic} introduced by the external force in a cycle and the total energy E_{Kmax} of the system itself. For $\Omega_E = \omega_n$ (excitation force frequency coincident with the natural frequency of the system) the value of r/r_c is given by:

$$\frac{r}{r_c} = \frac{1}{4\pi}\frac{E_{ic}}{E_{Kmax}} \tag{16.5}$$

Since the relationship between the energy E_{dc} dissipated by the damping r in a cycle and the maximum kinetic energy is:

$$\frac{E_{dc}}{E_{Kmax}} = \frac{\pi r \Omega_E |X_0|^2}{1/2 m\Omega_E^2 |X_0|^2} = \frac{2\pi r}{\Omega_E^2} \tag{16.6}$$

bearing in mind the expression of the critical damping:

$$r_c = 2m\omega_n \tag{16.7}$$

and since $E_{dc}=E_{ic}$, for $\Omega_E = \omega_n$ we have equation (16.5). The energy E_{ic} introduced by the excitation force is:

$$E_{ic} = -F_0 |X_0| \pi \sin\varphi \tag{16.8}$$

where φ represents the phase lag between the excitation force vector F_0 and the displacement vector X_0:

$$\varphi = \mathrm{arctg}\left(-\frac{2ar/r_c}{1-a^2}\right) \tag{16.9}$$

where $a = \Omega_E/\omega_n$. If the damping of the system is positive, eq. (16.8) assumes the positive value since $\sin\varphi < 0$. For $r<0$, i.e. for instability conditions, we have positive values of $\sin\varphi$. To define r/r_c with the forced method it is therefore possible to use equation (16.5) which in explicit form is:

$$\frac{r}{r_c} = \frac{1}{4\pi}\,\frac{F_0\,\pi\,|X_0|\sin\varphi}{1/2 m\Omega_E^2|X_0|^2} \tag{16.10}$$

being, as we remember, $\Omega_E = \omega_n$. From a pratical point of view the system is excited by an excitation force with a variable pulsation until we obtain the maximum of the amplitudes: in these conditions (resonance conditions) the ratio r/r_c is determined from equation (16.10) and from this the characteristic parameter that determines system stability α/ω_d which, as is known, is:

$$\alpha/\omega_d = -r/r_c \tag{16.11}$$

If r/rc>o the system is stable, otherwise it is unstable.

<u>Systems with several d.o.f.</u>

In the sytems with several d.o.f. instability conditions can arise due both to the effect of the active velocity coefficients, i.e. non dissipative, (due for e.g. to damping matrices [R] defined non positive) and to the positional forces field, in the case of fluid film instability or of hysteresis. The equations of free motion of a system with n d.o.f. in matrix form is:

$$[M]\ddot{\underline{X}}+[R]\dot{\underline{X}}+[K]\underline{X}=0 \tag{16.12}$$

where [M], [R] and [K] are the square matrices of order n and $\underline{X}$ the vector of n independent coordinates. The solution of eq. (16.12) is given by:

$$\underline{x}=\underline{X}\ e^{\lambda t} \tag{16.13}$$

with n values of λ_i equal to:

$$\lambda_i = \alpha_i + i\omega_i \qquad (i=1,2,3..,n) \tag{16.14}$$

if [R] is non symmetric or is either or non symmetric or defined non positive, at least one value of α_i can become positive and the system unstable. In systems with several d.o.f. the instability conditions are indicated by all the solutions λ_i with $\alpha_i>0$: the vibration mode defined by the complex eigenvector $\underline{X}^{(i)}$ and by its conjugate is an unstable mode. If the system is forced in a non nodal point of the K-th vibration mode, the amplitudes become maximum for $\Omega_E = \omega_K$, the lower the value of α_K the higher the resonance peak. Moreover, in resonance condition, the contribution of the other modes can be negligible and this enables us to consider, in these conditions, the system as if it had only one d.o.f. and therefore apply the concepts already seen. In particular:

$$\left(\frac{r}{r_c}\right)_K = \frac{1}{4\pi}\left(\frac{E_{ic}}{E_{Kmax}}\right)_K \tag{16.15}$$

where E_{ic} is either the energy introduced or dissipated by the excitation force in a cycle and E_{max} is the maximum value of the total energy relative to the K-th vibration mode. By varying the frequency Ω_E of the excitation forces it is possible to define the natural pulsations of the system ω_K and, from equation (16.15), obtain the non-dimensional damping relative to the generic K-th vibration mode. In this case the energy introduced by the excitation force is:

$$E_{ic} = -\pi F_0 |X_{of}| \sin\varphi \tag{16.16}$$

where $X_{o\varphi}$ and φ are respectively the vibration amplitude of the application point of the excitation force and the phase angle between the vibration and the excitation force evaluated at the same point. The so called "forced method" therefore enables us to evaluate the stability conditions of the system by evaluating the index $(r/r_c)_K$ relative to the generic natural frequency ω_K and associated to a vibration mode $\underline{X}^{(K)}$, using the response of the system to a harmonic excitation force in a frequency range in the neighbourhood of the frequency whose corresponding mode can become unstable. In the case of the rotors the instability manifests itself in correspondence to the lowest flexural natural frequency of that rotor of the shaft line placed on the two bearings which can cause instability to arise. Normally, these frequencies are known. By applying an excitation force (for e.g. horizontally) to that rotor which could become unstable, by suitably varying the forcing frequency it is possible to evaluate the natural frequency of the system. In these conditions we calculate the energy introduced by the excitation force and the total energy by assuming that the deformation of the K-th vibration mode is coincident with the deformation obtained by forcing the system in resonance conditions. Having evaluated the two types of energy it is possible to obtain the value of $(r/r_c)_K$ and therefore evaluate the instability index. In order to apply this method, it is necessary to repeatedly solve the system:

$$[M]\ddot{\underline{X}}+[R]\dot{\underline{X}}+[K]\underline{X}=\underline{F}_0 e^{i\Omega_E t} \tag{16.17}$$

where, as has been said, [M], [R] and [K], are the mass, damping and stiffness matrices of the rotor + oil film + foundation system already defined in Block 5 and $\underline{F} = \underline{F}_0 \exp(i\Omega_E t)$ is the harmonic excitation force. By substituting solution:

$$\underline{x}=\underline{X}_0 e^{i\Omega_E t} \tag{16.18}$$

equation (16.17) becomes (see Block 11):

$$\begin{gathered}(-\Omega_E^2[M]+i\Omega_E[R]+[K])\underline{X}_0=\underline{F}_0\\ [E(\Omega_E)]\underline{X}_0=\underline{F}_0\end{gathered} \tag{16.19}$$

The advantages of the method used lie in the fact that it is possible in this way to keep account of the foundation and of the carrying structure in general, through its mechanical impedances: àpropos of this it seems opportune to underline the difficulties of reproducing the rotor-foundation behaviour due to the difficulties of schematizing both the foundation and the casings of the machine. This can be obviated by experimentally defining, in correspondence to the supports, the response of the case and the foundation to a harmonic excitation force and from this to define the mechanical impedance matrices which can be used, if introduced in (16.19), both in the calculation of the frequency response as well as the stability analysis of the shaft line considered. For further details see ref. [22].

7.1.20 References

[1] Brown D.L. et al. - "Parameter Estimation Techniques for Modal Analysis" - S.A.E. Paper n. 790221

[2] Curami A., Vania A. - "An Application of Modal Analysis Techniques " L'Energia Elettrica, n.7-8, vol LXII, 1985

[3] Bishop R.E.D., Johnson D.C. - "Mechanics of Vibration" -Cambridge University Press, New York, 1960.

[4] Diana G.- "Appunti dalle lezioni di Dinamica e Vibrazioni delle Macchine" - Milano, 1983

[5] Pizzigoni B., Ruggieri G. - "Sulla Determinazione delle Caratteristiche Statiche e Dinamiche di Cuscinetti Lubrificati" - L'Energia Elettrica, n.2, vol. LV, 1978

[6] Przemieniecki J.S.- "The Theory of Matrics Structural Analysis"- Mc. Graw-Hill, New York, 1968

[7] Bathe K.J., Wilson E.L. - "Numerical Methods in Finite Elements Analysis"- Mc. Graw-Hill, New York, 1970

[8] Curami A., Pizzigoni B.- "Un Programma di Calcolo Automatico per l'Analisi Statica di una Linea d'ALberi" - L'Energia Elettrica, n. 12, vol. LVIII, 1981

[9] Bachschmid N., Pizzigoni B., Di Pasquantonio F. - "A Method for Investigating the Dynamic Behaviour of a Turbomachinery shaft on a Foundation" - Design Engineering Technical Conference, paper 77-DET-16, Chicago 1977

[10] Diana G., Bachschmid N.- "Influenza della Struttura Portante sulle Velocita' Critiche Flessionali di Alberi Rotanti" - L'Energia Elettrica, n.9, vol. LV, 1978

[11] Ruggieri G.- "Un Metodo Approssimato per l'Integrazione dell'Equazione di Reynolds" - L'Energia Elettrica, n. 2, vol. LIII, 1976

[12] Pizzigoni B., Ruggieri G. - "Caratteristiche di Funzionamento di Cuscinetti Lubrificati in Regime Turbolento" - V Congresso AIMETA, Palermo, 1980.

[13] Costantinescu V. N. - "On Turbulent Lubrication" - Proc. Inst. Mech. Engrs., London, vol. 173 n. 38, 1959.

[14] Chung Wan-Ng, C. H. T. Pan - "A Linearized Turbulent Lubrication Theory" - Journ. of Basic Engr., Sept. 1965

[15] Diana G., Borgese D., Dufour A. - "Experimental and Analytical Results on a Full Scale Turbine Journal Bearing" - Proc. 2nd Conf." Vibrations in Rotating Machinery ", Cambridge, 1980.

[16] Frigeri C., Gasparetto M., Vacca M. - "Cuscinetto Lubrificato in Regime Laminare e Turbolento : Parte 1" - Analisi Statica" - L'ENERGIA ELETTRICA, vol. LVII, 1980.

[17] Biraghi B., Falco M., Pascolo P., Solari A. - "Cuscinetto Lubrificato in Regime Laminare e Turbolento : Parte 2 - Lubrificazione Mista Idrostatica-Idrodinamica" - L'ENERGIA ELETTRICA, vol. LVII, 1980.

[18] Falco M., Macchi A., Vallarino G. - "Cuscinetto Lubrificato in Regime Laminare e Turbolento : Parte 3 - Analisi Dinamica"- L'ENERGIA ELETTRICA, vol. LVII, 1980.

[19] Pendered I.W., Bishop R.E.D.- "A Critical Introduction to Some Industrial Resonance Testing Techniques" - Journal of Mechanical Eng. Science, Vol. 5, n.4, 1963

[20] Thomson W.T. - "Vibrazioni Meccaniche" - Tamburini Editore, Milano, 1974

[21] Frigeri C., Gasparetto M., Pizzigoni B. - " Metodologia di Misura e di Elaborazione delle Vibrazioni Assolute di un Albero Rotante" - Atti V Congresso AIMETA, Palermo, 23-25 Ottobre 1980

[22] Cheli F., Curami A., Diana G., Vania A. - "On the Use of Modal Analysis to Define the Mechanical Impedances of a Foundation" -

Technical Rept. of Dipartimento di Meccanica del Politecnico di Milano, 1984

[23] Diana G., Massa E., Pizzigoni B. - "A forced Vibration Method to Calculate the Oil-film Instability Threshold of Rotor - Foundation Systems" - Proc. IFToMM Conference "Rotordynamic Problems in Power Plants", Rome 1982

CHAPTER 8.1

SENSOR TECHNOLOGY

J. Tonnesen
The Technical University of Denmark, Lyngby, Denmark

The basic concepts of sensors are described and discussed for non-contacting displacement types and piezo-electric type force and acceleration transducers.

8.1.1 Introduction

Sensors of some type must be used to measure vibrations when dealing with rotordynamic problems of turbomachinery. It is important to select the proper instrument transducer or even different types of transducers in order to cover a particular situation. One should consider the following questions: "What is to be measured?, "What are your looking for?, How is the machine built?, What access area(s) is available?" and also the follow-up questions like: What kind of display is wanted of the phenomenon?, and what is the time element?". That kind of question will almost eliminate the so called universal sensor, because there is no one sensor for all jobs.

In many instrumentation systems the sensor is still the limiting factor of accuracy, though not universally so, for the time-honoured barrier of 1 per cent is fast approaching 0,1 per cent. This trend has of course been matched by more accurate electronic and recording equipment and also more sophisticated calibration methods and equipment. Another general trend in sensor development is towards wider extremes of environmental conditions. Higher and lower ambient temperatures and more severe conditions of acceleration, vibration and shock have to be considered.

A final requirement of a sensor is invariably the necessity for small size and weight, for exceptional reliability, and for low cost. In the light of these exacting and often contradictary requirements it is difficult to understand why the effort extended on sensor selection is minor compared, for instance, with that allocated to the electronic sections of modern instrumentation systems. Data processing and data reduction always seem to carry greater weight and be of greater attraction than data acquisition, whether it be in the planning, the development or the production stage of an instrumentation system. After all, what is common knowledge in the computer age: "Garbage in, garbage out". This experience may be partly due to the fact that sensors, regardless of their high performance, are small and comparatively cheap items which people tend to overlook, but it may also be due to one of Parkison's laws that sensor being, as they are, on the borderline of mechanical and electrical technology, often do not get any support from either the mechanical or the electrical expert, because of their two-sided character. Sensor design and development assumes indeed as much expertise in the mechanical as in the electrical field, the former extending deeply into many branches of physics and the latter not excluding the art and science of electronics.

In the following it is attempted to demonstrate the diversity and interdepence of the various brances of science and technology which are employed in the design and the application of these unassuming and therefore often neglected "gadgets", sensors.

8.1.2 Displacement sensors

This type of sensor is also often called proximity transducer and the physical quantity to be measured can be made to vary the inductance, capacitance or eddy-current. The first two groups represent the sensor types used in the laboratory and the last group is used both in the laboratory and in the field (industrial application). A common feature of the three sensors is that a carrier signal (typically 8 kHz to 2 MHz) must be fed to them in order to operate, and this signal is modulated in proportion to the gap size. A demodulator then converts the transducer output into a voltage that many or may not be linearly proportional to the gap size. Thus both static and dynamic displacements can be measured and since it is basically a gap variance that causes the effect the sensors are also termed the non-contacting type.

8.1.3 Inductance sensors

As the variation of inductance is the property of such a sensor a coil arrangement is required which gives the largest possible inductance change for a given input quantity, and an ironcored coil will basically yield larger effects than an aircored coil. The variable air gap type with ferro-magnetic core is probably the most sensitive type if the fractional change of inductance for a given air gap change is considered. Inductances with ferro-magnetic cores have also the additional advantage, that in a closed magnetic circuit, as provided by a ferro-magnetic core with only a small air gap, it is the least affected by external magnetic fields.

The inductance of a coil is inversely proportional to the small air gap and hence the characteristic is non-linear. This is an undesirable feature of a displacement sensor and several corrective means must be taken. The simplest one is to arrange two coils in a push-pull arrangement (e.g. in a bridge circuit of the carrier amplifier). In such a set-up the non-linearity will probably be 1 to 3 per cent at full scale deflexion, depending on the individual design. A further advantage gained by this method is a higher sensitivity (or resultion). Another solution is to use the assosiated electronics to convert a non-linear input signal in the feed-back path to provide a linear output.

The physical dimensions are small of both inductance and eddycurrent sensors, the smaller industrial types going down to 3 mm in diameter and then standard connestors (plugs and sockets) are usually much too large and their contact properties are not reliable in the presence of shock and vibration. A convenient type of cable attachment is a permanently soldered and cleated cable, say 1 meter long with a connector socket attached at the free cable end, which can be protected from shock and vibrations. The connecting cable, by virtue of its low impedance, can be used in great lengths and a shielded cable is preferrable, but not necessary, however ground loops, caused by more than one grounding point along the line, must be avoided. The shunting capacity provided by the cable must be considered when the sensor is calibrated with a cable different in length from that required for the actual measurement. At high carrier frequencies the permissible cable length is limited by the cable capacitance.

The inductance and eddy-current displacement sensors are widely accepted due to their small size, high sensitivity and rugged construction. When used to measure rotating machinery they can be installed with a relatively large air gap, say 1 to 1.5 mm and are unaffected by lubricating oil and most gasses. On the other hand, one will have to accept that the sensors are susceptible to inhomogenities of the shaft like electrical and magnetical properties, coatings of different conductivity and shaft finish.

8.1.4 Capacitance sensors

By changing the distance between two parallel electrodes a variation in capacitance is obtained. The dielectric may be simply an air or gas gap (or even fluid filled gap). Most variable capacitance sensors can be presented by a pure capacitance and the computation of the "geometrical" capacitance of a given condenser configuration, neglecting fringing, can be found in many textbooks of physics or electrical engineering. Even at very high carrier frequencies of the order of several megacyles per second and with very small capacitance values, losses are normally negligible in capacitance sensors, possibly with the exception of some high-temperature sensors.

In practice one of the electrodes constitutes the moving part and the actual displacement sensor is the second electrode. The moving electrode is typically connected capacitatively to the rest of the electronic equipment.

The capacitance of a pair of electrodes is inversely proportional to the small air gap changes and just like previously mentioned under the inductance type sensor the same type of corrective measurements must be applied in order to have a linear output. Similarly, the connecting cable must be screened right down to the sensor housing without leaving an unscreened gap. For this reason a convenient length of cable is usually made an integral part of the sensor and when it is calibrated the entire cable must be included.

When discussing the construction of variable capacitances used in sensors it seems more difficult to divorce the essential variable capacitance unit from the specific sensor design than in the previous mentioned variable inductance sensors where coil and core were distinct elements. Essentially, a conductive electrode is required. The electrode must be well insulated from the mounting parts and still connected solid mechanically. The choice of a suitable insulation material is of great importance. It must be of sufficient mechanical strength and, even more important of an extremely high form of stability. Its thermal coefficient of expansion should be as low and as predictable as possible, because it will most likely affect the stability of the active air gap. In some cases its coefficient of expansion should match the coefficient of expansion of the other structural parts of the sensor, usually made of metal, so as to achieve improved zero stability by way of compensation. Ceramic insulation materials are generally a better choice than plastics or organic materials. The face of the electrode cannot normally be cleaned in use so it should be protected from humidity and condensation as well as from corrosion and rust. Metals and alloy combinations used in the design of the sensor must be chosen to avoid electrolytic corrosion, especially if the sensor cannot be effectively sealed against ambient influence.

The variable capacitance displacement sensor is mostly used in the laboratory because it can easily be designed to match a given test set-up. The major disadvantage is contamination of the air gap or a variable dielectric condition. A common disadvantage to all types of displacement sensors is their inherent characteristic of sensing geometric changes in a moving surface commonly termed the mechanical out-of-roundness.

8.1.5 Acceleration sensors

The acceleration sensor most used, both in the industry and in the laboratory is the piezoelectric type. Piezoelectric sensors are energy converters and are therefore generator-type or active type sensors and the generated quantity is an electrical charge. The indicated voltage thus depends on the capacitance of the sensor and indicator circuit. This fact, together with the high impedances normally prevailing in these circuits determines the approach in the design of piezoelectric sensors and their associated electronic equipment. Owing to the finite insulation resistance of the sensor circuit and the shunting effect of the load resistance, the generated charge gradually leaks away and there is therefore no steady-state response.

Piezoelectric sensors are force sensitive devices and are therefore employed for the measurement of physical properties which can be reduced to forces, such as pressure, stress or acceleration. Piezoelectricity occurs in crystals of certain configurations when exposed to compression or tension. Pyro-electricity is a closely related effect observed in some crystals when heated. In the design and use of piezoelectric sensors the pyro-electric effect is unwanted and in some modern piezoelectric ceramics the high voltages generated by temperature changes may in some applications become a serious source of errors.

The main attraction of piezoelectric sensors for the measurement of pressure or acceleration is their high mechanical input impedance (stiffness). They are genuine force-measuring instruments of negligible deformation under load. In the lanquage of sensor performance they are instruments with high natural frequencies. The output side normally represents a very high electrical impedance which often requires special precautions and slightly more complex circuits than sensors of low output impedance.

The other advantage of piezoelectric sensors is the possibility of designing instruments of exceptionally small dimensions. There are piezoelectric acceleration sensors weighing less than one gram. From the point of small size and weight these sensors are rivalled only by resistance strain-gauges, thermocouples, and thermistors.

The main disadvantages of piezoelectric sensors are their lack of steady-state response and their high electrical output impedance coupled with the need for low-noise cables of low capacitance value. The methods of mounting need also special care and attention.

The maximum working temperatures of polarized piezoelectric ceramics are now in the neighbourhood of $500^{o}C$ and this is a remarkable achievement considering that no less that four sensor parameters: charge sensitivity, permitivity, dissipation factor, and leakage resistance are variable with temperature.

Humidity may affect piezoelectric materials such that they suffer loss of insulation resistance when water vapour condenses on their faces. This effect applies equally well to other parts of high-resistance circuits of piezoelectric sensors, such as internal connections, terminals, transducer housing etc. It is thus important that condensation of water vapour be prevented from entering, or affecting, critical parts of piezoelectric sensors. Waterproof coatings should be applied to these areas and silicone grease filled into cavities of sensors and connectors. Cold curing silicone rubber has proven useful in places where some mechanical durability is required.

When a linear-acceleration sensor is exposed to a sinusoidal linear acceleration in a radial direction perpendicular to the main axis of polarization the directional characteristic of sensitivity assumes a two lobed shape with a transverse sensitivity between 2 to 5 per cent of the longitudinal sensitivity. Manufacturers are constantly trying to improve this by paying attention to the selection of homogenous piezoelectric ceramics and in careful machining, polishing and lining-up of the parts in such a fashion that the direction of the anticipated transverse acceleration run perpendicular to the main axis of the lobe. Similarly, connecting leads must not exert radial forces on the sensor and the optimum position of the leads may have to be found experimentaly.

Piezoelectric acceleration sensors may also be sensitive to airborne high-intensity noise, but if a sensor is designed so as to prevent admittance of sound pressure to the crystal faces, say the cylindrical faces of a disk enclosed in a sealed housing, a net pressure sensitivity may be experienced.

Electrical cable noise is primarily an environmental effect since it is caused by whipping and twisting of the cable during calibration and mesurement. The mechanism of generation of cable noise in piezoelectric circuits is frictional separation of conductors and insulation which cause an electrostatic induction at the point of separation and can generate appreciable voltages on the load. The remedy is to lead away the changes by a conductive film on the surface of the insulator. Piezoelectric sensors are now normally supplied by the manufacturers with a suitable length of low-noise cable.

The frequency response of a piezoelectric acceleration sensor is best when it is mounted with a steel stud into a clean flat surface. However less rigid mounting methods works just as well depending on the application, enviromental conditions and frequency range of interest. Several possibilities are available like magnetholder, wax, double sided adhesive tape or disc, dental cement, epoxy, and quick setting cyanoacrylate cement. By using some of these methods, and also with an insulated stud, ground loops are prevented which may otherwise interfere with measurement of low acceleration levels.

8.1.6 References:

1) H.K.P.Neubert: "Instrument Transducers", Oxford at the Clarendon Press, 1963.
2) C.Rohrbach: "Handbuch für elektrisches Messen mechanischer Grössen", VDI-Verlag, Düsseldorf, 1967.
3) C.Jackson: "The Practical Vibration Primer", Gulf Publishing Company Book Division, 1979.
4) M.Neales and Associates: "A Guide to Condition Monitoring of Machinery", Her Majesty's Stationery Office, London, 1979.
5) E.B.Magrab: "Vibration Testing-Instrumentation and Data Analysis, A.S.M.E., AMD-Vol.12, 1975.

CHAPTER 8.2

EXPERIMENTAL TECHNIQUES FOR ROTORDYNAMICS ANALYSIS

J. Tonnesen
The Technical University of Denmark, Lyngby, Denmark

Experimental test rigs are described for:

1: identification of bearing parameters.

2: the influence of mass unbalance on a rotor's instability threshold speed and the effect of adding external damping to the rotor-bearing system.

8.2.1 Introduction

Experiments will always be needed, first at all in the laboratory or at the manufacturer's facilities, in order to validate the theoretical model or to provide further physical insight and just as often to give vital feed-back for new and improved theoretical models.

In bearing parameter identification, and for our purpose here is referred only to fluid-film bearings, many manufacturers are now conducting experiments on full-size bearings. It is first of all the static properties like load carrying capacity versus film thickness that is being investigated but the equally important dynamic properties like stiffness, damping and stability are getting close attention. Although large discrepancies are found between theory and test the overall tendency is towards a much better understanding of identifying these parameters and especially with the aid of modal experimental techniques.

8.2.2 Bearing Identification Test Rig

A typical bearing test stand is shown in fig.1 and the main data are as follows:

Bearing diameter:	100 mm, D
Bearing length:	50 mm, L
L/D range:	0.3 to 0.6
Speed range:	0 - 20,000 RPM
Load range:	0 - 10,000 N, by applying force and/or mass to the bearing housing.

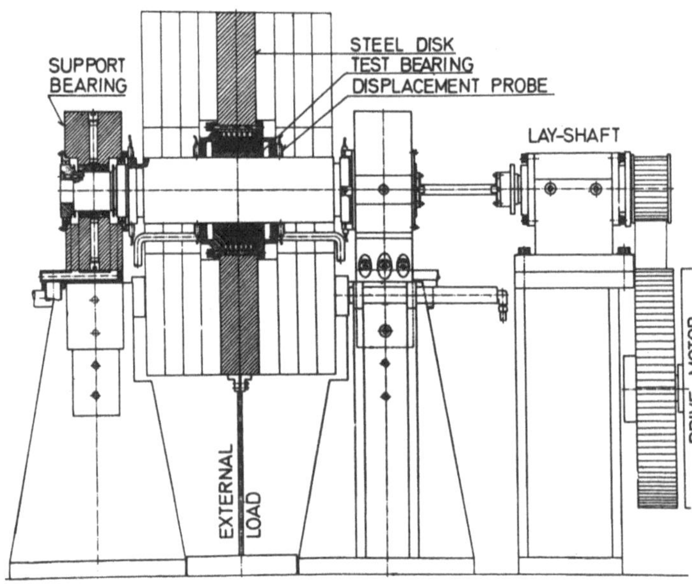

Fig.1 Bearing Test Rig.

The journal's position in the test bearing is measured by capacitance and eddy current displacement sensors. The capacitance displacement sensors are installed on both sides of the bearing house, and at each location there are four sensors, two for the vertical direction and two for the horizontal direction. The sensors are arranged in a push-pull connection, where the difference between the signals measures the relative motion of the parts and the sum of the signals gives the combined expansion of the parts, thus making it possible to partly eliminate the error caused by differential expansion of the parts. The measurement planes are located 75 mm from the bearing midplane and the position of the centre of the journal is obtained as the average of the two signals.

The eddy current sensors are installed in the test shaft in two planes each located 18.5 mm from the bearing midplane. There are two sensors in each plane and they are arranged to give both in-line and 180° apart readings. Figure 2 shows some details of the installation and in addition to the displacement sensors is also seen an oil pressure sensor and a thermocouple sensor, all mounted flush with the journal surface. The signals from the sensors are transmitted to the stationary parts via a mercury-cell slip-ring unit and thus simultaneous information is available about the absolute eccentricity, attitude angle, oil pressure profile and journal midplane surface temperature.

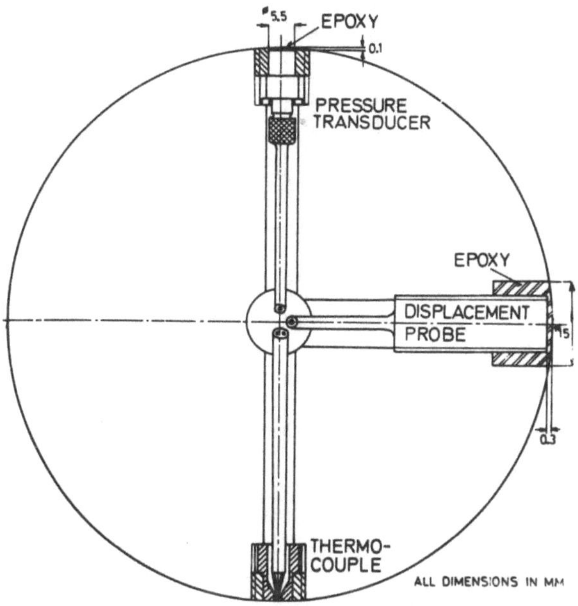

Fig.2 Cross-section of Test Shaft.

A broad range of tests must be conducted to provide reasonable data on the effect of the variable encountered in bearing parameter identification work. The static properties have three main variables: speed, load and oil temperature (viscosity), but design variables such as geometry and oil inlet conditions, like flow and supply pressure, may also have a noticeable influence on the behaviour. The test technique is straight forward: at a given speed the load is varied in increments large enough to give distinct changes in the displacement, oil pressure and temperature data. The test series are then repeated with a different oil viscosity or if the bearing geometry is of interest, a different bearing type is used.

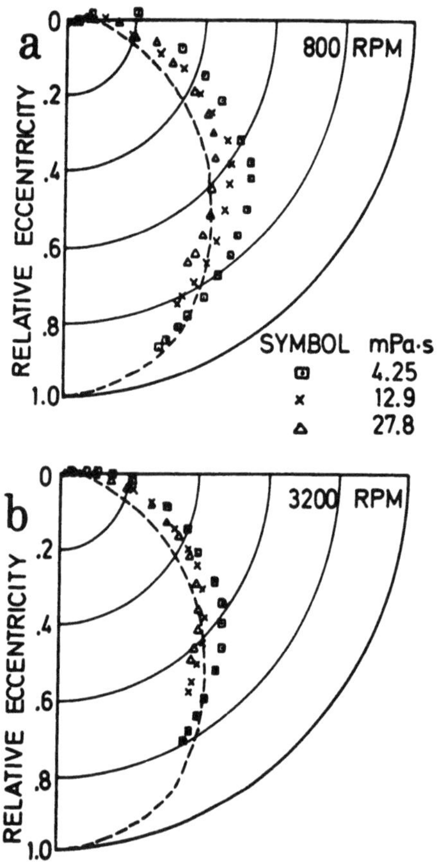

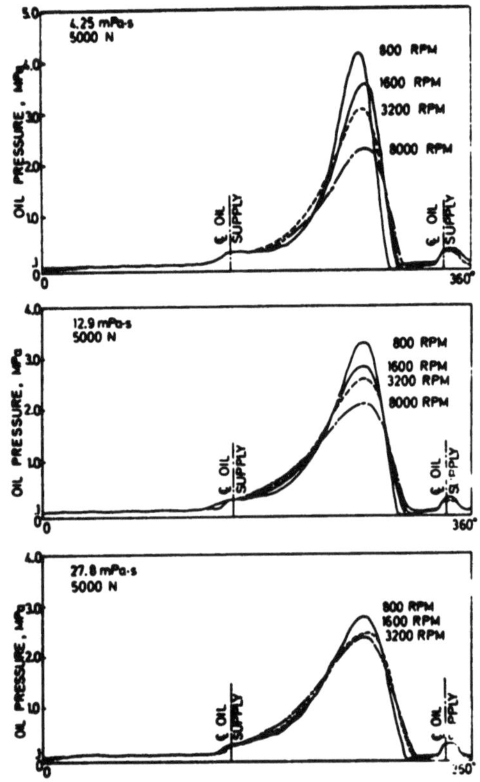

Fig.3 Locus Curves. Fig.4 Oil Pressure Profiles.

8.2.3 Static Properties of Bearings

Figures 3 - 6 show typical results obtained from the test rig. Figure 3 shows the locus curve as a function of speed and oil viscosity and interconnecting the individual test points will produce a smooth curve. For comparison is the theoretical locus curve shown as a dashed curve. Figure 4 shows the oil pressure as a function of speed and oil viscosity for a constant load. Integrating the areas under the curves give a check on the accuracy of the test method and in this case is the deviation less then 4 per cent from the applied load.

Figure 5 shows the measured and calculated journal surface temperature as a function of load and speed and as can be seen from the curve, the temperature is largely independent of load at any given speed, except for a rising trend at low loads caused by the reduction in the side flow of the oil. Figure 6 shows the

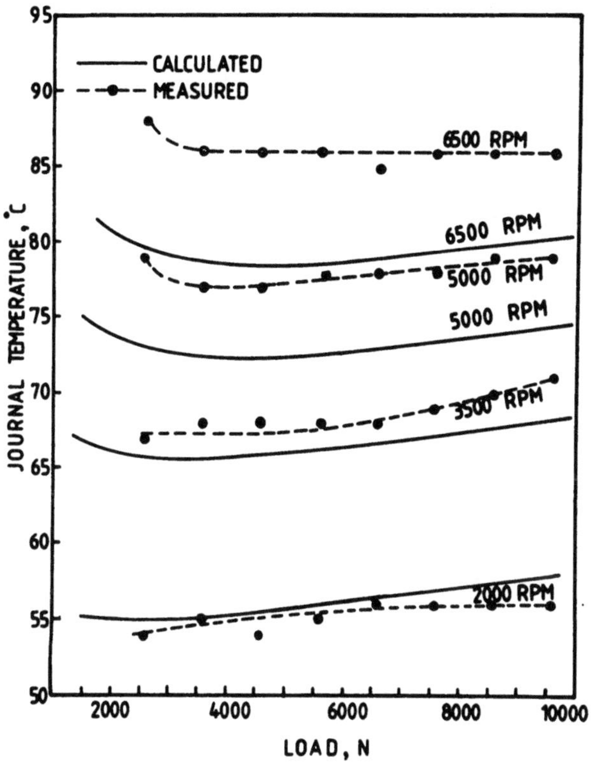

Fig.5 Journal Surface Temperatures.

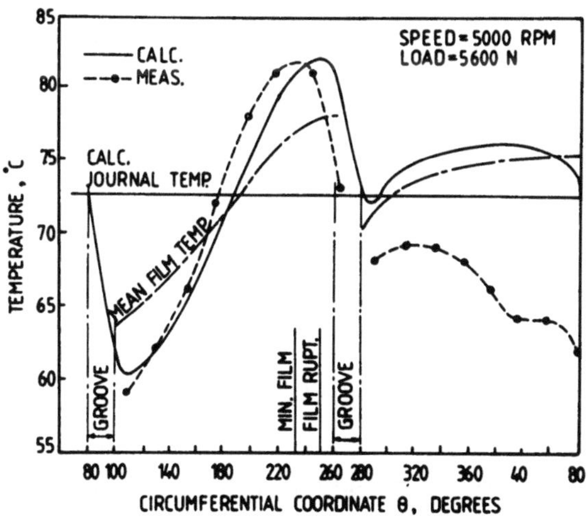

Fig.6 Bearing Bush Wall Temperature.

results for the maximum bearing bush wall temperature which occurs in the loaded portion of the bearing a little past the minimum oil film thickness. The agreement between calculation and measurement is good for the region where full oil film exists, however in the ruptured film zone there are large discrepancies. A possible explanation is that there may be a backflow of cooler supply oil from the grooves in the unfilled space in the striated film.

8.2.4 Dynamic bearing properties

The dynamic bearing parameters are obtained by applying a force function to the system. If only the stability threshold is wanted this is accomplished relatively simple by reducing the load or for a given load increasing the speed until whirl occurs. Under these condition is the whirl-frequency between 40 to 50 per cent of the running frequency. The force function used in order to determine the stiffness and damping coefficients of the bearing is either a sinusoidal, step or impulse force each of which have their own advantages.

8.2.5 Rotor-Bearing Tests

Once the bearing parameters have been identified in a test rig where the rotor dynamics influence is minimized as much as possible the next step is to investigate the combined rotor-bearing system. As the stiffness and damping coefficients of the bearings vary with speed, the natural frequencies of the system will also change and a speed may be reached where the damping becomes negative. At that speed, the threshold speed, the rotor becomes unstable in a self-exited whirl. Hence, from this point of view, it is important to be able to determine the damped natural frequencies of the system at the design stage. The information is, furthermore, of considerable value for monitoring purposes and in diagnosting vibration measurements, especially in connection with a FFT-analyzer. The present trend in experiments is towards measuring the onset of instability, methods of stabilization and the influence of mass unbalance as well as to determine which critical speeds must be avoided, and which of them may be safely ignored.

Figure 7 shows a test rig that effectively may be used to identify some of the before mentioned parameters. The main data of the test rotor is:

Rotor mass.	188 kg, 6 disks shrink fitted on a common shaft.
Rotor length:	1,20 m
Operating range:	0 - 20,000 RPM

The rotor is supported in fluid-film bearings with a bearing diameter of 62.7 mm and the L/D ratio is 0.3. The bearing arrangements are shown in figure 8 and it should observed that in the left hand bearing arrangement external damping can be added.

Fig.7 High Speed Rotor Test Rig.

The shaft whirl orbit is measured at ten locations along the shaft, namely at each of the six disks, at the two bearing pedestals on the outboard side and in the middle of the two bearings. At each location are capacitance displacement sensors for measuring in the vertical and horizontal directions. The sensors in the bearings are flooded by the supply oil, and to insure constant dielectric properties in the gap, they are additional flooded through separate supply holes around the tip of the sensor. From separate calibration tests the dielectric constant is found to be a factor of 2.3 less than that of air (at 50°) with a variation of ±3 per cent in the temperature range 20 to 80°C. It is found to be insensitive to pressures up to 6 MPa.

In addition to the displacement sensors there are also pressure tabs in the bearings' midplane for measuring the static and dynamic oil film pressures at two places in the bottom half, ±15 degrees relative to the vertical. 1 mm diameter holes drilled at these locations are connected by 600 mm long tubes to manometers and piezo-electric pressure sensors.

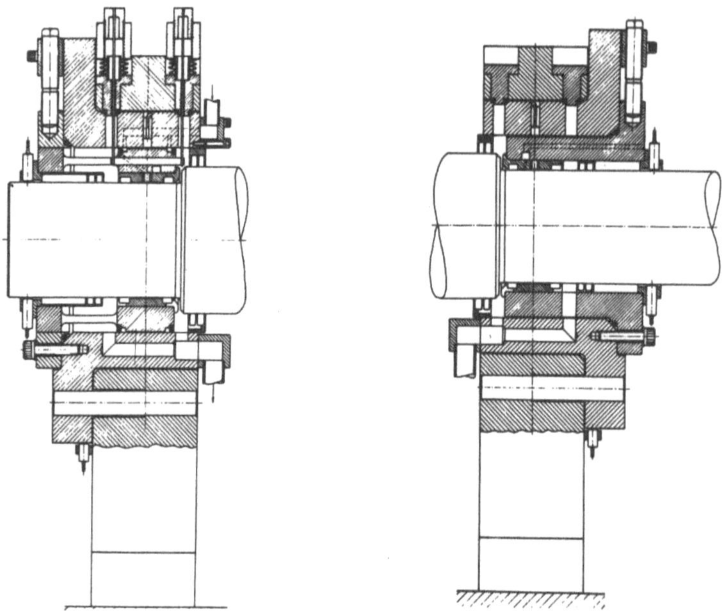

Fig.8 Bearing Arrangements.

The circumferential temperature distribution of the bearing bush wall temperature, which is an indication of the oil film temperature, is measured by eight thermocouples: two in the upper half of the bearing at ±45 degrees from vertical and six in the bottom half with 30 degrees spacing. The thermocouples are located in the centerplane of the bearing and are within 0.5-1 mm from the surface.

8.2.6 Squeeze Film Damper Performance

The vertical and horizontal motion of the squeeze film dampers, which contain the journal bearings, is measured by capacitance displacement sensors which are flooded by the squeeze film oil supply. The transmitted bearing forces are measured by force gauges made up of a solid ring connected by eight flex bars to the bearing housing. The force is measured using strain gauges attached to the flex bars and by calibrating the force gauge up to 2.000 N it is found that the non-linearity and cross-coupling does not exceed 3 per cent. Taking simultaneous measurements of the transmitted force and the squeeze film damper makes it possible to determine the system's external damping and stiffness.

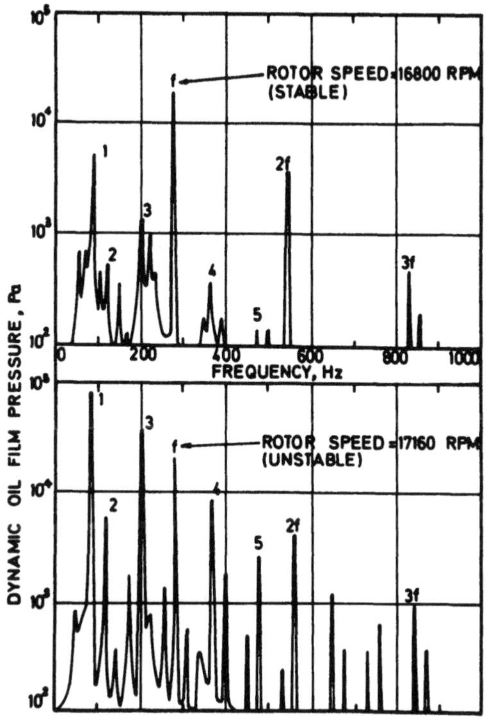

Fig.9 Test Rotor Frequency Response, with no External Damping.

8.2.7 Stability Test Data

Some typical test results from a stability test run are shown in figure 9. The test rotor is well balanced and should theoretically be stable up to the max. speed of 20.000 RPM,when the rotor-bearing system consists of a rotor without the overhung disk and with rigid bearing supports (no squeeze film damper). However the calculations did also indicate that above 12.000 RPM the damping is (log.decrement) small, typically below 0.08. As may be seen from figure 9 the rotor became unstable at 17.160 RPM. The residual unbalance in the rotor results in a bearing whirl orbit which occupies approximately 20 per cent of the clearance just before the instability sets in, say at 16.800 RPM. At the threshold speed this whirl orbit grows rapidly, occupying about 50 per cent of the clearance circle when the speed has been raised 50-100 RPM.

Inserting a weight of 25 g·cm in the disk next to the rotor's free end bearing brings the threshold speed down to 14.235 RPM and increasing the weight 3 times more results in a further decline to 11.300 RPM. This rotor-bearing system is thus sensitive to unbalance which lowers the instability threshold speed and the character of the instability onset is the same, a small speed increase results in a very rapid growth of the whirl orbit.

The same behavior is observed when the overhung disk is installed. Here the theory predicted the instability threshold speed to be at 18.400 RPM, however, with residual unbalance, the measured speed was 12.600 RPM, and this is caused by a small log.decrement at the higher speeds.

The rotor-bearing system is now stabilized by adding external damping at the bearing supports and it can now be operated safely up to 20.00 RPM even with a substantial residual unbalance resulting in a shaft orbit covering 40 per cent of the bearing clearance. However it is interesting to notice that a frequency analysis of the oil film dynamic pressure signal, in comparison with the displacement, did reveal most of the predicted damped natural frequencies, an example of this is shown in figure 10.

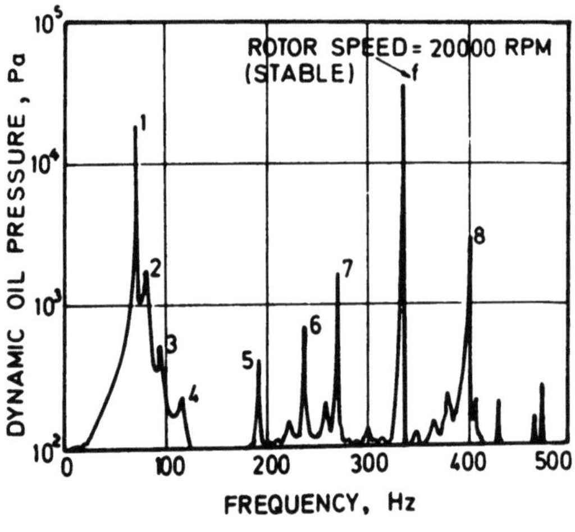

Fig.10 Test Rotor Frequency Response
with External Damping.

From the foregoing test it is evident that the bearings exert a controlling influence on the dynamics of the system. This influence is governed by the dynamic properties of the bearings, which again depends on the operating eccentricity ratio of the journal. A comparison between theory and tests is shown in figure 11 where the measured position of the journal center at various speeds is plotted against the calculated journal center locus. The overall curve fit is reasonable, but a point by point comparison shows an increasing descrepancy as the speed increases. It would require an unrealistic increase in the assumed oil viscosity value from

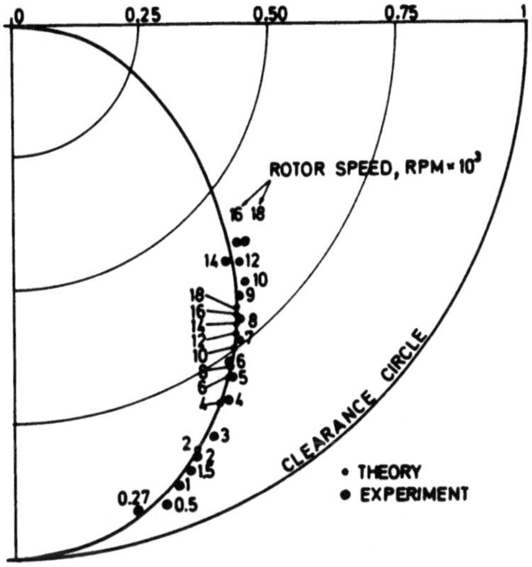

Fig.11 Locus Curve, Theory and Experiments.

which the theoretical points are obtained, in order to bring agreement and this would conflict with the measured temperatures in the bearings. The possibility of some systematic error in the instrumentation or the measurement technique cannot be ruled out entirely as it is notoriously difficult to insure accuracy of static measurements with displacement sensors. The measurements, however, repeat themselves whether the rotor is brought up to full speed rapidly or slowly, and reversing the direction of rotation yields a mirror image.

8.2.8 Recent Development

In recent years the technique of experimental modal analysis has been used to measure the dynamic performance of rotor-bearing system and with variable succes. It has so far been limited to simple test rigs, where the system damping was relatively small and where the rotor speed was below 6000 RPM.

In contrast to the previous example, where only synchronous force excitation was used, we now use several other techniques like swept sine, impact or random forces. The sinusoidal excitation is long and tedious but it has one subtle advantage which is often overlooked, namely it is a rather straight forward approach which allow you to get a good "physical feel" for the situation as you go through the analysis. For example, when the forcing frequency sweeps through a natural frequency the response is often quite visual or audible and as the response transducer is moved from point to point it is possible to feel (or easily see on the display media) the points of maximum response. Even an inexperienced person could basically "see" what is happening and if something did not look quite right one will find it fairly easy to use ones judgement to find the cause of the problem.

The advent of the mini-computer based digital analysis systems with its' tremendous capabilities of Fast Fourier Transforms (FFT), high speed computations, and data storage has drastically changed the methods used to determine the dynamic characteristics of structures. The excitation that is applied is often a transient (such as a pulse or step relaxation) or a force signal with energy at many different frequencies simultaneously (such as a random signal). The frequency response data is computed and can be automatically put into a disc storage. The mode shape analysis techniques and modal parameter extraction methods can be completely automated so that one never sees, handles or evaluates any of the data nor understands any of the preprogrammed functions that are being performed for you.

These techniques have revolutionised experimental modal analysis and it is no wonder, therefore, that mini-computer based analysis systems and these new analysis techniques have enjoyed such wide utilisation in a relatively short time. However, simply owning one of the new analysing machines does not gurantee good or useful results. It should be regarded as a tool and its characteristics must be specified properly by the manufacturer and you must understand it. One must always scrutinize ones results and have a good basic understanding of the theory that has been programmed and also if compromises have been made. If this is not done the results will often be worthless.

8.2.9 References:

1) Lund, J.W., "Stability and Damped Critical Speeds of a Flexible Rotor in Fluid-Film Bearings", Journ. of Eng. for Industry. Trans. ASME, Series B, Vol.96, No.2, May 1974, pp.509-517.
2) Thomsen, K.K., "Theoretical and Experimental Investigation of the Stability of Hydrodynamic Radial Bearings", PhD. thesis, Tech.Univ.of Denmark, Lyngby, Denmark, Sept.1975.
3) Christensen, E., Tonnesen, J., and Lund, J.W., "Dynamic Film Pressure Measurements in Journal Bearings for Use in Rotor Balancing", Journ.of Eng. for Industry, Trans.ASME, Series B, Vol.98, No.1, Feb. 1976, pp.92-100.
4) Glienicke, J., "Schwingungs- und Stabilitätsuntersuchungen an Gleitgelagerten Rotoren", Motortechnische Zeitschrift, Vol.33, No.4, April 1972, pp.135-139.
5) Morton, P.G., "The Derivation of Bearing Charachteristics by Means of Transient Excitation Applied Directly to a Rotating Shaft", IUTAM-Symposium, Dynamics of Rotors, Lyngby, Denmark, Springer Verlag 1975.

6) Nordmann, R., "Identification of Modal Parameters of an Elastic Rotor with Oil Film Bearings", Journ. of Vibration, Acoustics, Stress and Reliability in Design, Trans. ASME, VOl.106, No.1, Jan. 1984, pp.107-112.
7) Tonnesen, J., and Lund, J.W., "Some Experiments on Instability of Rotors Supported in Fluid-FIlm Bearings", Journ. of Mechanical Design, Trans. ASME, Vol.100, No.1, Jan 1978, pp.147-155.
8) Tonnesen, J., and Hansen, P.K., "Some Experiment on the Steady-State Characteristic of a Cylindrical Fluid-Film Bearing Considering Thermal Effects", Journ. of Lub. Technology, Trans. ASME, Vol.63, No.1, Jan 1981, pp.107-114.
9) Lund, J.W., and Tonnesen, J. "An Approximate Analysis of the Temperature Conditions in a Journal Bearing. Part II: Application", Journ. of Tribology, Trans. ASME, Vol.106, No.2, April 1984, pp.237-245.

CHAPTER 9

INTERACTION BETWEEN A ROTOR SYSTEM AND ITS FOUNDATION

L. Gaul
University of the Federal German Armed Forces, Hamburg, FRG

ABSTRACT: A theoretical approach is developed and programmed to analyze the three-dimensional dynamic response of machines on foundations interacting with soil. Structures and soil are coupled by means of a substructure technique. The substructure behaviour of soil is treated for rigid and flexible foundation slabs of arbitrary shape by superposition of semianalytical solutions of viscoelastic halfspace field equations. The interaction between a single turbomachinery frame foundation and soil as well as the interaction through the underlying soil between adjacent block foundations are considered. The assumptions of perfectly smooth and perfectly welded contact at the interface between soil and bases bound the influence of shear stresses. The impact of foundation flexibility with respect to rotor vibrations is discussed. Experimental studies describe the measured sine sweep response and vibration modes of a small scale frame foundation and a rigid circular block foundation on compressed sand.

9.1 Introduction

The prediction of machine vibrations by theoretical approaches as well as the modification of response after construction often require taking the interaction between machine, foundation-structure and subsoil into account. Three examples are given. Fig. 1 shows a discretized model of a drilling machine with long foundation slab on soil. The impact of static soil-structure interaction was calculated and measured by Thurat [22]. A base for the dynamic analysis is given by the soil model in the present paper.

A multi body model of a forging hammer (Fig. 2) is coupled with a viscoelastic truncated cone model of soil (Knobloch and Gaul [17]). Thurat [22] calculated and measured the transient response. Novak [19] treats a hammer foundation as a system of two masses on a viscoelastic halfspace including embedment effects.

The global response of turbomachinery frame foundations e.g. the low-tuned steel foundation with concrete raft (Dietz [7]) of Fig. 3, or the response of block foundations are calculated and studied experimentally by small scale models in the present paper. Dynamic

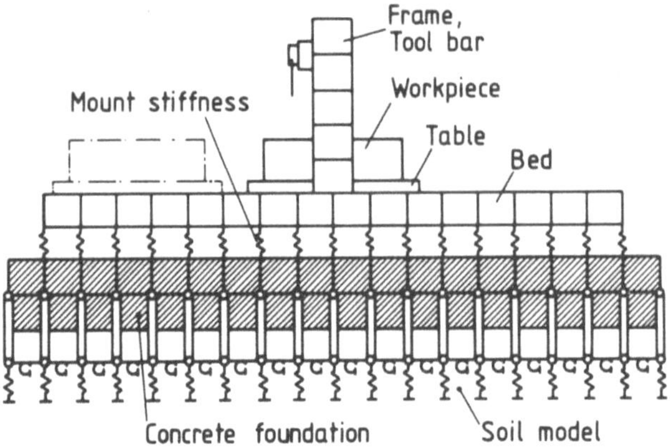

Fig. 1 Model of drilling machine foundation slab and soil

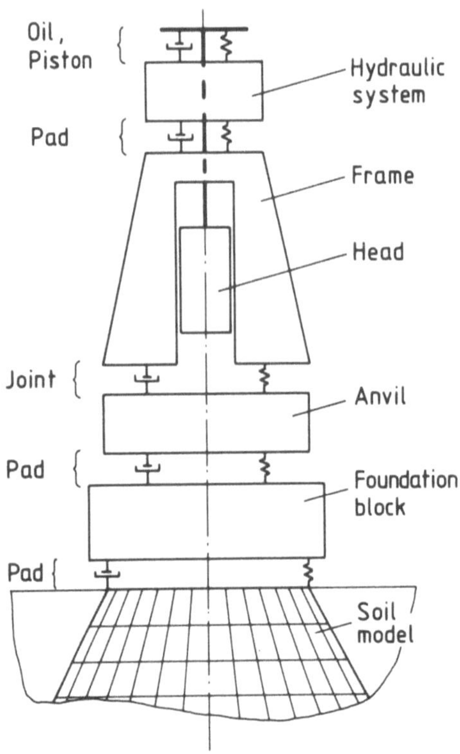

Fig. 2 Model of hammer, foundation and soil

Fig. 3 Low-tuned steel foundation with concrete raft

response results from active excitations by rotor unbalances, short circuit moments and shaft misalignement or by passive seismic excitation. The function of the foundation is not only to support the weight of the expensive equipment; the light upper steel plate on flexible columns (Fig. 3) minimizes the amplitudes of shaft whirling relative to the bearings. Although the tendency often prevails to treat the rotor, the frame and the foundation as if they were independent, actually all these substructures interact. This interaction was treated by Gasch and Sarfeld [8] for a Laval shaft on a block foundation and by Aboul-Ella and Novak [1] for a turbogenerator on a pile-supported frame foundation. The horizontal soil stiffness in the first paper was calculated by Gaul [10], the vertical soil stiffness matrix of the second paper by Gaul [9].

Methods for simulation of soil-structure interaction often take advantage of substructure techniques by coupling the model of structure and base plate with the model of the substructure soil. Structures are usually discretized by finite elements or can be treated in special cases by analytical dynamic stiffness matrices as in the present paper. Besides simplified soil models (Gaul and Plenge [14]) the substructure soil is usually described by finite elements (Waas [23]), halfspace theory (Holzlöhner [15], Gaul [11] or boundary elements (Ottenstreuer [20]).

Finite elements do also allow for a simultaneous discretization of structure and soil. The method is equally applicable to embedded foundations and inhomogeneous soil. It has however serious disadvantages when applied to three-dimensional problems since it requires extensive, complicated and expensive data mangement. Energy radiation travelling out to infinity by waves (geometrical damping) can be represented approximately by semi-infinite elements, which do only simulate the infinite extension in the horizontal direction (Waas [23]).

The halfspace theory presented here treats the substructure soil separately. The soil is assumed to be an elastic (Holzlöhner [15]) or viscoelastic (Gaul [11]) homogeneous halfspace. Dynamic stiffness matrices of the discretized soil surface can be coupled with rigid or flexible base plates of arbitrary shape (Sarfeld and Fröhlich [21], Gaul [11]). Three-dimensional motion of structures can be described even in the high frequency range. Soil inhomogenity has to be approximated by one or two layers or the concept of equivalent moduli. Embedment has to be approximated as well.

As another tool the boundary element method proved to be well suited to handle soil dynamics problems. It is possible to calculate embedded structures (Dominguez [24], Huh, Schmid and Ottenstreuer [16] as well as layered media.

Viscoelastic material properties and coupling effects of neighbouring foundations can be described by all three methods.

9.2 Coupling of Substructures

The neighbouring structures (Fig. 4) interact with soil. The transfer behaviour of the three substructures can be described in the frequency domain of Fourier transform by dynamic stiffness matrices $[K(i\omega)]$ including inertia, damping and stiffness properties.

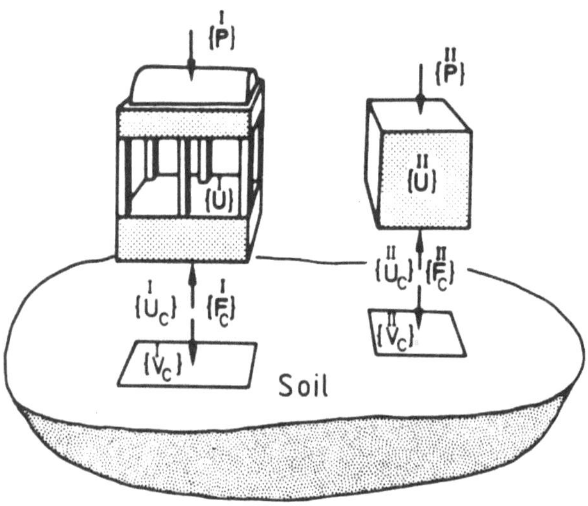

Fig. 4 Substructures of soil-structure interaction

The substructure matrices of soil $[\overset{S}{K}]$ and both structures $[\overset{I}{K}], [\overset{II}{K}]$ are coupled by compatibility requirements of generalized displacements $\{U_c\}$ and forces $\{F_c\}$ at the contact nodes of the interfaces I and II. Seismic excitation requires the input of generalized displacements $\{V_c\}$ at the unloaded interfaces generated by incoming waves. With the generalized forces of active excitation $\{P\}$ the substructure equations are given by

$$[\overset{I}{K}(i\omega)]\begin{bmatrix}\{\overset{I}{U}\}\\ \{\overset{I}{U}_c\}\end{bmatrix}=\begin{bmatrix}\{\overset{I}{P}\}\\ -\{\overset{I}{F}_c\}\end{bmatrix},\quad [\overset{II}{K}(i\omega)]\begin{bmatrix}\{\overset{II}{U}_c\}\\ \{\overset{II}{U}\}\end{bmatrix}=\begin{bmatrix}-\{\overset{II}{F}_c\}\\ \{\overset{II}{P}\}\end{bmatrix} \tag{1}$$

and

$$[\overset{S}{K}(i\omega)]\begin{bmatrix}\{\overset{I}{U}_c\}-\{\overset{I}{V}_c\}\\ \{\overset{II}{U}_c\}\quad\{\overset{II}{V}_c\}\end{bmatrix}=\begin{bmatrix}\{\overset{I}{F}_c\}\\ \{\overset{II}{F}_c\}\end{bmatrix}, \tag{2}$$

where the generalized displacements of structures $\{U\}$ are separated from those at the interfaces $\{U_c\}$. With given excitation data Eqs. (1) and (2) lead to the generalized displacement response of the coupled system by solving

$$\begin{bmatrix} [{}^{I}K] & & \\ & [{}^{S}K] & \\ & & [{}^{II}K] \end{bmatrix} \begin{bmatrix} \{{}^{I}U\} \\ \{{}^{I}U_c\} \\ \{{}^{II}U\} \\ \{{}^{II}U_c\} \end{bmatrix} = \begin{bmatrix} \{{}^{I}P\} \\ \{0\} \\ \{0\} \\ \{{}^{II}P\} \end{bmatrix} + {}^{S}[K] \begin{bmatrix} \{{}^{I}V_c\} \\ \{{}^{II}V_c\} \end{bmatrix} . \quad (3)$$

The solution of Eq. (3) leads to complex amplitudes $\{U\} = \{U_R\} + i\{U_I\}$ corresponding to real displacements $\{u(t)\} = \{U_R\} \cos \omega t - \{U_I\} \sin \omega t$ for time harmonic excitation or to Fourier transformed displacements, corresponding to transient excitation. Calculation of transient response requires the inverse transformation which can be computed efficiently by the fast Fourier transform algorithm.

9.3 Substructure Soil

9.3.1 Interaction of soil with rigid and flexible base plates.

The substructure behaviour of soil is calculated by the halfspace approach for idealized rigid base plates and for flexible plates. The plane interfaces of soil (Fig. 5) are loaded by forces F_i and moments M_i generated by the structures. Solutions of the field equations of soil have to fulfil mixed boundary values:

- rigid bases require plane displacement fields at the interfaces,
- the soil surface is stressfree elsewhere.

While rigorous formulations by dual integral equations (Gaul [12]) lead to approximate solutions only for simple base geometries, the presented superposition method provides solutions for arbitrary shapes and allows for taking flexible base plates into account.

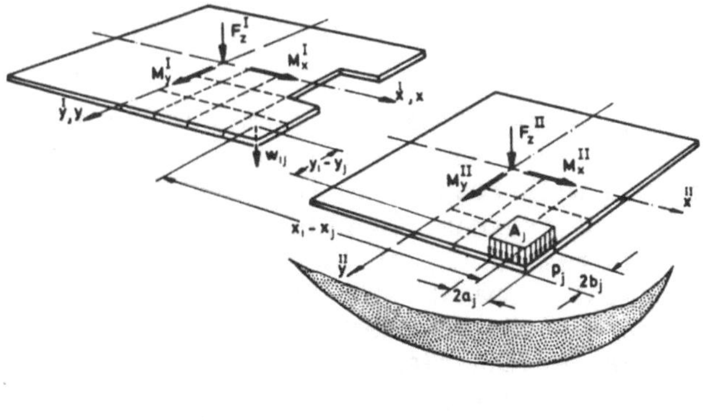

Fig. 5 Mixed boundary value problem of soil.
Stress boundary value problem of one interface soil element.

Arbitrary shapes are modelled by subdividing the interfaces into rectangular surface elements. The continuous stress distributions in the interfaces are discretized by locally constant pressures in each element, acting harmonically in time. Each loaded surface element in Fig. 5 defines a stress boundary value problem of the halfspace. The assumption of decoupled horizontal and vertical displacement fields simplifies the analysis. Only vertical displacements generated by the load components in Fig. 5 are calculated. The horizontal displacements (Gaul [9]) generated by the missing load components are superimposed. To bound the influence of shear stresses at the vertically moving interfaces

- smooth contact with vanishing shear stresses,
- welded contact with vanishing horizontal displacements

are compared. Semianalytical solutions of both boundary value problems lead to flexibility influence matrices. One complex frequency dependent matrix element $\bar{h}_{ij}$ relates the complex displacement $\bar{w}_{ij}$ in the middle of element i to the amplitude $(\bar{p}A)_j$ of the time harmonic force acting at element j.

Displacement superposition leads to

$$\bar{w}_i = \bar{h}_{ij}\,(\bar{p}A)_j \qquad (4)$$

written with the flexibility matrix [h]

$$\{w\} = [h]\,\{F\}$$

or with the inverse dynamic stiffness matrix $[K]^S$ of soil halfspace

$$\{F\} = [K]^H\{w\}.$$

For rigid bases the corresponding interface stress distribution is determined by requiring

- the displacement boundary conditions of the plane interface displacement fields to be fulfilled in the center of each element,
- the resultants of the interface stresses to be equivalent to the halfspace load resultants.

The interaction of soil with flexible base plates requires a plate discretization by finite elements compatible to the soil element discretization (Fig. 6).

The equations of motion of the discretized base plate, which is loaded by nodal external forces {P}, moments {T} and half-space reactions {F}, are partitioned with respect to the translational {w} and rotational {φ} degrees of freedom. Coupling of mass matrix [M] and viscoelastic stiffness matrix [K] of the base plate with soil is achieved by displacement compatibility at the plate nodes and soil element centers. Expressing the unknown displacements {w} by Eq. (4) avoids the inversion of the flexibility matrix [h] and leads to a linear set of complex equations

$$\left[\begin{array}{c|c} (-\omega^2[M_v]+[K_v])[h]+[E] & -\omega^2[M_{v\varphi}]+[K_{v\varphi}] \\ \hline (-\omega^2[M_{v\varphi}]^T+[K_{v\varphi}])\,[h] & -\omega^2[M_\varphi]+[K_\varphi] \end{array}\right] \left\{\begin{array}{c}\{F\}\\ \hline \{\varphi\}\end{array}\right\} = \left\{\begin{array}{c}\{P\}\\ \hline \{T\}\end{array}\right\} \qquad (5)$$

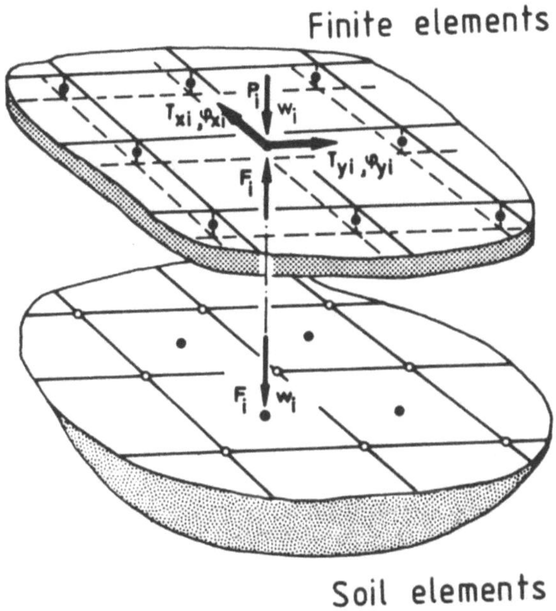

from which the halfspace reactions {F}, determining the soil pressure distribution, and the translational and rotational degrees of freedom {φ} and with Eq. (4) {w} follow. The results have to be interpreted as complex amplitudes for time harmonic excitations or as Fourier transforms of transient response.

Fig. 6 Base plate interacting with soil. Coupled finite elements and soil elements.

9.3.2 Flexibility matrix of soil halfspace, stress broundary value problems for smooth and welded contact.

The solutions of the stress boundary value problems of halfspace loaded vertically on one surface element (Fig. 5) are given semi-analytically by the Fourier integral for smooth and welded contact. Compared with elastic halfspace theories, a better approximation of the rheological properties of soil is given here by using visco-elastic constitutive equations. It turns out that the energy dissipation by viscoelastic damping is of considerable influence when the geometrical damping by wave radiation is of same order of magnitude. The equations of motion of the viscoelastic continuum in terms of displacements $u_i(x_j,t)$

$$\int_{-\infty}^{t} E_D(t-\tau)\,\frac{\partial u_{j,ji}}{\partial\tau}\,d\tau - e_{ijk}\,e_{k\ell m}\int_{-\infty}^{t} G(t-\tau)\,\frac{\partial u_{m,\ell j}}{\partial\tau}\,\delta\tau = \rho\,\frac{\partial^2 u_i}{\partial t^2} \qquad (6)$$

are decomposed by

$$u_i = \varphi,_i + e_{ijk}\,\psi_{k,j} \qquad (7)$$

in two wave equations for the scalar and the vector potential φ and ψ_k respectively

$$\int_{-\infty}^{t} E_D\, t-\tau\, \frac{\partial\varphi_{,\ell\ell}}{\partial\tau}d\tau = \rho\frac{\partial^2\varphi}{\partial t^2}$$
$$\int_{-\infty}^{t} G(t-\tau)\, \frac{\partial\psi_{k,\ell\ell}}{\partial\tau}d\tau = \rho\frac{\partial^2\psi_k}{\partial t^2} \quad . \tag{8}$$

This representation is complete (Gaul [11]) if the constraint condition $\psi_{k,k} = 0$ is satisfied. Steady state harmonic motion

$$u_i(x_j,t) = \mathrm{Re}\{\bar{u}_i(x_j)\exp(i\omega t)\} \tag{9}$$

leads to reduced wave equations for dilatation $\varepsilon = u_{\ell,\ell} = \varphi_{,\ell\ell}$ and rotation $2\omega_k = e_{kij}\, u_{j,i} = -\psi_{k,\ell\ell}$ with relaxation functions of plane $E_D(t)$ and shear $G(t)$ replaced by complex moduli

$$E_D^*(i\ ) = \lambda^*(i\omega) + 2G^*(i\omega) \text{ and } G^*(i\omega)$$
$$\bar{\varepsilon}_{,\ell\ell} + \frac{\rho\omega^2}{E_D^*}\bar{\varepsilon} = 0,\ \bar{\omega}_{k,\ell\ell} + \frac{\rho\omega^2}{G^*}\bar{\omega}_k = 0. \tag{10}$$

Excluding reflextions, solutions are given by

$$\bar{\varepsilon} = A\exp[-\alpha_D z + i(\beta x+\gamma y)],$$
$$\bar{\omega}_k = B_k\exp[-\alpha_S z + i(\beta x+\gamma y)], \tag{11}$$

where $\mathrm{Re}(\alpha_{D,S}) \geqq 0$, $\bar{\omega}_{k,k} = 0$. Displacement field and stress field are superimposed by these solutions

$$\bar{u}_i = -\bar{\varepsilon}_{,i} + 2e_{ijk}\,\bar{\omega}_{k,j}$$
$$\bar{\sigma}_{ij} = [E_D^*(i\omega) - 2G^*(i\omega)]\,\bar{u}_{,k,k}\,\delta_{ij} + G^*(i\omega)(\bar{u}_{i,j}+\bar{u}_{j,i}) \tag{12}$$

with A, B_k determined by the boundary conditions. The stress boundary value problem is solved by superposition of harmonic vertical displacement waves at the halfspace surface z = 0

$$w(x,y,0,t) = \bar{H}^{S,W}(\beta,\gamma,\omega)\,\bar{p}(\beta,\gamma)\exp.[i(\beta x+\gamma y+\omega t)] \tag{13}$$

generated by the stress wave

$$-\sigma_{zz}(x,y,0,t) = p(x,y,t) = \bar{p}(\beta,\gamma)\exp[i(\beta x+\gamma y+\omega t)] \tag{14}$$

propagating with phase velocity $v = \omega/\zeta$, where $\zeta = (\beta^2+\gamma^2)^{\frac{1}{2}}$. Real and imaginary part of the complex wave compliances $\bar{H}^{S,W}$ corresponding to smooth (s) and welded (w) contact are plotted in Fig. 7 versus the velocity ratio v/v_S.

The compliances show the features of a single-degree-of-freedom system. Welded contact leads to a resonant condition at the shear

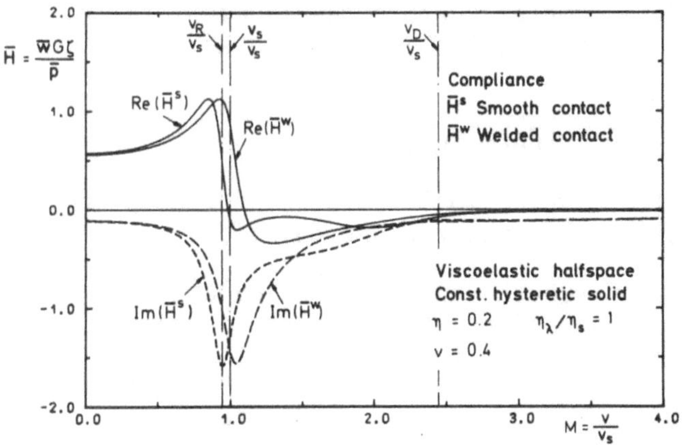

Fig. 7 Wave compliances of viscoelastic halfspace.

wave velocity $v_S = \sqrt{G/\rho}$, smooth contact leads to resonance associated with the slightly slower Rayleigh wave speed v_R. Fig. 8 shows an experimental set-up (Crandall et al. [4]) by which the damping factors $\eta_D(\omega)$, $\eta_S(\omega)$ of the complex moduli $E_D^* = E_D(1+i\eta_D)$, $G^* = G(1+i\eta_S)$ and Poisson's ratio can be measured. The damping factors are related by

$$\eta_D = \eta_\lambda + (\eta_S - \eta_\lambda)(1-2\nu)/(1-\nu),$$

with damping factor η_λ of Lamé modulus $\lambda^* = \lambda(1+i\eta_\lambda)$.

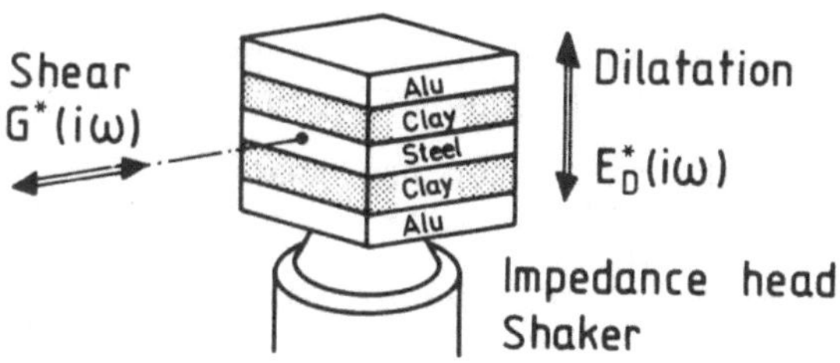

Fig. 8 Clay sandwich for alternating dilatation and shear tests.

Taking advantage of available measured data, results are presented for the constant hysteretic solid and Kelvin-Voigt-solid leading to damping factors $\eta_{S,\lambda}(\omega) \equiv \eta_{S,\lambda}$ and $\eta_{S,\lambda}(\omega) = a_0\, \xi_{S,\lambda}$ respectively, where $a_0 = \omega a/v_S$ is a frequency parameter with characteristic lenth a.

The obtained harmonic solutions are now superimposed by integrating with respect to the wave numbers β, γ in Fourier's integral theorem

$$w(x,y,0,t) = \frac{1}{2\pi} \iint_{-\infty}^{\infty} \bar{H}^{s,w}(\beta,\gamma,\omega)\bar{p}(\beta,\gamma) \times \exp[i(\beta x+\gamma y+\omega t)]d\beta\, d\gamma \quad (15)$$

with

$$\bar{p}(\beta,\gamma) = \frac{2p_j}{\pi}\left(\frac{\sin\beta a_j\ \sin\gamma b_j}{\beta\gamma}\right)$$

being the two-dimensional Fourier transform of the exciting stress field (Fig. 5) at one surface element of area $4a_j\, b_j$. The elastic halfspace leads to improper integrals due to poles of the compliance in Eq. (15) at the shear wave and the Rayleigh wave speed. Solutions can be obtained by choosing Cauchy's principal value and performing a contour integration in the complex plane. Here a different technique is used. Because the viscoelastic halfspace yields finite resonance amplifications instead of poles (Fig. 7), the integral of Eq. (15) is no longer improper with respect to the integrand and can be integrated directly.

By pointwise evaluation of the complex surface displacement field, the soil flexibility matrix [h] of Eq. (4) is obtained with elements

$$h_{ij} = \bar{w}(x_i, y_i)/[\bar{p}A(x_j, y_j)].$$

9.3.3 Flexibility matrix of excavated soil halfspace for coupling with embedded base plates by substructure deletion.

The dynamic stiffness matrix of substructure soil calculated analytically by solving the field equations of the halfspace can be applied not only for describing surface foundations but also for embedded foundations. The effect of embedment can be taken into account by substructure deletion utilizing the available continuum and discrete solution techniques.

Finite element discretizations of soil require to introduce artificial boundaries at some distance from the base plate although the foundation medium is geometrically unbounded. When applying the boundary element method, surface discretization too has to be truncated at some distance from the base implying a discretization error.

These difficulties are circumvented in the present approach by employing the substructure deletion technique (Fig. 9).

The dynamic stiffness matrix of the excavated halfspace $[\overset{H}{K}_e]$ is calculated from the known dynamic stiffness matrix of the halfspace obtained by the continuum approach $[\overset{H}{K}]$ and the dynamic stiffness matrix $[\overset{E}{K}]$ of the excavated domain.

Finite element discretization of the excavated domain (Dasgupta [6]) requires to condense the internal nodal degrees of freedom out and leads to $[\overset{E}{K}] = [K] - \omega^2[M]$ by static condensation.

The author applies surface discretization by the boundary element method, where condensation drops out.

The boundary nodes are divided in those at the ground level surface $\{\overset{E}{u}_s\}$ and those at the excavation surface $\{\overset{E}{u}_e\}$. Interface conditions at the ground level surface require identity of nodal forces and displacements at the halfspace surface and excavated domain surface:

$$\{\overset{H}{F}_s\} = \{\overset{E}{F}_s\}, \ \{\overset{H}{u}_s\} = \{\overset{E}{u}_s\}.$$

At the excavated surface equilibrium of nodal forces $\{\overset{E}{F}_e\} + \{\overset{H}{F}_e\} = \{0\}$ and displacement compatibility $\{\overset{E}{u}_e\} = \{\overset{H}{u}_e\}$ has to be fulfilled.

As simultaneous prescription of nodal forces and displacements on the common boundary points is required, Dasgupta [5] demonstrated that well posedness can be guaranteed if and only if the discrete models can reproduce those results which are the counterparts of Almansi's triviality theorem (Almansi [3]).

These requirements and the known dynamic stiffness matrices of halfspace and excavated domain lead to the dynamic stiffness matrix of the excavated $[\overset{H}{K}_e]$ (Fig. 9). Interaction with an embedded base plate can be calculated by coupling the dynamic stiffness matrices.

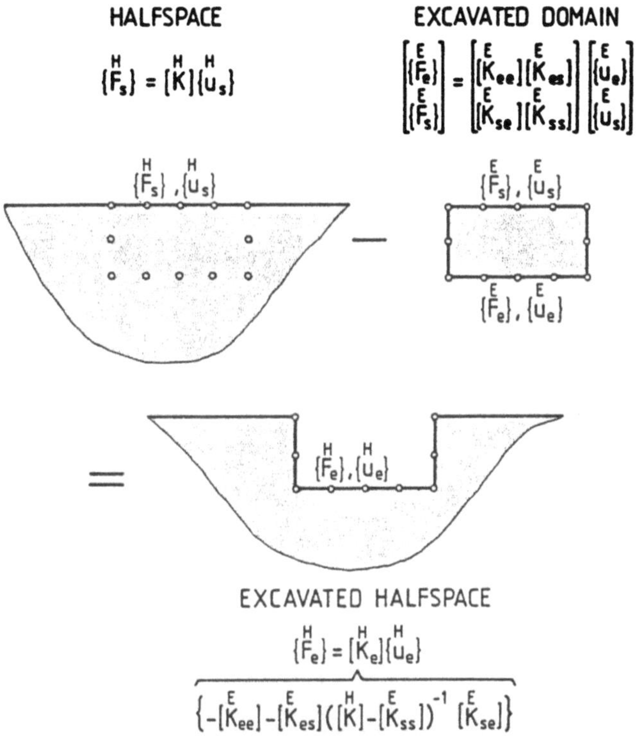

Fig. 9 Flexibility of excavated soil halfspace by substructure deletion.

9.4 Substructure Frame Foundation and Shaft.

As an example of a sensitive structure interacting with soil a frame foundation (Fig. 3) is considered. Only the global vibration behaviour in the low frequency range is treated on the bases of a simplified model (Fig. 9) which is suited for comparisons with experimental results from small scale frame foundations.

The rigid upper plate is excited harmonically by the force F_i and torque T_i^d generated e.g. by the unbalances of a Laval rotor with excentricity e. Upper plate and base plate are connected by flexible columns, rigid bearings and rotating disk are connected by the flexible shaft. The halfspace reactions are reduced to point B in the interface. The three-dimensional motion of upper plate, base plate and disk are described by displacement coordinates v_i, u_i, w_i and the angles of small rotations ϕ_i, φ_i, α_i respectively.

The geometrically linearized Newton-Euler equations of motion yield with symbols, coordinates and constraints from Fig. 10

- for the upper plate with mass M, inertia tensor I^D_{ij}

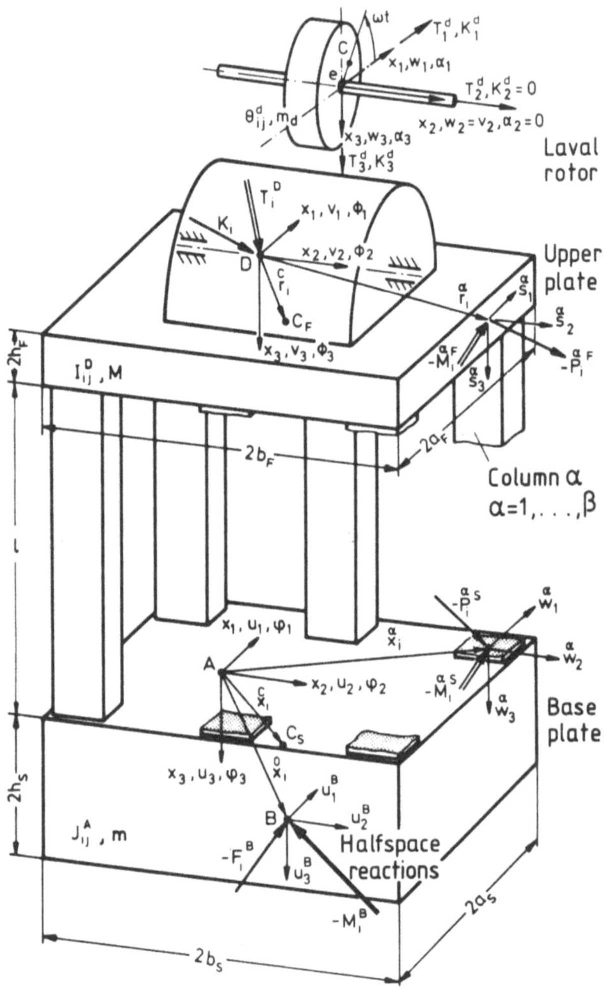

Fig. 10 Rotor on frame foundation interacting with soil.

$$I^D_{ij}\,\ddot{\phi}_j + M\,e_{ijk}\,\overset{c}{r}_j\,\ddot{v}_k = -\sum_{\alpha=1}^{\beta}\left(e_{ijk}\,\overset{\alpha}{r}_j\,P_k^{F} + \overset{\alpha}{M}_i^{F}\right) + T_i^{D}$$

$$M\left(\ddot{v}_i - e_{ijk}\,\overset{c}{r}_j\,\ddot{\phi}_k\right) = -\sum_{\alpha=1}^{\beta}\overset{\alpha}{P}_i^{F} + K_i^{D} \qquad (16)$$

- for the base plate with mass m, inertia tensor J^A_{ij}

$$J^{A}_{ij}\,\ddot{\varphi}_j + m\,e_{ijk}\,\overset{c}{x}_j\,\ddot{u}_k = -\sum_{\alpha=1}^{\beta}\left(e_{ijk}\,\overset{\alpha}{x}_j\,\overset{\alpha}{P}_k{}^{S} + \overset{\alpha}{M}{}_i^{S}\right) - \left(M_i{}^{B} + e_{ijk}\,\overset{o}{x}_j\,F_k{}^{B}\right)$$

$$m\left(\ddot{u}_i - e_{ijk}\,\overset{c}{x}_j\,\ddot{\varphi}_k\right) = -\sum_{\alpha=1}^{\beta}\overset{\alpha}{P}_i{}^{S} - F_i{}^{B}\,, \qquad (17)$$

- for the rotating disk with mass m_d , inertia tensor Θ^{d}_{ij}, angular velocity ω

$$\Theta^{d}_{ij}\,\ddot{\alpha}_j + \Theta^{d}_{11}\,\omega(\delta_{i1}\,\dot{\alpha}_1 - \delta_{i3}\,\dot{\alpha}_3) = -\,T_i{}^{d}$$

$$m_d\,\ddot{w}_i = -\,K^{d}_i + m_d\,e\,\omega^2(\cos\omega t\;\delta_{1i} - \sin\omega t\;\delta_{3i}). \qquad (18)$$

Flexible columns and flexible shaft are treated as space beams with distributed mass. The effects of shear, rotary inertia, static axial force and viscoelastic material properties are considered in dynamic, complex valued stiffness matrices (Aboul-Ella, Novak [2], Gaul [11]).

- column α

$$\begin{Bmatrix} M_i{}^{F} \\ P_i{}^{F} \\ M_i{}^{S} \\ P_i{}^{S} \end{Bmatrix}^{\alpha} = [K_{i\ell}]^{\alpha} \begin{Bmatrix} \phi_\ell \\ v_\ell - e_{\ell mn}\,\overset{\alpha}{r}_m\,\phi_n \\ \varphi_\ell \\ u_\ell - e_{\ell mn}\,\overset{\alpha}{x}_m\,\varphi_n \end{Bmatrix}\,, \qquad (19)$$

- shaft

$$\begin{Bmatrix} M_i{}^{d} \\ K_i{}^{d} \\ M_i{}^{D} \\ K_i{}^{D} \end{Bmatrix} = [K_{i\ell}]^{d} \begin{Bmatrix} \alpha_\ell \\ w_\ell \\ \phi_\ell \\ v_\ell \end{Bmatrix}\,. \qquad (20)$$

In the frequency domain Eqs. (16) to (20) lead to a substructure equation like Eq. (1)

$$[K(i\omega)]\begin{bmatrix} \{U\} \\ \{U_c\} \end{bmatrix} = \begin{bmatrix} \{P\} \\ -\{F_c\} \end{bmatrix}. \qquad (21)$$

9.5 Calculated Results of Soil-Structure Interaction

9.5.1 Lumped parameters of substructure soil.

The solution of the mixed boundary value problem, describing the interaction between one rigid base and soil leads to the complex elements of the soil stiffness matrix (Eq. 2) $K^S_{ij} = c_{ij}(a_0) + i\, a_0\, d_{ij}(a_0)$, which can be modelled as lumped parameters of soil. The spring and damping coefficients $c_z(a_0)$ and $d_z(a_0)$ corresponding to vertical vibration of rigid square base are plotted in Fig. 10 versus the frequency parameter $a_0 = \omega a/v_s$. The spring coefficient, describing elastic restoring forces and inertia forces, is slightly higher in the low frequency range for welded contact than for smooth contact. This is due to the displacement constraint at the halfspace surface. The same reason causes a higher geometrical damping primarily associated with the Rayleigh wave for smooth contact. Thus the damping coefficient d^S_z exceeds d^W_z. In the present paper all lumped parameters additionally depend on the energy dissipation of soil governed by viscoelasticity. One important result of the analysis with respect to the uncertainties of the contact boundary conditions is that it makes little difference whether the contact at the interface is smooth or welded.

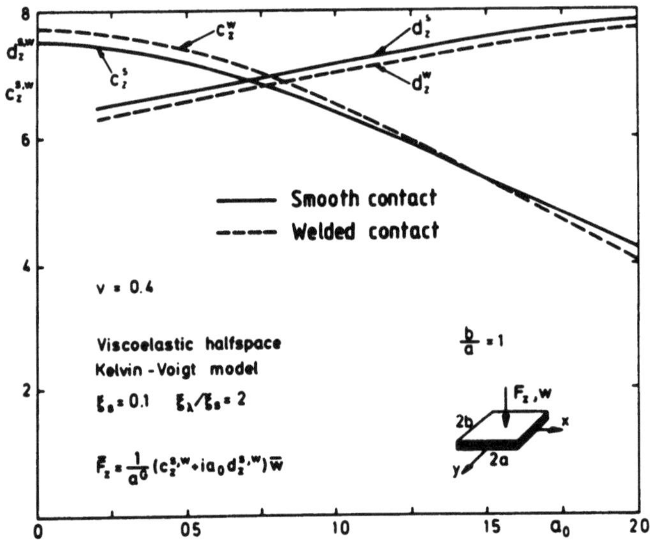

Fig. 11 Spring and damping coefficient of soil for vertical motion of a rigid base.

9.5.2 Interaction between two structures.

Real and imaginary parts of the complex interface pressure distribution are given in Fig. 12 corresponding to the interaction through the underlying soil between two rigid structures, which are excited by forces P^I_z, P^{II}_z and torques T^I_x, T^{II}_y due to rotating unbalanced masses, acting with a phase shift.

9.5.3 Dynamic response of machine foundations.

The dynamic response of a frame foundation (Fig. 3) with eight concrete columns is evaluated as an application of the presented substructure technique. The system is excited by an unbalanced rigid rotor. The magnification functions in Fig. 12 describe the amplitudes of horizontal displacements v_1, u_1 and vertical displacements v_3, u_3. According to Fig. 9 v_i belongs to the upper plate and u_i to the base plate. The coupled rocking and horizontal motion gives rise to two resonant amplifications indicated by the horizontal displacement amplitudes within the regarded frequency range. The resonant amplifications are affected by material damping of soil because rocking motion causes only small geometrical damping by wave radiation (Gaul [11]). This indicates vertical motion being associated with strong geometrical damping. Frequency independent static lumped parameters of soil lead to the compared deviations of response. The vibration modes in Fig. 14 show the coupling between a rocking and sliding motion.

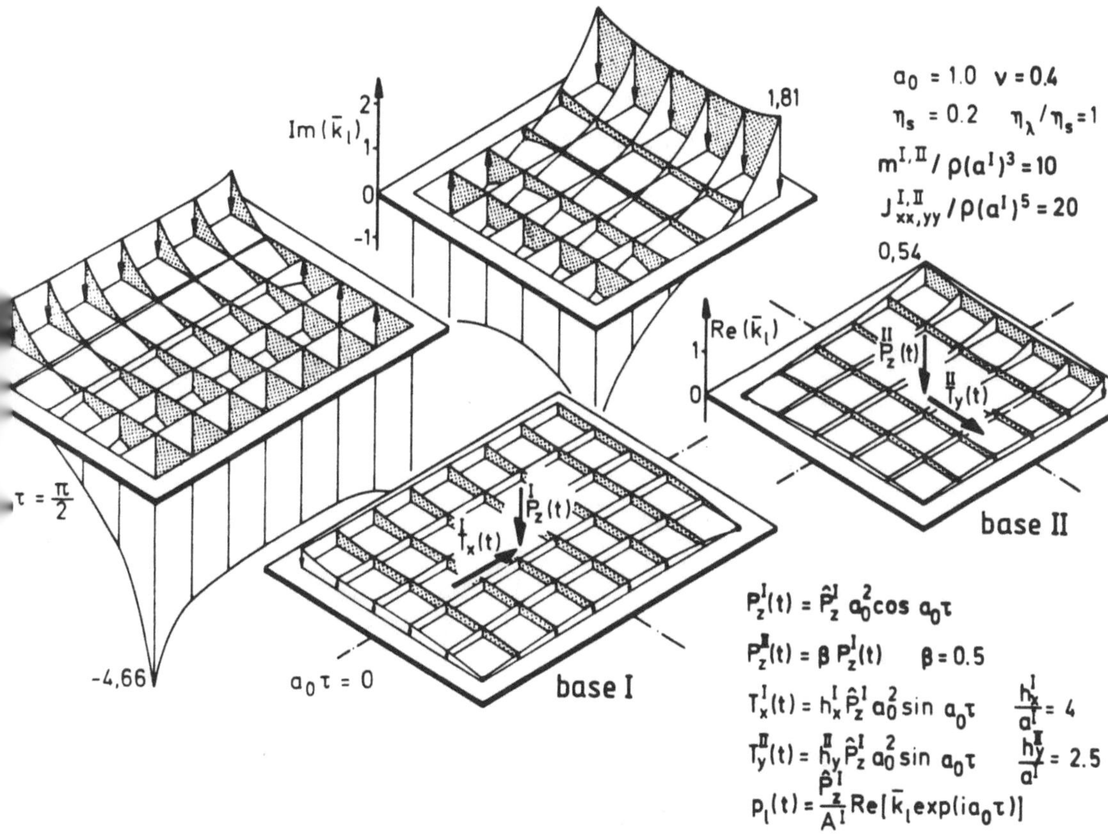

Fig. 12 Interface pressure distribution for adjacent excited bases.

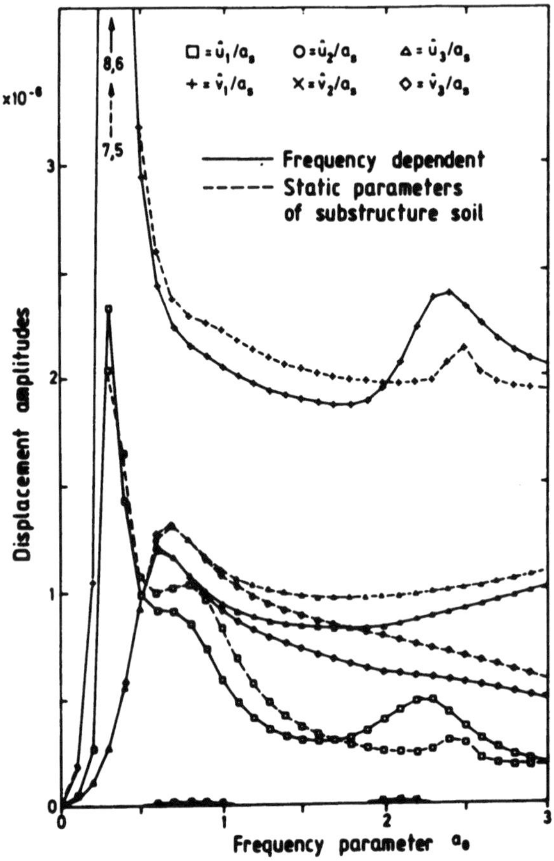

Fig. 13 Response of frame foundation on soil.

Rigid columns of the frame foundation simplify Eqs. (16) to (21) and lead to the description of a Laval rotor on a block foundation (Gasch and Sarfeld [8], Krämer [18]). If only a plane motion is considered, the system has 5 degrees of freedom. Fig. 15 compares the response of the rotor on a rigid foundation with the response corresponding to a flexible foundation.

Analogous to the frame foundation (Fig. 13) the first three resonant amplifications are predominately due to the foundation, while the last two are governed by the rotor. The foundation influence splits one resonant frequency of the rotor in two of the combined structure with lower amplitudes.

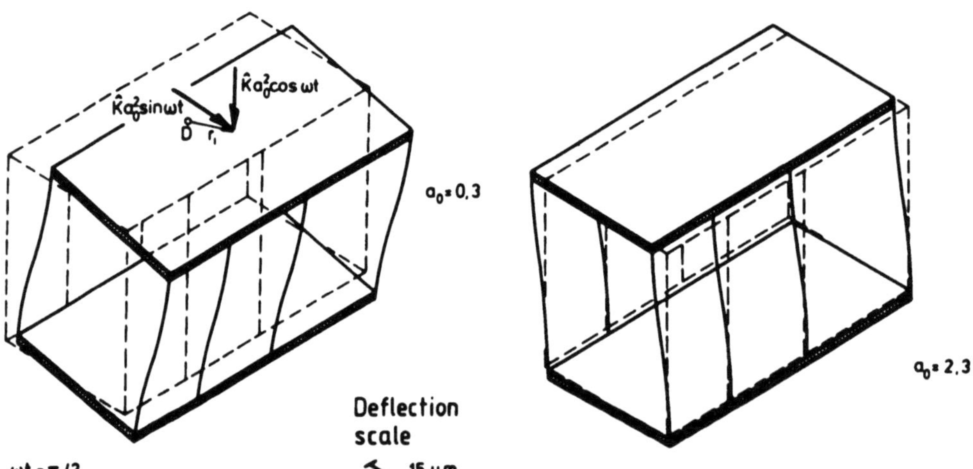

Fig. 14 Vibration modes of frame foundation on soil.

9.5.4 Turbomachinery frame foundation supported by piles.

Aboul-Ella and Novak [1] analyzed the dynamic response of turbomachinery frame foundations supported by piles or a foundation slab. Their study investigates interaction of all components of the system, i.e. flexible rotors, viscoelastic oil film, space frame, flexible mat, piles and soil (Fig. 16). The mat is composed of rectangular finite plate elements. The pile and soil resistance is included into mat element stiffness matrix. The dynamic complex soil stiffness matrix is obtained from Gaul [9].

In the study of Aboul-Ella and Novak [1] special attention is paid to the effects of soil structure interaction. It was found that this interaction markedly affects the response of the frame as well as the rotors in the lowest resonant regions. The interaction reduces rotor and frame amplitudes. This results from the increase in damping due to energy radiation in the soil and viscoelastic behaviour of soil and mat. The interaction reduces the frame vibration more than shaft vibration.

E.g. Fig. 17 compares the vertical response of frame under bearing pedestal corresponding to a rotor on elastic frame and rigid foundation with a rotor on elastic frame and elastic foundation.

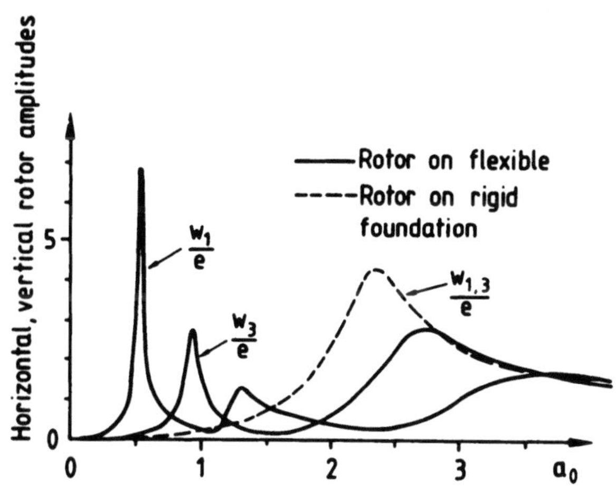

Fig. 15 Response of Laval rotor on rigid and flexible block foundation.

9.6 Experimental Investigation of Soil-Structure Interaction.

9.6.1 Measured response of a model frame foundation.

The response of a frame foundation according to the model of Fig. 10 was simulated by a small scale model (Fig. 18). Four rotating unbalances driven via a control gear allow for coupled and uncoupled excitation by horizontal and vertical forces and by torsion and rocking moments. The stiffness of coupling between upper plate and base plate can be varied by interchangeable columns. Rubber springs simulate the soil.

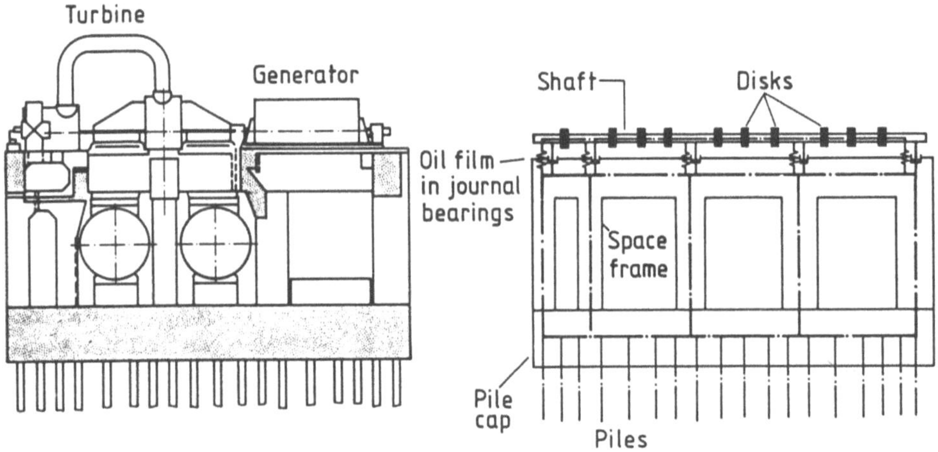

Fig. 16 Turbomachinery frame foundation and its model.

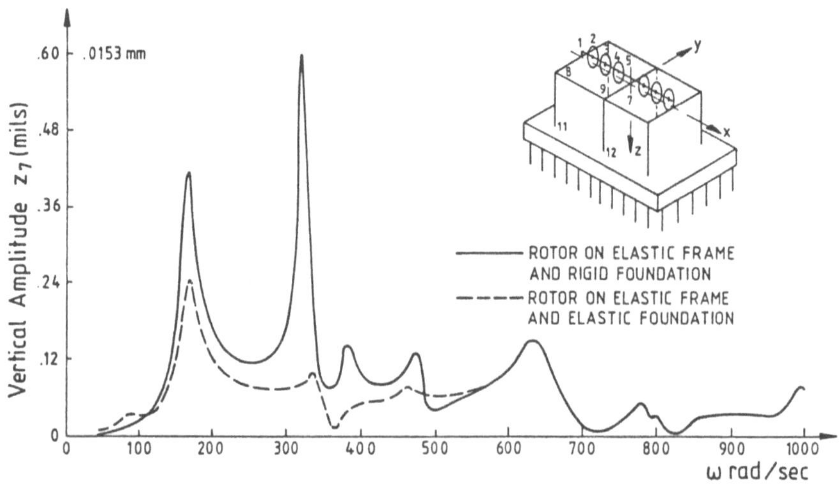

Fig. 17 Effect of foundation (piles and soil) flexibility on vertical response of frame under bearing pedestal.

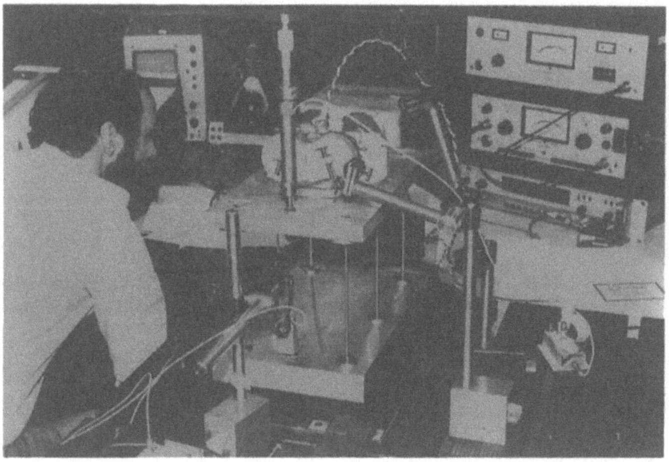

The sine sweep response of 12 degrees of freedom where measured by velocity pick-ups. The results are in good agreement with those calculated by Eqs. (16) to (21) (Gaul, Mahrenholtz [13]). The coupled horizontal and rocking modes in Fig. 19 correspond to the calculated modes in Fig. 14.

Fig. 18 Small scale model of frame foundation with measuring set-up.

9.6.2 Laval shaft on a model frame foundation.

For simulating the interaction between rotating shaft, frame foundation and subsoil on a model scale, the lab model of Fig. 20 has been built. A laval shaft with adjustable disc position and excentricity driven by a variable speed motor is supported by two ballbearings on the upper foundation plate.

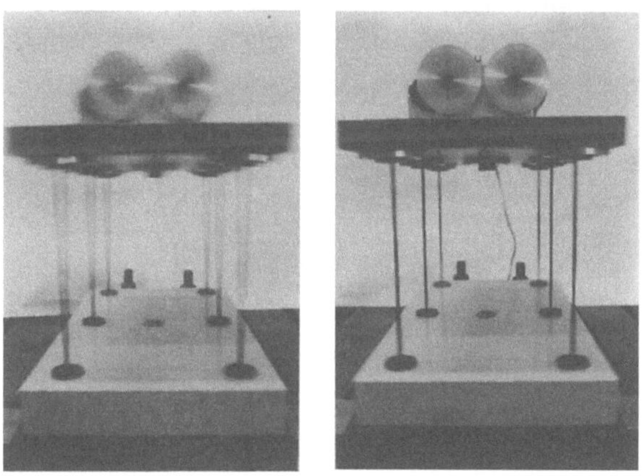

The experimental work in progress simulates the interaction effects of the three substructures where the base plate is bedded on rubber springs, on a foam layer or on a model sand foundation which is explained in the next chapter.

Fig. 19 Vibration modes of frame foundation on rubber springs.

Fig. 20 Lab model for measuring interaction effects between rotating shaft, frame foundation and subsoil.

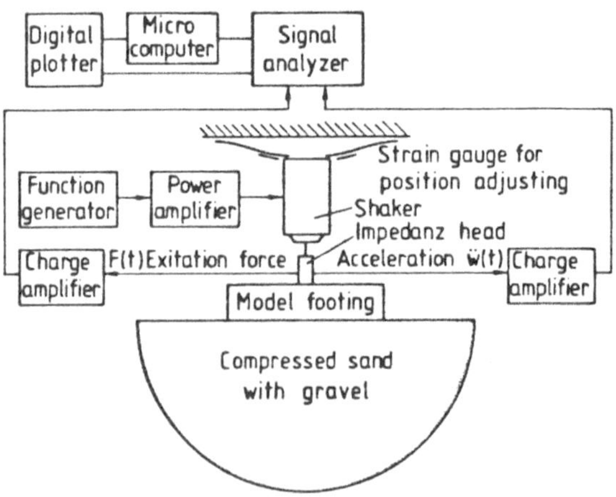

Fig. 21 Experimental set-up of shaker-driven model footing.

9.6.3 Steady-state vibrations of model footings.

The substructure behaviour of soil was measured on a model scale by shaker-driven footings at the surface of homogeneous or layered sand mixed with gravel (Fig. 21). The response of acceleration and phase angle versus frequency of the sine sweep (Fig. 22) as well as the deduced frequency dependent lumped parameters of soil are found in satisfactory agreement with calculated results (Gaul, Mahrenholtz [13]).

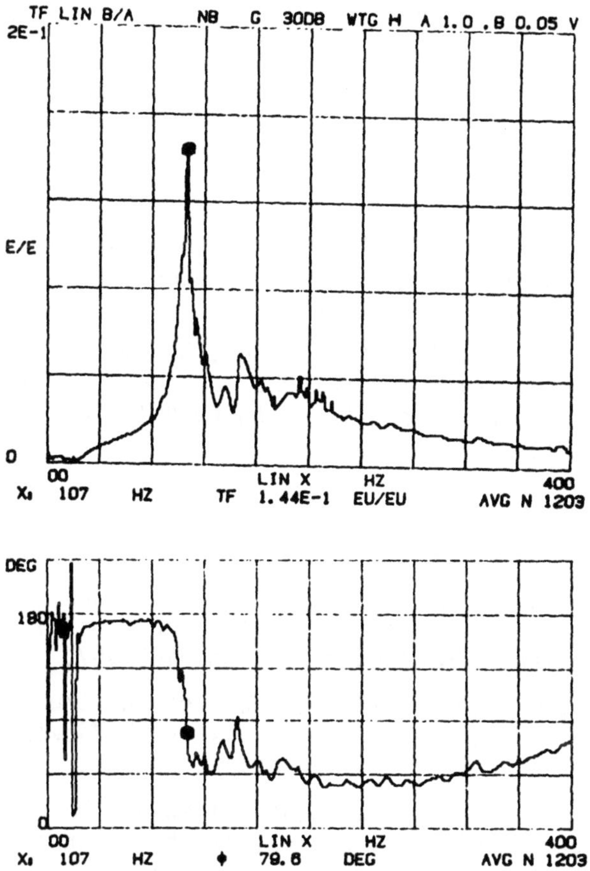

Fig. 22 Sine sweep response of vertical vibration for a circular model footing.

9.7 Summary and Conclusions.

On the basis of substructuring a theoretical approach has been formulated and programmed to analyze the three-dimensional dynamic response of machine foundations considering the interaction of the system components; namely, viscoelastic soil, frame and rotor, as well as the interaction through the underlying soil with an adjacent structure. Geometrical as well as material damping of soil are considered. Material damping is found to be of considerable influence for rocking motion. The influence of shear stresses at the interface between base and soil is limited. It makes little difference whether the contact at the interface is smooth or welded. The dynamic response of a model frame foundation and a model footing on compressed sand are measured. Both, theoretical approach and experiments, provide a good understanding of the basic interaction effects.

9.8 Acknowledgements

The experimental research with model footings performed by Dipl.-Ing. M. Plenge, University of Hanover, is gratefully acknowledged.

This study was supported by the Deutsche Forschungsgemeinschaft (German research council).

9.9 References

[1] Aboul-Ella, F.A. & M. Novak, 1978. Dynamic analysis of turbine-generator foundations. Presented at the 1978 Fall Convention, Housten, Tex., Oct.

[2] Aboul-Ella, F.A. & M. Novak, 1980. Dynamic response of pile-supported frame foundations. Journal of the Engineering Mechanics Division, Proc. ASCE, Vol. 106, No. EM6, Dec., pp. 1215-1232.

[3] Almansi, E., 1907. Un teorema Sulle Deformazioni Elastiche dei Solidi Isotropi, Atti della reale accademia dei nazionale Lincei, Vol. 16, pp. 865-868.

[4] Crandall, S.H., Kurzweil, L.G. & A.K. Nigam, 1971. On the measurement of Poisson's ratio for modelling clay. Experimental Mechanics, 11, pp. 402-407.

[5] Dasgupta, G., 1979. Wellposedness of Substructure Deletion Formulations. Proceedings, Sixteenth Midwestern Mechanics Conference, Vol. 10, Manhattan, Kansas.

[6] Dasgupta, G., 1980. Foundation Impedance Matrices by Substructure Deletion. Journal of the Engineering Mechanics Division, American Society of Civil Engineers, Vol. 106, No. EM3, pp. 517-523.

[7] Dietz, H., 1972. Stahlfundamente für Turbomaschinen. Beratungsstelle für Stahlverwendung, Düsseldorf, Merkblatt 146,3.

[8] Gasch, R & W. Sarfeld, 1980. Unwuchterzwungene Schwingungen des Systems Lavalläufer-Blockfundament - elastischer Halbraum, VDI-Berichte Nr. 381, pp. 129-138.

[9] Gaul, L., 1977. Dynamische Wechselwirkung eines Fundamentes mit dem viskoelastischen Halbraum. Ing.-Archiv, 46, pp. 401-422.

[10] Gaul, L., 1979. Dynamisches Halbraumverhalten infolge eines schuberregten Fundamentes. ZAMM 59, pp. 180-183.

[11] Gaul, L., 1980. Zur Dynamik der Wechselwirkung von Strukturen mit dem Baugrund. Habil. Univ. Hannover, June.

[12] Gaul, L., 1980a. Dynamics of frame foundations interacting with soil. Journal of Mechanical Design, Vol. 102, pp. 303-310.

[13] Gaul, L. & O. Mahrenholtz, 1981, 1984. Dynamische Wechselwirkung zwischen Maschine, Fundament und Baugrund. Arbeitsbericht zum DFG-Schwerpunktprogramm Betriebsverhalten dynamisch belasteter Maschinen (not published).

[14] Gaul, L. & M. Plenge, 1983. Ein Baugrundmodell für geschichtete Baugründe mit viskoelastischem Stoffverhalten. ZAMM 63, pp. T50 - T53.

[15] Holzlöhner, U., 1969. Schwingungen des elastischen Halbraumes bei Erregung auf einer Rechteckfläche. Ing.-Archiv, 18, pp. 370-379.

[16] Huh, Y., Schmid, G. & M. Ottenstreuer, 1983. Evaluation of kinematic interaction of soil-foundation systems by boundary element method. SMIRT 7 Chicago, August, Paper K814.

[17] Knobloch, W. & Gaul, 1975. Dämpfungs- und Federverhalten elastischer und viskoelastischer Gründungen bei harmonischer Erregung. Kolloq. Viskoelastische Systeme, VDI-GKE, TU Berlin, pp. 412-451.

[18] Krämer, E., 1984. Maschinendynamik. Springer Verlag, Berlin.

[19] Novak, M., 1982. Response of hammer foundations. Proc. soil dynamics and earthquake engineering conference Southampton, July, pp. 783-797.

[20] Ottenstreuer, M., 1982. Frequency dependent dynamic response of footings. Proc. soil dynamics and earthquake engineering conference, Southampton, July, pp. 799-809.

[21] Sarfeld, W. & C. Fröhlich, 1980. Dynamische Wechselwirkung von Gebäuden und Fundamenten auf dem elastisch-isotropen Halbraum, Bauingenieur 55, pp. 419-426.

[22] Thurat, B., 1978. Machine-Fundament-Baugrund. Diss. RWTH-Aachen.

[23] Waas, G., 1972. Linear two-dimensional analysis of soil dynamics problems in semi-infinite layered media. Ph.D. Thesis, University of California, Berkeley.

[24] Dominguez, J., 1978. Dynamic stiffness of rectangular foundations. Dep. of Civ. Eng. MIT R78-20.

CHAPTER 10.1

PROBLEMS OF TURBINE GENERATOR SHAFT DYNAMICS

D.W. King*, N.F. Rieger**
*Rochester Gas and Electric Corporation, Rochester New York, USA
**Stress Technology Incorporated, Rochester, New York, USA

ABSTRACT

The nature of faults and disturbances which can occur in an electrical power system are reviewed. The sequence of events to which a turbine-generator unit may be exposed from such events is described. Procedures for unit response analysis are mentioned, and current problems in predicting such response are discussed. The need for on-line shaft monitoring and cumulative damage analysis is mentioned. Several types of continuous monitoring and measurement equipment now in use on turbine-generators are described, with details of their performance. Conclusions are presented concerning the state-of-the-art in this area.

10.1.1 Introduction

Electrical faults and system disturbances can create severe transient torques in turbine generator (TG) shafts. TG shaft failures have been attributed to such transient torques in 1970 and 1971 by Jackson and Umans [1]. The IEEE Screening Guide for Planned Steady State Switching Operations to Minimize Harmful Effects on Steam Turbine Generators [2] indicates that the following electrical disturbances may create damaging transient torques in TG shafts.

a) Transmission line switching.

b) High speed reclosing of circuit breakers following fault-clearing on lines leaving power stations.

c) Single phase operation, such as caused by single pole operation of circuit breakers. This produces alternating torques at twice line frequency. While this frequency is generally above those of the lower modes of the TG set, there are many complex higher modes of vibration which involve internal deflections of the generator and turbine rotors. Further, individual discs and blade groups can respond to such pulsations.

d) Sub-synchronous resonance (SSR) in series-capacitor compensated systems.

e) Out-of-phase synchronization.

f) Line faults reaching the generator terminals.

g) Full load trips.

It is known that when series capacitors are used in transmission lines situated electrically close to the generator, steady state and transient currents may be generated at frequencies below the normal power system frequency. The existence of such currents causes alternating torques in the generator which can excite the natural vibration modes of the shaft, and cause significant dynamic torques at or near the shaft couplings.

From numerical calculations of such conditions, it has been determined that the above mentioned fault conditions typically occur in sequence:

The first impact on the generator might be a sudden short circuit in a transmission line close to the generator, which in electrical engineering is frequently called fault application.

The second impact results from electrical disconnection of the faulty transmission line, which is called fault clearing or fault removal. Usually this is all that is needed to extinguish the electrical arc, which is initiated by the fault, e.g., lightning, or by short circuits on the power line.

After a brief disconnection, the third impact is then applied when the line is electrically re-connected. This is also done automatically. The technical term for this operation is high speed re-closing.

High speed re-closing can be successful if the fault is effectively cleared by the disconnection of the line, or unsuccessful if the fault persists after re-closing. In this latter case a fourth impact is applied by a second clearing attempt.

These shocks are applied consecutively. TG shaft oscillations result from the initial shock, and the unit is re-shocked from the second, and then the third impacts. Depending strongly on the relative phasing of these shocks, the resulting dynamic strains can result in very high total dynamic shaft torques. These torques can be much higher than in the terminal short circuit case, which had been considered the worst design case until just a few years ago.

An evaluation of the effect of an electrical disturbance on the amount of fatigue damage per incident on the TG shaft was presented by Joyce and Lambrecht [3], and is shown here in Table 1. This chart was based on a large number of computer calculations for the fatigue damage incurred per transient on a four-flow 3600 rpm turbine with a 926 Mva two pole generator, and for a six-flow 1800 rpm turbine with a 1525 Mva four pole generator. These transients resulted from the successful and unsuccessful reclosure of system faults. The black sections of the chart represent the results of studies conducted on 45 machines of various design and sizes. From calculations it appears that the amount of TG shaft fatigue damage resulting from an electrical disturbance can vary by more than an order of magnitude. This is in part due to a lack of knowledge in certain critical areas of TG shaft vibrations and stress analysis. Some of these areas are:

a) Data on electrical disturbances with regard to their duration and transient amplitudes.

b) The effect of system size, impedance, load, and climate conditions.

c) The conversion of an electrical line transient to a mechanical torque transient within the generator is not satisfactorily understood.

d) The effect of turbine and generator construction, of blade-shaft interaction, and of modal damping on torsional vibration requires further clarification.

It is the purpose of this chapter to explore some of these areas, assessing what has been done in the past and what appears to be needed for the future.

10.1.2 Notation

C	Damping Coefficient
Hz	Frequency, Cycles/Second
I	Inertia
K	Spring Constant
T(t)	Time Varying Torque
θ	Angle of Twist
$\dot{\theta}$	Rate of Angle of Twist Angular Velocity
$\ddot{\theta}$	Angular Acceleration
ω_{shaft}	Angular Velocity

10.1.3 TG Shaft Vibration Analysis

In order to assess the fatigue damage incurred during a given transient, the resulting stress history within the shaft system must be determined from the imposed transient torques. Such transient torques can produce stress in two ways. The first is through an abrupt change in torque level which leads to torque magnification followed by a slow decay in stress amplitude which is inversely proportional to the damping magnitude. In Figure 1, M_m is the mechanical torque imposed on the TG shaft while M_e is the mechanical torque imposed on the TG shaft while M_e is the measured air gap torque imposed on the generator as a result of an electrical disturbance. The second manner in which transient stress is produced in a TG shaft is by sub-synchronous resonance (SSR). In this case the ratio of M_m to M_e may reach as high as 15 in the lower modes of the TG shaft system. SSR is typically of greater concern than step changes in torques, because of the greater rate at which significant damage can be acquired by the shaft.

TG shaft natural frequencies may be computed using a mass-elastic model of the shaft system. A simple example of such a model is shown in Figure 2. Actual analytical models commonly contain 200-500 inertia stations, and include several rows of LP blades for each rotor flow direction. Under transient loading conditions, the matrix equation of motion for such a system is:

$$[I]\ \{\ddot{\theta}\} + [C]\ \{\dot{\theta}\} + [K]\ \{\theta\} = [T(t)] \tag{1}$$

where the notation used is specified above. By integrating this equation using a time-marching technique in a finite element code, such as ANSYS or NASTRAN, the time-varying response of such a system may be readily obtained. Such programming may also be used to obtain mode shapes for the different natural frequencies of the TG system. These mode shapes are normalized displacement amplitudes (degree of twist in the shaft) and are helpful in predicting those shaft sections which are most susceptible to fatigue damage under transient conditions, or SSR. In order to realistically determine the TG shaft stresses by finite element calculations, information on the following must be available:

a) Transient time history of the disturbance reaching the generator rotor,

b) Transient load distribution within the generator,

c) TG shaft material strength,

d) Distribution of mechanical load on the TG shaft,

e) Shaft damping properties (magnitude, distribution for each mode).

Further information is needed in each of these areas, and this is of concern in the evaluation of stresses due to generator transients and SSR.

10.1.4 Continuous Monitoring and Measurement Equipment

In order to acquire data such as the time-variation and transient amplitudes during an electrical disturbance, continuous monitoring equipment has been developed. Continuous Monitoring Systems (CMS) are designed to start recording both the electrical disturbance and the mechanical transient torques when a threshold electrical disturbance is detected. Presently there is no common threshold value due to differences in the transmission line makeup and machine variances. Once triggered, the CMS records the electrical disturbance usually by measuring the three phase current and voltage at the generator terminals. In certain systems, generator currents are measured by use of a so-called Rogowski coil, as illustrated in Figure 3. Voltage is measured with potential transformers. The current and voltage values are then transmitted to a recording device and stored on magnetic tape (analog signals). The product of such voltages and torques gives instantaneous values for the generator air gap torque. In addition, devices such as shaft mounted toothed wheels with accompanying magnetic pickups are used to measure small variations in the shaft rotational speed. These are the two primary measurements that are used on TG systems to acquire basic data on the amount of fatigue incurred per transient incident.

Two such CMS units currently in use are the Torsional Stress Analyzer (TSA) developed by Kraftwerk Union AG, as described in reference [7], and the Torsional Vibration Monitor System (TVMS) developed at General Electric Large Steam Turbine

Generator Division, as described in reference [8]. Block diagrams for each unit are shown in Figure 4 and Figure 5. The first TSA was implemented in September 1977 (no location given) and several are now in service. The TSA uses the Rogowski coil to measure three phase current. An ironless core is used to eliminate inaccuracies due to core saturation in DC offsetting of fault currents. Current values are combined with three phase voltages by an analog computer to produce an instantaneous value for the generator air gap torque. Shaft angular velocity variations are measured as described previously, and these values along with the air gap torque are used as input signals to an analog shaft torsional model. Figure 6 shows this analog model. The turbine torques and electrical torques along with the shaft angular velocity enter the computer, are transformed into analog work, and are then transformed back into shaft torque exerted by all contributing modes. Measured shaft angular velocity is used as input to compensate for non-linear effects that machine damping could cause. The shaft torque figure derived from the model is used along with the properties of the shaft section to determine fatigue damage by use of the Rainflow Cycle Counting method. This method uses closed loop stress-strain cycles, which are determined from the succession of torque maxima and minima to obtain a cumulative figure for the fatigue damage resulting from each transient event. A typical stress-strain hysteresis damage loop is shown in Figure 7.

Sample results from the TSA are shown in Figure 8 for a TG set with four couplings, where the shaft motions are monitored. A code, shown in Figure 9, is used to describe the severity of the incident and recommended action, if any. The second column shows a figure for the shaft life expended in the incident, based on a 100% shaft life expenditure at first crack initiation. The third column shows the cumulative fatigue incurred since installation of the TSA. The accuracy of this model is evaluated in Figure 10 for a 970 Mw, 3000 rpm, six-flow TG undergoing a sudden three-phase short circuit (A), and for partial load rejection (B). The shaft torque between the last low pressure turbine and the generator is compared using the figures from the TSA and figures obtained throughout the application of strain gages to the subject shaft area. The results appear to endorse the TSA fatigue life model.

The GE TVMS was first installed in early 1981. The TVMS uses both air gap torque and angular velocity for its inputs and produces figures for fatigue life expended in much the same way as the TSA. The novel concept of the TVMS is the use of Data Acquisition Systems (DAS) which sends in information from the TG unit to a central computer to be analyzed. This permits centralized monitoring of many remote TG units, and tends to minimize the need for specialized test personnel at the TG locations.

10.1.5 Current Problem Areas

Due to the complex nature of this problem, many areas of uncertainty remain. One important concern is the prediction of remaining shaft fatigue life. This is a function of the complex state of stress that the shaft is exposed to, and it is also influenced by differences between actual shaft material properties and those derived from simple test, e.g., size and loading effects. Jackson and Umans [1] conclude that 'much more effort in this area (fatigue calculations) is required before a general consensus can be achieved. Predominant is the need for research in the area of torsional failure and fatigue modeling in order to resolve a major source of controversy which has arisen as to the loss of life predictions for torsional shaft transients.' Since this paper, programs to address these problems have been initiated under the sponsorship of the Electric Power Research Institute, Palo Alto, California.

Rusche and Mitsche in the Discussion of [1] state that 'Even its (fatigue and damage prediction) most basic aspects remain under continuing investigation by specialists.' The General Electric Company is presently working on a project under EPRI sponsorship to develop a torsional fatigue methodology for use by industry sources in fatigue calculations [8]. Similar developments concerning the fatigue properties of typical shaft materials are being undertaken by the Westinghouse Electric Corporation, again under EPRI sponsorship.

As can be seen in Figure 11, there exists many different types of mass-elastic models for the purpose of finding the modal properties of TG shaft systems. Both Joyce and Lambrecht [3] in Table 2, and Ramey and Kung [4] in Figure 12, show the difference between simple mass-elastic models (less than 10 masses) and complex models (over 200 masses). Both conclude that for basic calculations, adequate simple models can be developed. However, differences between the simple and complex models do not appear to have been explained in the literature, particularly in relation to the effect of such models on fatigue calculations. Adequacy in such cases means 'an appropriately calibrated modal model of the system,' and such a model must be defined using some proven system identification procedure.

Figure 1 shows that many TG shaft natural frequencies result from the use of such models. This is in accordance with practical observations on TG units. Correlation between observed and predicted torsional frequencies is typically close. Both Hizume [5] and Joyce and Lambrecht [3] have shown, in Figure 13 and Figure 14, that the torsional modes which contribute most significantly to transient torques are typically the first four modes. These four are therefore the primary modes which must be considered in such calculations.

The effect of machine damping on torque calculations has been examined by Joyce and Lambrecht in Figure 14. This figure shows the decaying torque oscillations for a 700 Mw TG unit over a period of ten seconds after clearing a close-in three-phase fault. Figure 15 shows the effect of damping on fatigue calculations for an undamped, a low damped, and an 'expected' damped shaft. The fatigue incurred by the shaft was inversely related to the damping value. Furthermore, the damping value does not appear to be linear (with vibration amplitude). Both CMS units described earlier use feedback of actual shaft angular velocity to compensate for such non-linear damping effects in their evaluations of shaft life.

Shaft torsional damping varies with the vibration modes excited, and with the amplitude of vibration in that mode. This has been shown by Joyce and Lambrecht [3] in Figure 16, and by Walker, Adams, and Placek [6] in Figure 17. Figure 16 shows the measured mechanical damping of the first three modes of torsional vibration of a 970 Mw turbine with a two pole generator after a terminal to terminal short circuit, in terms of reciprocal decay rate (seconds). Figure 17 uses a log. dec. vs. load graph, but it clearly shows the variation in damping values, as both a function of load and of model. Details of how this graph was obtained are given in the appendix to the paper. It is evident that the damping increases as a function of generator load, and the rate of increase depends on the mode of vibration. The need here appears to be for a calibratable computer procedure which will allow damping to be calculated for the relevant modes of a given TG system.

Current problems associated with the use of CMS units are the lack of a means for retrofitting the past fatigue history of the shaft into the cumulative fatigue calculations, plus a means for accounting for the effects of modal damping. A means for specifying an appropriate threshold strain value that will provide optimum use of these diagnostic tools is also needed.

10.1.6 Conclusions and Recommendations

Since the first reported shaft failure due to transient torques was disclosed, advances have been made as follows:

o Knowledge of these electrical disturbances which can cause transient torques in TG shafts is for the most part complete.

- o Vibration analysis can now be efficiently accomplished by use of tools such as the finite element method. A remaining question is: how accurate a calculation does one need?

- o Development of basic continuous monitoring systems to gather much needed data such as the frequency of an electrical disturbance, its duration, and its transient amplitudes have been accomplished.

- o Increased awareness exists concerning how system planning can reduce damage to the TG shaft from shaft, i.e., system disturbances.

- o Analytical refinements such as Rain Flow Cycle Counting technique, and of the Rogowski coil that increase the accuracy of calculations, and of data acquired during monitoring.

- o Increased knowledge of what direction further research should take.

Several areas exist in which further study of this problem is needed, i.e.:

- o Knowledge of shaft fatigue criteria must be increased in order to better predict shaft torque crack initiation time.

- o Effect of damping on shaft torque should be explored in greater detail.

- o A method to predict modal damping should be developed.

- o Continuous monitoring systems should be improved which can incorporate past damage into the life algorithm.

10.1.7 References

1) Jackson, M. C., Umans, S. D., 'Turbine Generator Shaft Torques and Fatigue: Part III-Refinements to Fatigue Model and Test Results,' IEEE Transactions Vol. Pas-99, No. 3., pp. 1259-1269, May/June 1980.

2) 'IEEE Screening Guide for Planned Steady-State Switching Operators to Minimize Harmful Effects on Steam Turbine Generators,' IEEE Transactions Vol. Pas-99, No. 4, pp. 1519-1521, July/August 1980.

3) Joyce, J., Lambrecht, D., 'Status of Evaluating the Fatigue of Large Steam Turbine Generators Caused by Electrical Disturbances,' IEEE Transactions Vol. Pas-99, No. 1, pp. 111-119, January/February 1980.

4) Ramey, D. G., Kung, G. C., 'Important Parameters in Considering Transient Torques on Turbine Generator Shaft Systems,' IEEE Transactions, 1978 IEEE ASME/ASLE Joint Power Generation Conference, Paper No. 7, November 1978.

5) Hizume, A., 'Transient Torsional Vibration of Steam Turbine Generator Shafts Due to High Speed Reclosure of Electric Power Lines,' ASME Transactions Paper No. 75-OCT-71, pp. 968-979.

6) Hammons, T. J., 'Stressing of Large Turbine Generator Shaft Couplings and LP Turbine Final Stage Blade Roots Following Clearance of Grid Systems Faults and Faulty Synchronization,' IEEE Transactions 1978 IEEE ASME/ASLE Joint Power Generation Conference, Paper No. 15, November 1978.

7) Fick, H., Stein, J., 'The Torsional Stress Analyzer for Continuously Monitoring Turbine Generators,' IEEE Transactions Vol. PAS-99, No. 2, pp. 703-708, March/April 1980.

8) Gibbs, E. E., Walker, D. N., 'Torsional Vibration Monitoring System,' General Electric Publication Presented at the Pacific Coast Electrical Association Engineering and Operating Conference, March 13-14, 1980.

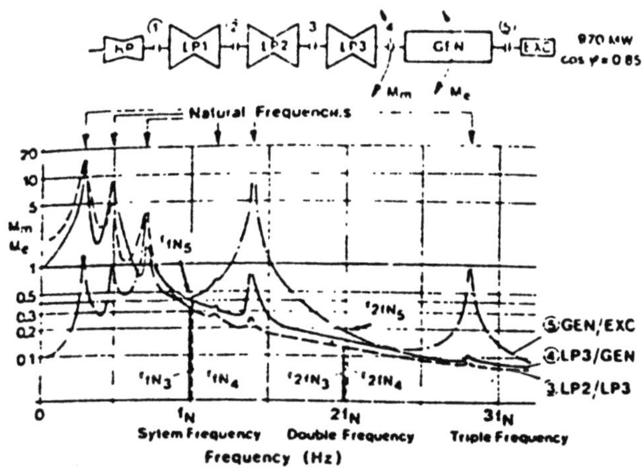

Figure 1 Response Characteristics of TG Shaft Sections [3]

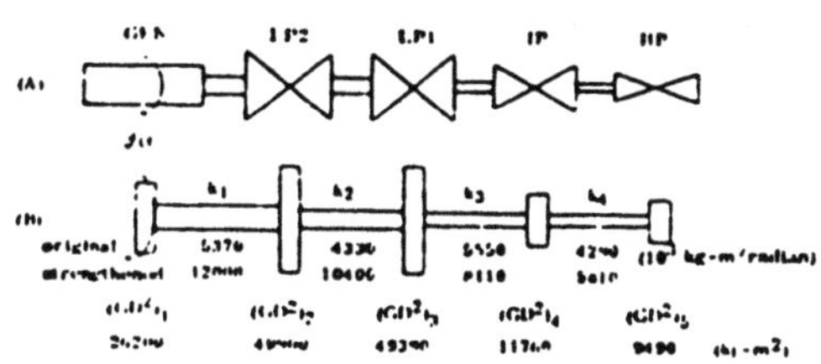

Figure 2 Simple Mass-Elastic System [3]

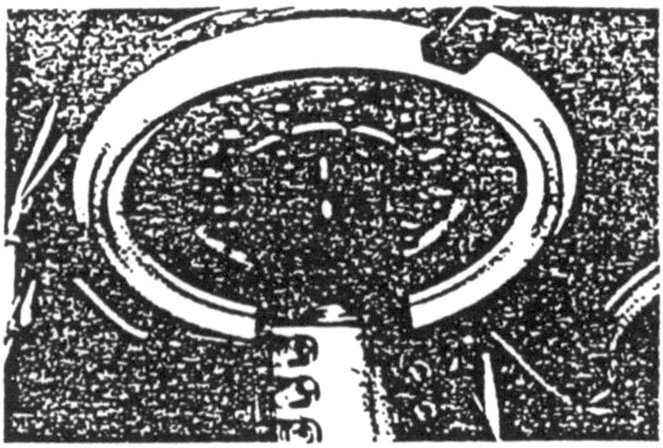

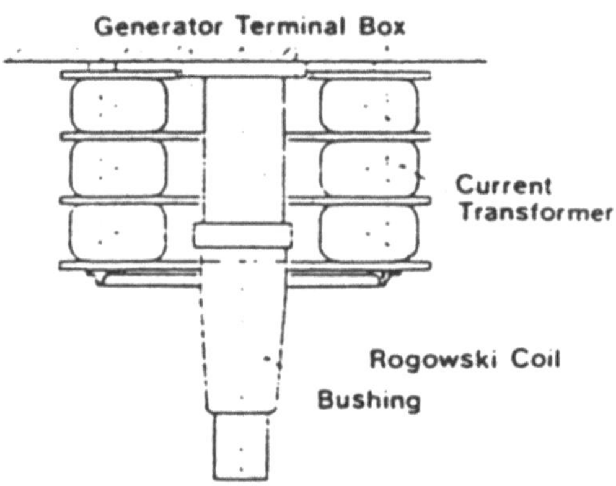

Figure 3 Rogowski Coil [7]

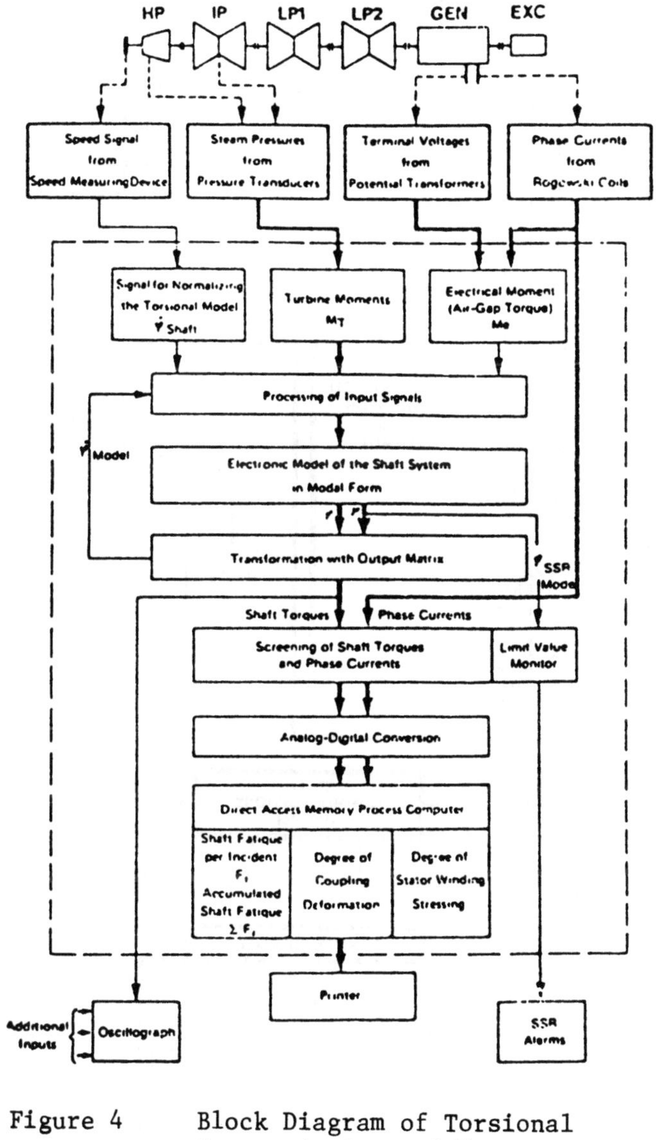

Figure 4 Block Diagram of Torsional Stress Analyzer [7]

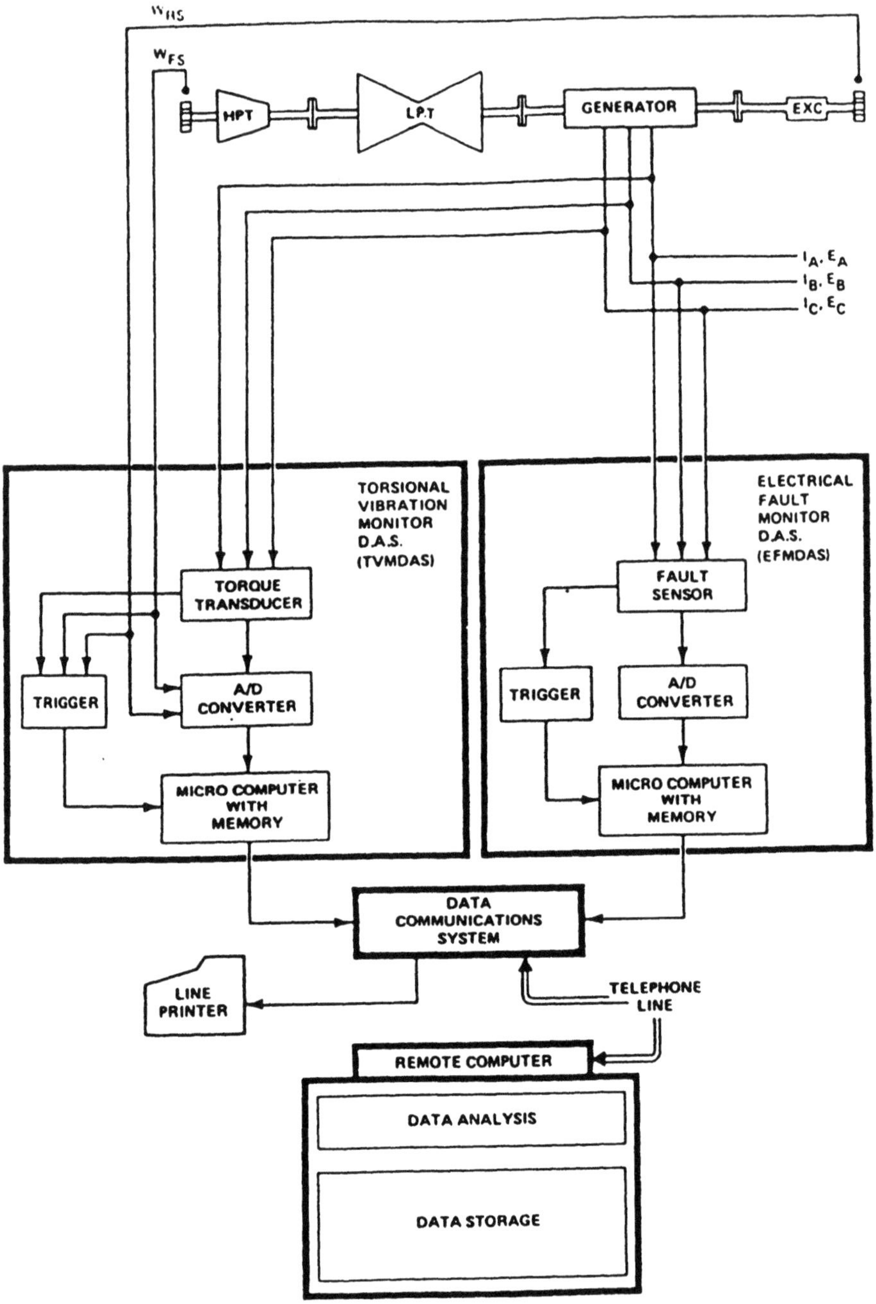

Figure 5 Block Diagram of Torsional Vibration Monitor System [8]

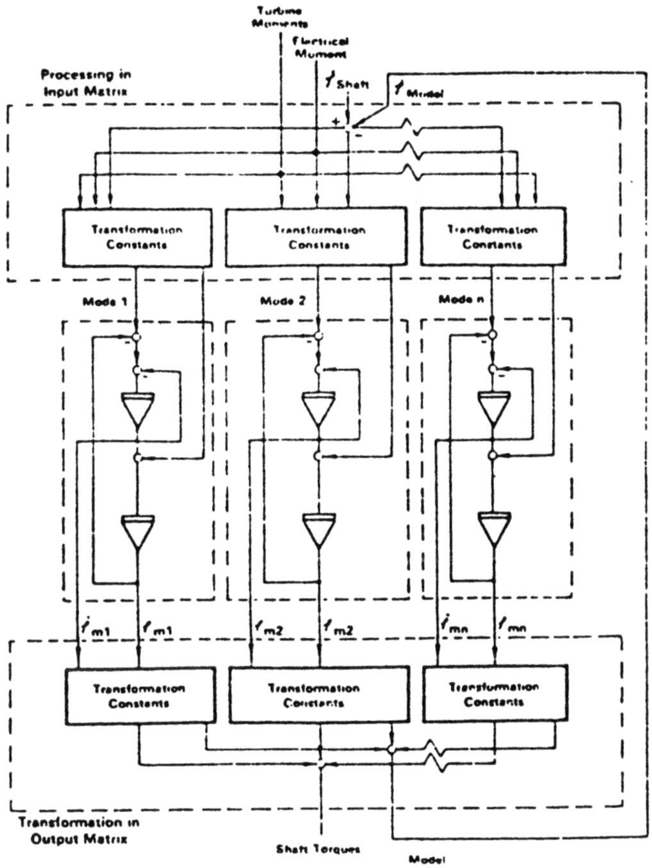

Figure 6 TSA TG Shaft Analog Model [7]

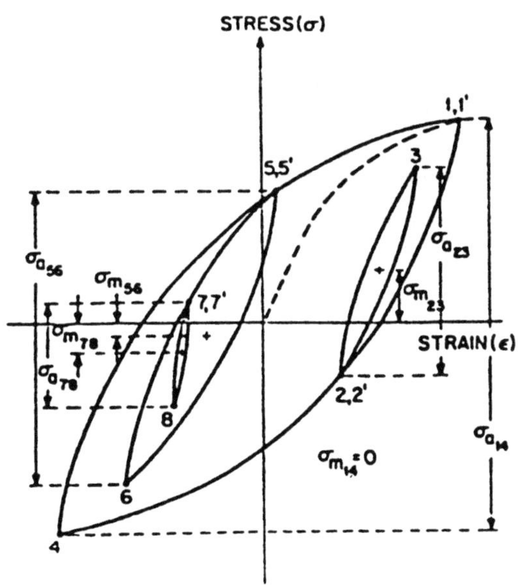

Figure 7 Typical Stress-Strain Diagram [1]

DATE: 10.12.77 HOUR: 11.25

TORSIONAL STRESS ANALYZER

C1:	MAXIMUM TORQUE LEVEL	0 REACHED	0.00005	99.9264
C2:	MAXIMUM TORQUE LEVEL	0 REACHED	0.24651	99.3258
C3:	MAXIMUM TORQUE LEVEL	1 REACHED	0.38194	99.1564
C4:	NO IMPACT		0.00000	99.7639

NO IMPACT ON STATOR END WINDING

Figure 8 Typical TSA Results [7]

Four Levels of Mechanical Impact of Operating Incidents on Turbine-Generator Shaft Couplings and Recommended Actions	
0	No permanent deformation.
1	Localized minor permanent deformations which bear no consequence for the operation of the machine.
2	Localized permanent deformations. Inspection should take place as soon as possible and convenient. If the running behavior deteriorates, immediate inspection is required. The coupling bolts should be replaced and the bolt holes reworked, if necessary.
3	Major deformation can be expected. Shut down for immediate inspection and repair of the couplings.

Figure 9 Code for TSA Printout [7]

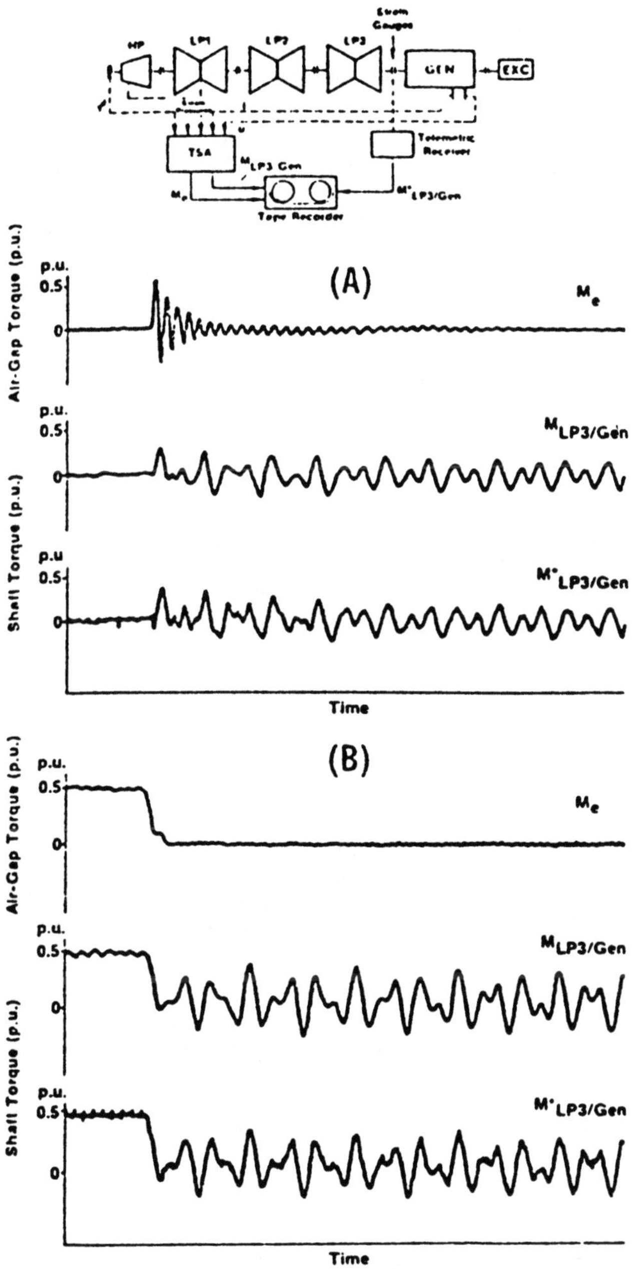

Figure 10 Comparison of Modeled vs. Measured Shaft Torque for (A) and Three-Phase Short and (b) Partial Load Rejection [7]

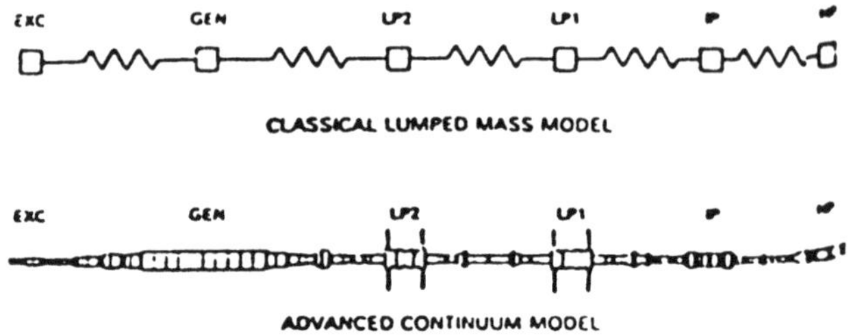

Figure 11 Different Types of Mass-Elastic Models [4] - Also See Figure 2

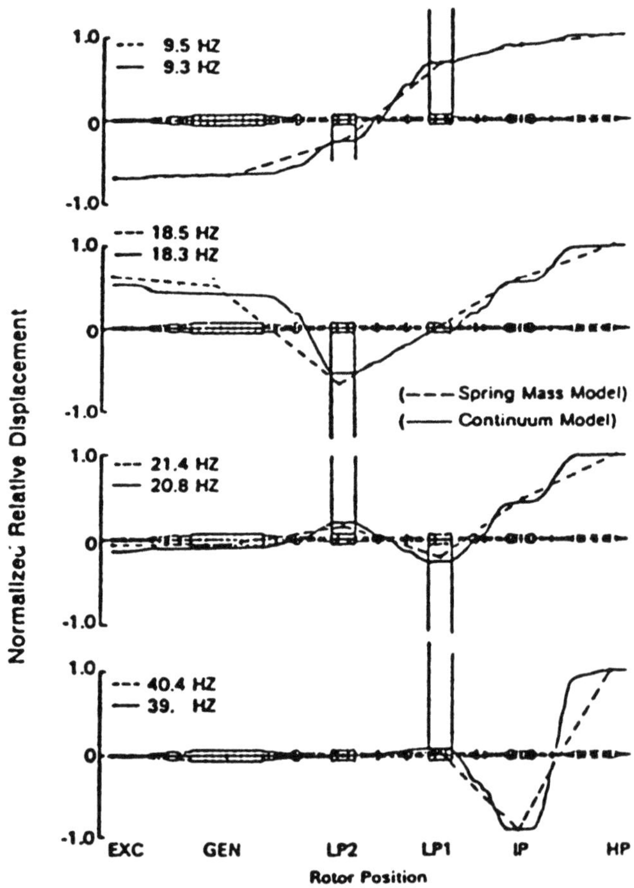

Figure 12 Comparison of Modes Between Simple Spring Mass Model and Complex Continuum Model [4]

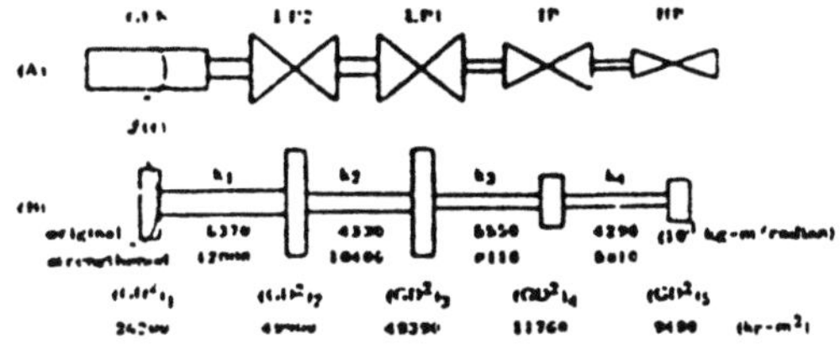

Shaft system of utility turbine and generator unit

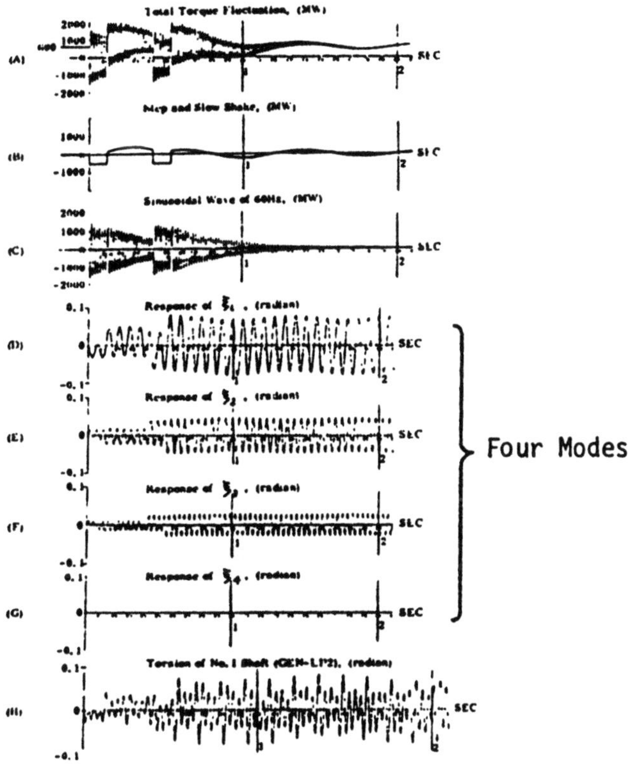

Figure 13 Effect of First Four Modes on Torques [5]

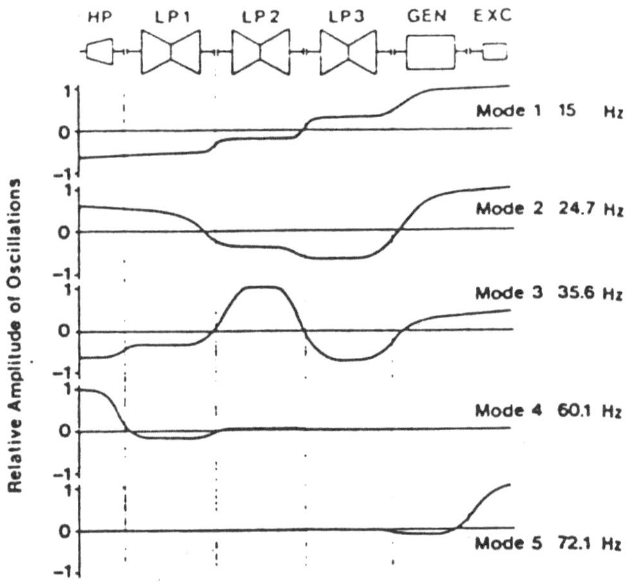

Torsional vibration modes of a 970 MW, 50 Hz nuclear-fueled turbine-generator.

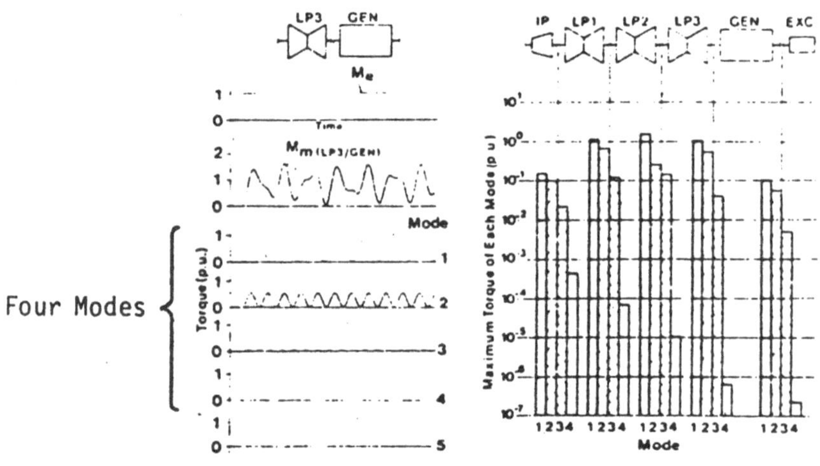

Torsional modal response at different shaft locations to a uni-directional electrical impulse. [3]

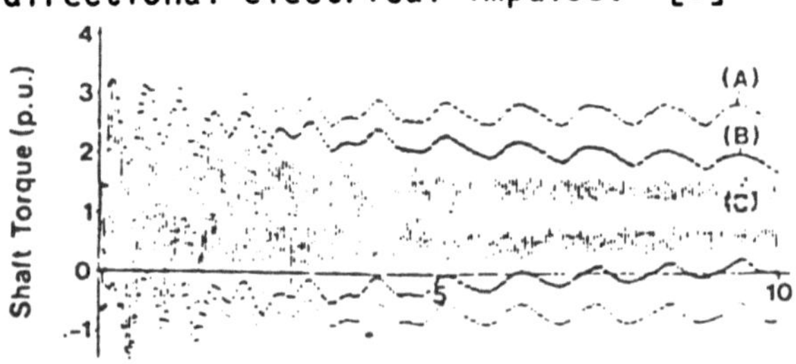

Decaying torque oscillations [3]

Figure 14

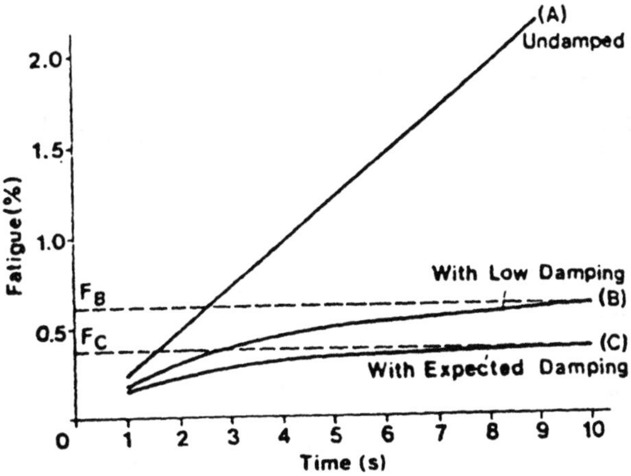

Figure 15 Relation Between Fatigue Calculations and Shaft Damping Values [3]

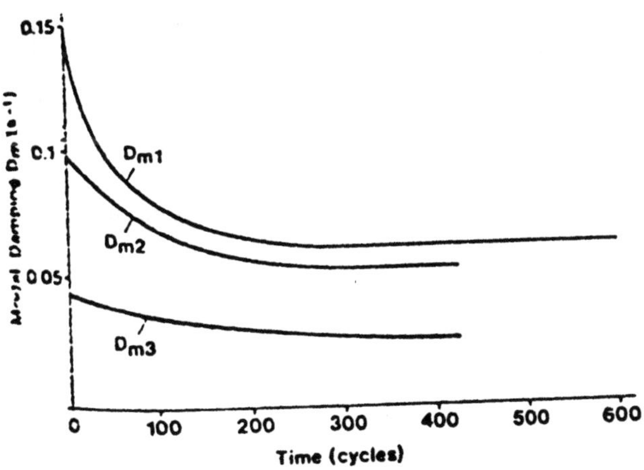

Figure 16 Measured Mechanical Damping of Torsional Oscillations in Three Different Vibration Modes [3]

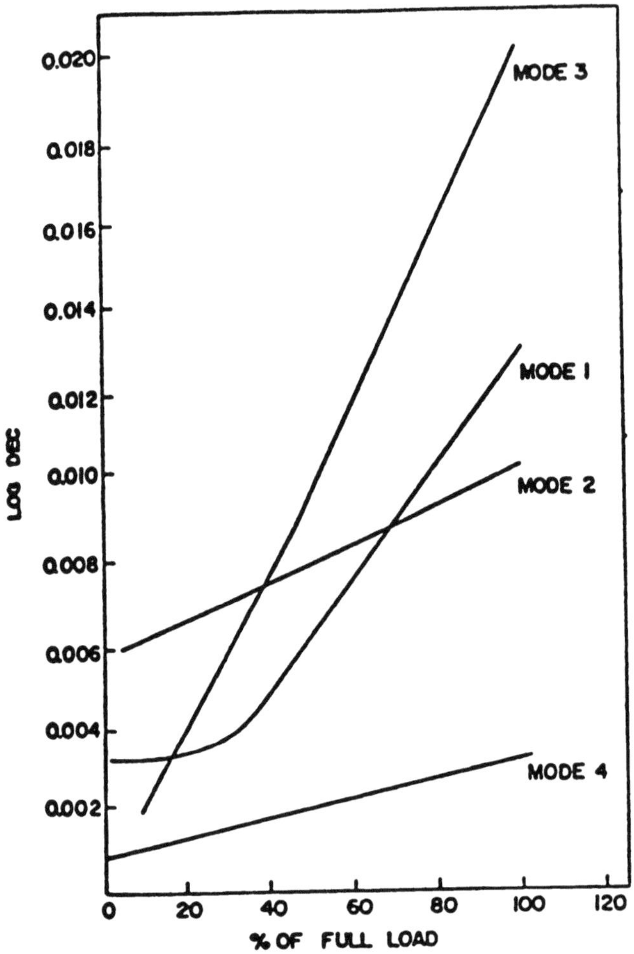

Figure 17 Typical Modal Damping Characteristics

Table 1 Effect of Different Electrical Disturbances on the Amount of Fatigue Incurred per Incident [3]

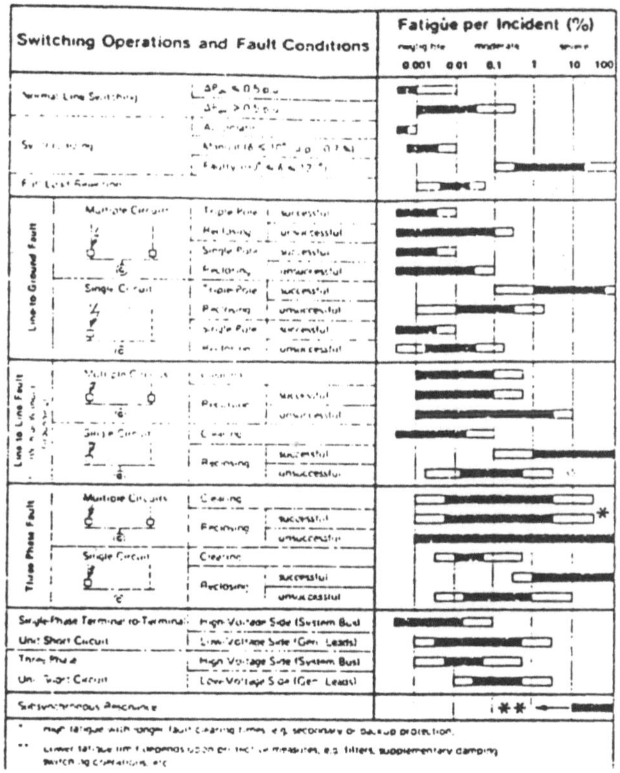

Table 2 Calculated Torsional Natural Frequencies for Simple (10 Masses) and Complex Shaft Models (200 Masses) [3]

Unit Rating (MW)	Number of Generator Poles	Vibration Mode	Calculated Frequency (Hz) Number of Masses in System n < 10	n > 200	Measured Frequency (Hz) by Double-Frequency Excitation (S.C.Test)	by Frequency Analysis from Sudden Short-Circuit Test Data
630	2	1	17.0	16.5 *	–	16.7
660	4	1	7.8	8.2	–	8.1
1200	4	1	6.5	6.5	–	6.3
970	2	1	14.9	14.9	14.8	14.9
		2	24.3	24.7	24.7	24.5
		3	34.9	34.6	34.3	34.3

CHAPTER 10.2

TORSIONAL SYSTEMS: VIBRATION RESPONSE BY MEANS OF MODAL ANALYSIS

P. Schwibinger, R. Nordmann
University of Kaiserslautern, Kaiserslautern, FRG

ABSTRACT

Large steam-turbine generators in operation may be stimulated to torsional vibrations by dynamic moments at the generator due to electrical system transients. To solve the torsional vibration problem the turbogenerator shaft is modelled by the finite element method. The equations of motion are solved for the transient vibrations with the modal analysis technique - 'Time-History-Method'. If the designer is interested in an approximation for the maximum response the 'Response-Spectrum-Method' proves to work very effectively. This paper discusses 'Time-History-' and 'Response-Spectrum-Method' and presents the results for a 600 MW and a 722 MW turboset.

10.2.1 Introduction

Large steam-turbine generators in operation may be stimulated to torsional vibrations by dynamic moments at the generator due to electrical system transients. The induced torsional stresses in the shaft have drawn growing attention over the past few years /1-4/, Fig. 1.

For the solution of the torsional vibration problem it is essential to find an appropriate torsional model for the turbine-generator shaft. A common approach is to model the torsional system by the finite element method. As a result one obtains the linear equations of motion for the rotor. The solutions of the ho-

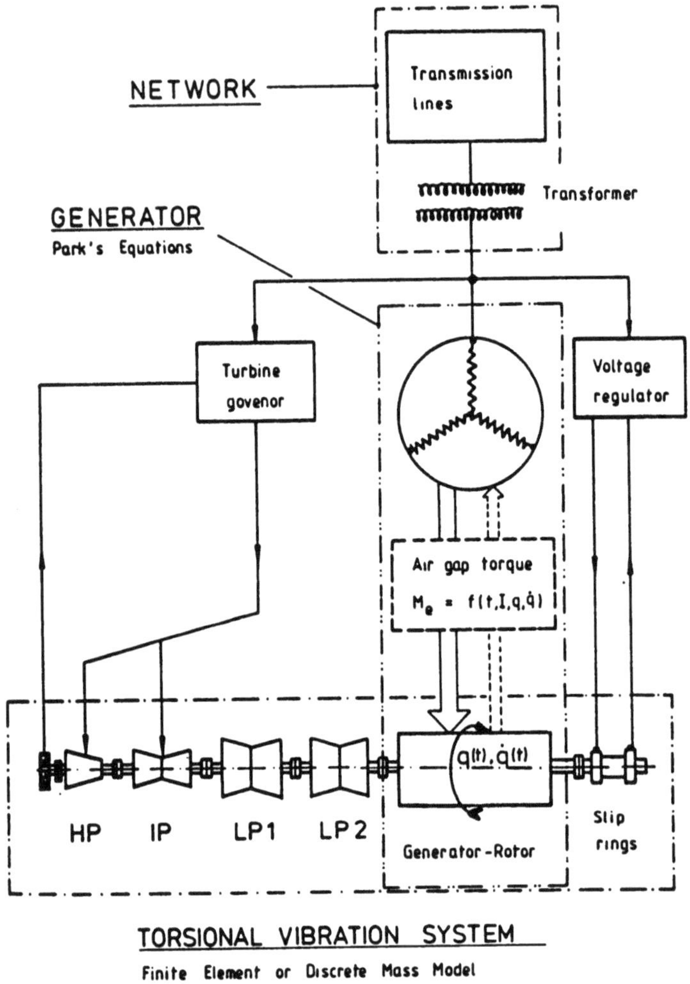

Fig. 1 Subsystems for calculating the effects of electrical system faults

mogeneous problem are the eigenfrequencies and modes to which correspond modal internal moments.

For arbitrary transient excitation of the system due to electrical moments the equations of motion are solved by <u>modal analysis:</u>
The decoupled equations of motion are integrated numerically in the time domain. But this so called 'Time-History-Method' needs extensive computer time. Often the designer is only interested

in the maximum of a displacement or a tension. In this case an approximation is sufficient which calculates only the maxima and works therefore more economically. In this paper the application of the 'Response Spectrum Modal Analysis' to calculate transient torsional vibrations of turbogenerators is discussed.

10.2.2 Modelling of the Turbogenerator Shaft

The first step in the analytical solution of a vibration problem in the field of machinery dynamics is to find a mechanical model for the built machine, which describes its dynamic characteristics adequately. For the solution of our torsional vibration problem it is a common approach to model the shaft finely by the finite element method. In this case the total shaft is subdivided in N-1 finite 'torsional elements', Fig. 2.
If modelled in this way a 600 MW turbogenerator has about 250 DOF.

We consider small - that means linearized - torsional vibrations about a stationary rotation of the shaft. By using the principle of virtual work we can describe the dynamic behavior of our model mathematically; we get the mathematical model.

$$\delta W = \underbrace{\sum_{e=1}^{N-1} \delta W_i^{\,e}}_{\substack{\text{virtual work} \\ \text{of elastic moments}}} + \underbrace{\sum_{e=1}^{N-1} \delta W_T^{\,e}}_{\substack{\text{virtual work} \\ \text{of inertia moments}}} + \underbrace{\sum_{e=1}^{N-1} \delta W_F^{\,e}}_{\substack{\text{virtual work} \\ \text{of external moments}}} = 0 \qquad (1)$$

If we approximate the unknown torsional displacement in an element with static deflection functions, we obtain a 2x2 mass- and stiffness-matrix. Each element has two nodes with one DOF each. Superposition of the element matrices yields the mass-matrix $\underline{M}$ and the stiffness-matrix $\underline{K}$ for the complete system, Fig. 3.

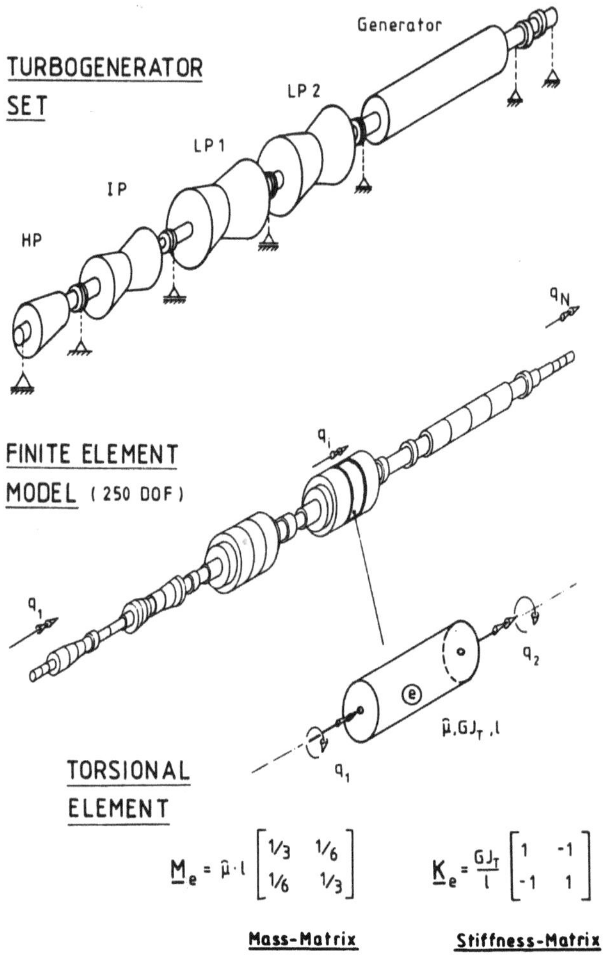

Fig. 2 Finite element model for the turbogenerator shaft

In a similiar manner we can transform the distributed external moments to the nodes and obtain the excitation vector $\underline{F}(t)$. The damping matrix is assumed as proportional to $\underline{M}$ respectively $\underline{K}$.

$$\underline{D} = \alpha\,\underline{M} + \beta\,\underline{K} \tag{2}$$

The result are the equations of motion for the turbogenerator shaft: a linear, inhomogeneous system of differential equations of order N.

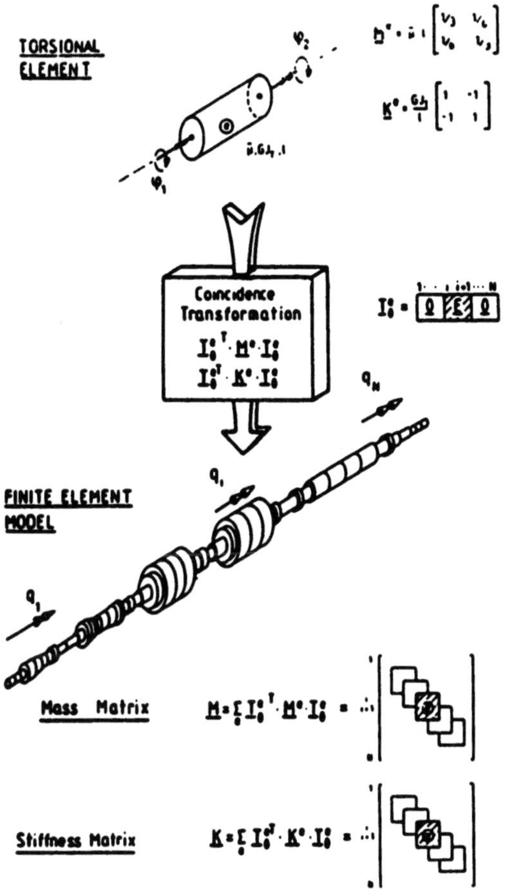

Fig. 3 Superposition of element matrices

$$\underline{M}\,\ddot{\underline{q}} + \underline{D}\,\dot{\underline{q}} + \underline{K}\,\underline{q} = \underline{F}(t) \tag{3}$$

To solve (3) we'll use the modal analysis method; but before we study the corresponding homogeneous problem ($\underline{F}(t) = \underline{0}$).

10.2.3 Modal Parameters of the Torsional Problem

Eigenfrequencies and Modes. First the equations of motion are solved for the undamped natural vibrations, we get from the homogeneous equations.

$$\underline{M}\,\ddot{\underline{q}} + \underline{K}\,\underline{q} = \underline{0} \tag{4}$$

Assuming a solution of the form

$$\underline{q}(t) = \underline{\varphi}\, e^{\lambda t} \tag{5}$$

we obtain the eigenvalue problem

$$(\underline{K} - \overline{\lambda}\, \underline{M})\, \underline{\varphi} = \underline{0} \quad \text{with} \quad \overline{\lambda} = -\lambda^2 \tag{6}$$

Eigenvalues $\overline{\lambda}_i$ Modes $\underline{\varphi}_i$

with N eigenvalues, $\overline{\lambda}_i$ respectively eigenfrequencies f_i,

$$f_i = \frac{1}{2\pi}\sqrt{\overline{\lambda}_i} \tag{7}$$

and corresponding modes $\underline{\varphi}_i$.
The lowest eigenvalue we obtain from (6) is zero and corresponds to the rigid body mode of our unbounded torsional system. Fig. 4 shows the next five elastic modes of a 600 MW turboset. In the first mode HP-, IP- and LP1-turbine oscillate with 18,19 Hz against LP2-turbine and generator. In the first, second and third mode the whole turbogenerator vibrates, whereas in the higher modes only parts oscillate, e.g. the slip ring shaft in the fifth mode.

Modal elastic moments. The natural vibrations in one mode cause elastic moments in the shaft - so called modal elastic moments $\underline{f}_m$. If we apply the principle of virtual work to one element we get, Fig. 5,

$$\underline{f}_m^{\,e} = \begin{bmatrix} -f_{m1} \\ f_{m2} \end{bmatrix} = \underline{M}^e\, \ddot{\underline{q}}^e + \underline{K}^e\, \underline{q}^e \tag{8}$$

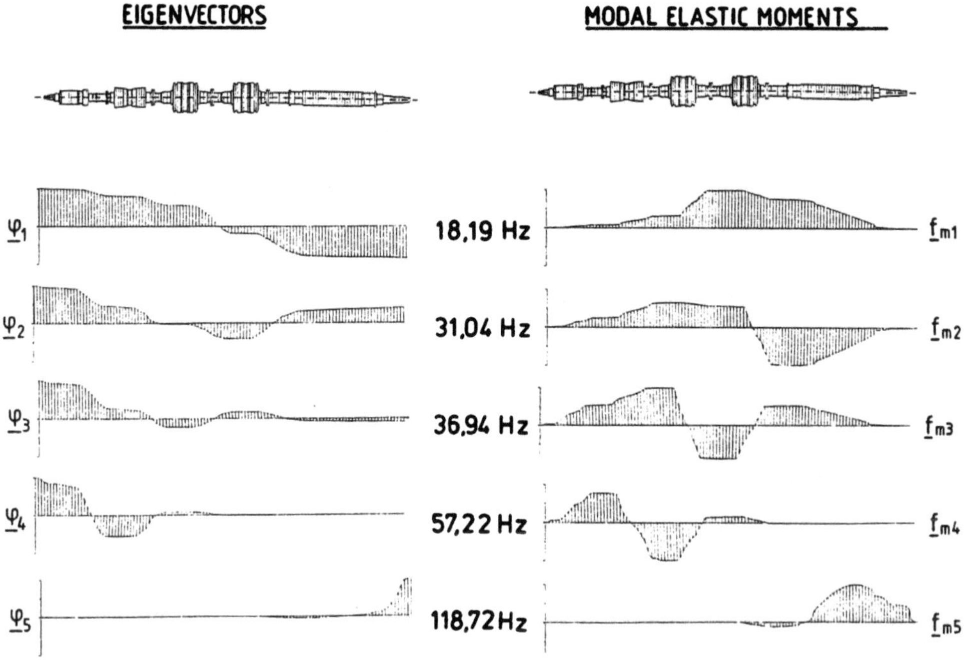

Fig. 4 Modes of a 600 MW turbogenerator with the corresponding modal elastic moments

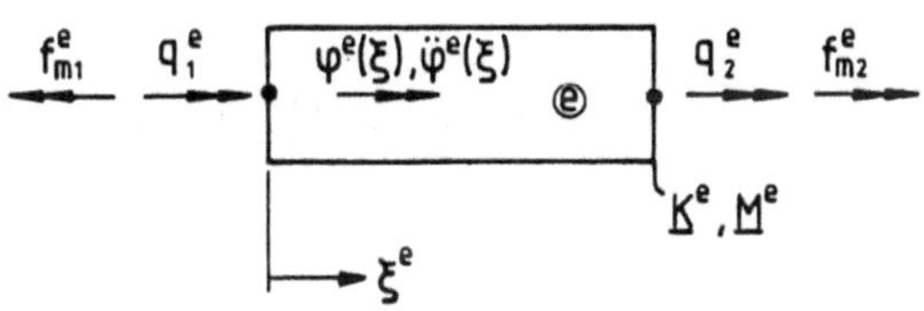

Fig. 5 Modal elastic moments at element e

If the rotor vibrates in a mode $\underline{\varphi}_i$ with the frequency ω_i it is

$$\underline{q}_i^e = \underline{\varphi}_i^e$$

resp. (9)

$$\ddot{\underline{q}}_i^e = -\omega_i^2 \underline{\varphi}_i^e$$

and we obtain from (8)

$$\underline{f}_{mi}^{e} = \left[\underline{K}^{e} - \omega_i^2 \underline{M}^{e}\right] \underline{\varphi}_i^{e} \tag{10}$$

In the slender shaft regions (couplings, slip-ring shaft) the contribution of the inertia term $\underline{M}^e\ddot{\underline{q}}^e$ in (8) is small compared to the elastic moments $\underline{K}^e\underline{q}^e$. In that case we can reduce (10) to

$$\underline{f}_{mi}^{e} \approx \underline{K}^{e}\, \underline{r}^{e} \tag{11}$$

In Fig. 4 the modal elastic moments which correspond to the lower modes are shown. The maximum amplitudes are normalized to equal 1.0. This maximum occurs mainly in the couplings where the gradient of the mode is high. Because the modes are normalized, we cannot infer from the modal elastic moments to the absolute stresses in the shaft. But they show clearly, where the maximum moments in the shaft occur, when it is excited in one natural mode. In the next point we'll expand the internal moments due to transients with the modal analysis method and use the modal elastic moments as ingredient.

10.2.4 Calculation of the Transient Vibrations with Modal Analysis - Time History Method

The transient vibrations due to dynamic electrical moments are calculated from the inhomogeneous equations of motion (3). Let us start with the virtual work of the complete system.

$$\delta W = \delta\underline{q}^T \{-\underline{M}\,\ddot{\underline{q}} - \underline{D}\,\dot{\underline{q}} - \underline{K}\,\underline{q} + \underline{F}(t)\} = 0 \tag{12}$$

The numerical solution of the resulting system of differential equations is very expensive, because the individual equations are coupled. We can solve the equations of motion economically with the modal analysis method. For this we expand the unknown torsional displacements $\underline{q}(t)$ linearly with the known elastic modes $\underline{\varphi}_i$.

$$\underline{q}(t) = \sum_{i=1}^{N} \underline{\varphi}_i \eta_i(t) = \underline{\Phi}\,\underline{\eta}(t) \tag{13}$$

The modes are arranged in the modal matrix $\underline{\Phi}$,[1])

$$\underline{\Phi} = \left[\underline{\varphi}_1, \ldots \underline{\varphi}_{N-1}\right] \tag{14}$$

expansion parameters are the modal displacements $\eta_i(t)$.

$$\underline{\eta}(t)^T = \left[\eta_1(t), \ldots, \eta_{N-1}(t)\right] \tag{15}$$

From the energy equation (12) we get with (13)

$$\delta\underline{\eta}^T\{-\underbrace{\underline{\Phi}^T\underline{M}\,\underline{\Phi}}_{\text{diag}\left[m_i\right]}\,\ddot{\underline{\eta}} - \underbrace{\underline{\Phi}^T\underline{D}\,\underline{\Phi}}_{\text{diag}\left[d_i\right]}\,\dot{\underline{\eta}} - \underbrace{\underline{\Phi}^T\underline{K}\,\underline{\Phi}}_{\text{diag}\left[k_i\right]}\,\underline{\eta} + \underbrace{\underline{\Phi}^T\underline{F}(t)}_{g_i(t)}\} = 0 \tag{16}$$

Because the modes are orthogonal to mass-matrix $\underline{M}$ and stiffness-matrix $\underline{K}$ and on the condition that (2) is valid, we get N-1 decoupled equations of the type of a single DOF system, Fig. 6,

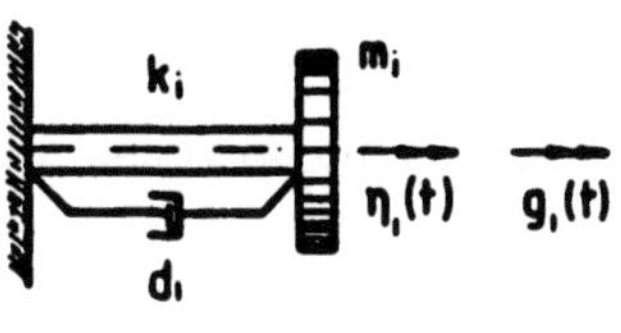

Fig. 6 Modal one degree of freedom system

$$m_i\ddot{\eta}_i + d_i\dot{\eta} + k_i\eta_i = g_i(t) \tag{17}$$

[1])Without the rigid-body mode $\underline{\varphi}_o$.

with

$$\begin{aligned} m_i &= \underline{\varphi}_i^T \underline{M}\, \underline{\varphi}_i && \text{- Modal Mass} \\ d_i &= \underline{\varphi}_i^T \underline{D}\, \underline{\varphi}_i && \text{- Modal Damping} \\ k_i &= \underline{\varphi}_i^T \underline{K}\, \underline{\varphi}_i && \text{- Modal Stiffness} \\ g_i(t) &= \underline{\varphi}_i^T \underline{F}(t) && \text{- Modal Force} \end{aligned} \tag{18}$$

Equation (17) can be solved for η_i analytically or for arbitrary modal forces numerically with suitable algorithms. The unknown torsional displacements follow with (13) from a superposition of the modal solutions. For technical applications it is mostly sufficient to consider only the lowest 1,...,n elastic modes, to obtain accurate results.

$$\underline{q}(t) = \sum_{i=1}^{n} \underline{\varphi}_i\, \eta_i(t) \tag{19}$$

Besides the decoupling of the equations through this the computer time can be further reduced.

Solution of the single degree of freedom system (SDOF). The response of a SDOF to an arbitrary excitation can be written with the Duhamel-Integral

$$\eta_i(t) = \frac{1}{m_i \omega_i} \int_0^t g_i(\tau)\, e^{-D_i \omega_{io}(t-\tau)} \sin\omega_i(t-\tau)\, d\tau \tag{20}$$

with the initial conditions

$$\eta_i(t{=}0) = 0 \qquad \dot{\eta}_i(t{=}0) = 0 \tag{21}$$

the damping factor

$$D_i = \frac{d_i}{2m_i\omega_{io}} \tag{22}$$

and the eigenfrequencies

$$\omega_{io} = \sqrt{k_i/m_i} \qquad \omega_i = \omega_{io}\sqrt{1-D_i^2} \tag{23}$$

The numerical solution of (20) is very uneconomical. Therefore other more effective algorithmus should be used. A linear interpolation of the excitation function proves to be most effective.

In this approximation method the given modal force $g_i(t)$ is replaced by a polygonal course, which is formed by the discrete values g_i^k at the times t^k, Fig. 7.

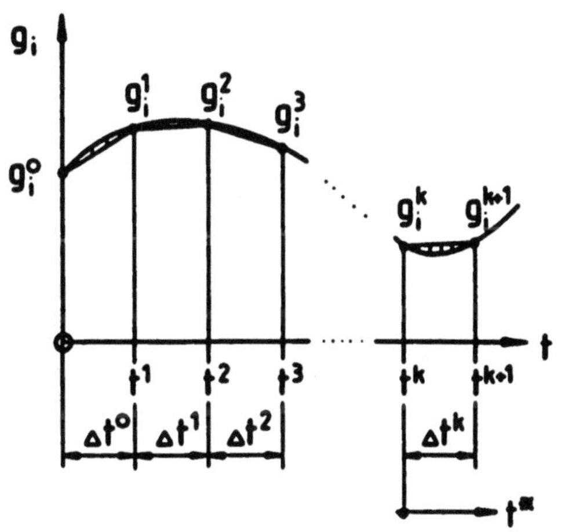

Fig. 7 Linear Interpolation of the modal force $g_i(t)$

For the approximated modal force the equation of motion can be solved piecewise exactly. For the k-th interval is the equation of motion for $\eta_i(t^*)$

$$m_i\ddot{\eta}_i + d_i\dot{\eta}_i + k_i\eta_i = g_i^k + s_i^k t^* \quad \text{with} \quad t^* = t - t^k \tag{24}$$

and

$$s_i^k = \frac{g_i^{k+1} - g_i^k}{t^{k+1} - t^k} = \frac{\Delta g_i^k}{\Delta t^k}$$

If we restrict the solution to the discrete times t^k and assume constant intervals Δt^k, the solution of (24) is

$$\begin{bmatrix} \eta_i^{k+1} \\ \dot{\eta}_i^{k+1} \end{bmatrix} = \begin{bmatrix} a_{11} & a_{12} \\ a_{21} & a_{22} \end{bmatrix} \begin{bmatrix} \eta_i^{k} \\ \dot{\eta}_i^{k} \end{bmatrix} + \begin{bmatrix} b_{11} & b_{12} \\ b_{21} & b_{22} \end{bmatrix} \begin{bmatrix} g_i^{k} \\ g_i^{k+1} \end{bmatrix} \tag{25}$$

whereas the a_{mn}, b_{mn} $(m,n=1,2)$ are constants which must only calculated once /5/.

The only error of the method lies in the approximation of the modal force. Our experience for a periodic excitation function is: If the period of the highest harmonic is approximated by 30 intervals the error in amplitude and phase is smaller than one percent.

As an example we take the response of the 600 MW turboset to a short circuit excitation, Fig. 8. The torsional displacement at a global DOF q_k superimposes with (13) from the contributions of the modal SDOF η_i multiplied with the corresponding eigenvectorcomponent φ_{ik}

$$q_k(t) = \sum_{i=1}^{n} \varphi_{ik}\, \eta_i(t) \tag{26}$$

In Fig. 8 the contributions of the individual modes to the angular displacement of the LP2-Generator coupling are shown. From the first three modes only the two lower ones contribute a considerable amount to the displacement q_k.

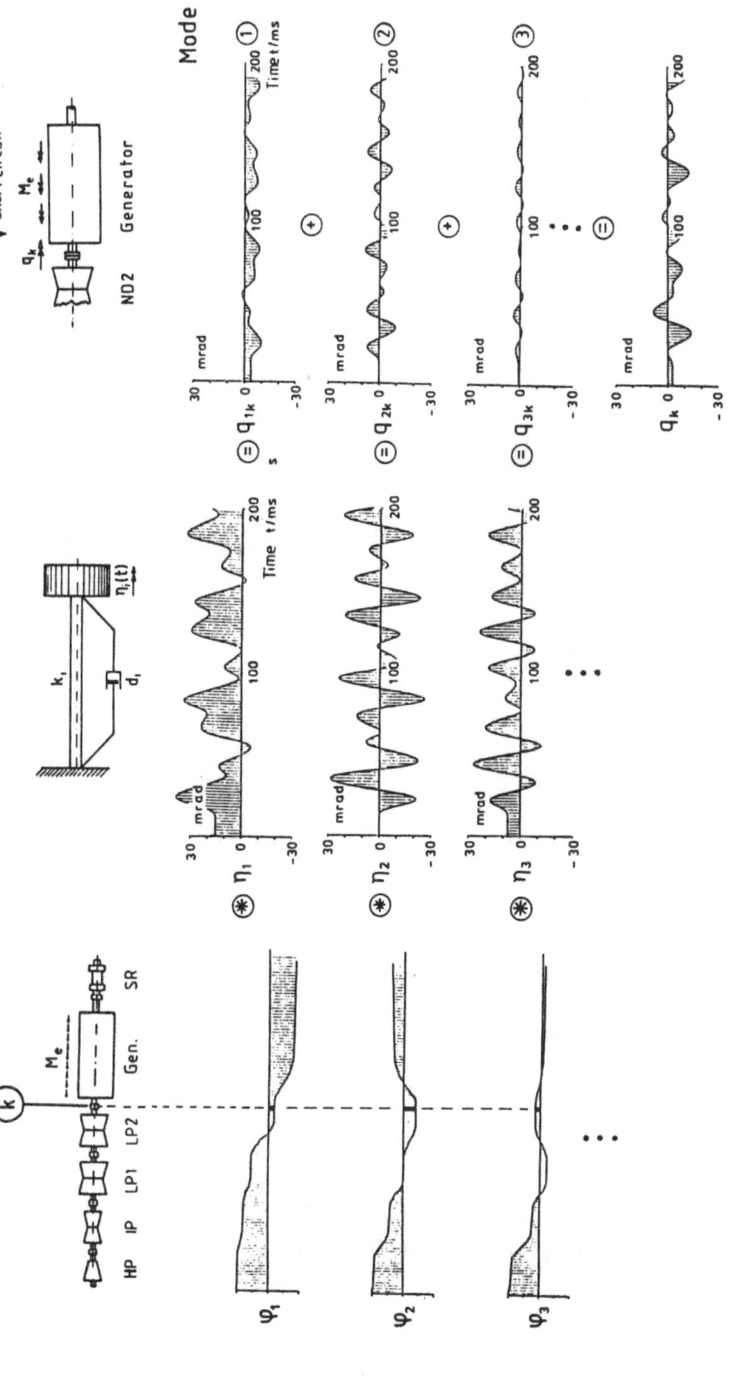

Fig. 8 Modal expansion of displacement

The amplitude of the third mode at the coupling is small, its contribution to the displacement q_K therefore insignificant.

Calculation of internal moments. The designer will ask: How are the interval moments and shear forces in the shaft due to electrical disturbances?

For the transient vibration we can develop the elastic forces at an element e in a similiar way as we have done it above with the principle of virtual work, Fig. 9.

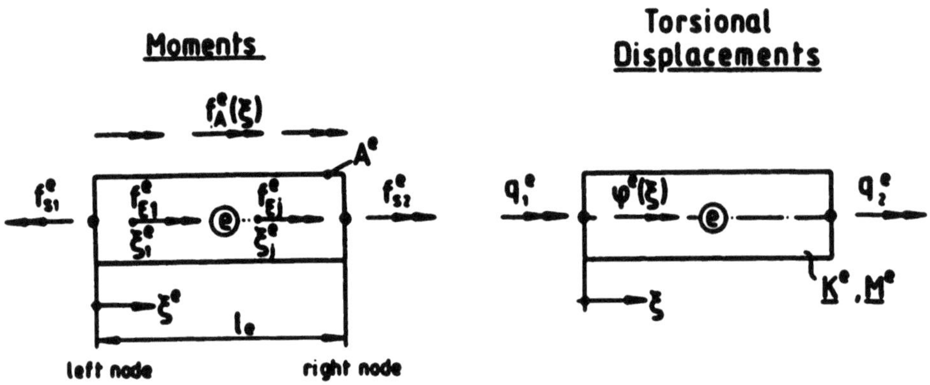

Fig. 9 Exitation and torsional displacements to calculate the elastic moments

At a torsional element act the distrubuted moment f_A^e, the discrete moments f_{Ei}^e, the inertia moments and the internal moments at the nodes. Assuming the distributed moment as constant in an element, we get for the internal moments

$$\underline{f}_s^e = \begin{bmatrix} -f_{s1} \\ f_{s2} \end{bmatrix} = \underline{M}^e\ddot{\underline{q}}^e + \underline{K}^e\underline{q}^e - \frac{A^e f_A^e}{2}\begin{bmatrix} 1 \\ 1 \end{bmatrix} - \sum_{i^e} f_{Ei}(\xi_i^e)\begin{bmatrix} 1-\xi_i \\ \xi_i \end{bmatrix}^e \qquad (27)$$

As already mentioned at the modal elastic moments, the maximum stress occurs in the slender couplings. Here no external moments act on the shaft, in addition is the contribution of the inertia terms in (27) small compared to the elastic terms. For the couplings we may therefore simplify (27) to

$$\underline{f}_s^e(t) = \begin{bmatrix} -f_{s1} \\ f_{s2} \end{bmatrix}^e \approx \underline{K}^e \, \underline{q}^e(t) \tag{28}$$

or written as a superposition of the individual modes, compare (11), (13)

$$\underline{f}_s^e(t) \approx \sum_{i=1}^{n} \underline{K}^e \, \underline{\varphi}_i^e \, \eta_i(t) = \sum_{i=1}^{n} \underline{f}_{mi}^e \, \eta_i(t) \tag{29}$$

For the 600 MW turboset Fig. 10 shows the calculated coupling moment between LP2-turbine and generator.
At time t=0 the shaft is tensioned by the constant turbine and generator moments. At t=10 ms a two phase short circuit occurs and excites the generator with net- and double net frequency components. For the modal damping factors values were taken from literature ($D_i \approx 0{,}01-0{,}02$) /6,7/. The highest amplitude of the resulting coupling moment is four times the static moment. Because of the small damping the response decreases slowly.

As in the case of the displacements, Fig. 8, we discuss the contributions of the modes to the maximum coupling moments, Fig. 11. We see, e.g. for the LP2-Generator coupling, how the individual modes superimpose to the coupling moment, from which we find the maximum. In the diagram on the right hand side the contributions for the different couplings are plotted. The lowest three up to four modes have a determining influence on the maximum at the couplings - except for the slip ring shaft, the elastic moment of which is determined by the fifth mode. One compare with the

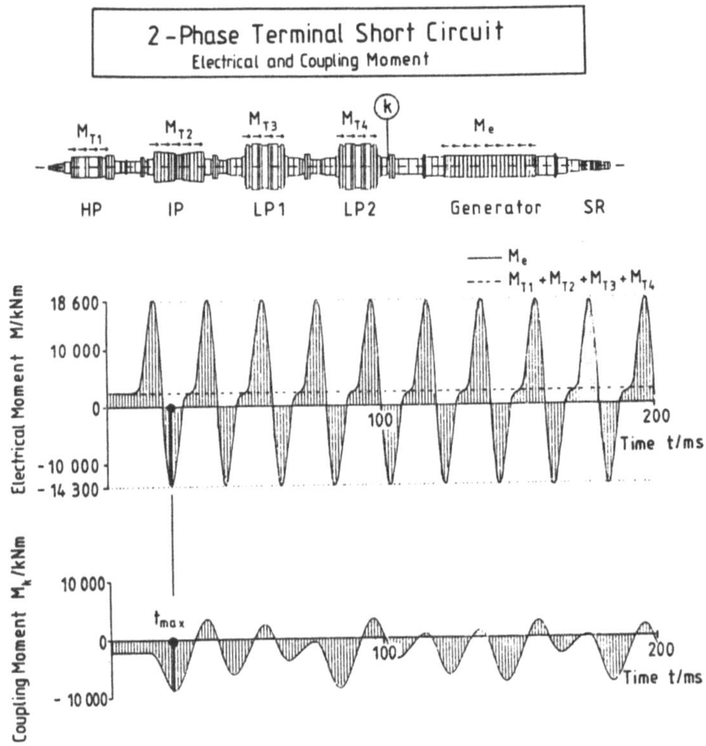

Fig. 10 Excitation moments and corresponding coupling moment at LP2-Generator

corresponding modal elastic moment in Fig. 4.

So when we use modal analysis we get some information on the composition of the systems response. If we are only interested in the maximum response we can also use a technique based on modal analysis to save further computer time - this technique is called 'Response-Spectrum-Method' (RSM).

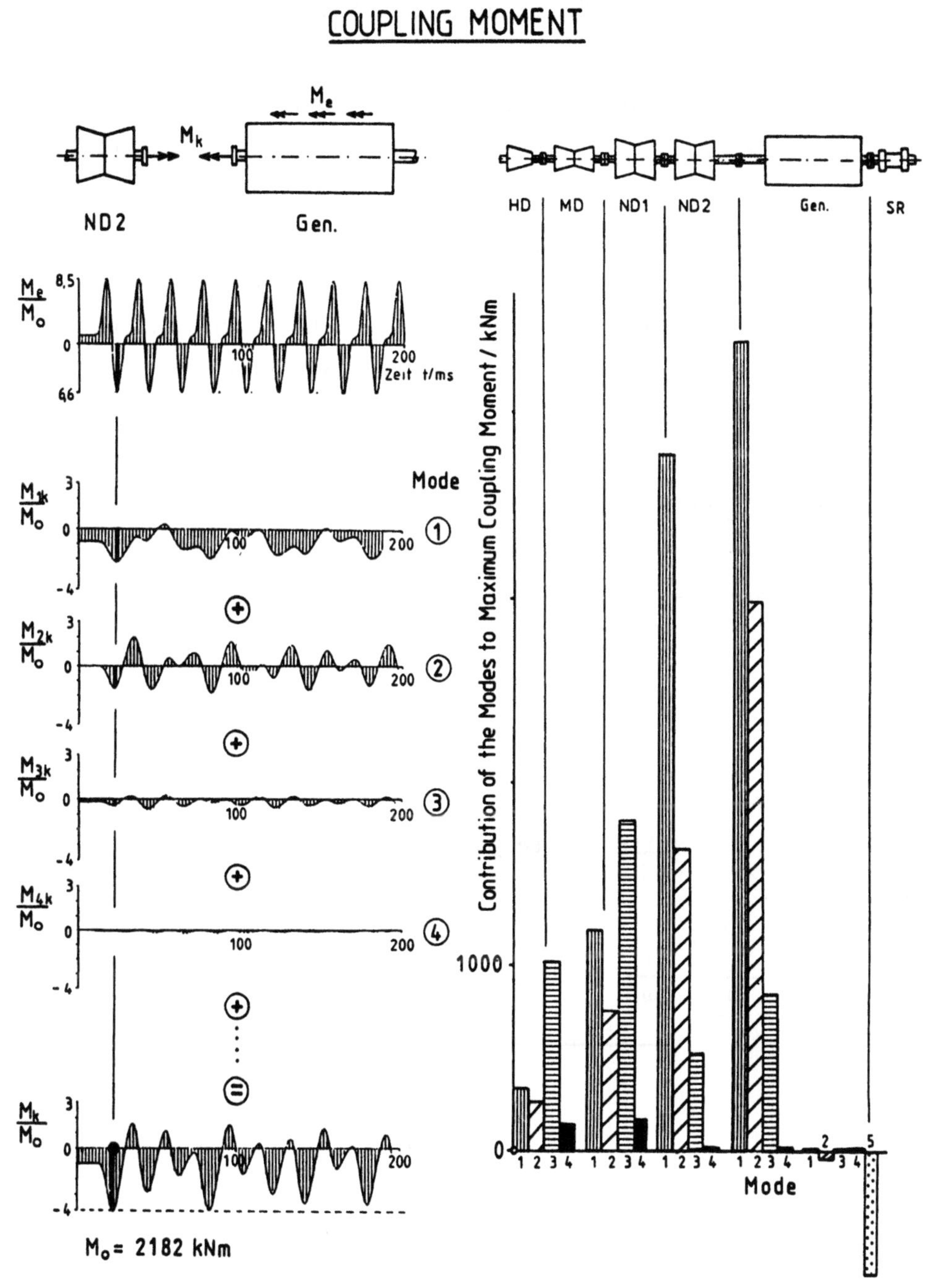

Fig. 11 Contribution of the modes to the maximum coupling moment

10.2.5 Calculation of the Transient Vibrations With Modal Analysis - Response Spectrum Method (RSM)

In this point we apply the RSM to approximate the torsional response of a turbogenerator set to transient excitation.

Basic idea and theory. The response of our linear system superimposes from the contributions of modal SDOF, Fig. 6. Equation (20) gives the solution of a modal SDOF to an arbitrary excitation. We consider the case that the excitation vector $\underline{F}(t)$ in (16) may be written as

$$\underline{F}(t) = \hat{\underline{F}} \cdot h(t) \tag{30}$$

and therefore the modal forces (18)

$$g_i(t) = \underline{\varphi}_i^T \hat{\underline{F}} \cdot h(t) \tag{31}$$

where h(t) is an arbitrary 'time function'.

If (31) is valid the response $\eta_i(t)$ from (20) can be written as

$$\eta_i(t) = \frac{\underline{\varphi}_i^T \hat{\underline{F}}}{m_i} \underbrace{\frac{1}{\omega_i} \int_0^t h(\tau)\, e^{-D_i \omega_{io}(t-\tau)} \sin \omega_i (t-\tau)\, d\tau}_{\eta_i^{(g)}(t)} \tag{32}$$

or

$$\eta_i(t) = \frac{\underline{\varphi}_i^T \hat{\underline{F}}}{m_i} \eta_i^{(g)}(t)$$

where $\eta_i^{(g)}(t)$ is the modal response of the i-th SDOF to the excitation h(t), Fig. 12, and the term

$$W_i = \frac{\underline{\varphi}_i^T \underline{F}}{m_i} \tag{33}$$

is called 'participation factor'.

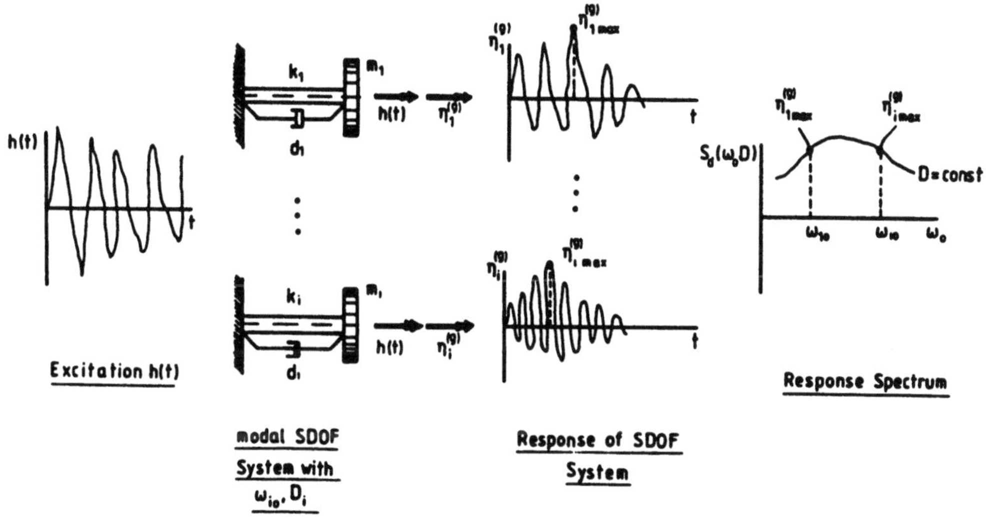

Fig. 12 Response of modal SDOF to excitation h(t)

Instead of the 'Time-History' that means the complete time function we work further on only with the maxima of $\eta_i^{(g)}(t)$.

$$\eta_{i\max}^{(g)} = \max\{\frac{1}{\omega_i} \int_0^t g(\tau)\, e^{-D_i \omega_{io}(t-\tau)} \sin \omega_i (t-\tau) d\tau\} = S_d(\omega_{io}, D_i) \tag{34}$$

This value we plot in a diagram as the maximum response of a modal SDOF, characterized by ω_{io} and D_i, to the excitation h(t). If we examine other modal SDOF with varied eigenfrequencies ω_{io} and modal damping D_i, and store always the maximum in a 'Response-Spectrum', we obtain the following set of curves

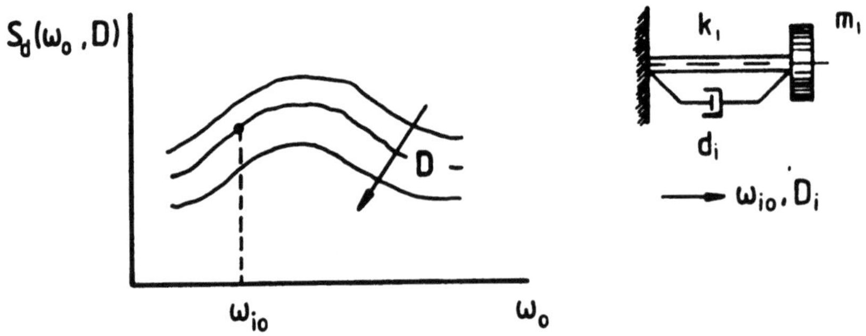

Fig. 13 Response-Spectrum

Besides the 'Response-Spectrum' for the displacement also spectras for the velocity resp. acceleration can be calculated.

We know the maximum of η_i, which is

$$\eta_{imax} = \eta_{imax}^{(g)} \frac{\underline{\varphi}_i^T \hat{\underline{F}}}{m_i} = \frac{\underline{\varphi}_i^T \hat{\underline{F}}}{m_i} S_d(\omega_{io}, D_i) \tag{35}$$

and therefore the maximum of the i-th modal contribution to the displacement $\underline{q}$ (13)

$$|\underline{q}_i|_{max} = |\eta_{imax} \underline{\varphi}_i| = \frac{\underline{\varphi}_i^T \hat{\underline{F}}}{m_i} S_d(\omega_{io}, D_i) \tag{36}$$

and to the coupling moments (29)

$$|\underline{f}_{si}^e|_{max} = |\eta_{imax} \underline{f}_{mi}^e| = |\frac{\underline{\varphi}_i^T \hat{\underline{F}}}{m_i} S_d(\omega_{io}, D_i) \underline{f}_{mi}^e| \tag{37}$$

<u>An example</u>

We apply the RSM to approximate the transient response of a 722 MW turbogenerator[1)] to a 2-phase terminal short circuit excitation. The approximations with the RSM are compared with the 'exact' solutions of the THM.

First the 'Response-Spectrum' for a excitation function correlated to a 2-phase terminal short circuit, Fig. 9, was calculated. The corresponding electrical moment contains mainly the single and double net frequency (= 60 Hz resp. 120 Hz in the United States). We have assumed in (30) that only one excitation function works at the system. But our shaft is excited with the constant steam forces <u>and</u> the generator moment, which is constant first and starts oscillating when the short circuit occurs. In this special case of excitation, the static preload condition can be transformed to an equivalent system with zero initial conditions and an electrical moment to which the static part is added /8/. Therefore we have a system which is excited by only one excitation function and (30) is valid.

Fig. 14 shows the calculated corresponding response spectrum of the displacements for different modal damping factors. It is important to note, that this response spectrum is independent of the considered system!
After the response spectrum has calculated once, the modes $\underline{\varphi}_i$ and the participation factors W_i are calculated for the considered 722 MW turboset - they are also marked in Fig. 14. The maximum of the i-th modal contribution $|q_i|_{max}$ can therefore with (36)be easily obtained from Fig. 14.

[1)] built for the United States

The question is now: How to superpose the contributions of all individual modes to get the total system response? The simplest approach is simply to add all the maxima of the individual modes, e.g. for the displacements

$$|\underline{q}|_{max} \approx \sum_{i=1}^{n} |\underline{q}_i|_{max} \tag{38}$$

But because the maxima in the different modes don't occur at the same time this approximation overestimates the solution in general. The experience shows that a mean square approximation yields much better results. We use

$$|\underline{q}|_{max} \approx \sqrt{\sum_{i=1}^{n} |\underline{q}_i|^2_{max}} \tag{39}$$

if the eigenfrequencies don't lie closely together, that means

$$\frac{\omega_{jo}-\omega_{io}}{\omega_{io}} > 0,1 \quad (1 \leq i < j \leq n) \, .$$

For eigenfrequencies lying closely together the 10%-method is used.

$$|\underline{q}|_{max} \approx \sqrt{|\underline{q}_k|^2_{max} + 2 \sum_{i \neq j} |\underline{q}_i|_{max} \, |\underline{q}_j|_{max}} \tag{40}$$

whereas i,j are indices of the eigenfrequencies for which

$$\frac{\omega_{jo}-\omega_{io}}{\omega_{io}} \leq 0,1 \tag{41}$$

is valid.

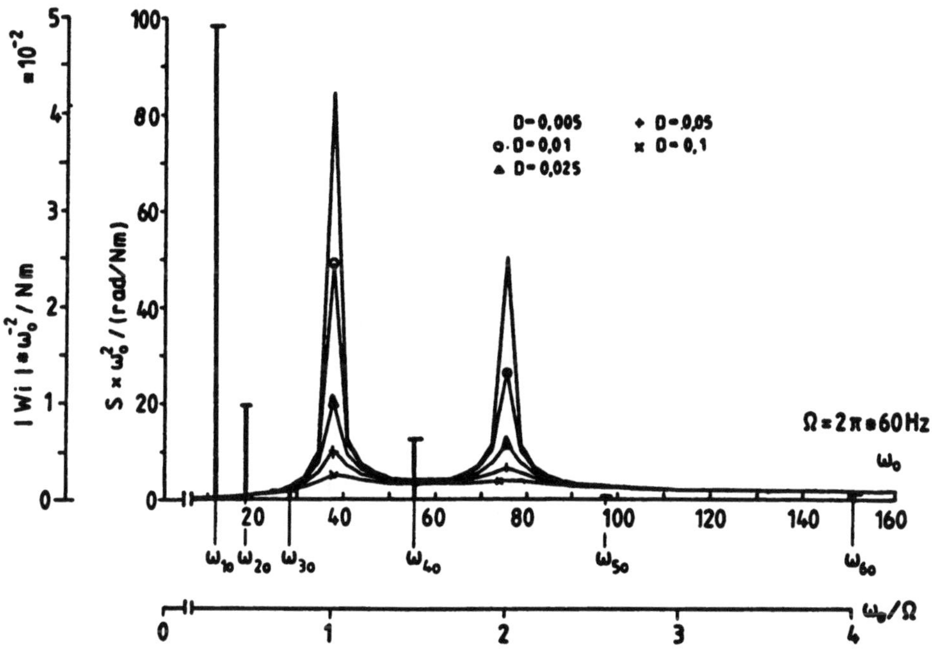

Fig. 14 Response-Spectrum for a two-phase short circuit and participation factors of a 722 MW turboset

The results for the maximum displacements are presented in Fig. 15, calculated with

- Time-History-Method
- 10% Response-Spectrum-Method
- to check - the sum of maximum response of all modes

The sum of maximum response of all modes overestimates the solution all over the shaft. The 10% RSM using mean square averaging

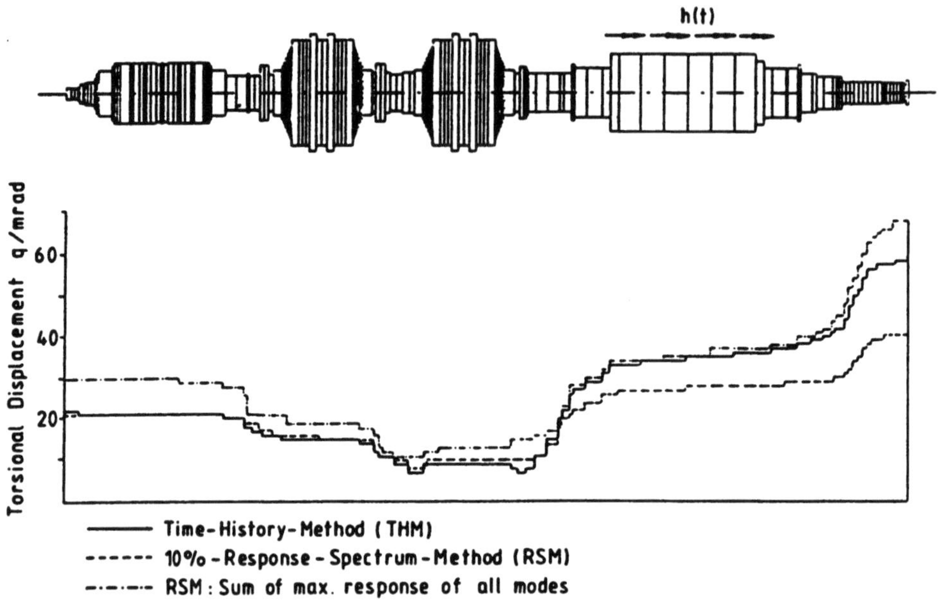

Fig. 15 Maximum displacements of a 722 MW turboset calculated with Time-History-Method, 10% Response-Spectrum-Method (40) and sum of maximum response of all modes (38)

yields quite good results in all shaft regions except the generator. Here the approximation yields to small displacements compared to the TH-Method. Probably in the response the maxima of two or several modes supperpose here. It is important to note that near the couplings, where the maximum twisting and the maximum stresses occur the RSM apprcximation is fairly good.

The intention of Fig. 16 is, to show the saving in computer time by using the RSM. In the THM about 16% of solution time is spent solving the eigenvalue problem and 84% calculating the response of the modal SDOF-Systems and internal forces. In comparison to that and under the condition the adequate 'Response-Spectrum' is already calculated the RSM needs 16% time for the eigenvalue problem and only 1% calculating the maximum response for the modal SDOF and the approximation of the internal forces. That

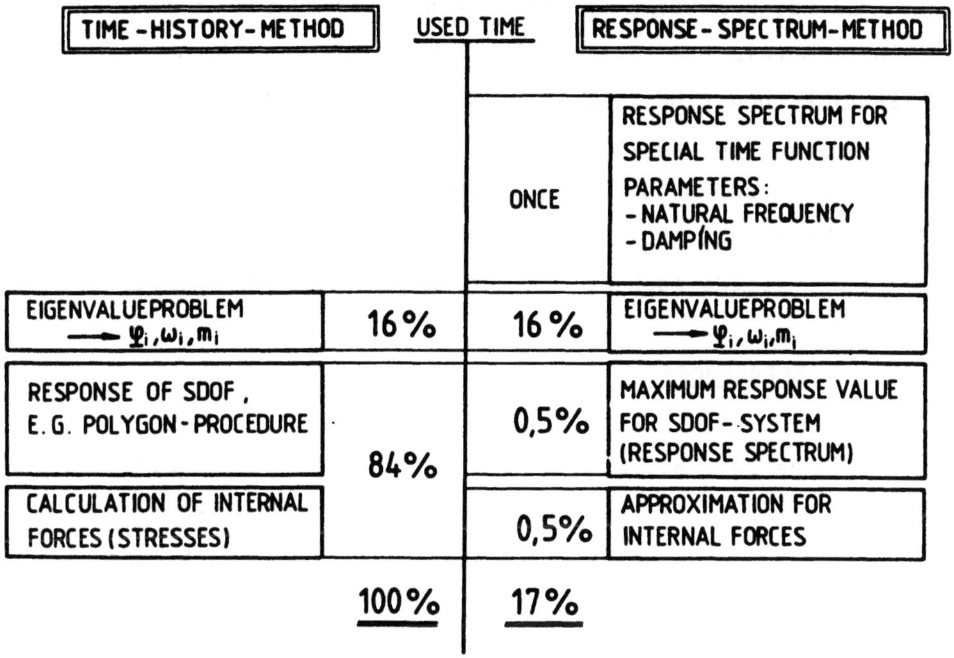

Fig. 16 Comparision of computer time used by 'Time-History'- and 'Response-Spectrum-Method'

means 83% time saved.

10.2.6 Conclusions

1. To calculate the torsional vibrations of turbine-generator sets the dynamic behavior of the shaft system is modelled by the finite element method. The result are the linear equations of motion. The solution for the natural vibrations yields the eigenfrequencies and modes together with the corresponding modal elastic moments.

2. For transient electrical excitation the equations of motion are solved with the modal analysis technique. The equations of motion are decoupled with the modes. As result we get n modal single degree of freedom systems (SDOF). For the solution of the SDOF an algorithm based on a linear interpolation of the excitation function proves to be most effective.

3. If the solutions of the SDOF are superposed in the time domain we obtain the total system response. We call this Time-History-Method (THM). If we are only interested in the maximum response we can use very effectively the Response-Spectrum-Method (RSM) as an approximation. The THM and the RSM are applied to a 722 MW turboset. It is shown that the RSM saves 83% of computer costs and yields good approximations in the coupling regions.

10.2.7 References

1. Berger, H., Kulig, T.S.: Simulation Models for Calculating the Torsional Vibrations of Large Turbine-Generator Units after Electrical System Faults, Siemens Forschungs- und Entwicklungsberichte, Band 10, 1981, Nr. 4.

2. Gonzales, A.J., Kung, G.C., Raczkowski, C., Taylor, C. W., Thomm, D.: 'Effects of Single- and Three-Pole Switching and High-Speed Reclosing on Turbine-Generator Shafts and Blades', IEEE Trans., Vol. PAS-103, 1984.

3. Canay, M., Rohrer, H.J., Schnirel, K.E.: 'Effect of Electrical Disturbances, Grid Recovery Voltage and Generator Inertia on Maximization of Mechanical Torques in Large Turbogenerator Sets', IEEE Trans., Vol. PAS-99, No. 4, 1980.

4. Schwibinger, P.: 'Torsionsschwingungen von Turbogruppen und ihre Kopplung mit den Biegeschwingungen bei Getriebemaschinen', PhD-Thesis, Univ. Kaiserslautern, 1986.

5. Krämer, E.: 'Maschinendynamik', Springer-Verlag, Berlin, 1984.

6. Lambrecht, D., Kulig, T.: 'Torsional Performance of Turbine Generator Shafts Especially Under Resonant Excitation', IEEE Trans., Vol. PAS-101, No. 10, 1982.

7. Hammons, T.: 'Electrical Damping and its Effect on Accumulative Fatigue Life Expenditure of Turbine-Generator Shafts Following Worst-Case Supply System Disturbances', IEEE-Trans., Vol. PAS-102, 1983.

8. Clough, R.W., Penzien, J.: 'Dynamics of Structures', Mc Graw-Hill, 1975.

CHAPTER 10.3

TORSIONAL DYNAMICS OF POWER TRANSMISSION SYSTEMS

N.F. Rieger
Stress Technology Incorporated, Rochester, New York, USA

ABSTRACT

The sources of torsional vibration in geared power transmission systems are reviewed. The modal analysis of typical geared systems is described, with useful formulas for natural frequency analysis. Damper design principles are reviewed, with criteria for suppression of troublesome modes. Two Case Histories from practice are discussed, indicating the diagnostic procedure used and the vibration suppression methods in each case.

10.3.1 Sources of Torsional Vibrations

Torsional vibrations occur in many machine systems. Rotating power transmission machinery is particularly susceptible to torsional vibration problems because of the long shaft sections and concentrated disk construction. Torsional pulsations arise in practice from typical sources such as: reciprocating mechanisms, e.g., internal combustion engines; impulsive excitation of rotating equipment, e.g., punch press, generator electrical transient, and in a wide range of other machine systems. A general classification of torsional excitation sources related to machine types is shown in Table 1.

Increased power transmission requirements are usually accompanied by an increase in machine size and an increase in operating speed. Increased power frequently means an increase in the magnitude of the applied fluctuating torque, in addition to an increase in the steady torque. Increased machine size means lower natural frequencies for the system torsional modes, and increased speed means more modes are possible within the machine operating range. Each factor associated with increased power output therefore tends to increase the vibration potential of the torsional system.

10.3.2 Theory of Torsional Vibrations

The simplest torsional systems are shown in Figure 1. The simple torsional oscillator, Figure 1 (a) obeys the same vibration laws as the simple linear oscillator: the analogous relations are shown in Table 2.

The two inertia semi-definite model shown in Figure 1 (b) is more representative of torsional systems, which usually have disks at their ends and are free to oscillate at these locations. Machine drive trains are typical of such systems.

The natural frequencies of such a system are readily found. Applying Newton's Law for free vibrations of the disks gives:

$$I_1\ddot{\theta}_1 = q(\theta_2-\theta_1)$$

$$I_2\ddot{\theta}_2 = -q(\theta_2-\theta_1) \tag{1}$$

In matrix form:

$$\begin{bmatrix} I_1 & 0 \\ 0 & I_2 \end{bmatrix} \begin{Bmatrix} \ddot{\theta}_1 \\ \ddot{\theta}_2 \end{Bmatrix} + \begin{bmatrix} q & -q \\ -q & q \end{bmatrix} \begin{Bmatrix} \theta_1 \\ \theta_2 \end{Bmatrix} = \begin{Bmatrix} 0 \\ 0 \end{Bmatrix}$$

For free harmonic vibrations assume solutions:

$$\theta_1 = Ae^{i\omega t} \; ; \; \theta_2 = Be^{i\omega t} \tag{2}$$

On substitution these expressions lead to the frequency equation:

$$I_1 I_2 \omega^4 - q(I_1+I_2)\omega^2 = 0 \tag{3}$$

The roots are:

$$\omega_1 = 0$$

$$\omega_2 = \left[\frac{q(I_1+I_2)}{I_1 I_2}\right]^{1/2} \tag{4}$$

The zero root is characteristic of semi-definite systems. It arises from the free-end conditions, and is referred to as a 'rolling mode,' in which both disks move (roll) in phase. This mode is only of significance in shock induced vibration problems.

The second mode is important. It consists of the disks vibrating against each other. Typically this mode is the lowest observed mode in all free-end torsional systems.

Modes of torsional systems may be calculated as follows: From equations 1 and 2 we have:

$$\theta_2 = \left(\frac{q - I_1\omega^2}{q}\right) \cdot \theta_1 \tag{5}$$

Knowing the natural frequency values and the system constants allows the coefficient of θ_1 to be found. By allowing θ_1 = 1.0, the value of θ_2 may be found, for ω_1 and ω_2.

Example: A turbocharger consists of two steel disks 6.0 inches diameter by 2.0 inches thick, connected by a uniform solid circular steel shaft 1.0 inch diameter and 10 inches long, as shown in Figure 2. Find the two lowest frequencies of this system.

o Disk inertia:

$$W = \pi/4\ D^2L \cdot \omega = \pi/4\ (36)(2)(0.283)$$

$$= 16.00 \text{ lb.}$$

$$I_p = \frac{1}{8}\,\frac{WD^2}{g} = \frac{(36)}{8(386.4)}$$

$$= 0.186 \text{ lb. sec. in.}^2$$

o Shaft stiffness:

$$q = \pi/32\ (12.10^6)\ 6^4/10$$

$$= 152.68 \times 10^6 \text{ lb. in./rad.}$$

o Natural frequency:

$$\omega_2 = \sqrt{\frac{q(I_1+I_2)}{I_1 I_2}} = \sqrt{\frac{(152.7)(10^6)(0.372)}{(0.186)^2}}$$

$$= 40.52 \times 10^3 \text{ rad./sec.}$$

$$f_2 = \omega_2/2\pi = 7449 \text{ hz}$$

o Angular relation:

$$\theta_2 = \left[1 - \frac{I_1\omega^2}{q}\right]\theta_1$$

$$= \left[1 - \frac{(0.186)\omega^2}{152.7 \times 10^6}\right]\theta_1$$

$$= [1 - 0.00122\omega^2]\ \theta_1$$

Mode 1: $\omega_1^2 = 0$ $\theta_2 = 1.0\ \theta_1 = 1.0$ (rolling mode)

Mode 2: $\omega_2^2 = 1641.87$ $\theta_2 = [1-2.00]\theta_1 = -1.0\ \theta_1$

These modes are illustrated in Figure 3. Note that the undamped mode analysis shown here is normalized on the displacement $\theta_1 = 1.0$ (selected arbitrarily). No knowledge of applied excitation or damping is required for this result.

The analysis of other torsional systems proceeds in a similar manner. The general torsional system shown in Figure 4 is found frequently in practice. With θ_1 and θ_4 for the disk displacements, θ_2 and θ_3 for the gear displacements, and q_1 and q_2 for the shaft torsional stiffness, the equations of motion for the end disks are:

$$I_1\ddot{\theta}_1 + q_1\theta_1 - q_2\theta_2 = 0$$

$$I_4\ddot{\theta}_4 + q_2\theta_4 - q_2\theta_3 = 0$$

Assuming the shafts have zero inertia (or that this has been included in the disks), and neglecting the inertia of the gears (close to node in mode 2), the equations of equilibrium at the gears are:

$$-q_1\theta_1 + q_1\theta_2 = FR_2$$

$$q_2\theta_3 - q_2\theta_4 = -FR_3$$

where F is the gear contact force, and R_1, R_2 are the gear radii. From these equations the lowest non-zero natural frequency is:

$$f_1 = \frac{1}{2\pi}\left[\frac{q_1 q_4}{I_1 I_4}\cdot\frac{I_1 + r^2 I_4}{q_1 + r^2 q_4}\right]^{1/2} \quad \text{hz}$$

and:

$$r = \frac{R_2}{R_3} = \frac{N_3}{N_2} = N,$$

when N is the shaft speed ratio.

Assuming that system excitation is caused by gear two only, the following relationship applies between the gear cycles at any time:

$$\theta_3 = r\theta_2 + ra_2\sin\theta_2$$

and:

$$\alpha_r = \frac{a_2}{R_2}$$

is the run-out angle for gear two, where a_2 is the eccentricity of the pitch circle about the center of rotation, as shown in Figure 4. For constant angular velocity (ω_2) of gear two:

$$\theta_2 = \omega_2 t$$

Solving the above equations, the dynamic force F between the gears at frequency ω_2 is:

$$F = \frac{-\alpha_2 q_1 q_4 I_1 I_4 \omega_2^2}{I_1 q_1 (q_4 - I_4 \omega_2^2) R_3^2 + I_4 q_4 (q_1 - I_1 \omega_2^2) R_2^2}$$

The response amplitudes at this frequency are:

$$\theta_1 = \frac{\alpha_2 K_1 K_4 I_4 R_2}{I_1 q_1 (q_4 - I_4 \omega_2^2) R_3^2 + I_4 q_4 (q_1 - I_1 \omega_2^2) R_2^2}$$

$$\theta_4 = \frac{\alpha_2 K_1 K_4 I_4 R_2}{I_1 q_1 (q_4 - I_4 \omega_2^2) R_3^2 + I_4 q_4 (q_1 - I_1 \omega_2^2) R_2^2}$$

It is important to note that in geared system cases where the shafts of a system rotate at m different speeds, the number of resonant (or critical) speeds of the system is then the number of shafts m times the number of natural frequencies n of the system, i.e., mn. Thus for the above case there are two shaft speeds (m = 2) and one natural frequency (ignoring $\omega_1 = 0$). Thus mn = 2, and the system has two resonant speeds. The reason for this is shown in Figure 5. The two per-rev excitation lines cross the natural frequency line at f_1 and f_2, the two resonant speeds.

10.3.3 Natural Frequencies of Shaft Systems

Formulas for calculating the natural frequencies of several shaft systems are given in Table 3. Each of these systems is semi-definite in nature. Thus, a three inertia system may be solved for the two significant eigenvalues directly. The frequency matrix for four order and higher systems must usually be solved by some root-seeking procedure.

Table 3 gives frequency equations for branched systems and for looped systems. Many cases frequently encountered in practice may be solved conveniently using these results. There appears to be no information in the open literature on looped systems, though branched systems are occasionally discussed. The formulas relate to both harmonic oscillations and to transient system dynamics, in which the first step in a modal analysis is to determine the natural frequencies and modal vectors of the system. Some experiences with branched and looped systems are described in Case Histories 1 through 4 herein.

10.3.4 Viscous Damper Analysis and Design

The following general rules have been shown to be generally effective for the suppression of torsional vibrations:

- Eliminate the source of vibration.
- Attenuate the vibration response.
- Provide additional damping in the responsive modes.
- Improve the fatigue resistance of affected components.

This section is concerned with rules (b) and (c). Consider the undamped single degree of freedom system shown in Figure 6 (a) which has a single natural frequency ω_o. If a second inertia is added as shown in Figure 6 (b), the system then has two natural frequencies ω_1 and ω_2. These frequencies are displaced about ω_1 as shown in Figure 6 (c). Evidently, the addition of an additional inertia is a method for changing the dynamic characteristics of a simple system. However, without the addition of damping, the resonant peaks will still involve very large amplitudes of vibration.

Note that this de-tuning has been achieved at the potential price of a second system resonant frequency.

This above detuning procedure is a method for reducing the vibration amplitudes of certain systems which receive excitation at some constant frequency which is situated close to a natural frequency of the system. In practice it is usually only possible to add small inertias I_2, e.g., the inertia ratio (I_2/I_1) frequently lies in the range $0.1 < I_2/I_1 < 0.3$. Such attenuation is usually possible only where the source of excitation is steady. Variable excitation would evidently re-introduce resonance problems.

A vibration damper can also reduce resonant vibration amplitudes. Again, consider an undamped single mass system, Figure 7 (a), to which a viscous untuned damper is added, as shown diagrammatically in Figure 7 (b). The damper has the effect of: (a) reducing the vibration amplitude, and (b) reducing the system natural frequency. (This may also detune the resonance). The reason is readily seen by considering two extreme cases:

Case 1: Damper coefficient $c = 0$. Zero damping, single degree of freedom system is unaffected, amplitude response as previously shown in Figure 7 (c).

Case 2: Damper coefficient $c = \infty$. Zero effective damping, no relative motion. Single degree of freedom system natural frequency reduced to $\omega_1 = \{K/(I_1 + I_2)\}^{1/2}$, as in Figure 7 (d). Amplitude response again high.

The practical case is the well-designed damper, which causes a reduced system natural frequency, and the relative motion $(\theta_2 - \theta_1)$ between the inertias causes energy dissipating slippage (damping) to occur.

It can be shown that the degree of damper effectiveness is related to the inertia ratio $\nu = I_2/I_1$. For optimum damper performance the required damping ratio ζ_D is given by:

$$\zeta_D = \frac{1}{\sqrt{2(2+\nu)(1+\nu)}}$$

where:

$$\zeta_D = \frac{C}{2I_2\omega_o} \qquad \text{and} \qquad \omega_o = \left(\frac{q_1}{I_1}\right)$$

The damped amplitude ratio is then given by:

$$\frac{\theta_1^D}{\theta_S} = \frac{2+\nu}{\nu} \quad ; \quad \theta_S = \frac{T}{q}$$

This may be compared with the amplitude ratio without the damper which is:

$$\left(\frac{\theta_1^u}{\theta_S}\right) = \frac{1}{2\zeta_u}$$

where ζ_u is the original system damping ratio. This factor was ignored in the damped design because it is almost negligible for many torsional systems, and $\zeta_D >> \zeta_u$.

For a typical inertia ratio $\nu = 0.2$, the amplitude ratio is $(2.2/0.2) = 11.0$, compared with an undamped amplitude ratio, which for $\zeta_u = 0.005$ is 100.0. The introduction of a viscous damper in such a case would therefore lead to a vibration amplitude reduction of 9:1 in the original system.

It has been found that optimum viscous damping occurs at the frequency where the two undamped response curves intersect, Figure 7 (b). This is called the 'tuning' point. The design condition is that the damper responsive curve has its peak amplitude at this tuning frequency. It is seen that the larger the inertia ratio ν, the greater the distance between the undamped peaks, and the lower the value of the damped peak amplitude.

From a practical standpoint, as the damper inertia I_2 increases, the inertia of the surrounding casing I_2^c also increases. This in turn, increases I_1 to $(I_1 + I_2^c)$ as I_2^c is attached to I_1. The actual inertia ratio is then:

$$\nu = \frac{I_2}{I_1 + I_2^c}$$

which being less than I_2/I_1 lessens the damper effectiveness. Lightweight casings which minimize I_2^c/I_2 are therefore another important aspect of damper design. Details of a practical viscous damper are illustrated in Figure 8.

10.3.5 Tuned Damper Analysis and Design

The effectiveness of the torsional dampers may be further improved if a suitable stiffness (or tuning) element is inserted in parallel with the dashpot, as shown in Figure 9 (a). The effect of this element is shown in Figure 9 (b). The original system response is shown, together with the two mode response of the corresponding undamped two inertia system. This response intersects the single inertia system response at two tuning points, as shown. By a suitable choice of damping constant, the damper response can be tuned to have local maximum values at the tuning points. Two cases must be considered when this is done, as follows:

Case 1: Response peaks equal at tuning points, Figure 9 (c).

Required inertia ratio:

$$\nu = \frac{I_2}{I_1}$$

Required damper ratio:

$$\zeta = \frac{C}{2I_2\omega_2} \text{ (see below)}$$

Required spring ratio:

$$Q = \frac{q_2}{q_1} = \frac{\nu}{(1+\nu)^2}$$

Amplitude ratio:

$$\frac{\theta_1}{\theta_S} = \sqrt{\frac{2 + \nu}{\nu}} , \quad \theta_S = \frac{T_o}{q_1}$$

Case 2: One response peak minimized, other disregarded, Figure 9 (d).

Required inertia ratio:

$$\nu = \frac{I_2}{I_1}$$

Required damper ratio:

$$\zeta = \frac{C}{2 I_2 \omega_2} \quad \text{(see below)}$$

Required stiffness ratio:

$$Q = \frac{q_2}{q_1}$$

These notations are defined in Figures 9 (a) through 9 (d).

Stiffness ratio calculation: Nestorides [1] discusses the following procedure for calculation of the stiffness ratio q_2/q_1. The required tuning condition reduces to the expression:

$$\omega^4 - \omega^2 \cdot \frac{2}{2+\nu} \{\omega_1^2(1+\nu)+\omega_2^2\} + \frac{2}{2+\nu}\,\omega_1^2\omega_2^2 = 0$$

or

$$(\omega^2-\omega_a^2)(\omega^2-\omega_b^2) = 0$$

The sum of the roots of this factorized expression $(\omega_a^2+\omega_b^2)$ is equal to the coefficient of the ω^2 term in the first expression times (−1), i.e.,

$$\omega_a^2 + \omega_b^2 = + \frac{2}{2+\nu} \cdot \{\omega_1^2(1+\nu)+\omega_2^2\}$$

This expression allows the required stiffness ratio to be found by specifying the frequencies ω_1 and ω_2 at which the system resonances are desired. These frequencies are substituted in the above biquadratic, which is solved for ω_a^2 and ω_b^2. These terms then give the amplitude ratios at the resonant peaks:

$$\left(\frac{\theta_1}{\theta_S}\right) a = \frac{-q_1}{q_q - (I_1+I_2)\omega_a^2}; \quad \left(\frac{\theta_1}{\theta_S}\right) b = \frac{q_2}{q_1 - (I_1+I_2)\omega_b^2}; \quad \theta_S = \frac{T_o}{q}$$

The required damping ratio does not influence these peak amplitudes, but it affects the off-peak amplitudes. Nestorides [1] shows that the required damping ratio ζ can be obtained from the following equation:

$$\left|\left(\frac{\theta_1}{\theta_S}\right)^2\right| = \frac{1}{\Delta}\left[\{\left(\frac{\omega_1}{\omega_2}\right)^2 - \left(\frac{\omega}{\omega_2}\right)^2\}^2 + 4\left(\frac{\omega}{\omega_2}\right)^2 \zeta^2\right]$$

$$\Delta = \left\{\left[\left(\frac{\omega_1}{\omega_2}\right)^2 - \left(\frac{\omega}{\omega_2}\right)^2\right]\left[1 - \left(\frac{\omega}{\omega_2}\right)^2\right] - \nu\left(\frac{\omega}{\omega_2}\right)^2\left(\frac{\omega_1}{\omega_2}\right)^2\right\}^2 + 4\left(\frac{\omega}{\omega_2}\right)^2 .$$

$$\zeta^2\left[1 - \left(\frac{\omega}{\omega_2}\right)^2 - \nu\left(\frac{\omega}{\omega_2}\right)^2\right]^2$$

Examples of use of the above procedures are described by Nestorides [1], Ker Wilson [2], and others.

10.3.6 Case History No. 1

Methanol Compressor Drive Train Torsional Vibrations

Problem Details:

The methanol compressor drive train shown in Figure 10 was exhibiting the following vibration symptoms:

- Excessive gearbox noise level, 103-105 dB.
- Rapid erosion of gear surface at pitch line (clearly evident after 72 hour operation).
- Intermittent sharp metallic knocks from within gearbox.
- Noise and knocks increased in severity with increased process loads.
- One gear pair removed from service due to pitch line wear six months after startup.

Machine Specifications:

Motor type:	AC squirrel cage rotor, 4 pole
Speed:	1780 rpm
Power:	1250 hp
Frequency:	60 Hz
Gearbox type:	Hobbed double helical
Ratio:	1.685:1
Quality:	AGMA Grade 8
Face Width:	10.875 inch
Pinion Teeth:	92
Gear Teeth:	155
Pitch:	9.148
Compressors:	Methanol gas
Axial Flow:	10 stage
Inlet Temp:	220°F
Outlet Temp:	650°F

Inspection Details:

- Impact damage was clearly evident upon inspection as a bright band at the pitch line after 72 hours of operation with replacement gears.
- The sharp intermittent (impact) noise from the gearbox was apparent, over the plant background noise (also high, around 90 dB).
- Couplings were gear tooth type.

- Compressors were rigidly secured to the poured concrete foundation. No special provisions for thermal growth.

Diagnostic Procedure:

- Preliminary diagnosis suggested that gear tooth impacts due to torsional vibration were the most likely cause. The bright metal surface at the pitch line was though to be due to plastic flow.

- The torsional natural frequencies and mode shapes shown in Figure 12 were calculated to provide data on the dynamic properties of the torsional system.

- It is evident from Figure 12 (b) that the second torsional natural frequency (1728 cpm) and the motor speed (1780 rpm) are very close so that a one-per-rev gear excitation could strongly excite the second mode.

- Since the gearbox is nearly at a node of the system, relatively small excitations result in large amplitude response.

- The most probable source of excitation was one-per-rev excitation from the low speed gear. This could arise from machining tolerances or from non-concentric mounting of the gear on its shaft.

- The motor is also a possible source of one-per-rev torsional excitations, due to air gap and electrical waveform non-uniformities.

Proposed Remedy:

- New gears were manufactured with a pitch of approximately 6 to AGMA 10 grade. This probably reduced the excitation and increased the impact wear resistance.

- An elastomer coupling was inserted between the drive motor and the gearbox. The intention was to detune the second torsional mode from resonance with rotational speed. The elastomer also provides a small amount of torsional damping.

- This remedy utilized off-the-shelf components, for which the torsional stiffness was known (manufacturer's data).

- The plot of system natural frequencies vs. coupling stiffness shown in Figure 13 was prepared to determine the effectiveness of the proposed remedy. The specific stiffness was selected from a range of available coupling options.

o Other possible remedies were:

a) Replace gearbox rolling element bearings with fluid film bearings. This would significantly increase system torsional damping. Rejected because of (a) possible center distance variation of gears, (b) machinery costs, and (c) delay involved.

b) Fluid coupling. Rejected because of costs and associated inconvenience (unnecessarily complex).

c) Torsional (Holset) damper. Rejected because detuning is preferred to absorbtion where possible.

Effectiveness of Fix:

Trouble-free torsional performance was observed twelve months after installation of the new coupling.

Note: System torsional measurements were made before and after the fix was installed. A CEC torsiograph (velocity transducer) was used to obtain waveforms for frequency analysis. The rotor of this device was conveniently attached to a stub shaft mounted to the end of the motor shaft. The stator is free in rolling element bearings. Signals are obtained directly without slip rings or telemetry.

10.3.7 Case History No. 2

Branched Compressor Drive Vibrations

Problem Details:

The branched compressor drive train shown in Figure 14 consists of a hot gas expander turbine driving two axial flow screw compressors. The drive input shaft carries through to the LP compressor. The gearbox power take-off involves a line of five gears driving the HP compressor. The screw compressor consists of two meshing helical groove rotors inter-connected by timing gears.

The machine exhibited vibrations throughout its structure from the time of startup. The problem became critical when the HP compressor slipped its drive timing gears, allowing the rotors to impact together torsionally during rotation. This caused impact damage to the helical surfaces, and an alarming noise. The timing gears were replaced and doweled into place, and the machine operation was measured with an IRD hand-held pickup. Unacceptable vibration levels remained as follows:

- o Torsional vibrations of HP compressor rotor.
- o Axial and lateral vibrations of the turbine.
- o Many harmonics of rotational frequency present in the LP compressor vibration signature, including half running speed (0.5 x)
- o Some indication of compressor surge was detected in the HP output pressure trace.
- o Vibration was also strong on the gearbox casing.

System Parameters:

Turbine:	4500 rpm, 3000 HP
1st Compressor:	4500 rpm, 1500 HP
2nd Compressor:	3000 rpm, 1500 HP
Gearbox:	Hobbed double helical, ratio (a) 1:1.000, (b) 1:0.57, 7.0 diametral pitch, plain cylindrical sleeve bearings, 10,400 rpm at 4500 rpm surface speed

Inspection Details:

Several possible sources of vibration existed:

- o Unbalance vibrations of the turbine (frequency 75 Hz).
- o Unstable turbine bearing operation (frequency 38 Hz or bending critical frequency, depending on nature of system, rigid, or flexible rotor).
- o Torsional vibrations of system. Several possible excitation causes exist.
- o Coupling misalignment or foundation settlement.
- o Observed gearbox casing frequency of 0.47 times turbine speed, i.e., 35.3 Hz or 2120 rpm.
- o Some 2x vibration evident on bearings of both shafts.
- o Thermal distortion of the long gearbox.

Possible Causes:

- o Unbalance vibrations of the turbine could cause lateral vibrations of the turbine bearings and casing. Strong vibrations could occur if the turbine was running close to its critical speed.

- o Unstable whirling of the turbine rotor in its bearings could occur above twice the rotor critical speed, or around $N_u = 500/\sqrt{c}$, where c is the bearing radial clearance.

- o Torsional vibrations could arise from drive gear run-out, timing gear run-out, and from gas pressure pulsations on the rotor helices.

- o Thermal distortions of the casing could unload bearings and make them unstable.

- o Gas pulsations could cause the axial vibrations observed at the turbine couplings.

Diagnostic Procedure:

The primary cause (or causes) of the timing gear slippage was sought. Incidental possibilities were attended to as part of the correction.

- o Balance of the turbine rotor was checked. Bearing gap accelerations were high but not excessive. Fix: balance the rotor at the next maintenance shutdown.

- o The turbine critical speed for the rotor in its bearings was calculated to be 3100 rpm. Rotor critical speed vibrations were rejected as the primary cause.

- o Hot distortion of the gearbox was checked optically (surveyors theodolite) and found to be small: good foundations and bolting.

- o Bearing instability was rejected as (a) the turbine runs below 6200 rpm, and (b) the rigid rotor instability speed is $N_u = 500/\sqrt{c} = 500/\sqrt{0.002}$ = rpm.

- o Torsional natural frequencies for this system were calculated to be 12.7, 17.7, 34.7, and 43.3. Hz. Corresponding critical speeds are shown in the resonance diagram, Figure 15.

- o Gear run-out pin checks were available: see Figures 16 and 17. All gears had substantial 1x run-out. The second idler gear had 3x run-out. This 3x corresponded to the 0.47x turbine speed detected on the casing.

- o The torsional forced response of this system was calculated, Figure 18. Measured torsional amplitudes agreed well with the predicted response.

Primary Remedy:

A viscous shear damper was first proposed to suppress the compressor peak amplitudes, but it was shown that this would increase the non-resonant vibrations to unacceptable levels.

A quill shaft between the gearbox and the LP compressor was then studied and chosen to detune the torsional vibrations. The quill shaft detuned the compressor from the gearbox excitation without increasing the non-resonant vibrations.

Secondary Remedies:

Several additional steps were taken to reduce the overall machine vibrations, as follows:

- The turbine vibration problem was eliminated by re-aligning the turbine casing and re-balancing the turbine rotor with the coupling sleeve attached.

- All gears were machined to AGMA 9 standards and balanced to reduce the torsional excitation.

10.3.8 References

1) Nestorides, E., Torsional Vibration Analysis, British Internal Combustion Engine Research Institute Handbook, Cambridge University Press, 1958.

2) Ker Wilson, W. J., A Handbook of Torsional Vibration Analysis, Volumes 1 through 5, John Wiley and Sons, New York, 1958, Second Edition.

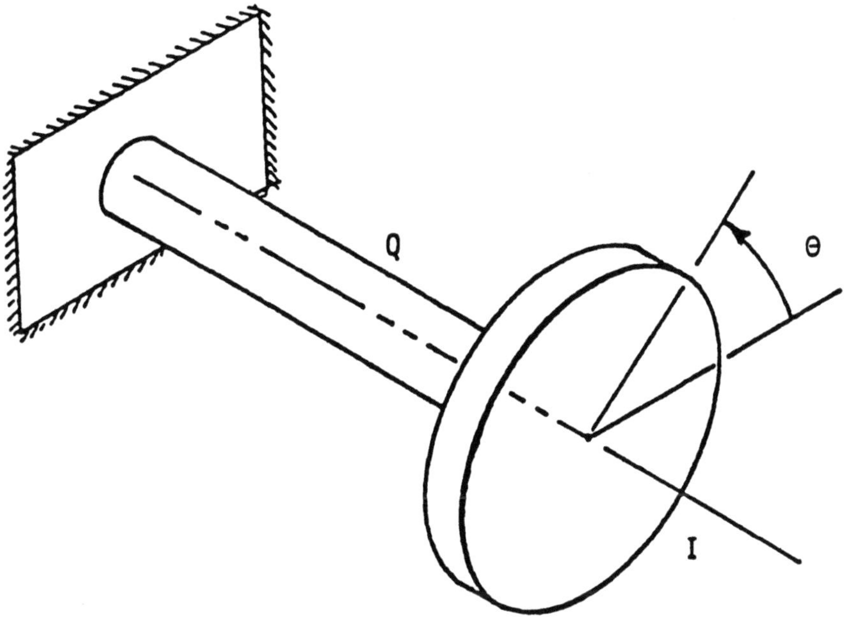

(a) Simple torsional oscillator

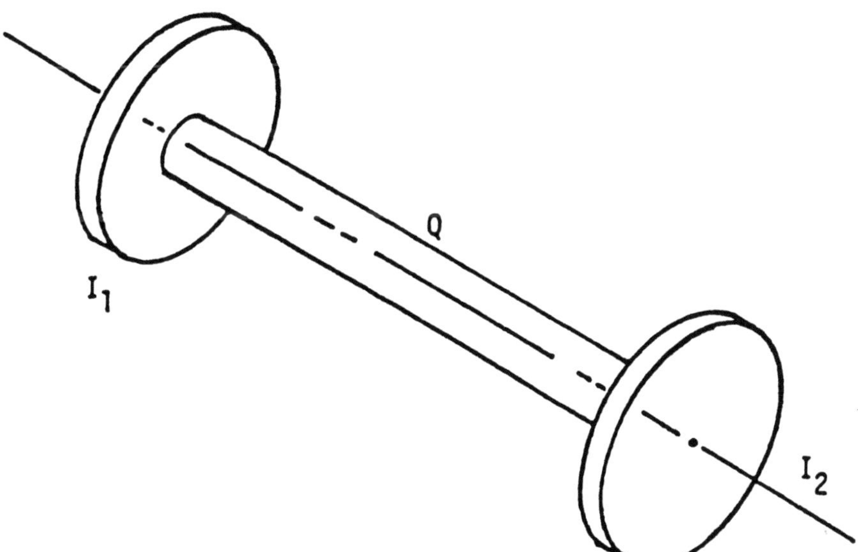

(b) Two inertia semi-definite system

Figure 1 Simple Torsional Systems

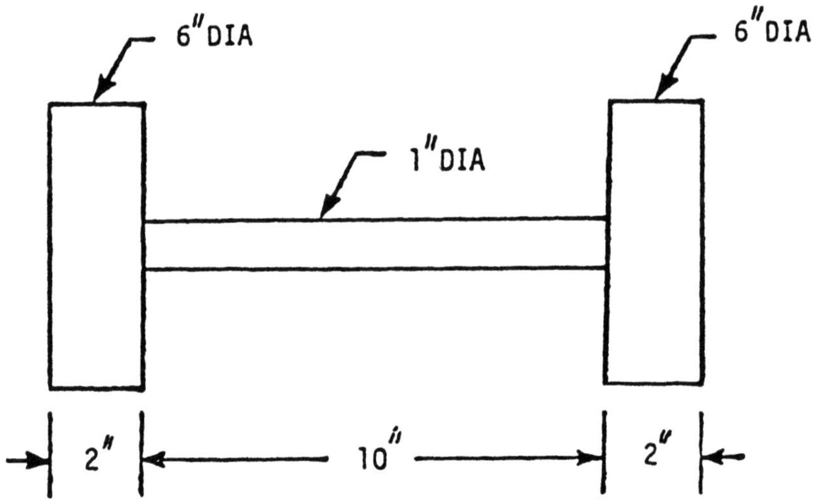

Figure 2 Dimensions of Turbocharger Rotor

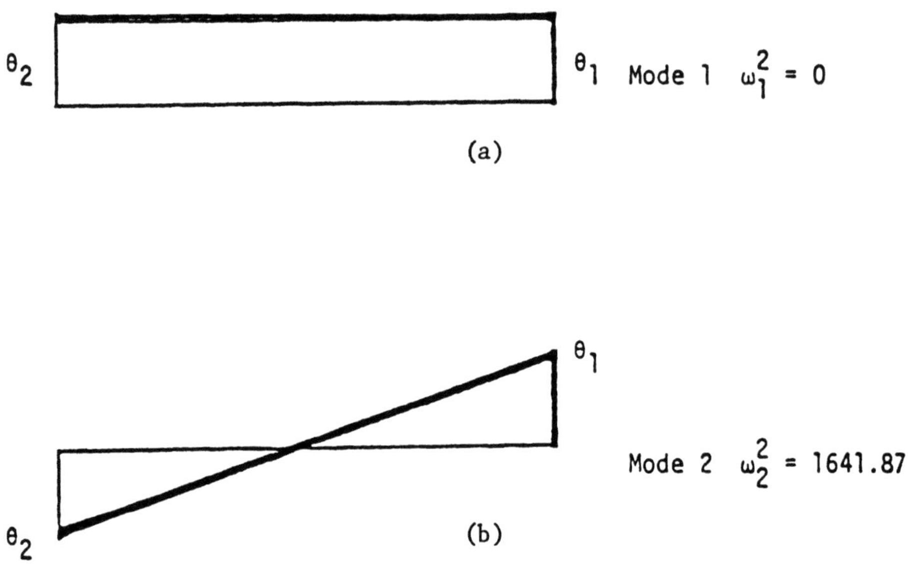

Figure 3 First Two Mode Shapes and Corresponding Natural Frequencies for Turbocharger

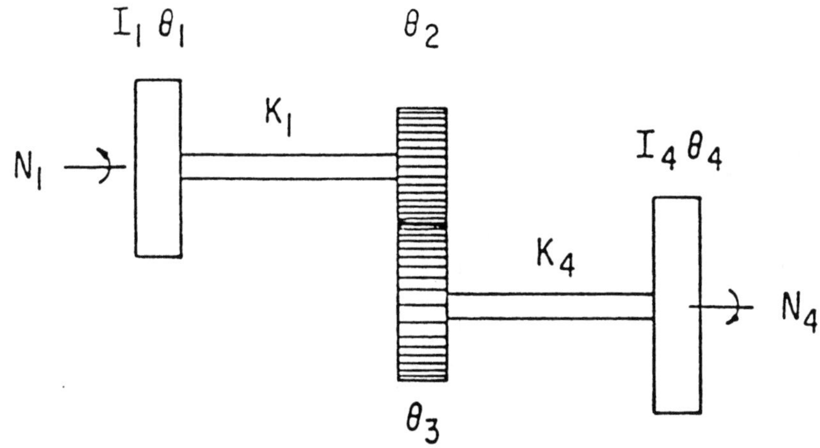

(a) Simple torsional system

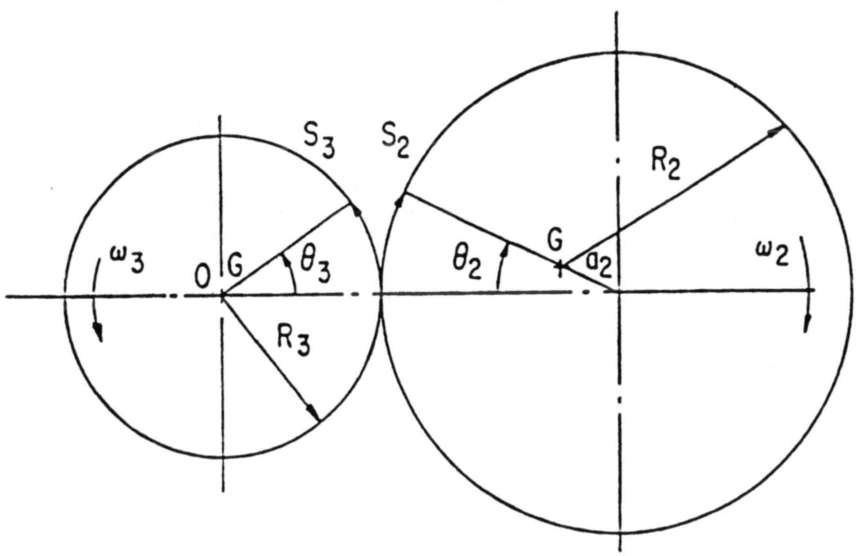

(b) Gear error geometry

Figure 4 Simple Geared Torsional System Showing Geometry of Gear Runout Error

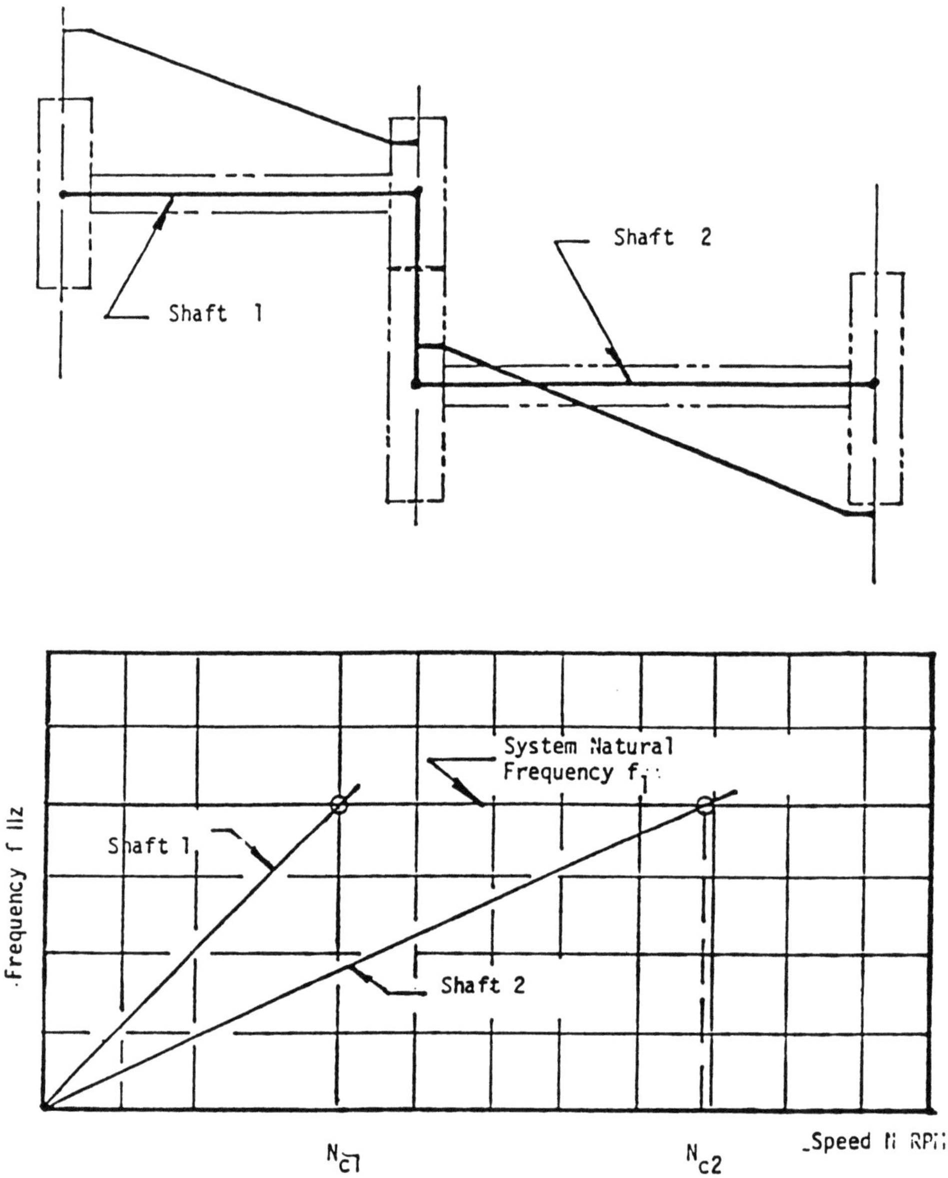

Figure 5 Resonance Diagram and Corresponding Mode for Geared Line Drive

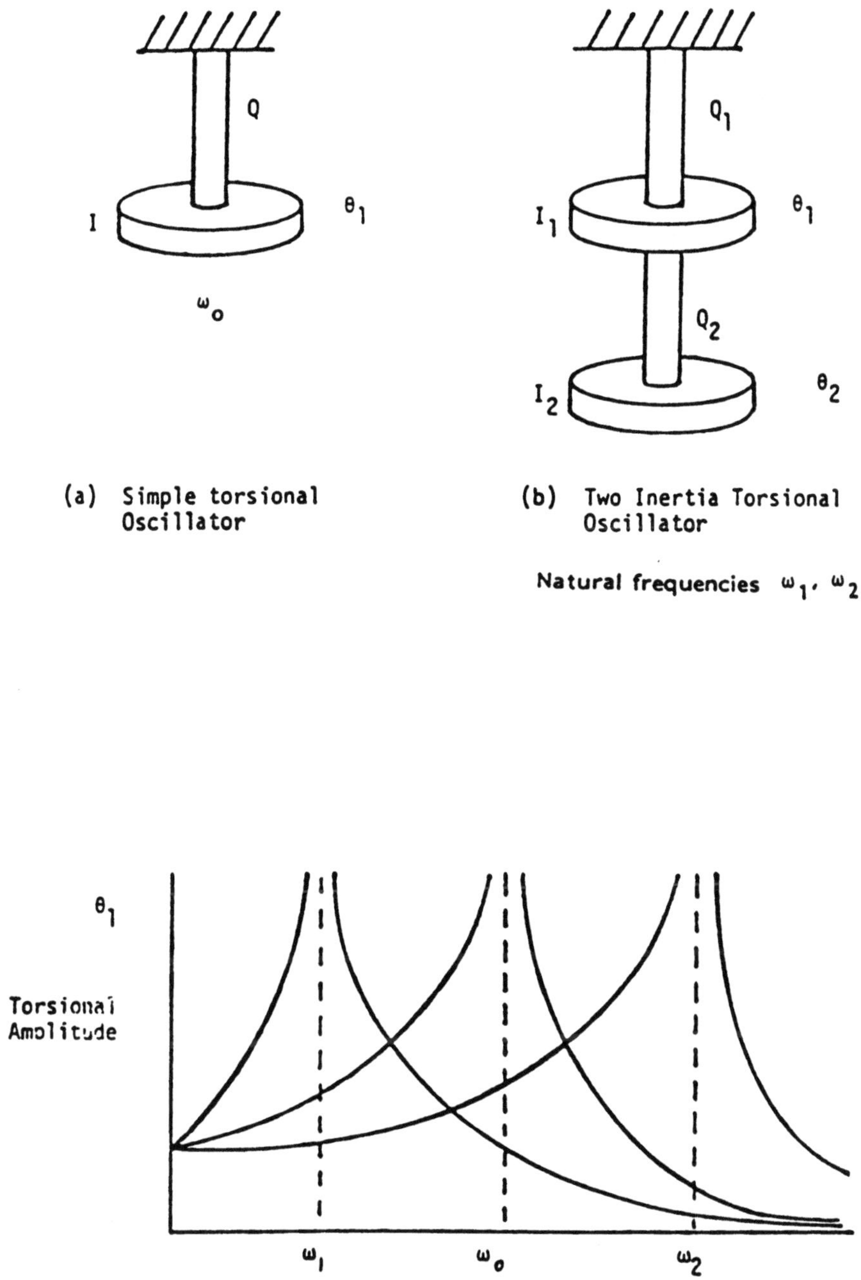

Figure 6 Torsional Response Attenuation at Frequency ω_o by the Addition of a Second Inertia

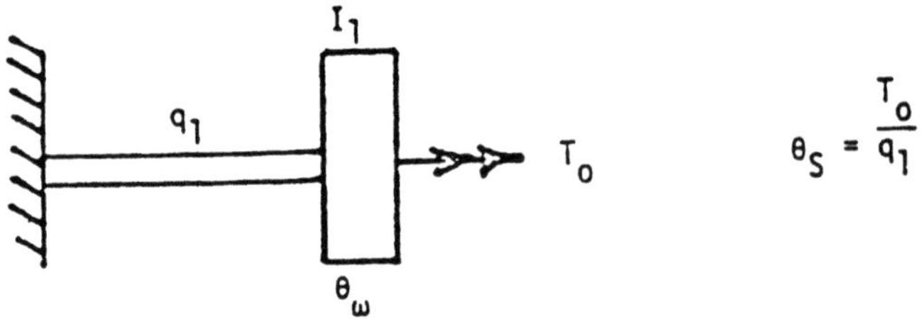

(a) Undamped single degree of freedom torsional system

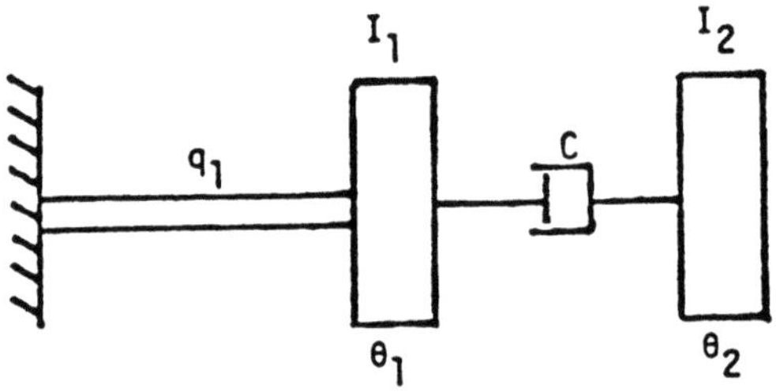

(b) Damped untuned torsional system

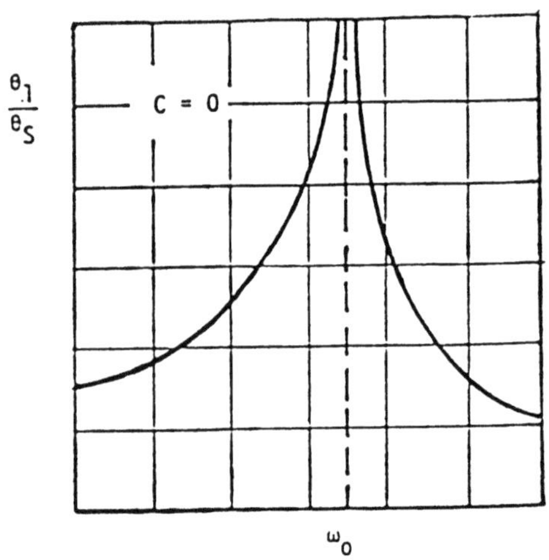

(c) Response of single degree System to T_0

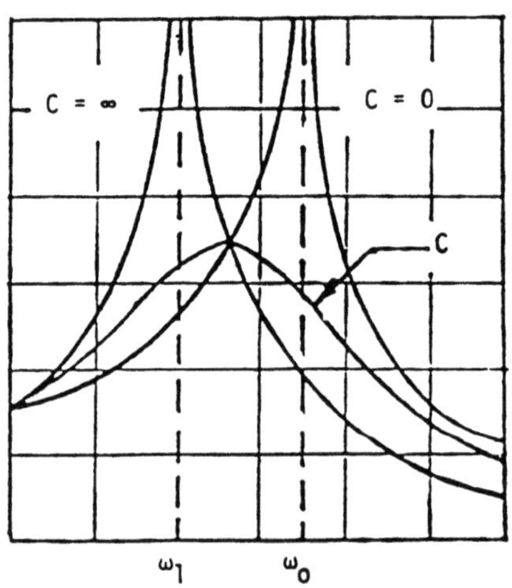

(d) Effect of damping on response

Figure 7

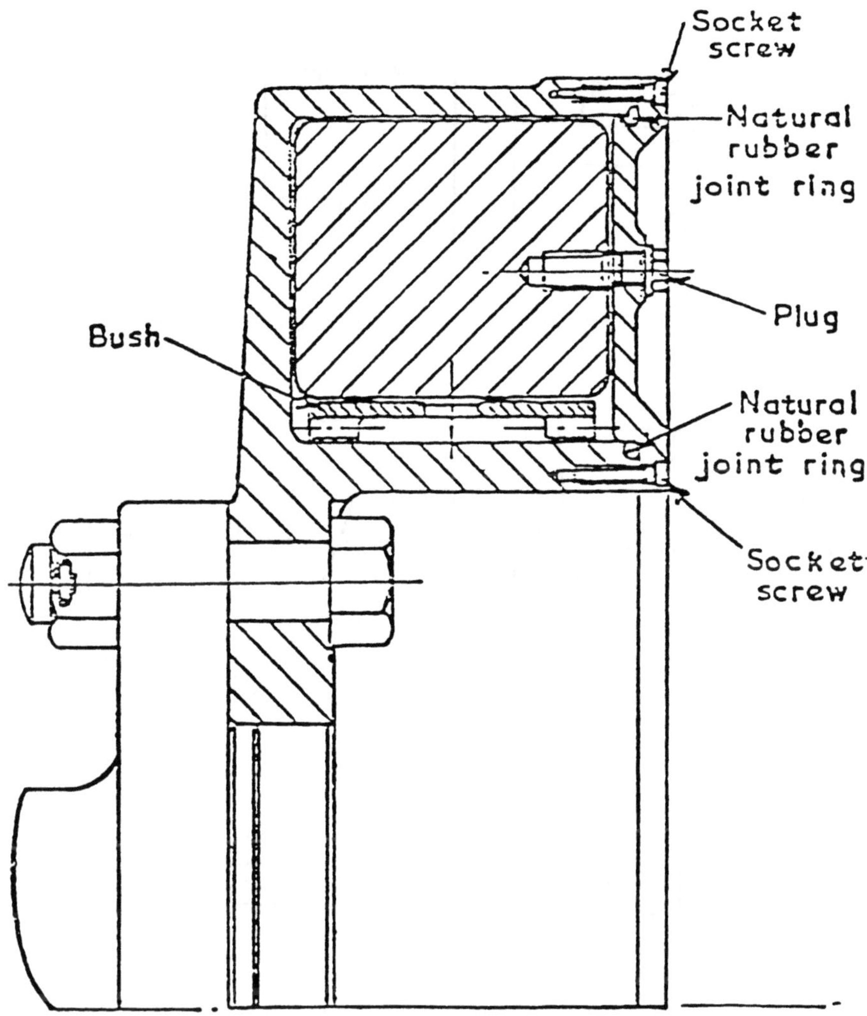

Figure 8 Viscous Shear Damper Details

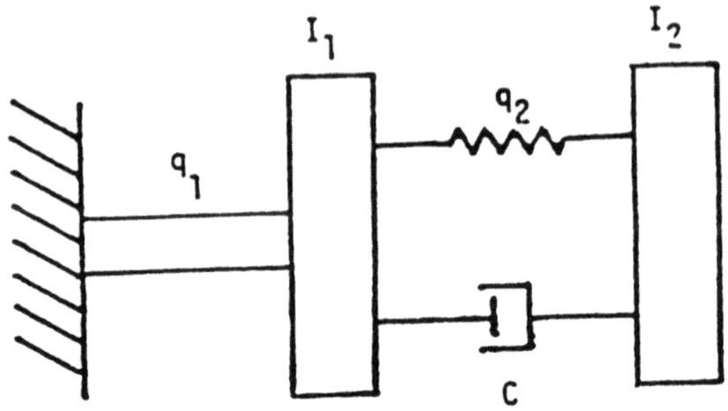

(a) Simple torsional system with tuned damper

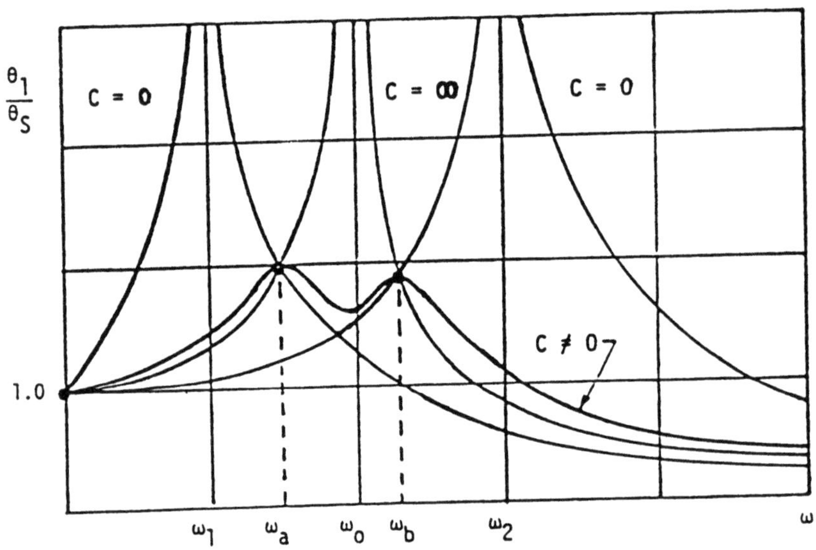

(b) Response of simple system for various amounts of damping

Figure 9

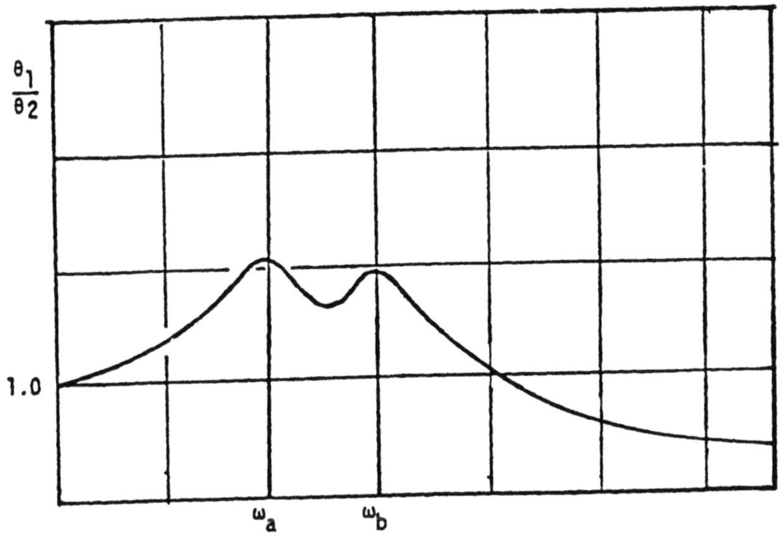

(c) Damper response with both peaks tuned equally

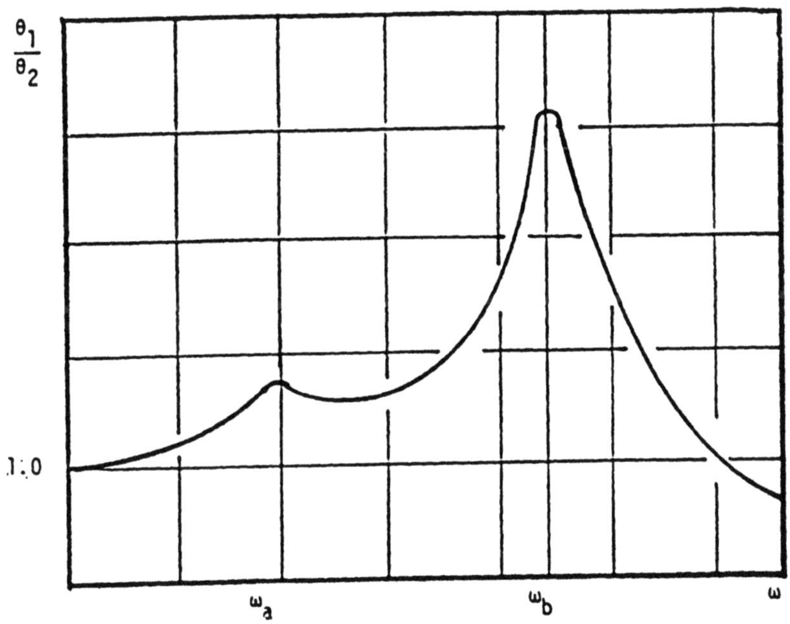

(d) Damper response with one peak tuned to low response at selected frequency ω_a

Figure 9 (cont')

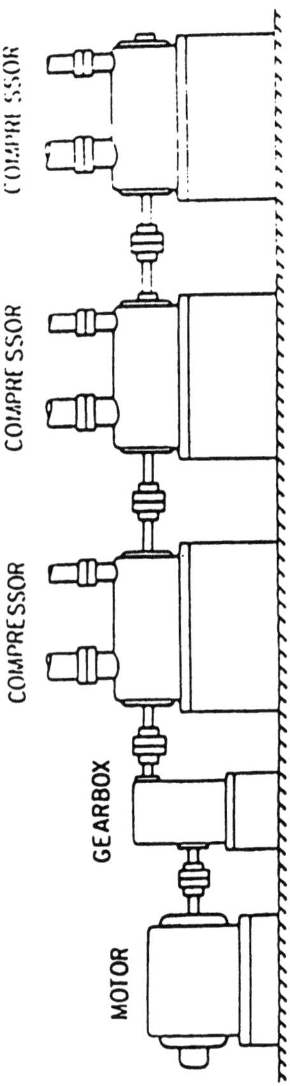

Figure 10 Methanol Compressor Drive Train Schematic

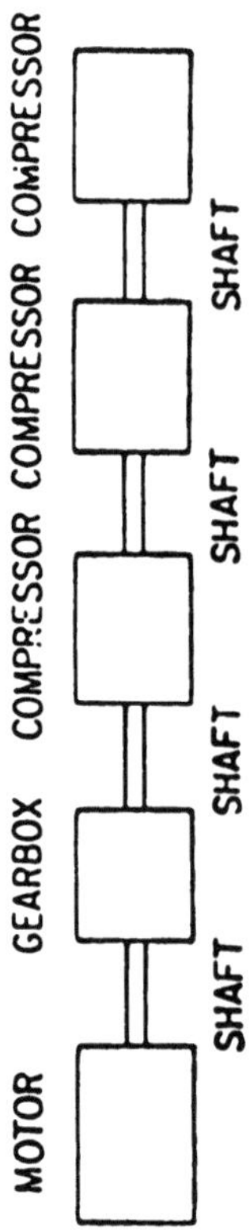

Figure 11 Block Diagram of Torsional System

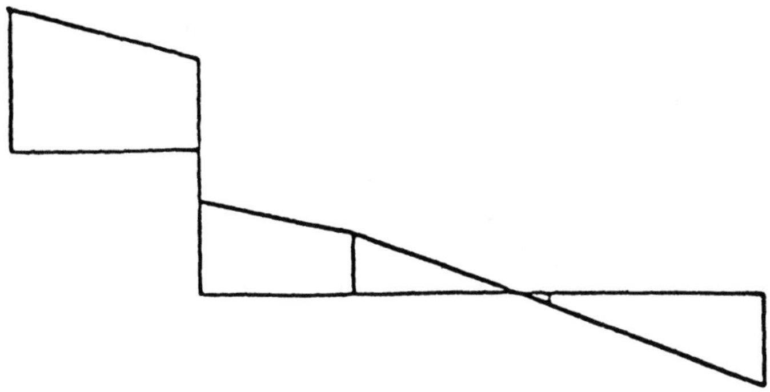

(a) First torsional mode. $f_1 = 960$ cpm

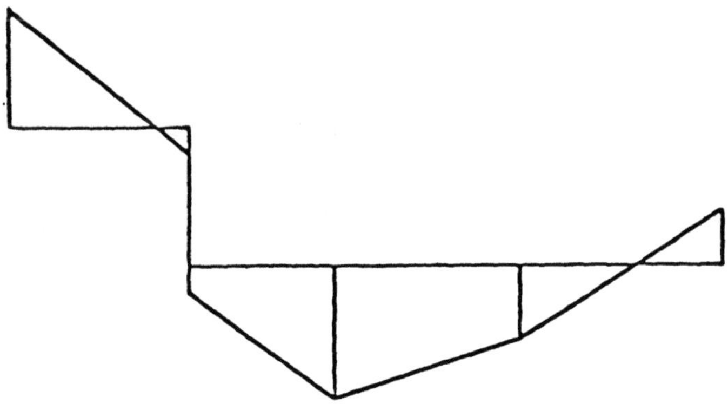

(b) Second torsional mode. $f_2 = 1728$ cpm

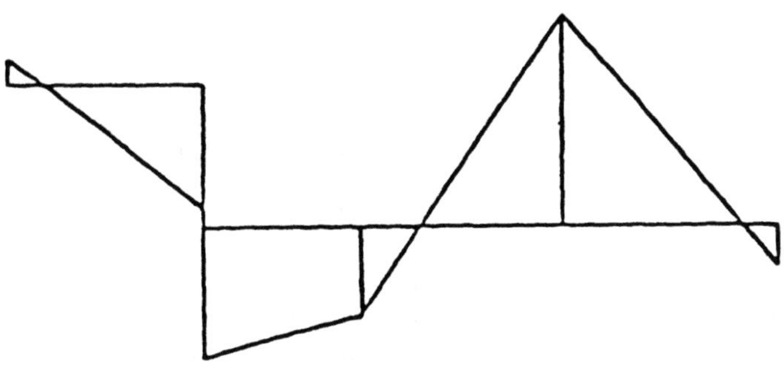

(c) Third torsional mode. $f_3 = 3792$ cpm

Figure 12 Torsional Mode Shapes and Natural Frequencies for Compressor Train

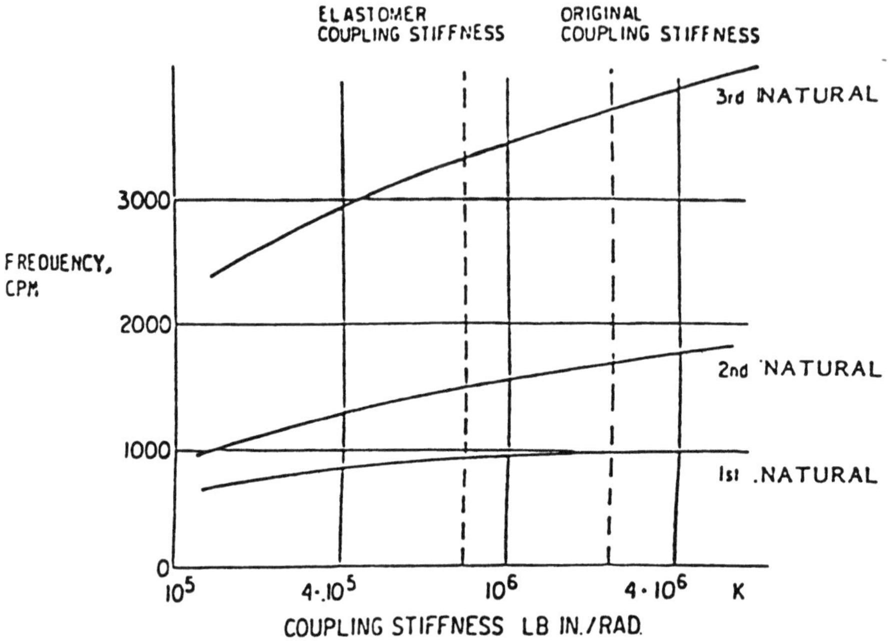

Figure 13 Variation of Natural Frequency vs. Coupling Stiffness

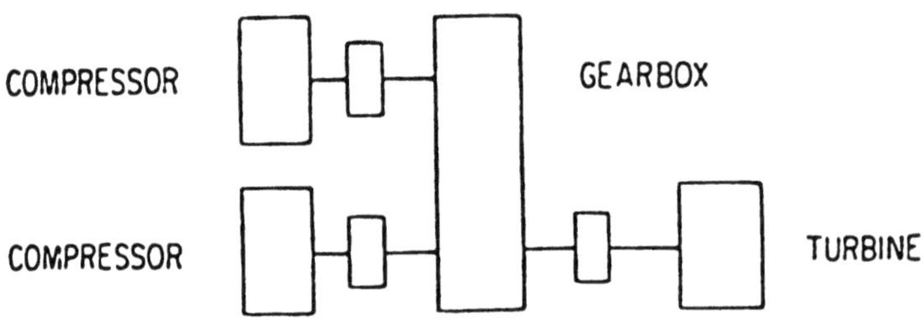

Figure 14 Schematic of Compressor Drive Train

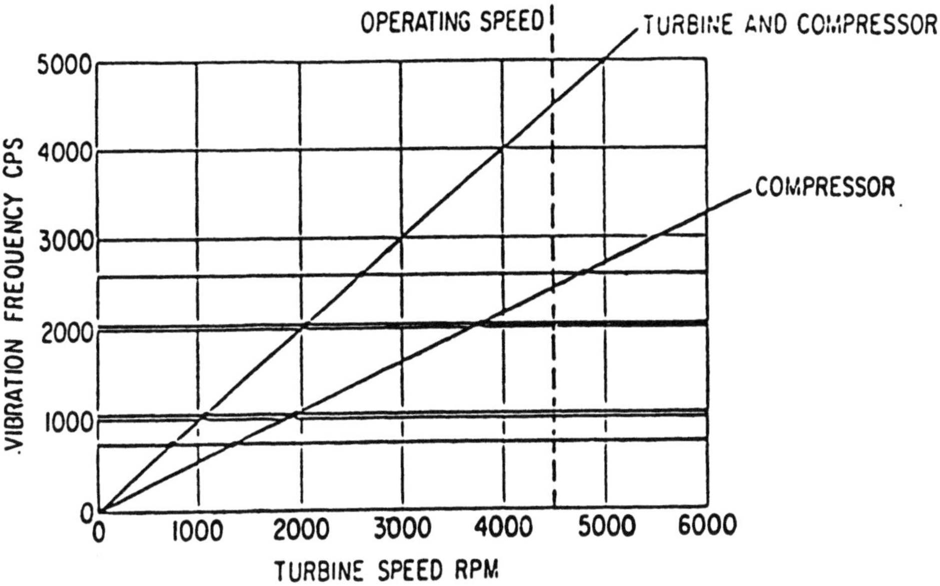

Figure 15 Resonance Diagram, Branched Compressor Drive

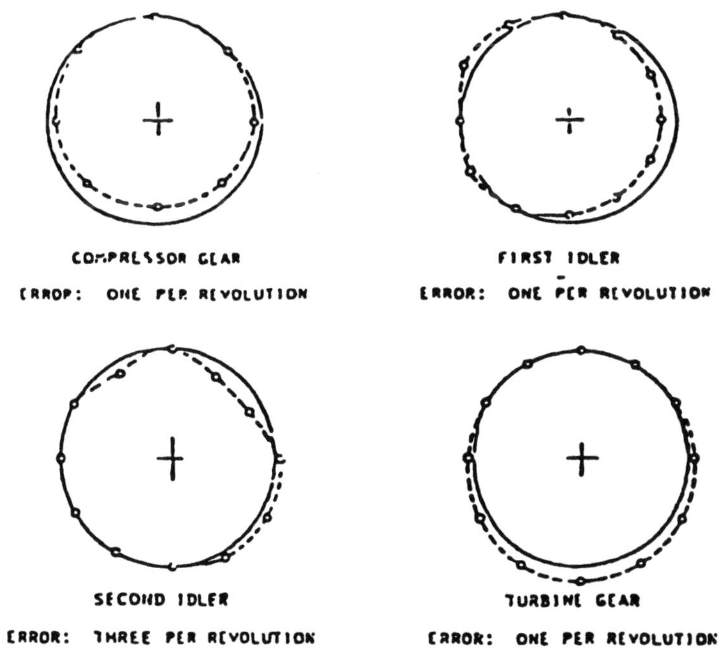

Figure 16 Gear Profiles, Branched Compressor Drive

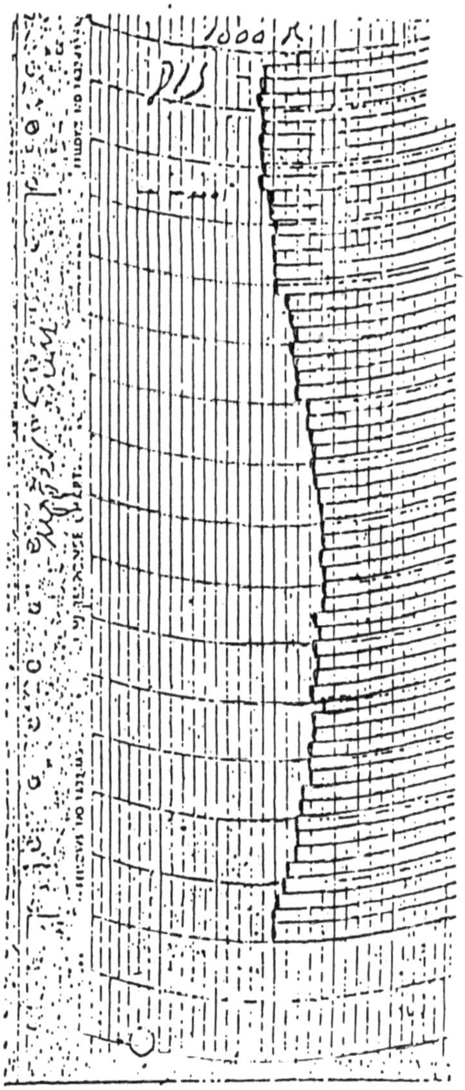

SUN GEAR INSPECTION TRACE
FOURIER AMPLITUDE

First 2.35×10^{-4} in. Second 0.39×10^{-4} in.
Third 0.27×10^{-4} in. Fourth 0.03×10^{-4} in.

57 T

Figure 17

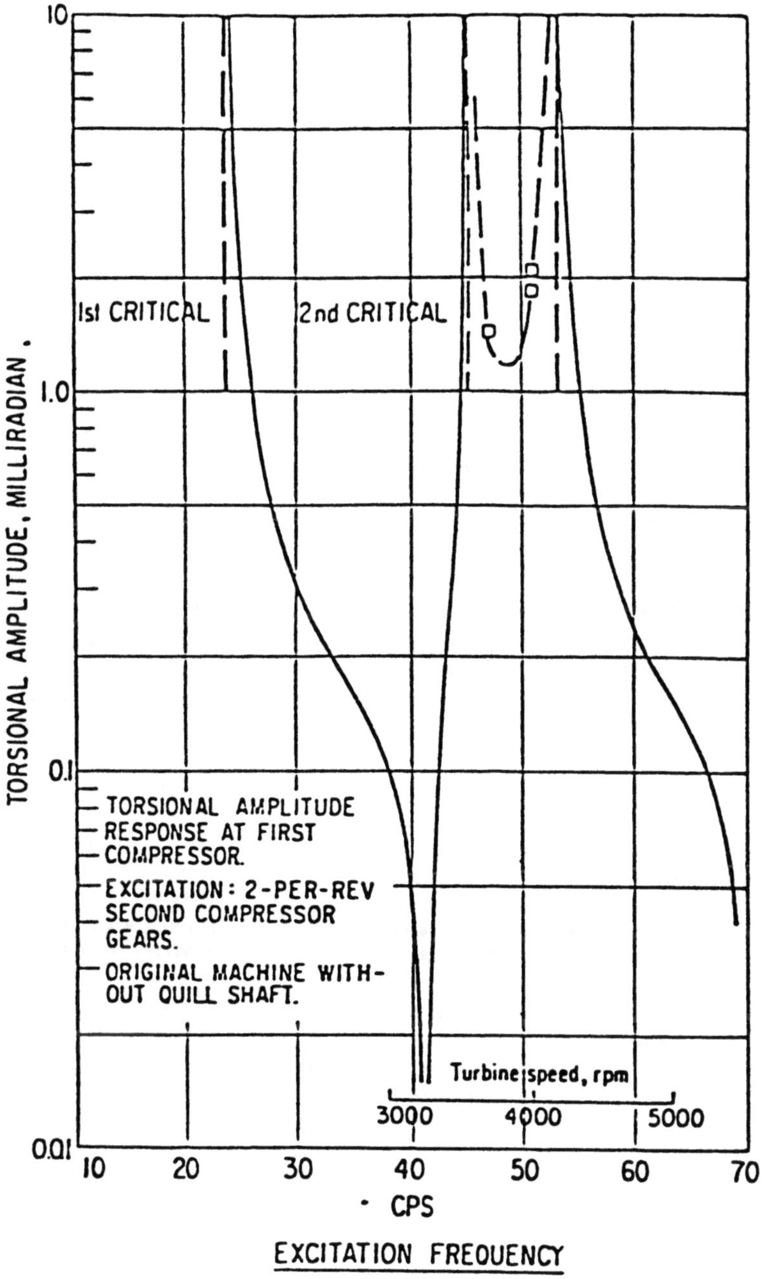

Figure 18 Amplitude Response vs. Frequency and Speed

Table 1 Typical Causes of Torsional Vibration

Class	Cause	Machine System	Typical Fix
1. Simple Harmonic	Gear runout	Compressor drive	Replace gear and balance. Avoid resonance.
	Electromagnetic	Turbine-generator set	Balance electrical loads. De-tune shaft.
	Hydrodynamic interactions	Marine propeller Ventilating fan	De-tune shaft with flexible coupling. Add viscous damper.
2. Complex Harmonic	Cylinder gas forces	Reciprocating engine Reciprocating compressor	Flywheel, viscous damper.
	Gear inaccuracies	Printing rollers	Stiffen shafts. Double gearing.
		Helicopter gearbox	Isolate gearbox. Fluid-film bearings. De-tune shafts.
	Turbocompressor blade discharge	Wind tunnel drive	De-tune shafts with viscous coupling.
3. Transient Shock	Electrical starter adjustment	Paper machine	Smooth starter. Strengthen shaft. Flexible input coupling.
	Start of load cycle	Blooming mill Punch press	Fluid-film bearings. Eliminate backlash. Flexible coupling de-tune.
	Coupling, gear backlash	Rolling mill	Eliminate backlash. Flexible, coupling de-tune.
	Indexing motion	Automatic coining press (laminates)	Tuned viscous damper.

Table 2 Analogy Between Rectilinear and Torsional Vibrations

	Rectilinear vibration		Torsional vibration	
	Symbol	Unit	Symbol	Unit
Time	t	sec	t	sec
Displacement	x	in	θ	rad
Velocity	$\dot{x}$	in/sec	$\dot{\theta}$	rad/sec
Acceleration	$\ddot{x}$	in/sec²	$\ddot{\theta}$	rad/sec²
Spring constant	k	lb/in	K	in-lb/rad
Damping coefficient	c	lb-sec/in	η	in-lb-sec/rad
Damping factor	ζ	dimensionless	ζ	dimensionless
Mass	m	lb-sec²/in	J	lb-in-sec²/rad
Force or torque	$F = m\ddot{x}$	lb	$T = J\ddot{\theta}$	in-lb
Momentum	$m\dot{x}$	lb-sec	$J\dot{\theta}$	in-lb-sec
Impulse	Ft	lb-sec	Tt	in-lb-sec
Kinetic energy	$\frac{1}{2}m\dot{x}^2$	lb-in	$\frac{1}{2}J\dot{\theta}^2$	lb-in
Potential energy	$\frac{1}{2}kx^2$	lb-in	$\frac{1}{2}K\theta^2$	lb-in
Work	$\int F\,dx$	lb-in	$\int T\,d\theta$	lb-in
Natural frequency	$\omega_n = \sqrt{k/m}$	rad/sec	$\omega_n = \sqrt{K/J}$	rad/sec
Equation of motion	$m\ddot{x} + c\dot{x} + kx = F_0 \sin \omega t$		$J\ddot{\theta} + \eta\dot{\theta} + K\theta = T_0 \sin \omega t$	
Initial conditions	$x(0) = x_0,\quad \dot{x}(0) = \dot{x}_0$		$\theta(0) = \theta_0,\quad \dot{\theta}(0) = \dot{\theta}_0$	
Transient responses	$x_c = Ae^{-\zeta\omega_n t} \sin(\omega_d t + \phi)$ $\omega_d = \sqrt{1-\zeta^2}\,\omega_n$		$\theta_c = Ae^{-\zeta\omega_n t} \sin(\omega_d t + \phi)$ $\omega_d = \sqrt{1-\zeta^2}\,\omega_n$	
Steady state responses	$x_p = X \sin(\omega t - \psi)$ $X = \dfrac{F_0}{\sqrt{(k - m\omega^2)^2 + (c\omega)^2}}$		$\theta_p = \Phi \sin(\omega t - \psi)$ $\Phi = \dfrac{T_0}{\sqrt{(K - J\omega^2)^2 + (\eta\omega)^2}}$	

Table 3 Torsional Natural Frequency Equations for Simple Systems

	CASE	FREQUENCY EQUATION
1	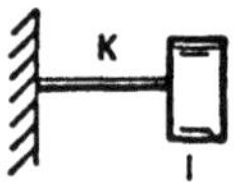	$\omega_n = \sqrt{\frac{k}{I}}$
2	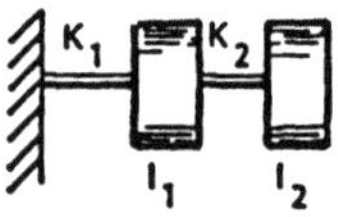	$\omega = \frac{1}{2}\left[\omega_1^2 + I\omega_2^2 \pm \sqrt{(\omega_1^2 - I\omega_2^2)^2 - 4\omega_1^2\omega_2^2}\right]$ $\omega_1^2 = K_1/I$; $\omega_2^2 = K_2/I_2$; $I = (I_1+I_2)/I_1$
3	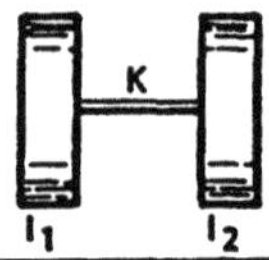	$\omega_1 = 0$; $\omega_2 = \omega_N = \sqrt{\frac{K(I_1+I_2)}{I_1 I_2}}$
4	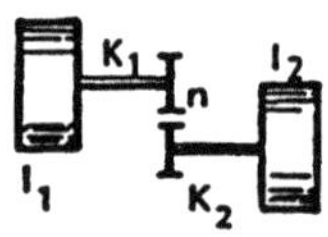	$\omega_1 = 0$; $\omega_2 = \sqrt{\frac{K_1K_2(I_1 - n^2 I_2)}{I_1I_2(K_1 - n^2K_2)}}$
5	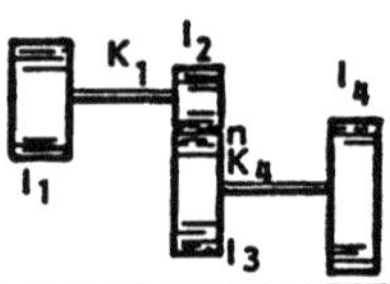	$\omega^4\left[I_1I_4(I_2+n^2I_3)\right] - \omega^2\left[K_4I_1(I_2+I_3n^2+I_4n^2) + K_1I_4(I_1+I_2+n^2I_3)\right] + K_1K_4(I_1+I_2+n^2I_3+I_4n^2) = 0$
6	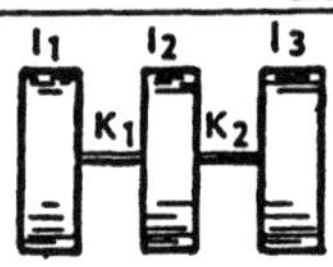	$\omega^4(I_1I_2I_3) - \omega^2\left[K_2I_1(I_2+I_3)+K_1I_3(I_1+I_2)\right] + K_1K_2(I_1+I_2+I_3) = 0$
7	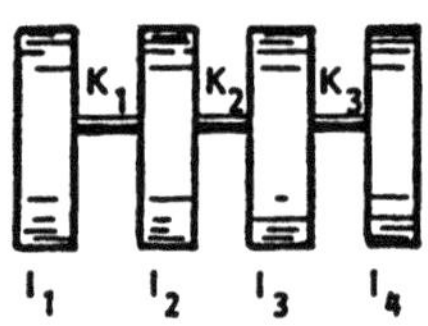	$\begin{vmatrix} K_1 - I_1\omega^2 & -K_1 & 0 & 0 \\ -K_1 & K_1+K_2-I_2\omega^2 & -K_2 & 0 \\ 0 & -K_2 & K_2+K_3-I_3\omega^2 & -K_3 \\ 0 & 0 & -K_3 & K_3-I_4\omega^2 \end{vmatrix} = 0$

Table 3 (cont') Torsional Natural Frequency Equations for Simple Systems

	CASE	FREQUENCY EQUATION
8	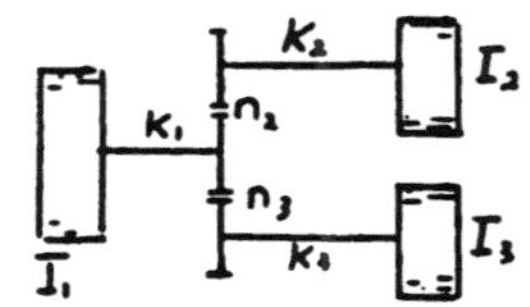	$\begin{vmatrix} K_1-I_1\omega^2 & -K_1 & 0 & 0 \\ -K_1 & K_1+n_2^2K_2+n_3^2K_3 & K_2n_2^2 & K_3n_3^2 \\ 0 & n_2^2K_2 & K_2n_2^2-I_2\omega^2 & 0 \\ 0 & n_3^2K_3 & 0 & K_3n_3^2-I_3\omega^2 \end{vmatrix} = 0$
9	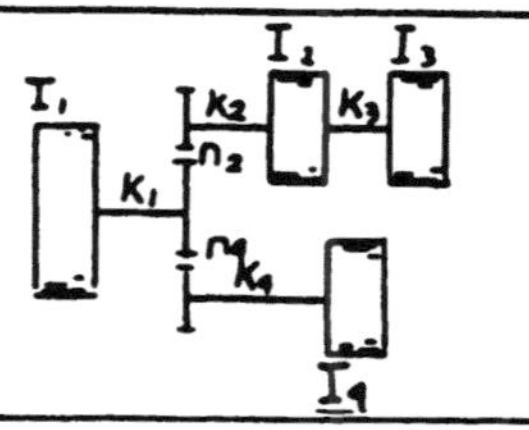	$\begin{vmatrix} K_1-I_1\omega^2 & -K_1 & 0 & 0 & 0 \\ -K_1 & K_1+n_2^2K_2+n_4^2K_4 & n_2^2K_2 & 0 & n_4^2K_4 \\ 0 & n_2^2K_2 & n_2^2(K_2+K_3-I_2\omega^2) & -K_3n_2^2 & 0 \\ 0 & 0 & -K_3n_2^2 & n_2^2(K_3-I_3\omega^2) & 0 \\ 0 & n_4^2K_4 & 0 & 0 & n_4^2(K_4-I_4\omega^2) \end{vmatrix} = 0$
10	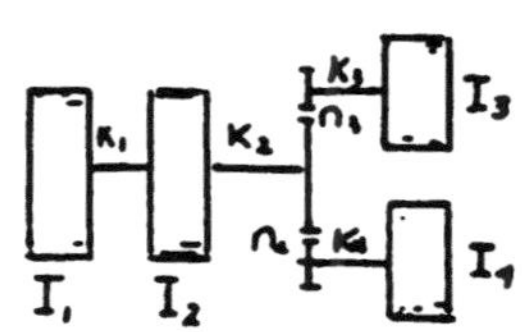	$\begin{vmatrix} K_1-I_1\omega^2 & -K_1 & 0 & 0 & 0 \\ -K_1 & K_1+K_2-I_2\omega^2 & -K_2 & 0 & 0 \\ 0 & -K_2 & n_3^2K_3+n_4^2K_4+K_2 & K_3n_3^2 & K_4n_4^2 \\ 0 & 0 & K_3n_3^2 & n_3^2K_3-n_3^2I_3\omega^2 & 0 \\ 0 & 0 & K_4n_4^2 & 0 & K_4n_4^2-I_4n_4^2\omega^2 \end{vmatrix} = 0$
11	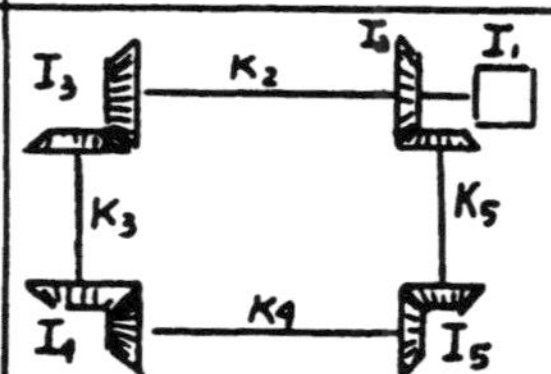	$\begin{vmatrix} K_1-I_1\omega^2 & -K_1 & 0 & 0 & 0 \\ -K_1 & K_1+K_2+K_5-I_2\omega^2 & -K_2 & 0 & -K_5 \\ 0 & -K_2 & K_2+K_3-I_3\omega^2 & -K_3 & 0 \\ 0 & 0 & -K_3 & K_3+K_4-I_4\omega^2 & -K_4 \\ 0 & -K_5 & 0 & -K_4 & K_4+K_5-I_5\omega^2 \end{vmatrix} = 0$

FREE AND FORCED VIBRATIONS OF TURBINE BLADES

H. Irretier
University of Kassel, Kassel, FRG

ABSTRACT

The free and forced vibrations of turbine blades are considered. First, various models for the description and the calculation of their dynamical behaviour are presented and methods of solution are discussed. Second, results and parametric studies are considered which show the influence of the most important blade parameters on the vibrations.

1. INTRODUCTION

One of the important components of steam and gas turbines are the blades. They're attached to the rotor surface to convert the fluid energy of the gas or the steam into rotational energy of the rotor. A typical bladed rotor is shown in Fig. 1

In the following paper the vibrational behaviour of turbine blades is considered whereas it is assumed that the blade vibration is uncoupled from the disc and the shaft. This assumption is valid in the most real cases. Otherwise, the coupled vibration of the blade and the disc or the complete system of the turbine rotor must be considered.

Fig. 1: Rotor of the 1200-MW-Turbine (KKW Biblis, FRG)

2. SOURCE OF BLADE VIBRATION PROBLEMS

The main source of turbine blade vibration problems arising in turbomachines is the fact that the distribution of the gas or steam load on the blade is not constant with respect to the rotational angle but varies during one rotation of the blade. This angular dependence of the blade load is caused by various circumstances. The main reason for that is the fact that the rotor blades run behind a row of still standing stator blades which are fixed in the casing of the machine (Fig. 2). Thus, the rotor blade is loaded in tangential and circumferential direction by fluid forces which are lower than the average value in the wakes of the stator blades and higher than the average value in the nozzles between the stator blades.

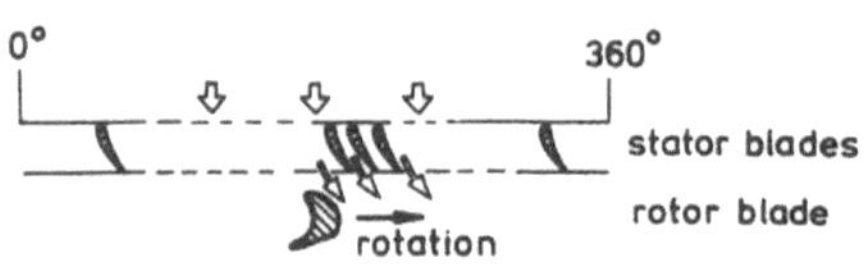

Fig. 2: Nozzle excitation of a rotor blade

Another reason for a non-uniform blade load during one revolution can occur by partial admission e.g. the loading of the rotor blades by fluid forces in one or more sectors around the machine while the other sectors remain unloaded. In both cases, the load on the rotating blade is periodic and can be described by the Fourier-series

$$p = p(t) = \sum_{\nu=0}^{\infty} P_\nu \cos(\Omega_\nu t - \psi_\nu) \qquad (2.1)$$

where t is the time, P_ν are the Fourier-coefficients, which can vary along the blade axis, ψ_ν is a phase angle and Ω_ν are the excitation frequencies of the periodic blade force. These excitation frequencies are given by

$$\Omega_\nu = \nu z \Omega \quad ; \quad \nu = 1, 2, 3 \ldots \qquad (2.2)$$

where z is the number of periodics of the blade load during one revolution (for the nozzle excitation for instance z equals the number of the stator blades) and Ω is the angular speed of the machine. The lines of excitation frequencies corresponding to equation (2.2) are plotted in the Campbell diagram which is shown in Fig. 3.

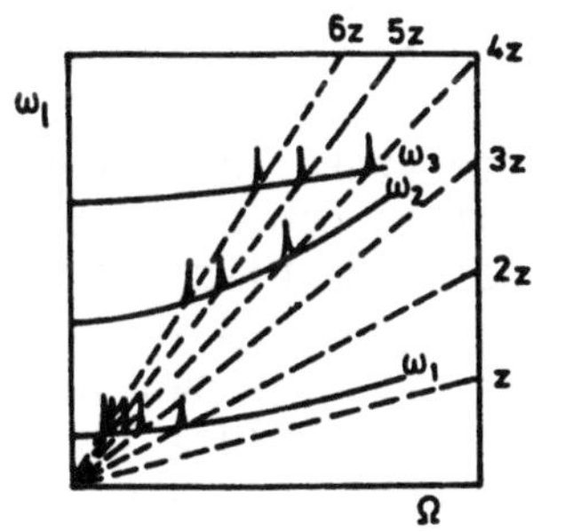

Fig. 3: Campbell-diagram for a rotating blade

On the other hand, each rotor blade possesses a spectrum of eigenfrequencies which generally depends on the angular speed because of a stiffening effect due to the centrifugal force field. Thus, the eigenfrequency-angular speed curves are parabolic functions as they are shown in Fig. 3, too. At each point of intersection of one eigenfrequency curve with one of the excitation lines in the Campbell-diagram resonance vibrations occur. These resonances can only be avoided if the corresponding Fourier-coefficient P is very small and/or the damping is high enough. Both requirements are mostly not given in real turbomachines. Thus, the only way to avoid strong resonance vibration is the possibility to keep the eigenfrequencies of the blade sufficient far away from the excitation frequencies. However, in many turbomachines it is not possible to avoid resonance because the blades run with varying operational speeds and various resonance vibrations occur during the run-up or the shut-down of the machine. The described circumstances show that the design of turbomachine blading from the vibration point of view must include three different considerations:

(1) The determination of the amplitudes and the frequencies of the aerodynamic forces acting on rotor blades,
(2) the determination of damping values and
(3) the investigation of the eigenfrequencies and the mode shapes of the rotor blades.

While the first two points are discussed in detail in other chapters this paper deals predominantely with the third point and additionally with the calculation of the forced response of turbine blades due to excitations as described above.

3. MECHANICAL MODELLING AND MATHEMATICAL DESCRIPTION OF TURBINE BLADES

The first problem to solve in the determination of the eigenfrequencies and the mode shapes of turbine blades is to find a sufficiently accurate mechanical model. This model depends on the shape and the size of the considered blade. In Fig. 4 blades are shown which are marked by a typical beam-like shape. The lower natural vibration of such blades may be described by more or less extended beam theories. In contrast to the blades shown in Fig. 4, other turbine blades may have typical shell-like shapes or can only be described by three-dimensional models. Such blade types often occur in modern gas turbines with very high inlet temperature so that the blades must be cooled. For this purpose the blades are designed as composed shell structures or, as shown in Fig. 5, in a monoblock form which is penetrated by cooling drills.

Fig. 4: Beam-like turbine blades/1/

The examples of turbine blades shown in the Fig. 4 and 5 indicate that the right mechanical model to describe their vibrational behaviour may be quite different and depends substantially on the shape of the considered blade. Consequently, the mathematical description is also dependent on the considered type of blade.

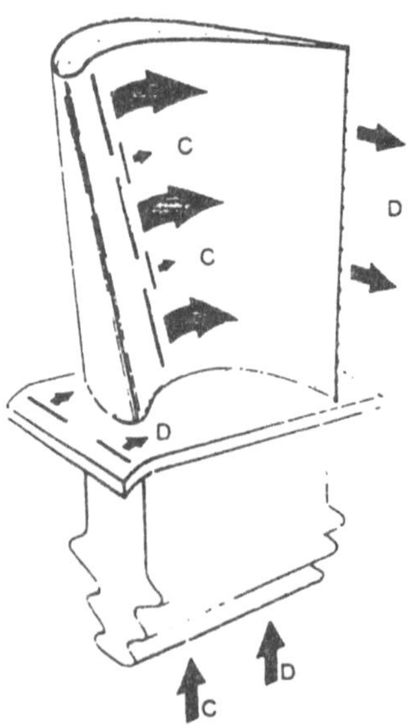

Fig. 5: Cooled, three-dimensional blade of a gas turbine

In the following chapters various kinds of mechanical models of blades are discussed and the corresponding mathematical description is given. While the beam-like blades are considered in detail, the shell-like and three-dimensional blades can only be discussed shortly. This is also valid for extended blade row constructions e.g. shrouded blade stages or blade packets etc.

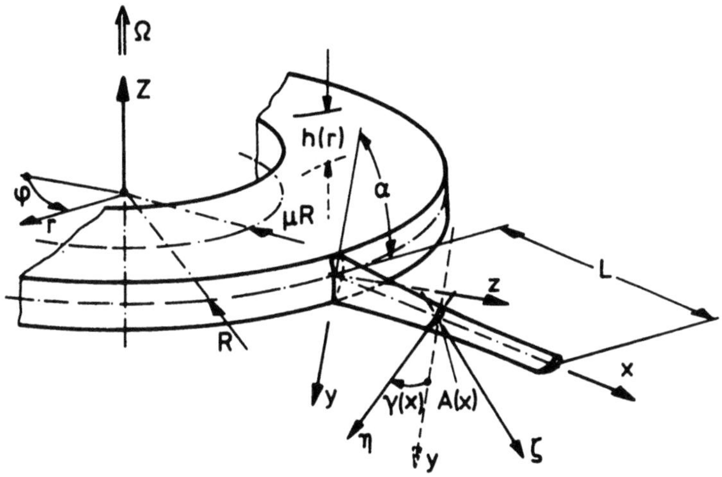

Fig. 6: Beam-like blade; notations

The blade can perform bending vibrations in the plane y, z and torsional vibrations around the x-axis. In the case, that the x-axis of the centre of gravity of cross-section is equal to the centre of shear or if the distance between both is neglectable - which in the most practical cases holds true - the bending vibration and the torsional vibration are uncoupled from each other. Otherwise, the bending and the torsional vibration are coupled.

Depending on the fact which constructive properties of the blade (tapering, twisting, etc.) are included in the mathematical description of the mechanical model shown in Fig. 6, various types of equations have been developed in the past to calculate the natural frequencies, the mode shapes and the forced response of beam-like turbine blades considered in this chapter. A complete survey on that is given in /2/ and /3/.

3.1.1 BENDING VIBRATIONS

Depending on the blade properties included in the model, various mathematical descriptions of the bending vibrations have been developed and applied in the past. The simplest model which neglects the rotation, the twisting, the shear deflection and the rotary inertia are of course the differential equations

$$- \frac{\partial^2}{\partial x^2} (EI_z \frac{\partial^2 v}{\partial x^2}) = \rho A \frac{\partial^2 v}{\partial t^2} \tag{3.1a}$$

$$- \frac{\partial^2}{\partial x^2} (EI_y \frac{\partial^2 w}{\partial x^2}) = \rho A \frac{\partial^2 w}{\partial t^2} \tag{3.1b}$$

of the classical Bernoulli-Euler beam theory. However, these equations do not take into account the important effects of rotation, twisting, rotary inertia and shear deformation. If in addition these influences are also included in the model, the complete set of equations of motion is /4/

$$\frac{\partial}{\partial x}\left[\kappa GA(\frac{\partial v}{\partial x}+\gamma_z)\right]+\frac{\partial}{\partial x}(n_x^* \frac{\partial v}{\partial x})+\rho A\Omega^2(v\cos\alpha-w\sin\alpha)\cos\alpha= A\frac{\partial^2 v}{\partial t^2}$$

$$\kappa GA(\frac{\partial v}{\partial x}+\gamma_z)-\frac{\partial}{\partial x}(EI_z\frac{\partial\gamma_z}{\partial x}+EI_{yz}\frac{\partial\gamma_y}{\partial x})=-\rho(I_z\frac{\partial^2\gamma_z}{\partial t^2}-I_{yz}\frac{\partial^2\gamma_y}{\partial t^2})$$

$$\frac{\partial}{\partial x}\left[\kappa GA(\frac{\partial w}{\partial x}+\gamma_y)\right]+\frac{\partial}{\partial x}(n_x^* \frac{\partial v}{\partial x})-\rho A\Omega^2(v\cos\alpha-w\sin\alpha)\sin\alpha=\rho A\frac{\partial^2 w}{\partial t^2}$$

$$\kappa GA(\frac{\partial w}{\partial x}+\gamma_y)-\frac{\partial}{\partial x}(EI_{yz}\frac{\partial\gamma_z}{\partial x}+EI_y\frac{\partial\gamma_y}{\partial x})=-\rho(-I_{yz}\frac{\partial^2\gamma_z}{\partial t^2}+I_y\frac{\partial^2\gamma_y}{\partial t^2}),$$

(3.2a) - (3.2d)

where

$$n_x^* = \rho\Omega^2\int_x^L A(R+\xi)d\xi \tag{3.3}$$

is the static normal force in the blade caused by rotation, v and w are the blade displacements, γ_y and γ_z are the angles of cross-section in the sense of Timoshenko's beam theory and the other parameters are as shown in Fig. 6.

The set of equations (3.2a) to (3.2d) for the twisted, staggered and rotating blade including the shear deformation and rotary inertia effect is the mostly developed mathematical model to describe the bending vibrations of turbine blades when there is no coupling to the torsional vibrations. This set of equations was first given in /4/ and proved to be very reliable in comparison to test results.

Introducing the displacement vector

$$\underline{w} = \left[v\ \gamma_z\ w\ \gamma_y\right]^T \quad , \tag{3.4}$$

the equations of motion (3.2) for the vibrating blade can be written as

$$\underline{M}\,\ddot{\underline{w}} + \underline{K}\,[\underline{w}] = \underline{0} \tag{3.5}$$

where $\underline{M}$ is the mass matrix including the transverse and rotary inertia terms and $\underline{K}$ is a matrix of differential operators with respect to x. This stiffness matrix $\underline{K}$ can be divided into two terms

$$\underline{K} = \underline{K}_{EI} + \Omega^2 \underline{K}_\Omega \tag{3.6}$$

where $\underline{K}_{EI}$ describes the stiffness properties of the bent, non-rotating blade whereas $\underline{K}_\Omega$ includes the stiffening effect of the centrifugal force field due to rotation.

Equation (3.5), which is the differential equation of the continuous, non-discretized turbine blade, describes the natural vibrations of the blade. Its ℓ-th natural vibration is given by

$$\underline{w} = \underline{W}_\ell \exp (j\omega_\ell t) \tag{3.7}$$

where $\underline{W}_\ell = \underline{W}_\ell(x)$ denotes the mode shape and ω_ℓ the corresponding circular frequency. Both are solutions of the eigenvalue problem

$$- \omega_\ell^2 \, \underline{M} \, \underline{W}_\ell + \underline{K}\left[\underline{W}_\ell\right] = \underline{0} \tag{3.8}$$

which follows from equation (3.5) where it must be payed attention to the fact that the stiffness matrix $\underline{K}$ is a function of the angular valocity Ω of the blade. Thus, the eigenfrequencies ω_ℓ and, on principle, even the mode shapes $\underline{W}_\ell$ are dependent on the rotational speed of the machine.

3.1.2 FREE TORSIONAL VIBRATIONS

If the assumption that the centroid and the centre of shear of the blade cross-section are equal to each other is still valid, the torsional vibrations are uncoupled from the bending vibration considered above. The simplest model to describe the torsional vibration of the blade is the well-known equation

$$\frac{\partial}{\partial x} (GI_T \frac{\partial \varphi}{\partial x}) = \rho I_p \frac{\partial^2 \varphi}{\partial t^2} \tag{3.9}$$

which is a wave equation of a drilled shaft. I_T is the torsional rigidity of the cross-section while I_p denotes its polar moment of inertia. Of course, this equation neglects all refined effects on the problem like the increase of the rigidity of the blade due to the centrifugal force field, additional inertia forces because of the formulation of Newton's law in the rotating reference system and the twisting effect. If these influences are taken into consideration, too, the equation of motion of the torsional blade vibrations becomes :

$$\frac{\partial}{\partial x} \left[GI_T + \sigma_x^* \, I_p + E\{\frac{d\gamma}{dx}\}^2 I_\gamma) \frac{\partial \varphi}{\partial x}\right] - \rho\Omega^2 (I_z - I_y) \cos 2(\alpha + \gamma) \cdot = \rho I_p \frac{\partial^2 \varphi}{\partial t^2} \tag{3.10}$$

where $\sigma_x^* = n_x^*/A$ is the static stress in the blade due to rotation

$$I_\gamma = \int_A (\eta^2 + \zeta^2)^2 \, dA - \frac{I_p^2}{A} \tag{3.11}$$

is an area moment which describes the increase of the torsional

rigidity of the blade due to twisting /5/.
Introducing a displacement vector $w = [\varphi]$ the equation of motion for the torsional vibration of the blade also can be written in the form of equation (3.5) and for harmonic natural vibrations the angle φ of torsion is of the form which is given in equation (3.7) which again yields an eigenvalue problem of the form shown in equation (3.8). It is obvious that the eigenfrequencies ω_ℓ and the mode shapes W_ℓ of the free torsional blade vibration are also dependent on the rotational speed Ω of the system.

3.1.3 FREE COUPLED BENDING AND TORSIONAL VIBRATIONS

In the two preceding chapters it was assumed that the centroid and the centre of shear of the cross-section are equal to each other or that the distance between both is small in comparison to the other sizes of the cross-section.
In this case the both bending vibrations and the torsional vibrations are decoupled and the eigenfrequencies and the mode shapes of one of the natural vibrations can be calculated independent from the other. However, in some types of turbomachines blades are used which have a large curvature so that this assumption is no more valid. Now, the bending and the torsional vibration are coupled and influence each other. The equations of motion must be extended to take this coupling effect into consideration.

There are a lot of models developed in the past to include this coupling effect. Probably the first set of equations including this effect was applied in /6/ and reads

$$- \frac{\partial^2}{\partial x^2} \left(EI_z \frac{\partial^2 v}{\partial x^2} \right) = \rho A \frac{\partial^2}{\partial t^2} (v + c_z \varphi) \qquad (3.12)$$

$$- \frac{\partial^2}{\partial x^2} \left(EI_y \frac{\partial^2 w}{\partial x^2} \right) = \rho A \frac{\partial^2}{\partial t^2} (w - c_y \varphi) \qquad (3.13)$$

$$\frac{\partial}{\partial x} \left(GI_T \frac{\partial \varphi}{\partial x} \right) = I_p \frac{\partial^2}{\partial t^2} \left(\varphi + \frac{c_z v - c_y w}{i_p^2} \right) \qquad (3.14)$$

where c_y and c_z are the coordinates of the centre of shear of the cross-section and i_p is its radius of gyration ($I_p = i_p^2 A$). Of course, these equations neglect the effects of shear deformation, rotary inertia, rotation of the blade and twisting. Refined models to include also these effects where given in several references like /2/, /3/, /5/, /7/, /8/ and /9/.

3.1.4 FORCED VIBRATIONS

Following the considerations in the preceding chapters we recognize that the equations of motion of the free decoupled and even coupled bending and torsional vibrations of beamlike blades can be described by the standardized equation (3.5). If now, in addition, also the excitation of the turbine blades as described in chapter 2 should be taken into account to calculate the stress response of blades, external transverse loads acting in the y- and z-direction must be added up in the equations for the bending vibrations and an external torsional moment along the x-axis for the torsional vibrations. This leads to the general equation of motion

$$\underline{M}\,\ddot{\underline{w}} + \underline{C}\left[\dot{\underline{w}}\right] + \underline{K}\left[\underline{w}\right] = \underline{p} \tag{3.15}$$

in which in addition a damping term is included which is assumed to be proportional to the velocity.

The general method to solve this linear differential equation is the modal transformation technique /10/, /11/. This technique transforms the displacement vector $\underline{w}$ in the space of mode shapes by the relation

$$\underline{w} = \sum_{k=1}^{m} \underline{W}_k\,(x)\;q_k(t) \tag{3.16}$$

where q_k are the time-dependent modal coordinates of the problem. Introducing this modal series and paying attention to the orthogonality relations of the mode shapes, the general equation of motion (3.15) is transformed into

$$\ddot{q}_\ell + \sum_{k=1}^{m} c_{\ell k}\,\dot{q}_k + \omega_\ell^2\,q_\ell = r_\ell \tag{3.17}$$

where

$$c_{\ell k} = \int_0^L \underline{W}_\ell^T\,\underline{C}\left[\underline{W}_k\right]\,dx \tag{3.18}$$

and

$$r_\ell = \int_0^L \underline{W}_\ell^T\,\underline{p}\,dx \tag{3.19}$$

describe the damping coefficient and the transformed blade load, respectively. From equation (3.18) it is obvious that generally the damping couples the differential equations (3.17) for the modal coordinates q_ℓ if no special assumptions are introduced at this point of consideration. The basic idea of these assumptions is that in the most practical cases the damping effect is so small that there is no coupling between the various types of natural vibrations of the system. Under these circumstances it is possible to replace equation (3.17) by the uncoupled equation

$$\ddot{q}_\ell + 2\zeta_\ell\,\omega_\ell\,\dot{q}_\ell + \omega_\ell^2\,q_\ell = r_\ell \tag{3.20}$$

where the introduced factor ζ_ℓ is the viscous damping factor. If structural damping is used, where the damping matrix is given by

$$\underline{C} = \frac{\eta}{\Omega}\,\underline{K}, \tag{3.21}$$

where

η is the loss factor of material and Ω the frequency domain parameter, the transformed equations of motion are

$$\ddot{q}_\ell + \omega_\ell^2 \left(1 + \frac{\eta}{\Omega}\frac{\partial}{\partial t}\right) q_\ell = r_\ell \quad . \tag{3.22}$$

To calculate the vibration response of the blade the next step is to consider the solution of the non-homogeneous, linear differential equation (3.20) or (3.22). This solution depends on the form of the right hand side r_ℓ. In a lot of practical cases - of course there are also important exceptions of this - the external blade load can be separated in a product form

$$\underline{p} = \underline{P}(x)\ \underline{f}(t) \tag{3.23}$$

where $\underline{P}(x)$ is a diagonal matrix which includes the distributed load on the blade and f(t) is a vector which contains the time-dependent functions of excitations which come into the problem as considered in chapter 2. Thus, the transformed blade loads as pointed out by equation (3.19) are

$$r_\ell = \underline{R}_\ell\ \underline{f}(t) \tag{3.24}$$

where

$$\underline{R}_\ell = \int_0^L \underline{W}_\ell^T\ \underline{P}\ dx \quad . \tag{3.25}$$

We notice that the character of time-dependence of the blade load is completely described by the vector $\underline{f}(t)$.

The most convenient way to solve the equations (3.20) and (3.22) for various types of excitation (3.24) is the frequency domain method. This method starts from the Fourier-spectrum

$$\overline{R}_\ell(\Omega) = \frac{1}{2\pi} \int_{-\infty}^{\infty} r_\ell(t)\ e^{-j\Omega t}\ dt \tag{3.25}$$

of the time-dependent load function r_ℓ. In a lot of practical cases these spectra are given directly. In other cases, they must be calculated from the time-dependent blade load using a Fast-Fourier-algorithm. Transforming the equation of motion (3.20) or (3.22) also into the frequency domain yields

$$\left[-\Omega^2 + j\cdot 2\zeta_\ell\omega_\ell\,\Omega + \omega_\ell^2\right] \overline{Q}_\ell(\Omega) = \overline{R}_\ell(\Omega) \tag{3.26}$$

and

$$\left[-\Omega^2 + \omega_\ell^2\,(1 + j\eta)\right] \overline{Q}_\ell(\Omega) = \overline{R}_\ell(\Omega) \quad , \tag{3.27}$$

from which the vibratory response

$$\overline{Q}_\ell(\Omega) = \overline{H}_\ell(\Omega)\cdot \overline{R}_\ell(\Omega) \tag{3.28}$$

for the ℓ-th modal coordinate of the blade can be calculated using the transfer functions

$$\overline{H}_\ell = \frac{1}{\omega_\ell^2 - \Omega^2 + j \cdot 2\zeta_\ell \, \omega_\ell \, \Omega} \tag{3.29}$$

or

$$\overline{H}_\ell = \frac{1}{\omega_\ell^2 (1 + j\eta) - \Omega^2} \quad , \tag{3.30}$$

for viscous or structural damping, respectively. If necessary, the modal coordinates q_ℓ in the time domain follow from the backward Fourier-transformation

$$q_\ell(t) = \int_{-\infty}^{\infty} \overline{Q}_\ell(\Omega) \, e^{j\Omega t} \, d\Omega \quad . \tag{3.31}$$

With $\overline{Q}_\ell(\Omega)$ or $q_\ell(t)$ the displacements of the blade follow from equation (3.16) in the frequency or time domain. Using the constitutive equation relating stresses to displacements yield the blade stresses due to the forced vibrations, from which a life fatigue analysis can be performed.

It is important to discuss one point at this stage of consideration. It was pointed out in the chapters above that generally the natural frequencies of turbine blades depend on the rotational speed. Thus, if the rotational speed of the machine changes for instance during a run-up or a run-down of the machine the natural frequencies of the blade change, too. If forced vibrations are considered during such periods of blade operation this changing of the natural frequencies must be taken into account and the application of equation (3.28) should be restricted only on sufficiently short time periods whereas in each period the actual eigenfrequency ω_ℓ must be considered to calculate the transfer function $\overline{H}_\ell$. An alternative possibility to overcome this difficulty is to solve the equations of motion (3.20) or (3.22), respectively, not in the frequency domain as described above but in the time domain by a direct numerical integration using a set of Runge-Kutta-integration formulars. During this time integration it is possible to take into account the actual eigenfrequency ω_ℓ for each time t because the dependence $\Omega(t)$ is a given parameter of the problem.

3.2 SHELL-LIKE BLADES

A beam-like model for a turbine blade is only sufficient for the consideration of the lower modes of blades which have cross-sections much smaller than the blade length. Otherwise, extended theories must be applied. For such types of blades, for which still one dimension, the thickness h, is small in camparison to the two other dimensions, shell theories have been successfully applied to compute the free vibration behaviour. It would exceed this chapter to discuss in detail all the approximations which have been applied in the past. The models start from simple cantilever cylindrical shells with constant thickness and end with complicated finite element models for rotating blades with non-constant thickness and a twisted shape. The influence of prestressing due to the rotation is more or less accurately included in the applied models.

A complete survey on the subject is given in /13/. Details considering various shell theories are discussed in /14/ and in /15/ a comparison of beam and shell theories for the vibrations of blades is given.

In comparison to the free vibration problem of shell-like blades the response of such blades due to forced vibrations was considered much less in the literature. The convenient method to consider this problem is also the technique of modal transformation which was described in detail in the chapter 3.1.4 by using the eigenfrequencies and the mode shapes from a free vibration analysis.

3.3 THREE-DIMENSIONAL BLADES

The only suitable method to deal with free and forced vibrations of three-dimensional blades is the method of finite elements. After intensive investigations of this method in the sixties and seventies today the finite element approximation is a well-established technique in the dynamic analysis in mechanical engineering sciences. Many efficient programs have been developed since then and are available for the application of a vibration analysis of three-dimensional blades. In the most cases shell or tetrahedron elements are used to calculate the eigenfrequencies and mode shapes of such complicated blades. Some programs also allow to take into account the important effect of the centrifugal force field. By the steps of a finite element procedure, the problem is formulated in a form given in equation (3.5) and the eigenfrequencies are calculated from equation (3.8). To determine the eigenfrequencies and mode shapes from this equation a lot of efficient methods of eigenvalue calculation have been developed parallel to the finite element method. Details on both are given in /16/, /17/ and /18/.

In addition to the free vibration analysis, programs also have been used to perform a forced vibration analysis. The most adequate method is again the modal analysis technique which is described in chapter 3.1.4 for the beam-like blade. Applications of this method for the analysis of three-dimensional blades are given in /19/ and /20/.

4. METHODS OF ANALYTICAL AND NUMERICAL SOLUTIONS

The various types of mechanical and mathematical models require different techniques of solution. Closed form solutions are only known for the very simple case of the free bending and free torsional vibrations of beam-like blades without the consideration of twisting, tapering and rotation i.e. for the equations (3.1a) and (3.1b) for the bending and (3.9) for the torsional vibrations, respectively.

If the refined effects for the beam-like blade models are taken into account, only approximate solutions are available. The classical procedures are the Rayleigh-Ritz and Galerkin-method. They are described in /4/ and /15/. In the last twenty years, however, computer-aided numerical methods are preferred to calculate the natural frequency and mode shapes of beam-like blades. One classical procedure is the method of transfer matrices while more modern techniques are the method of finite elements and a direct numerical integration of the differential equations by a Runge-Kutta-method and a shooting technique. A large amount of references to the first method is given in /3/ while the second technique was used in /5/ and /21/.

The suitable method for the calculation of the free vibration of shell-like and three-dimensional blades is today the method of finite elements which was described in the chapters 3.2 and 3.3.

For all three types of blades the technique of modal transformation is the most applied method to calculate forced response of blades due to the various types of excitation. Details on this have been described in the chapters 3.1.4, 3.2 and 3.3. For details of the application and numerical problems is referred to the numerous literature in this field.

5. RESULTS AND PARAMETRIC STUDIES

General statements concerning the dependence of the vibrational behaviour of turbine blades on the constructive parameters are only possible for the eigenfrequencies of beam-like blades. For their forced vibrational behaviour and in particular for shell-like and three-dimensional blades the connections are so complicated that only some typical results can be discussed here.

5.1 BEAM-LIKE BLADES

Besides the mode shapes of the bending and torsional vibrations of blades, the influence of the constructive parameters on their eigenfrequencies is considered in particular. Moreover, two recent results concerning the forced vibration response of the beam-like blades subjected to nozzle excitation and partial admission, respectively, are presented.

5.1.1 FREE BENDING VIBRATIONS

5.1.1.1 MODE SHAPES

As pointed out in the chapter 3.1.1 the most simple model for the free bending vibrations of a turbine blade with constant cross-section, without twisting and rotation and neglecting the shear influence and the rotary inertia effect are the differential equations (3.1a) and (3.1b) for an Euler-Bernoulli-beam. The corresponding mode shapes are those of a cantilever and consist of harmonic and hyperbolic functions /10/.

If the refined effects are included in the consideration, the blade mode shapes change more or less considerably. The influence of the rotational speed and therefore of the centrifugal force field on the mode shapes is small in a wide range of operational speeds. Even the influence of shear deformation and rotary inertia affect the mode shapes only a little except the higher vibration modes where these influences dominate. In contrast to that, the tapering of the blade and in particular the twisting have a large influence on the blade mode shapes.

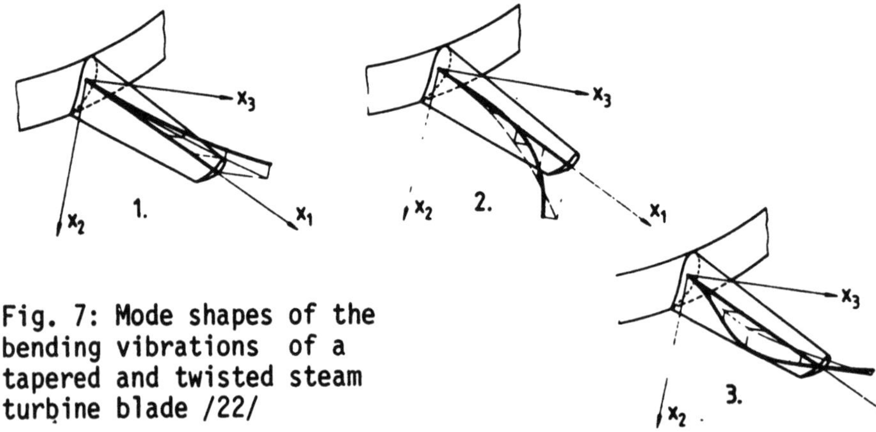

Fig. 7: Mode shapes of the bending vibrations of a tapered and twisted steam turbine blade /22/

From the mode shapes of a tapered and twisted turbine blade shown in Fig. 7 these effects are visible. While the first mode is similar to that one of a cantilevered Euler-Bernoulli-beam the mode shapes two and three are characterized by complicated spatial curved lines. This result has particular consequences when the forced vibrations are considered because an excitation in the direction x_2 can cause strong vibrational response in the perpendicular direction x_3 and vice versa. This phenomenon must be taken into account for refined constructions of turbine blades with damping wires or shrouding.

5.1.1.2 EIGENFREQUENCIES

In contrast to the mode shapes of beam-like blades, the eigenfrequencies are considerably influenced by a stiffening effect of the blade rotation. This is shown in Fig. 8 where the first three eigenfrequencies of a slender rotating beam are plotted as a function of the angular velocity Ω. Parameters on the curves are the stagger angle α and the ratio ε of disc radius to blade length. We notice from the curves that there is a parabolic increase of the eigenfrequencies versus the angular velocity. This effect is well-known from Southwell's theorem which yields for the rotating beam

$$\omega_\ell^2 = \overset{o}{\omega}{}_\ell^2 + \beta_\ell \Omega^2 \tag{5.1}$$

where $\overset{o}{\omega}_\ell$ is the ℓ-th eigenfrequency for the non-rotating case and β_ℓ is a centrifugal force factor which considers the stiffening effect of rotation. As investigated in /4/ this factor can be approximated for $\ell = 1$ by

$$\beta_1 = 1.54\varepsilon + 1.18 - \sin^2\alpha \tag{5.2}$$

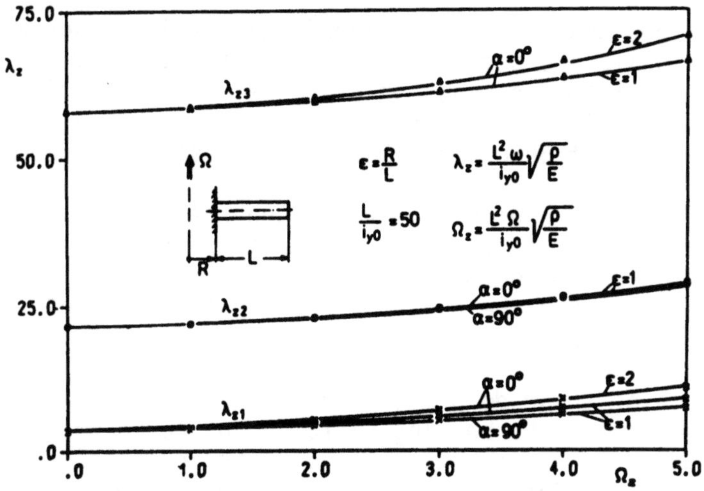

Fig. 8: Eigenfrequencies of the bending vibrations of a slender rotating beam as function of angular velocity

for the first eigenfrequency of bending vibrations of slender untwisted and even twisted beams. For the higher eigenfrequencies the relationship (5.1) can only be used in a qualified sence. For this case, useful approximated values for β_ℓ are given in /23/.

Transversed to real turbine blades with beam-like shape the results of Fig. 8 indicate that there is a large influence of the rotation in particular for long and slender blades which are fixed on a disc with large radius. For short blades on a small disc this influence decreases.

The influence of taper on the eigenfrequencies of an Euler-Bernoulli-beam-like blade on the eigenfrequencies is considered in Fig. 9.

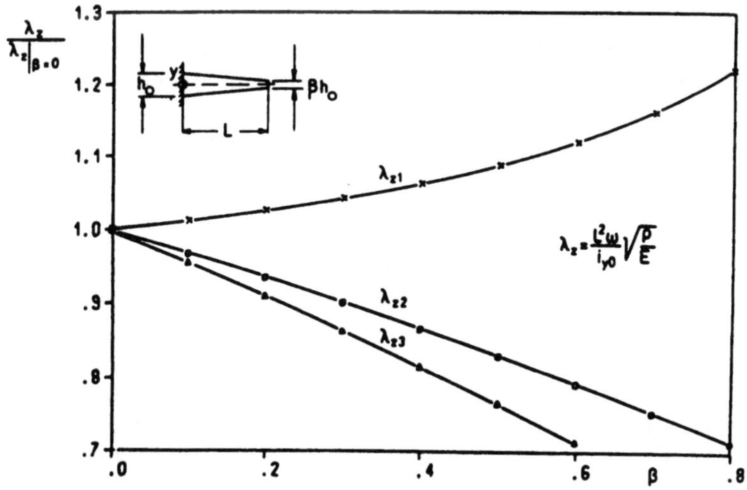

Fig. 9: Eigenfrequencies of tapered beam-like blades as a function of tapering parameter

The curves show that the first eigenfrequency increases with blade taper while the second and third eigenfrequency decrease. The magnitude of the increase or decrease of the eigenfrequencies can be considerable in practice because blades with a decrease of cross-section about 50 or more percent are used in turbomachinery.

If in addition to the effect of taper the rotational influence on the eigenfrequencies is considered it can be shown that the eigenfrequency of the tapered beam increases less than for the untapered one which has the same uniform cross-section as the tapered beam at its root.

The influence of shear deformation and rotary inertia on the eigenfrequencies of the bending vibration of beams had been frequently studied in the past initiated by the classical work in/24/. Introducing this effect in the consideration of beam-like blades fixed at the root and free at the end, numerical results show that this effect can be often neglected for the first eigenfrequency but is of importance for the higher ones.

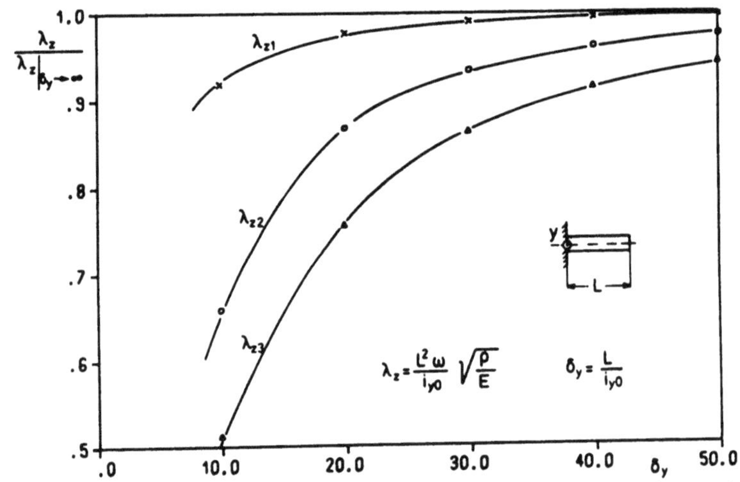

Fig. 10: Influence of shear deformation and rotary inertia on the eigenfrequencies of bending vibration of beam-like blades

In Fig. 10 the eigenfrequencies of a cantilever blade with these effects in account related to the eigenfrequencies neglecting both effects are plotted as function of the ratio of the blade length L to the radius of gyration $i_y = \sqrt{I_y/A}$. We notice from the curves that the influences of shear deformation and rotary inertia are relatively small for L/i_y-ratios greater than 30 but become more and more important for shorter blades.

A further effect which is very important for the eigenfrequencies spectra of beam-like turbine blades is the influence of twisting. The curves in Fig. 11 show the change of the first three eigenfrequencies of non-rotating blades without shear deformation, rotary inertia and tapering as a function of the angle of twist between the blade root and the blade tip while a linear twisting along the blade is assumed. The plot shows that the influence of

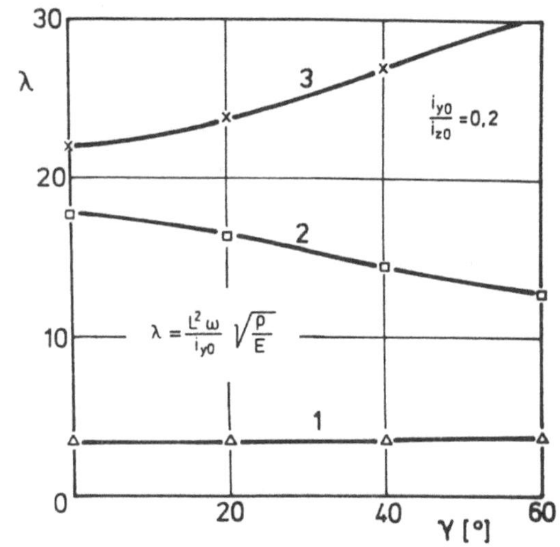

twisting is completely neglectable for the first eigenfrequency while the second one decreases and the third one increases considerably. This result is a consequence of the increasing coupling between the two bending vibrations in y- and z-direction with increasing twisting angle.

Fig. 11: Influence of twisting on the eigenfrequencies of the bending vibrations of beam-like blades

5.1.2 FREE TORSIONAL VIBRATIONS

5.1.2.1 MODE SHAPES

The mode shapes of a cantilever with constant cross-section performing free torsional vibrations are described by harmonic functions which are the general solutions of equation (3.9). For turbomachinery blades the torsional mode shapes are affected by tapering, twisting and the other refined influences in a complicated manner. A typical mode is shown by the hologram in Fig. 12 which is taken from /25/. We notice that the lines of equal displacements are no more parallel to the edge of the blade.

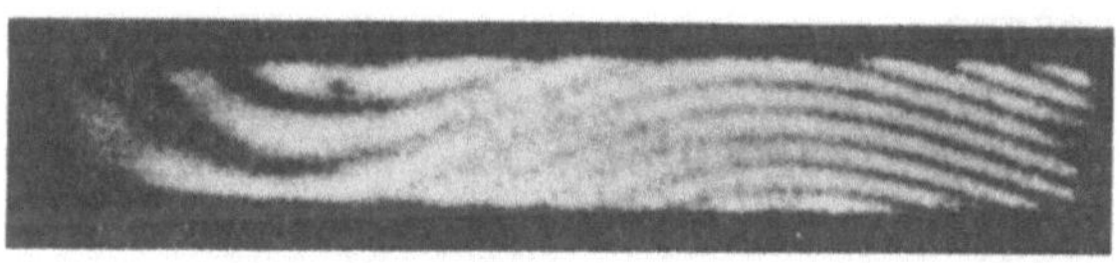

Fig. 12: Mode shape of the first torsiona vibration of a twisted turbine blade /25/

If, in addition, the torsional vibrations are coupled to the bending vibrations as described in chapter 3.1.3, the mode shapes are even more complicated and no general description of the influence of the blade parameters on its mode shapes is possible.

5.1.2.2 EIGENFREQUENCIES

As for the bending vibrations also the eigenfrequencies of the torsional vibrations of turbine blades are considerably influenced by the effects of twisting, tapering and rotation. As an example, Fig. 13 shows the influence of the rotation on the fundamental eigenfrequency of the torsional vibration of a beam-like blade. We notice again a parabolic increase of the eigenfrequency. However, in comparison to the bending vibrations the increase of the torsional vibration eigenfrequencies is much less

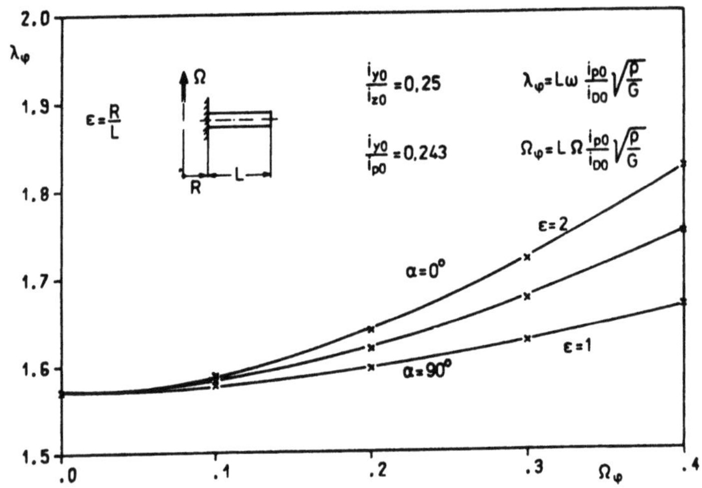

Fig. 13: First eigenfrequency of the torsional vibration of a slender rotating beam as function of angular velocity

for real turbomachine blades. Therefore, this influence is often neglected in practical calculations.

As to the influence of other properties of the blade on its torsional frequencies spectrum is referred to the numerous literature in particular to /2/, /3/, /4/ and /21/.

5.1.3 FORCED VIBRATIONS

In contrast to the free vibration analysis of turbine blades, the forced vibration problem was considered much less in the literature. Some typical results of the calculations of the blade response due to nozzle and partial admission excitation are considered in the following chapters.

5.1.3.1 PERIODIC EXCITATION AT CONSTANT ROTATIONAL SPEED

The usually considered problem of forced blade vibration is the dynamical behaviour of the blade running with constant rotational speed. The blade response is calculated as described in chapter 3.1.4 where various models for the description of the bending and torsional vibrations of the blade are used. In /27/ a numerical model for a beam-like blade is applied based on a Timoshenko-beam-theory where the coupling between the bending and torsional vibrations is taken into account. For a typical steam turbine blade the eigenfrequencies are calculated and plotted in a Campbell-diagram which is shown in Fig. 14. For the first three coupled modes various resonance points in the rotor speed range between 400 and 2000 rpm are visible. The cross-sections at different distances from the blade root are also plotted in Fig. 14.

For an excitation by z = 24 nozzles - the detailed data of the forces are given in /27/ - the blade response is shown in Fig. 15. It can be seen that the amplitudes reach their peak values at the points when the nozzle passing frequency lines intersects the

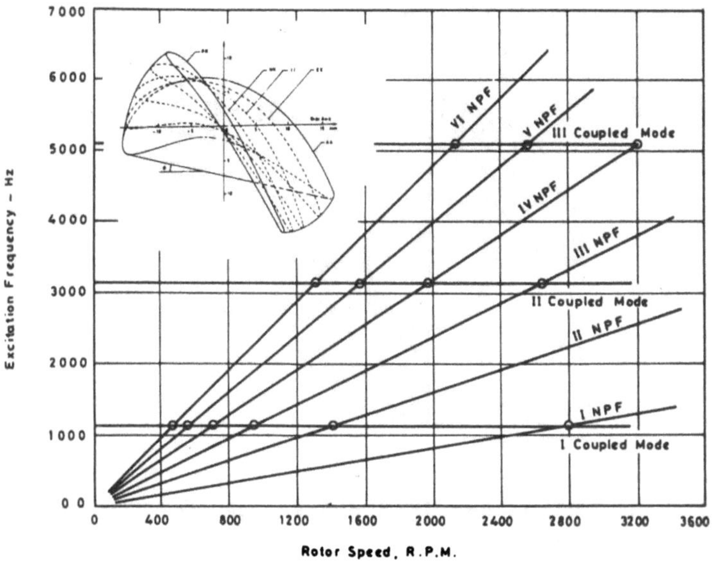

Fig. 14: Turbine blade and Campbell-diagram /27/

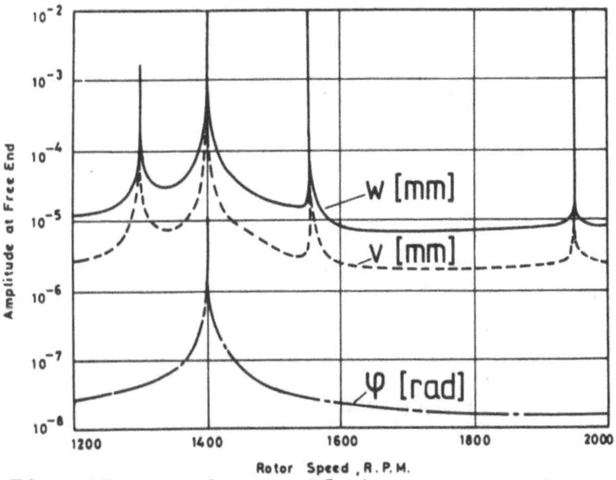

Fig. 15: Stationary blade response due to nozzle excitation /27/

curves of natural frequencies. It can also be observed from the curves that the intensity of resonance depends on the order of the nozzle passing frequency. We observe that the renance with the first coupled mode at 1400 rpm yields a strong torsional response while at the other resonances in the considered speed range only bending responses occur. This result is caused by the special geometry of the blade shown in Fig. 14.

The most important problem in the calculation of blade responses due to forced vibrations is the determination of the exciting forces and of the damping values. However, in recent times sufficiently accurate data for these parameters are given in numerous papers for a lot of practical cases.

5.1.3.2 TRANSIENT EXCITATION DURING RUN-UP AND RUN-DOWN

As an example for a turbine blade, which is subjected to transient excitation, a rotating beam is considered which runs up from one rotational speed to another while it is loaded by partial admission. The considered system is shown in Fig. 16 from which two 90°-sectors of partial admission and the blade load in tangential direction are visible. The blade runs up from a rotational fre-

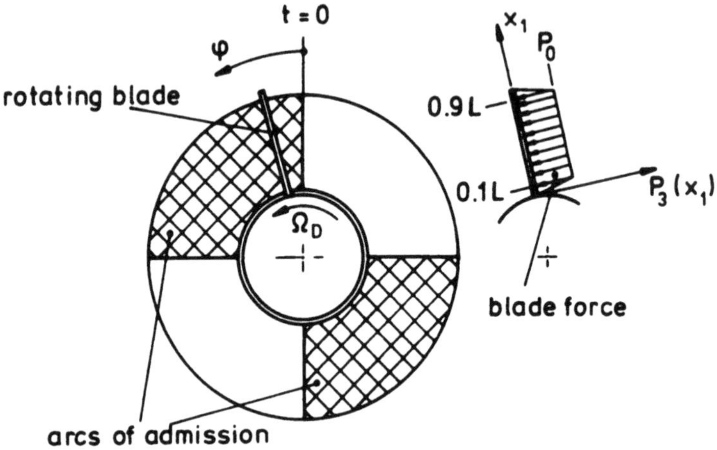

Fig. 16: Rotating blade and load due to two arcs of partial admission /26/

quency of f_D = 10 Hz up to a rotational frequency of 60 Hz as shown in Campbell-diagram in Fig. 17 where the resonance points during this run-up are indicated for the first two eigenfrequencies of the blade. The stress response of the blade at its root

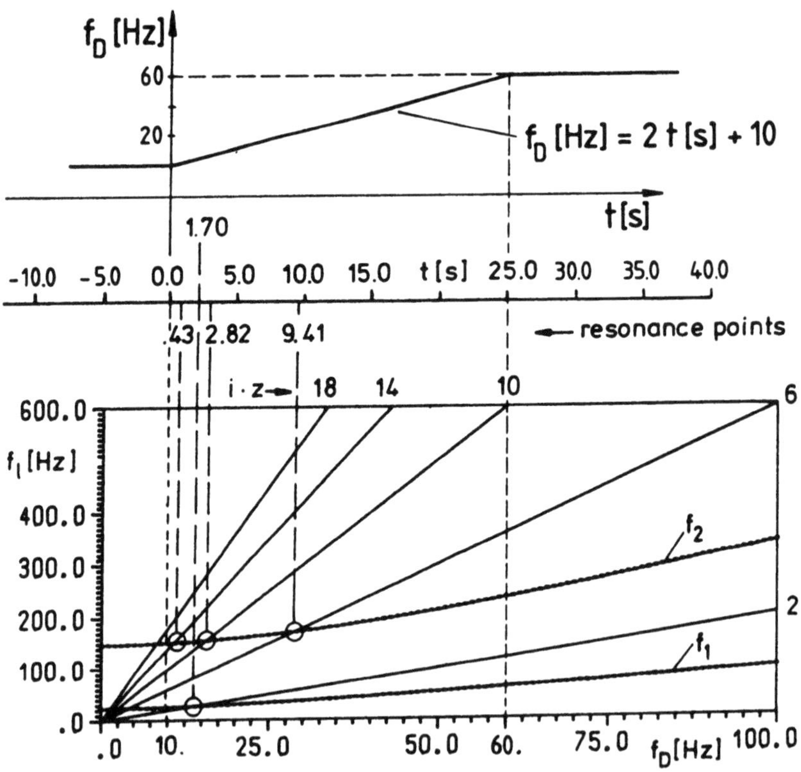

Fig. 17.: Campbell-diagram of a simulated run-up of a turbine blade loaded by partial admission /26/

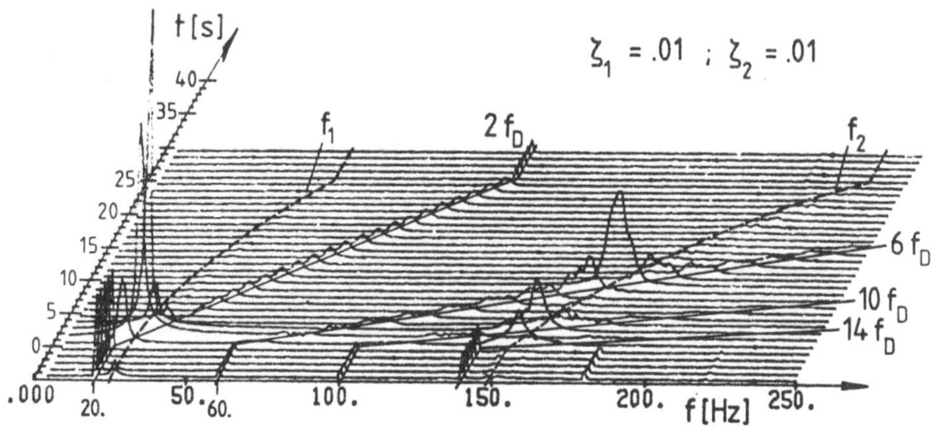

Fig. 18: Stress response of a blade at its root during a run-up loaded by partial admission /22/, /28/

plotted in Fig. 18 which shows for various time windows of the run-up the blade response spectrum up to 250 Hz. We notice the various harmonics of excitation which increase from the values before the beginning of the run-up to the values of its end. These harmonics cut at different times the parabolically-growing eigenfrequency lines of the blade. Thus, at these points resonance vibrations occur which are indicated by the significant responses.

The results plotted in Fig. 18 are taken from the Turbine Blades Simulation Program TUBSIM which is under development at the University of Kassel /22/. For more detailed information, also for the considered example of this chapter, the reader is referred to /26/ and /28/.

5.2 SHELL-LIKE BLADES

As to the vibrations of shell-like blades only a few results can be given here. As a typical example Fig. 19 shows the Campbell-diagram and the mode shapes of a blade of the final stage of a steam turbine. We notice the more or less complicated shape of the mode of the blade especially the higher bending modes. Strong distinction between bending and torsional modes is no more possible because of the complicated shape of the blade. The eigenfrequency curves show that those frequencies corresponding to more or less typical bending modes increase much more with the rotational speed than the torsional frequency does. This is also

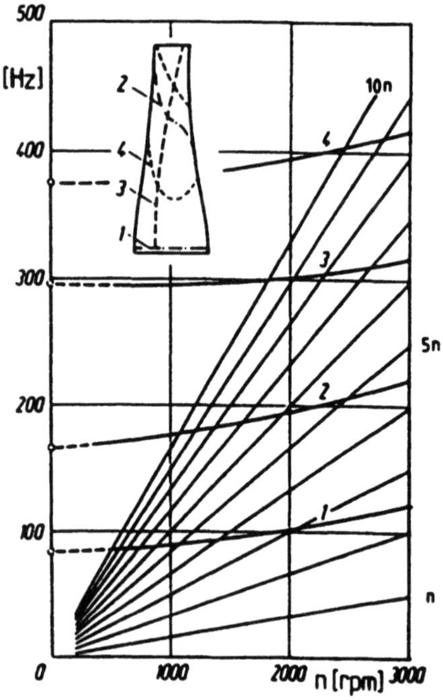

Fig. 19: Campbell-diagram and mode shapes of a shell-like blade of the final stage of a steam turbine /29/

a typical result which can be generalized for all beam- or shell-like blades.

5.3 THREE-DIMENSIONAL BLADES

As for the shell-like blades in the previous chapter also for three-dimensional blades only a few typical results can be described here. Fig. 20 shows a cylindrical hollow blade with aerodynamic profile which was investigated numerically as well as experimentally /30/. For the calculations the method of finite element was used. The applied model is shown in Fig. 20 and has about 600 degrees of freedom. Some calculated and measured eigenfrequencies and mode shapes are shown in Fig. 20, too. We notice for them a good agreement between the numerical and experimental results.

Finally, a complicated finite element model using the NASTRAN

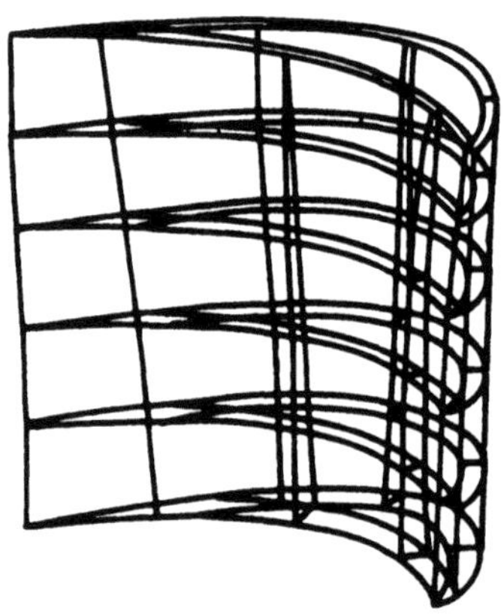

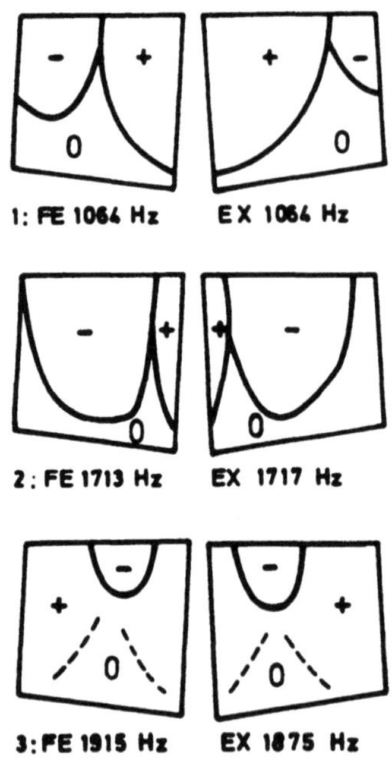

Fig. 20: FE-model, eigenfrequencies and mode shapes of a cylindrical hollow blade /30/

computer program of a twisted hollow blade is shown in Fig. 21. The calculated eigenfrequencies as function of the rotational speed are shown in the Campbell-diagram which is plotted in Fig. 21, too. Again, the typical effect of the rotation on the

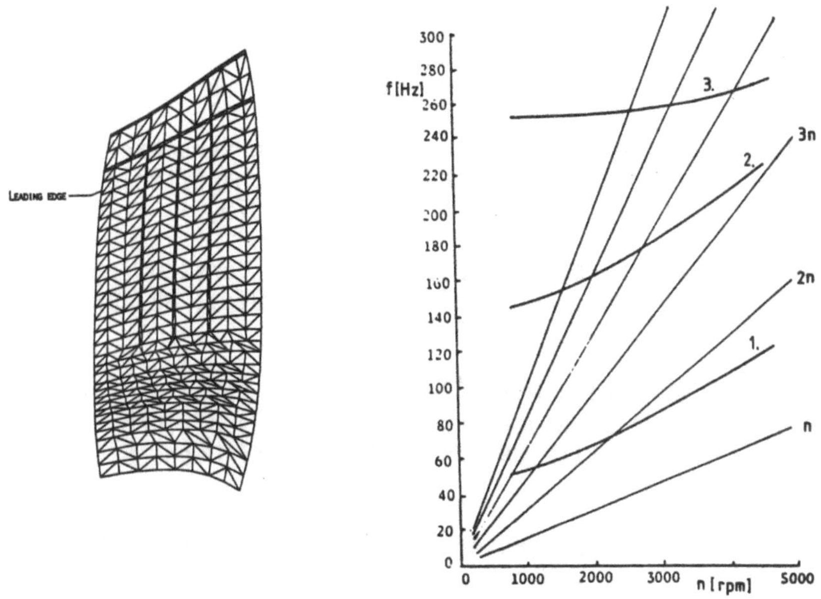

Fig. 21: FE-model (NASTRAN) and Campbell-diagram of a twisted hollow blade /31/

eigenfrequencies is obvious. From the results described in this chapter we can conclude that today it is possible to give numerical results for the natural vibrations of turbine blades even for very complicated shapes and for refined influences for instance like the effect of rotation. The most adequate method to consider these problems is the finite element technique as shown with the results in this chapter.

6. CONCLUSIONS

Free and forced vibrations of turbine blades had been considered. Various types of models were introduced for beam-like, shell-like and three-dimensional blades. It it obvious from these models that even complicated shapes of turbine blades can be considered for calculations of the eigenfrequencies and the corresponding mode shapes. As to the forced vibration problem of turbine blades much less results are known in the literature. However, the development of reliable numerical models is part of several investigations today. Some of the early results had been described in this paper.

7. REFERENCES

/ 1/ BLOHM & VOSS-Prospekt "Schaufeln für Dampfturbinen", Blohm & Voß, Hamburg, 1976

/ 2/ Rao, J.S.: Turbine Blade Excitation and Vibration. Shock and Vibration Digest 9 (1979) 15 - 23

/ 3/ Rao, J.S.: Turbomachine Blade Vibration. Shock and Vibration Digest 12 (1980) 19 - 26

/ 4/ Böhm, J.: Theoretische und experimentelle Parameterstudien an schwingenden Turbinenschaufeln im Fliehkraftfeld. Fortschr.-Ber. VDI-Zeitschrift 11, 29 (1979) 1 - 136

/ 5/ Montoya, J.G.: Gekoppelte Biege- und Torsionsschwingungen einer stark verwundenen rotierenden Schaufel. BBC Brown Boveri Mitteilungen 53 (1966) 216 - 230

/ 6/ Rao, J.S.; Carnegie, W.: Solution of the Equations of Coupled Bending-Bending-Torsion Vibrations of Turbine Blades by the Method of Ritz-Galerkin. Int. J. of Mech. Sci. 12 (1970) 875 - 882

/ 7/ Carnegie, W.: Vibrations of Pre-twisted Cantilever Blading. Proc. of the Instn. Mech. Engrs. 173, 12 (1959) 343 - 374

/ 8/ Carnegie, W.: Vibrations of Pre-twisted Cantilever Blading Allowing for Rotary Inertia and Shear Deflection. J. of Mech. Engng. Sciences 6, 2 (1964) 105 - 109

/ 9/ Fu, C.C.: Computer Analysis of a Rotating Axial Turbomachine Blade in Coupled Bending-Bending-Torsion Vibrations. Int. J. Num. Meth. Engng. 8 (1974)

/10/ Meirovitch, L.: Analytical Methods in Vibrations. Collier-Macmillan Ltd.; London, 1967

/11/ Bishop, R.E.C.; Gladwell, G.M.L.; Michaelson, S.: The Matrix Analysis of Vibration. Cambridge University Press, Cambridge/London/New York/Melbourne, 1965

/12/ Hurty, W.C.; Rubinstein, M.F.: Dynamics of Structures. Prentice-Hall Inc., Englewood Cliffs N.J. (USA) 1964

/13/ Leissa, A.W.: Vibrations of Turbine Engine Blades by Shell Analysis. Shock and Vibration Digest 12, 11 (1980) 3 - 10

/14/ Vogt, H.J.: Zur Berechnung der Eigenfrequenzen und Eigenformen von Schaufeln thermischer Turbomaschinen mit Hilfe eines gekrümmten finiten Schalenelementes. Dissertation, TU Hannover (1974) 1 - 87

/15/ Leissa, A.W.; Ewing, M.S.: Comparison of Beam and Shell Theories for the Vibrations of Thin Turbomachinery Blades. ASME Int. Gas Turbine Conf.,London (England), April 1982; ASME Publication 82-GT-223 (1982) 1 - 12

/16/ Zienkiewicz, O.C.: The Finite Element Method. McGraw-Hill Book Company Ltd., London (England), 1977

/17/ Meirovitch, L.: Computational Methods in Structural Dynamics. Sijthoff and Moordhoff, Alphen aan den Rijn (The Netherlands)/ Rockville MD(USA), 1980

/18/ Bathe, K.J.; Wilson, E.L.: Numerical Methods in Finite Element Analysis. Prentice-Hall Inc., Englewood Cliffs N.J. (USA), 1976

/19/ Steele, J.M.; Lam, T.C.T.: Stress and Fatigue Analysis of Steam Turbine Blades with ANSYS. Proc. ANSYS Conf., Pittsburgh PA (USA), April 1983, 4.48 - 4.65

/20/ Lam, T.C.T.: Computer Simulation of Fatigue Damage. ASME Conf. on Computers in Mech. Eng., Chicago (USA), August 1983

/21/ Irretier, H.: Die Berechnung der Schwingungen rotierender, beschaufelter Scheiben mittels eines Anfangswertverfahrens. Dissertation, TU Hannover (1978), 1 - 129

/22/ Irretier, H.: Turbine Blade Simulation Program TUBSIM. Institut für Mechanik, Universität - Gh Kassel, 1985 (unpublished)

/23/ Traupel, W.: Thermische Turbomaschinen, Bd. 2. Springer-Verlag Berlin/Heidelberg/New York, 1977

/24/ Timoshenko, S.; Young, D.H.; Weaver, W.: Vibration Problems in Engineering. John Wiley & Sons Inc., New York, 1974

/25/ Vogt, E.: Hybride Schwingungsformanalyse an Turbinenschaufeln. Fortschr.-Ber. VDI-Zeitschrift 11, 59 (1984) 1 - 118

/26/ Irretier, H.: Computer Simulation of the Run-up of a Turbine Blade Subjected to Partial Admission. ASME Conf. Mech. Vibr. and Noise, Cincinnati OH (USA), Sept. 1985; ASME Publication 85-DET-128 (1985) 1 - 12

/27/ Rao,J.S.; Jadvani, H.M.: Free and Forced Vibrations of Turbine Blades. Proc. ASME Conf. Mech. Vibr. and Noise, Dearborn MI (USA), Sept. 1983, 11 - 24

/28/ Irretier, H.: Transient Vibrations of Turbine Blades due to Passage through Partial Admission and Nozzle Excitation Resonances. IFToMM Int. Conf. on Rotordynamics, Tokyo (Japan) Sept. 1987

/29/ Pan-Report "Institute of Fluid-Flow Machinery", Polish Academy of Sciences, Gdansk (Polen), 1978

/30/ Gill, P.A.T.; Ucmaklioglu, M.: Isoparametric Finite Elements for Free Vibration Analysis of Shell Segments and Non-Axisymmetric Shells. Journal of Sound and Vibration 65, 2 (1979) 259 - 273

/31/ Aiello, R.A.; Hirschbein, M.S.; Chamis, C.C.: Structural Dynamics of Shroudless, Hollow, Fan Blades with Composite In-Lays. ASME Publication 82-GT-284, (1982) 1 - 7

CHAPTER 11.2

FLOW PATH EXCITATION MECHANISMS FOR TURBOMACHINE BLADES

N.F. Rieger
Stress Technology Incorporated, Rochester, New York, USA

ABSTRACT

The sources of non-steady forces in a turbine stage are reviewed. Procedures for line-vortex stage flow analysis and for actuator disk analyses are described, with details of analytical contributions to these areas. Experimental procedures for measurements of blade non-steady forces in the laboratory and in the field are described. Selected test data are included. Recent developments using the rotating water table approach are reviewed.

11.2.1 Introduction

Non-steady forces arise from local changes (distortions) of the fluid flow, from trailing wakes, and from flow instabilities such as Karman vortices and rotating stall. In a turbomachine stage such disturbances are commonly associated with flow obstructions such as nozzles, guide vanes, diaphragm joints, and certain downstream pressure conditions. During rotation, such stationary flow disturbances may cause harmonic forces to act on the moving blades. Figure 1 illustrates the interaction between stationary guide vanes and a moving blade row, showing the manner and sources from which flow excitations may arise. Table 1 lists several important flow excitation sources. One major source of excitation is passage of the moving blades through the nozzle wakes. This gives rise to blade excitations at nozzle passing frequency (NPF) and at higher multiples (2x, 3x, etc.) of NPF. The strength of these excitation harmonics depends on factors such as the pressure drop across the flow guides, ratio of number of guides to number of blades, axial spacing between the flow guides and blades, guide vane trailing edge thickness, local flow Mach number, and so on. Such excitations have been observed on the moving blades in the tangential, axial, and torsional directions. Similar excitations also act on the non-rotating guide vanes.

Low pressure stage excitations arise mostly from low frequency harmonic distortions of the flow field. These excitations are shown in Figure 1. Harmonic flow distortions may arise from the inlet flow guide passage spacing, e.g., from guide vane pitching and/or gaging errors during manufacture or assembly of the guide rows. These excitations also occur from construction features such as diaphragm joints, and from radial or angular eccentricity of the moving row. Low pressure moving blades are airfoil-shaped, i.e., relatively thin with small camber (and hence small turning angle), as in Figure 1. Low frequency per-rev excitations may be of the same magnitude as the high frequency NPF excitations for these stages. LP blade excitation spectrum data may be obtained using a rotating pressure transducer on a moving blade, or by analog sumulation of the stage flow using a rotating water table, as described in Section 11.2.4.

At present, pratical computer methods for predicting magnitude of the excitation on turbine blades are in a developmental stage. The fundamental problem is highly complex: the stage geometry creates a three-dimensional flow field in which interactions occur between the nozzle flows and the flow in the moving blade passages. Flow-field analyses of this problem involve severe analytical difficulties. Up to the present such analyses have mostly been restricted to two-dimensional potential flow studies. The wake/boundary layer interaction is thereby excluded from such formulations of this problem, though such effects may be significant. Further difficulties exist with available methods for handling effects from the high turning angles of impulse blading. Localized transonic flow conditions may occur in HP blade passages, and such effects are beyond the scope of all but research analysis. Details of some approaches which have been developed for non steady flow calculations are given in Section 11.2.3.

The calculation of LP stage exciting forces is simpler than for an HP stage because smaller blade turning angles are involved, and the stage pressure ratios may be lower. Non-steady line vortex theory may then be used as an approximate method for evaluating the high frequency exciting forces on low camber airfoils, while actuator disk techniques (and others) have been applied to predict the low frequency per-rev harmonic forces.

The inherent danger from non-steady forces acting on the moving blades is evident from the Campbell diagram shown in Figure 2. This diagram is used to identify potential resonant operating conditions. Blade frequency is plotted as ordinate against machine rpm as abscissa. Blade natural frequencies are plotted as characteristic lines. Speed-related excitation harmonics are shown as the 1x, 2x,..., per-rev excitation lines. The intersection of any per-rev harmonic with any natural frequency is potentially dangerous, because it represents a resonant condition. Sustained resonant operation under such conditions constitutes a possible failure hazard. Whether danger actually exists depends on three additional factors: (a) excitation magnitude, (b) damping associated with the excited mode, and (c) the phasing of the exciting forces relative to the blade group. Non-steady flow studies are directed toward defining the magnitude of the exciting forces in the spectrum of a given stage, for nozzle wake harmonics, and for the lower per-rev harmonics, together with details of their phasing.

A further source of excitation in HP turbine stages arises from partial admission operation. Here, the inlet flow enters via one or more nozzle sectors around the inlet circumference: see Figure 3. As the blades pass through the flow admission arcs, they are subjected to transient loading from the inlet gas jet, followed by full unloading in the inactive (non-flow) sectors, one or more times per rotor revolution. The steady-state blade load is readily determined, but details of the associated time-dependent transient loadings are more difficult to obtain, because impulse harmonics are involved. An experimentally-determined load profile for a partial admission stage is shown in Figure 4. Until recently, there has been little research information available on partial admission blade loadings, due to the obvious difficulties of making force measurements in high-pressure, high-temperature turbine stages. Blade strain gage data is more readily obtained, and this has been used to calibrate the force-time profile. Some comments and information on partial admission loads on turbine blades are given in Section 11.2.4.

11.2.2 Notation

A	tan θ
a	sonic velocity in gas ($\sqrt{\gamma g\ RT}$)
B	constant in reference [19] see below
B_s	constant in Whitehead's analysis [13]
b	axial distance between blade rows

b_s	constant in Whitehead's analysis [13]
c	blade chord
$C(\sigma)$	Theodorsen function (defined in text)
C_L	lift coefficient normal to chord
C_F	dimensionless force (defined in text)
C_M	dimensionless force (defined in text)
D	constant in reference [19] see below
F	constant in reference [19] see below
Fr	Froude Number
G_m	Glauert expansion (defined in text)
h	constant in reference [17]. Also water depth
i	$\sqrt{-1}$
$J(\sigma)$	Bessel function of first kind
$K(\sigma)$	modified Bessel function of second kind
k	constant in reference [8]
l	disturbance in wavelength $2\pi V_s/l$
L_x	non-steady axial force
L_y	non-steady tangential force
L_o	steady lift
L	unsteady lift
M	Mach number
N	rpm
p	pressure
P	constant in reference [19] see below
q	blade velocity due to vibration
r_a	blade mean radius
R	gas constant
$S(\sigma)$	Sears' function (defined in text)
s	blade pitch
t	time
T	temperature
$T'(\sigma)$	Holmes function (defined in text)
$T(\sigma)$	Horlock function (defined in text)
u	non-steady axial velocity of gas
U	mean axial velocity of gas
v	non-steady tangential velocity of gas
V	mean tangential velocity of gas
V_s	blade tangential velocity
w	general non-steady gas velocity
W	mean gas velocity relative to blade
$X(\sigma)$	function in reference [11]
x	axial direction
y	tangential direction
$Y(\sigma)$	function in reference [11]
α	angle of attack
β	mean flow angle relative to axial direction
γ	specific heat ratio (Cp/Cv)
ε	constant in Whitehead's analysis [13]
θ	stagger angle
θ_{LE}	lead edge blade surface angle relative to axial direction

θ_{TE} trail edge blade surface angle relative to axial direction
$\mathcal{K}$ constant in reference [19] see below
λ Whitehead's reduced frequency parameter, $\nu c/U \sec \beta$
ν disturbance frequency $2\pi V_s/1$
ξ_p location of center of pressure of blade
ξ_T location of center of twist of blade
π 3.14159
ρ mass density
σ reduced frequency parameter relative to semi-chord $\nu c/2W$
Φ potential function
ψ constant in reference [19] see below
ω reduced frequency parameter relative to pitch $2\pi s/1$

Constants:

Horlock, Greitzer, and Henderson [19]

$$B = \mathcal{K}^2 V_s \sec\theta\{[2(A^4-2A^2-1) + D - 2\psi A(1+A^2)] + i[DA+(1+A^2)(\psi-4A^3)]\}$$

$$D = 4A^2 - \psi^2$$

$$F = [(2A^2+\psi A-D) + i(2A^3-\psi)]$$

$$P = [(2A^2+\psi A) + i(2A^3-\psi)]$$

$$\psi = (1+A)\ [1+2\sigma(1-A)]$$

$$\mathcal{K} = 2\pi/1$$

Whitehead [13]

$$B_s = (1+b_s)\ /\ (1-b_s)$$

$$s = \text{arc tan } U/V_s$$

Henderson - Horlock [17]

Let $\tan\theta_{LE} = g_1$ $\quad$ $\tan\theta_{TE} = g_2$

$$h_1 = g_2 - g_1$$

$$h_2 = \frac{(1+g_1^2)}{2} + \frac{g_1 h_1}{3} + \frac{h_1^2}{12}$$

$$h_3 = \frac{g_1 h_1}{2} + \frac{h_1^2}{6}$$

$$h_4 = g_1 + \frac{h_1}{2}$$

$$h_5 = \frac{g_1}{2} + \frac{h_1}{3} + \frac{g_1}{2} + \frac{2g_1^2 h_1^2}{3} + \frac{g_1 h_1^2}{3} + \frac{h_1^3}{15}$$

$$h_6 = \frac{g_1 h_1}{2} + \frac{g_1 h_1^2}{2} + \frac{h_1^3}{8}$$

$$h_7 = g_1^2 + g_1 h_1 + \frac{h_1^3}{3}$$

$$h_8 = g_2^2 - g_1^2$$

11.2.3 Non-Steady Force Theories

Information Sources:

Several recent reviews of the excitation literature have been published. Sisto [1] has summarized the status of non steady flow analyses, with emphasis on vortex theories. This is a brief but useful introduction to the excitation literature of turbine LP stage flows. Rao [2] [3] has catalogued vortex analyses with reference to turbine blade vibration, in two general survey papers. Samoylovich [4] has published (in Russian) a monograph concerned with turbine blade excitation and vibration, which reviews excitation sources and describes blade measurement and test procedures for dynamic pressure distributions and for blade response. Osborne [5] has given an excellent comparative review of developments in non-steady compressible flow theories. Existing procedures for calculating non-steady interactions between blade rows are reviewed and compared with results from the matched asymptotic expansion approach for high subsonic flows (M>0.9). Gostelow [6] has reviewed steady state compressible theories for potential flow through cascades with reference to the transonic flow problem, and has indicated some promising numerical techniques for its solution. Crofoot [7] has presented a comprehensive state-of-the-art review for non-steady forces on turbomachine blades.

Vortex Theories:

Kemp and Sears [8] represented lifting surfaces by line vortices to predict the non-steady forces acting on both the moving blades and the stationary blades of a turbomachine stage such as that shown in Figure 5. The flow through the cascade is assumed to be inviscid and incompressible. The airfoils are assumed to be isolated i.e., the non-steady circulation of neighboring blades is neglected when calculating the non-steady effects on a particular airfoil. For rotor blades however, the effects of the vortex wakes shed by the stators are included. The rotor and stator blades are assumed to be flat plates though they may be thin, slightly cambered, lightly loaded airfoils. The implication of these assumptions is that the flow can tolerate only small turning angles as it passes through the blade rows. This analysis is therefore more suited to low solidity LP turbine stages than HP turbine stages.

The stator wakes are modeled as gusts which decay in amplitude as they propagate downstream. The velocity of the gust perpendicular to the chord is assumed to be of the form,

$$w = w_o e^{-i\nu t} e^{-iky/W} \tag{1}$$

where ν is the stator vane passing frequency, W is the free stream relative velocity, and k is an arbitrary constant. The lift fluctuation is then given by:

$$L = \pi \rho c W w_o S(\sigma,\lambda) e^{i\nu t} \tag{2}$$

where ρ is the fluid mass density, c is the chord length, and $S(\sigma,\lambda)$ is the modified Sears function, given by:

$$S(\sigma,\lambda) = J(\lambda)\, C(\sigma) + i(\sigma/\lambda)\, J_1(\lambda) \tag{3}$$

where the $J_n(\lambda)$ are Bessel functions of the first kind, and $C(\sigma)$ is the Theodorsen function. The Theodorsen function is given by:

$$C(\sigma) = K_1(i\sigma) \,/\, [K(i\sigma) + K_1(i\sigma)] \tag{4}$$

where $K_n(i\sigma)$ are modified Bessel functions of the second kind. The arguments of the modified Sears function are the reduced frequency ($\sigma = \nu c/2W$) and the frequency associated with the decay term ($\lambda = Kc/W$).

The results shown in Figure 6 are for the first two harmonics ($m = 1,2$) of the lift ratio (unsteady lift/steady lift) plotted against the axial spacing ratio (b/c) and the pitch ratio (S_r/S_s). The dashed curves correspond to an elliptical steady load distribution on both the rotor and stator blades, while the solid curves are for an elliptical distribution on the stator blades, and a flat-plate distribution on the rotor blades. The authors concluded that 'the non-steady part of the lift may be as large as 18% of the steady lift and therefore may be of practical importance.' This low value is a feature of the simple model chosen. It is by no means an upper limit for practical blade conditions.

In a subsequent study [9] Kemp and Sears considered the effect of the viscous wakes of an upstream stator on a downstream rotor blade as the blade passes through the stator wakes. The assumptions of thin airfoil theory were again imposed. The wake behind each stator blade is the same as the wake behind an isolated airfoil. The wake consists of an inviscid, symmetrical, shear perturbation of the undisturbed stream. If the wake velocity in the streamwise direction is assumed to be constant, the non-steady lift on the rotor due to viscous wakes is given by:

$$L = \pi c\ W\ w_o\ S(\sigma)e^{i\nu t} \tag{5}$$

where ρ is the mass density, W is the free stream velocity relative to the rotor blade, and c is the chord length. The non-steady lift coefficient C_L is given as:

$$C_L = \sum_{m=1}^{\infty} G_m\ S(m\sigma)e^{im\nu t}$$

where G_m is the Glauert expansion for the non-steady velocity normal to the chord, $S(m\sigma)$ is the Sears function for reduced frequency ($\sigma = \nu c/2w$), and ν is the disturbance (stator passing) frequency. Comparing the results of this analysis with those obtained previously suggests that the viscous-wake effects on a rotor blade may approach the magnitude of the circulation-induced non-steady lift.

Horlock [10] [11] noted that the Kemp-Sears analyses considered only velocity perturbations normal to the chord. He then derived an expression for fluctuating lift due to a streamwise velocity perturbation. The Kemp-Sears assumptions for thin airfoil theory were again used. Since the flow inlet velocity is not necessarily parallel to the chord, small angles of attack (α) were thereby included in this analysis. The non-steady velocity parallel to the chord is assumed to be of the form:

$$u = u_0 \; e^{i\nu(t - \frac{x}{U})} \qquad (7)$$

where ν is the stator passing frequency, and the coordinate x is in the direction of the chord. The fluctuating lift is then given by the expression:

$$L = 2\pi \; U \; u_o \; \alpha \; \rho \; e^{i\nu t} \; T(\sigma) \qquad (8)$$

where ρ is the mass density, w_c is the maximum value of the non-steady chordwise velocity, α is the angle of attack, and $T(\sigma)$ is the Horlock function:

$$T(\sigma) = X(\sigma) + iY(\sigma) \qquad (9)$$

where, $X(\sigma) = (2-a) \; J(\sigma) - b \; J_1 \; (\sigma)$

$Y(\sigma) = (a+1) \; J_1 \; (\sigma) - b \; J(\sigma)$

and $K_o/(K_o+K_1) = a + ib$

where $J_n(\sigma)$ are Bessel functions of the first kind and K_n are modified Bessel functions of the second kind. The results of this analysis may be combined with the results of the first Kemp-Sears analysis to obtain the total circulation-induced lift fluctuation on a blade due to a general periodic disturbance of the free stream flow. This disturbance w is resolved into components perpendicular to the chord w_t and parallel to the chord w_c as shown in Figure 7. In general, the velocity of the disturbance can be expressed as a Fourier series:

$$w = \sum_n w_{on} \; \sin \frac{2\pi n x}{1} \qquad (10)$$

where 1 is the wavelength of the disturbance. For the nth component of this series, the fluctuating lift is:

$$L = 2\pi \; \rho \; U\sec\beta \; w_{on} \; e^{i\nu t}[S(\sigma) \; \sin\beta - aT(\sigma) \; \cos\beta_m] \qquad (11)$$

where U is the axial velocity, β is the mean flow angle, ν is exciting frequency, and σ is reduced frequency. The above expression for fluctuating lift applies to flat plate airfoils. To account for camber Horlock included the Holmes function, $T'(\sigma)$, which was derived for an airfoil of parabolic camber. The total lift fluctuation for the nth component of the Fourier series including all the above effects is:

$$L=\pi \rho U 1 e^{i\nu t} v_o S(\sigma)+u_o[\alpha T(\sigma)+(\frac{2y_{max}}{1})T'(\sigma)] \qquad (12)$$

where y_{max} is the maximum airfoil thickness and c is the airfoil chord. Naumann and Yeh [12] independently derived an expression relating the unsteady lift of a cambered airfoil to a general flow perturbation.

Actuator Disk Method:

Whitehead [13] introduced a procedure for calculating the non steady blade force resulting from a given harmonic disturbance profile in the incoming flow. Figure 8 shows the blade row considered as a thin 'actuator disk,' with the associated velocity changes resulting from the momentum transfer through the disk. It is also assumed that the blades vibrate nearly in-phase, and that the blades are flat plates at stagger angle Θ to the axial flow. The blade row is considered to be narrow, such that the time for the fluid to pass through is small compared with the disturbance frequency acting on the blades. This assumption is expressed in terms of the reduced frequency parameter $\lambda = \nu c/U\sec\beta$, where ν is the frequency of the non-steady excitation, c is the chord length, and U is the mean velocity of flow in the axial direction. The reduced frequency is assumed to be close to zero in this analysis. Using these assumptions, Whitehead obtained a simple expression for the dimensionless non-steady force acting on any blade.

Whitehead's procedure depends only on a knowledge of the blade/cascade proportions and geometry, and the steady and non-steady flow velocities. The above theory is restricted to cases where the reduced frequency λ is small, i.e., $\lambda \ll 1$. It is therefore well-suited to the analysis of low frequency harmonics of high inlet velocity machines with narrow blades, but it is not suited to analysis of nozzle wake excitations from low inlet velocities. To relax this restriction Whitehead [14], Horlock and Danyshar [15], and Smith [16] have developed more advanced procedures which are suitable for $0<\lambda<1.0$.

Henderson and Horlock [17] have approximated the unsteady lift on airfoils for cascades having low pitch-to-chord ratios. The pitch is also assumed to be small with respect to the wavelength of the inlet disturbance, Figure 9. The rotor blades are thin and may have considerable camber but the lift coefficient is assumed to be not large due to the low pitch to chord ratio. The flow is assumed to be two-dimensional, inviscid, and incompressible. Sinusoidal axial disturbances are considered, of finite frequency parameter σ, based on the blade chord, but of low frequency parameter ω, based on blade pitch.

The basis of this theory is the so-called pitch-averaging technique due to Horlock and Marsh [18] applied to the unsteady equations of motion for the fow through the moving blade passage. The total force in the tangential and axial directions was obtained by integrating the pressure difference between the pressure and suction faces of the blade along the chord. For an inlet velocity of:

$$u = U + u_o \sin\nu(t - y/V) \tag{14}$$

the unsteady tangential component of lift is:

$$\frac{L_y}{\rho c V^2} = \left\{[2 \frac{c}{s} \frac{u_o}{V} \frac{h_2}{\sec^2\beta}(1-\cos\omega) - \frac{u_o S}{V\omega c}\sin\omega - \frac{2uoUs}{\omega V^2 c} h_1 \sin\omega]\right\}\sin\nu t$$
$$+\left\{[\frac{4u_o U}{\omega V^2 \sec\beta} h_3 (1-\cos\omega) + \frac{u_o h_4}{V\sec\beta}\sin\omega]\right\}\cos\nu t \tag{15}$$

A similar expression was obtained for the unsteady axial component of lift, $L_y/\rho c V^2$.

The symbols used are defined in the Notation section. The total unsteady lift for a turbine cascade is the vector sum of the tangential lift and the axial lift components:

$$\frac{L}{\rho c V^2} = \frac{L_x \sin\beta - L_y \cos\beta}{\rho c V^2} \tag{16}$$

A comparison of results obtained using the above theories is shown in Figure 10.

The above calculation procedures have been further developed by Horlock, Greitzer, and Henderson [19], who presented an alternative 'semi-actuator disk' method. This procedure assumes that the blade row is made up of a closely-spaced set of flat plates set at an angle Θ to the axial direction, and that:

a) Far upstream there is a harmonic disturbance in the axial velocity,

b) The stream function is continuous at the leading edge of the blades with a discontinuity in slope,

c) Continuous stagnation pressure exists at the leading and trailing edges,

d) Continuity of velocities exists at the blade trailing edge.

These authors obtained the following expression for non-steady lift coefficient:

$$C_L = \frac{L_1 + L_2}{\pi\rho W w_o e^{i\omega t}} = \frac{s}{\pi c D} \{P(2\sigma^2 \sec\theta \operatorname{cosec}\theta - 1) + i(\frac{B 2\sigma\cos\theta}{V_s \mathcal{K}^2 \tan\theta} - F)\} \quad (17)$$

This result applies for the case of a general stagger angle θ, where:

$$L_1 = -\rho V_s s \mathcal{K} \sec\theta\ (F - iP) \exp i(\nu t - \mathcal{K} y)$$

$$L_2 = \frac{1}{2}\rho s \cos\theta\ \{2Bc - i(\mathcal{K}^2 c^2 \nu \sec^2\theta)P\} \exp i\nu t \quad (18)$$

and where P, F, $\mathcal{K}$, and B are defined in the Notation section.

Results obtained by the semi-actuator disk method have been compared with data from Smith [16] in Figures 11 and 12, from [19]. It is evident that the semi-actuator disk theory gives good results when the space/chord ratio is low, and when the reduced frequency σ is fairly small.

11.2.4 Experimental Studies

Types of Blade-Force Tests and Experiments:

Measurements for non-steady blade forces and stresses have been conducted as follows:

a) In-situ tests on turbine and compressor blades, rotating and non-rotating, using strain gages.

b) Experimental turbine stage studies in wheelboxes using strain gages and developmental holographic techniques.

c) Wind tunnel tests on blade cascades.

d) Water table measurements using strain gages.

e) Conducting-sheet analogy and electrolytic tank analogy.

These studies may be divided into several categories:

a) Direct measurement of blade response in a rotating stage flow test.

b) Laboratory measurement of blade response under controlled stage flow conditions.

c) Quasi-static studies of stage flow for specified nozzle-blade static configurations.

d) Nozzle cascade wind tunnel measurements.

Direct measurement of blade response to exciting forces is a meaningful procedure for non-steady force estimation only where the blade dynamic response properties do not substantially interfere with the blade excitation frequencies. Any tendencies toward blade resonant response will give incorrect excitation data. The strain gages used in such tests must be calibrated to give load readout. Direct readout of forces from strain gages is difficult to achieve because the blades themselves are very stiff and the strain gages have limited sensitivity and signal/noise properties. Good signals may be achieved under blade resonance conditions, but non-resonant strain gage data from blades are rarely taken in practice. For this reason in-situ blade testing is most commonly used to test/define the Campbell diagram over the speed range under operating load conditions.

Laboratory measurement of blade response. Such measurements may be undertaken in several different ways. Heater box tests are common blade verification procedures. They are usually conducted in vacuum (to minimize power and heating), without gas flow. Blade natural frequencies can be excited under such circumstances, usually with electromagnets. Blade tests in experimental turbines have also been conducted e.g., by Wagner [20] with water jet excitation to measure blade group damping, and by Partington [21] to measure blade excitation and stresses: see later. Other laboratory tests have been conducted on a rotating water table model, using the hydraulic analogy to model the flow through one or more turbine stage geometries. These tests have provided data on non-steady force (and torque) spectra, for a wide range of flow parameters, and are discussed later in this section.

Stationary blade-nozzle stage tests. Several stationary stage tests have been reported in the literature e.g., Kearton [22], Malavard [23]. Some indication of the flow through a stage is usually sought by measuring blade forces at series of increments across the nozzle exit. Quasi-static testing of this type does not match the velocity conditions in an actual stage, and the results are therefore considered to be unrepresentative of stage flow conditions, and of wake effects.

Nozzle cascade wind tunnel tests. Some indication of the excitation likely to arise from a given nozzle row or guide row is sometimes sought using a model row of guide vanes in a wind tunnel (frequently installed as corner turning guides). Tests of this type do not give blade forces, but they can give guidance of HP stages where the pressure drop occurs primarily in the nozzle and not in the blades. Velocity traverses from such models have been reported by several authors. Samoylovich [4] has discussed wake properties in detail.

Non-steady Blade Stress Measurements:

Measurement of blade dynamic stresses is now a relatively straightforward procedure. In the field such measurements are conducted on a 'need' basis because of the high costs involved in monitoring the number of channels required to obtain representative information. Relatively few blade stress reports of general usefulness have been published despite years of such testing. Partington [21] used an experimental turbine test to determine the sources of low pressure blade excitation and stress. Excitations arising from the following factors and others were investigated:

a) Obstructions in inlet and exhaust flow fields.

b) Nozzle and blade exit profile changes.

c) Axial distance between guides and blades.

d) Gaging modifications in nozzles and blades.

The blades and nozzles of the test turbine were instrumented with strain gages, and a slip ring assembly at the end of the rotor was used to obtain the blade strain gage signals. Partington found that:

a) Large differences in stress at similar blade locations were found between corresponding blade groups.

b) Small differences in stress were found between blades of the same group (presumably for in-phase modes).

c) Stress generally increases with output torque, but decreases with increasing exhaust pressure.

d) Plots of stress amplitude vs. rpm near resonance showed the first mode response to be complicated by multiple stress peaks. This was apparently caused by coupling between different blade groups around the disk. Each blade group is evidently not an independent vibration system.

This latter point has been observed by many other investigators. Partington also made blade group damping tests and found group logarithmic decrements in the order of 0.017 and 0.025. Heyman [24] conducted similar tests to determine the influence of axial gap between blades and nozzles at various wheel speeds on blade dynamic stresses. Strain gages were again used to measure the stress response.

Test Turbine Studies of Blade Forces:

Test turbines are used frequently in blade development work. Pressure profiles upstream and downstream of blades and nozzles have been measured by pitot nozzles and by pressure transducers. Descriptions of test turbines and flow measuring equipment have been given by Renaudin and Somm [25] and others. Samoylovich [4] shows drawings of two test turbines and gives details of pressure instrumentation and strain gage results obtained, with details of local pressure vs. time at various locations across the blade wake. The Westinghouse test turbine facility in Lester, Pennsylvania is designed to operate with either steam or air as the working fluid.

Studies of Flow in Turbomachines by the Hydraulic Analogy:

The earliest investigation of turbine stage flow using the hydraulic analogy appears to have been that of Preiswerk [26] in 1942. This paper contained photographs of subsonic, low supersonic, and high supersonic flows through a stationary cascade. For twenty years, the stationary cascade represented the state-of-the-art for hydraulic analogy studies of flow in turbines and compressors. Results from several investigations of this type have been compared favorably with cascade wind tunnel test results: see Hoyt [27]. More recent studies have used both stationary blades and moving blades to more accurately model the fluid-structure interaction in a turbine stage. Heen and Mann [28] studied flow in a partial admission turbine stage using a track and carriage apparatus, Figure 13. The water depth was measured along a blade passage. A theoretical depth at each position was obtained from one-dimensional water flow theory. Typical results plotted against theoretical values are shown in Figure 14. Johnson [29] has described a two stage axial flow compressor study made at the General Electric Research Laboratory.

Johnson's model employed two rows of rotor blades which moved relative to adjacent stator rows. The apparatus was used to study off-design conditions as well as the effects of varying rotor and stator angles and spacings. Slow-motion moving pictures were taken during testing. Similar tests have been performed on turbine stages using a similar apparatus. Rhomberg [30] made a qualitative comparison of results obtained from a water table model with results obtained from a rotating transonic air cascade. Shadowgraphs of the flow downstream of a transonic air cascade are given. Good correlation was found for the relative locations of the shock waves obtained by each method. Owczarek [31] investigated a periodic wave phenomenon occurring from stage interactions between turbine blades and stators. This phenomenon occurs when a pressure wave is generated on the leading edge of the rotor blades. Owczarek constructed a rotating radial inflow water table model of a suitable configuration for this phenomenon, according to theory. Photographs were taken of the flow between the stator and rotor which showed waves (analogous to gas shocks) which propagated as predicted by theory.

Rotating Water Table Tests of Blade Forces:

Non-Steady Forces from Nozzle Wakes. A turbine stage was constructed [32] and tested [33] to determine the magnitude of the exciting forces associated with a certain nozzle design which was associated with in-service failures. The stage geometry is shown in Figure 15. A typical force response amplitude vs. frequency spectrum from the moving blade strain gage output is shown in Figure 16. A large amplitude spike is evident at nozzle passing frequency (NPF). Such spikes represent the harmonic force which is applied to the moving blades at NPF. Harmonics of the NPF are also present. The test results are shown as a plot of dimensionless non-steady force amplitude vs. velocity ratio for several pressure ratio tests in Figure 17.

Transient Forces From a Partial Admission Stage. This test was conducted to measure the transient load variation on a moving blade as it passed through the flow from a nozzle arc in a partial admission stage. The general form of the load transient on a moving blade in an actual partial admission stage is well known from prototype strain gage tests. This water table test showed the degree to which the water table load transient resembled the load transient obtained from actual blading tests. The stage parameters and typical results are shown in Figure 4. The following similarities can be observed:

a) Relative slope of the inlet response

b) Inlet response spike magnitude

c) Outlet response spike magnitude

d) Relative magnitude of main curve

Subsequent partial admission tests have demonstrated that this correlation is typical, and that quantitative comparison (e.g., peak/average) is representative of that observed in practice.

11.2.5 References

1) Sisto, F., 'Nonsteady Flow Excitation in Steam Turbine Stages,' Proceedings, Workshop on Improved Turbine Availability, Electric Power Research Institute, Palo Alto, California, January 1977.

2) Rao, J. S., 'Turbine Blade Excitation and Vibration,' Shock and Vibration Digest, Vol. 9, No. 3, pp. 15-22, March 1977.

3) Rao, J. S., 'Turbomachine Blade Vibration,' Shock and Vibration Digest, Vol. 19, No. 5, pp. 3-10, May 1987.

4) Samoylovich, G. S., ''Vibration Problems in Rotating Turbomachines,'Machinestroya Publishing House, Moscow, 1975 (in Russian).

5) Osborne, C., ''Compressible Unsteady Interactions Between Blade Rows', AIAA Journal, Vol. 11, No. 3, pp. 340-346, March 1973.

6) Gostelow, J. P., 'Review of Compressible Flow Theories for Airfoil Cascades,' Trans. ASME, Journal of Engineering for Power, Series A, Vol. 95, No. 4, pp. 281-292, October 1973.

7) Crofoot, J. F., 'Theories and Experiments for Determination of Non-steady Loads on Turbomachine Blades,' M.S. Thesis, Rochester Institute of Technology, Rochester, New York, February 1979.

8) Kemp, N. H., Sears, W. R., 'Aerodynamic Interference Between Moving Blade Rows',' Journal of the Aeronautical Sciences, Vol. 20, No. 9, pp. 585-597, 1953.

9) Kemp, N. H., Sears, W. R., 'The Unsteady Forces Due to Viscous Wakes in Turbomachines,' Journal of the Aeronautical Sciences, Vol. 22, No. 7, pp. 478-483, 1955.

10) Horlock, J. H., 'Unsteady Flow in Turbomachines,' Proceedings, Conference on Hydraulics and Fluid Mechanics, Institution Of Engineers, Australia, pp. 221-227, November 1968.

11) Horlock, J. H., 'Fluctuating Lift Forces on Aerofoils Moving Through Transverse and Chordwise Gusts,' Trans. ASME, Journal of Basic Engineering, Series D, pp. 494-500, December 1968.

12) Naumann, H., Yeh, H., 'Lift and Pressure Fluctuations of a Cambered Airfoil Under Periodic Gusts and Applications to Turbomachinery,' Trans. ASME, Journal of Engineering for Power, Series A, pp. 1-10, January 1973.

13) Whitehead, D. S., 'Vibration of Cascade Blades Treated by Actuator Disk Methods,' Proceedings, IMechE, Vol. 173, No. 21, pp. 555-574, 1959.

14) Whitehead, D. S., 'Force and Moment Coefficients for Vibrating Aerofoils in Cascade,' Aeronautical Research Council, R and M 3254, 1960.

15) Horlock, J.H., Daneshyar, H., 'Stagnation Pressure Changes in Unsteady Flow,' Aeronautical Quarterly, Vol. 22, Part 3, August 1971.

16) Smith, S., 'Discrete Frequency Sound Generation in Axial Flow Turbomachines,' Aeronautical Research Council, R and M 3684, 1972.

17) Henderson, R. E., Horlock, J. H., 'An Approximate Analysis of the Unsteady Lift on Airfoils in Cascade,' Trans. ASME, Journal of Engineering for Power, Series A, pp. 233-240, October 1972.

18) Horlock, J. H., Marsh, H., 'Flow Models for Turbomachines,' Journal of Mechanical Engineering Science, Vol. 13, No. 5, pp. 358-368, 1971.

19) Horlock, J. H., Greitzer, E. M., and Henderson, R. E., 'The Response of Turbomachine Blades to Low Frequency Inlet Distortions,' Trans. ASME, Journal of Engineering for Power, Series A, 1976.

20) Wagner, J. T., 'Blade Damping Tests,' Westinghouse Engineering Report EC401, NOBS N00024-67-C-5494, May 1969.

21) Partington, A. J., 'Experimental Study of the Sources of Blade Vibration Excitation in a Low Pressure Turbine,' Westinghouse Electric Corporation Development Engineering Dept., Lester, PA, Contract NOBS-94380, Report No. EC 384, April 1971.

22) Kearton, W. J., Steam Turbine Theory and Practice, Sir Isaac Pitman and Sons Ltd., London, England, 1958.

23) Malavard, L., 'The Use of Rheo-Electrical Analogies in Certain Aerodynamical Problems,' Journal of Royal Aeronautical Sciences, Vol. 51, pp. 739, 1947.

24) Heyman, F. J., 'Turbine Blade Vibrations Due to Nozzle Wakes,' ASME Publication, Paper No. 68-WA/PWR-1, Presented December 1-5, 1968.

25) Renaudin, A., Somm, E., 'Quasi-Three-Dimensional Flow in a Multistage Turbine Calculation and Experimental Verification,' Flow Research on Blading, Elsevier Publishing Company, London, England, pp. 51-88, 1970.

26) Preiswerk, E., 'Some Applications of the Method of Hydraulic Analogy,' Publ. Scientifiques et Tech. du Ministere de l'Air, Paris, France, 1942.

27) Hoyt, J. W., 'The Hydraulic Analogy for Compressible Gas Flow,' Applied Mechanics Review, June 1962.

28) Heen, H. K., Mann, R. W., 'The Hydraulic Analogy Applied to NonSteady Two-Dimensional Flow in the Partial-Admission Turbine' , Trans. ASME, Journal of Basic Engineering, Series D, pp. 408-421, September 1961.

29) Johnson, R. H., 'The Hydraulic Analogy and its Use with Time Varying Flows,' GE Research Laboratory Report, 64-RL-(3755 C), Schenectady, NY, August 1964.

30) Rhomberg, E., 'Investigations into Rotating Blade Cascades for Transonic Flow,' The Brown Boveri Review, Vol. 51, No. 12, pp. 762-773, December 1964.

31) Owczarek, J. A., 'On a Wave Phenomenon in Turbines,' Trans. ASME, Journal of Engineering for Power, Series A, pp. 262, July 1966.

32) Rieger, N. F., Wicks, A. L., 'Design and Development of Water Table Analog for Dynamic Forces in Turbine Stages,' ASME Paper 78WA-DE16.

33) Rieger, N. F., Wicks, A. L., 'Non-Steady Force Measurements and Observations of Flow in Three Turbine Stage Geometries,' Trans. ASME Vol. 100, pp. 525-532, October 1978.

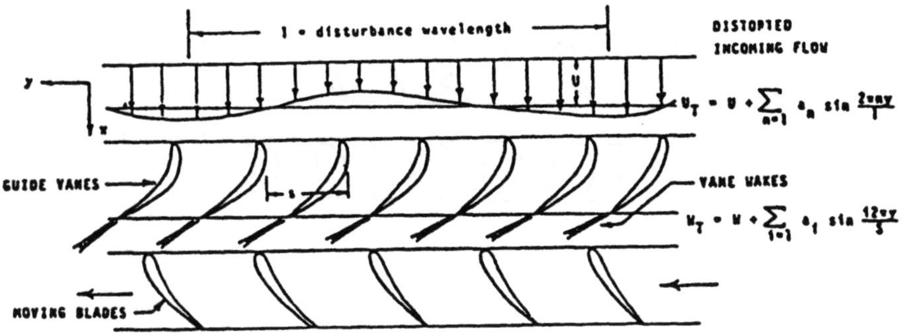

Figure 1. Interaction Between Guide Vanes and Moving Blade Row Showing Sources of Flow Induced Excitation

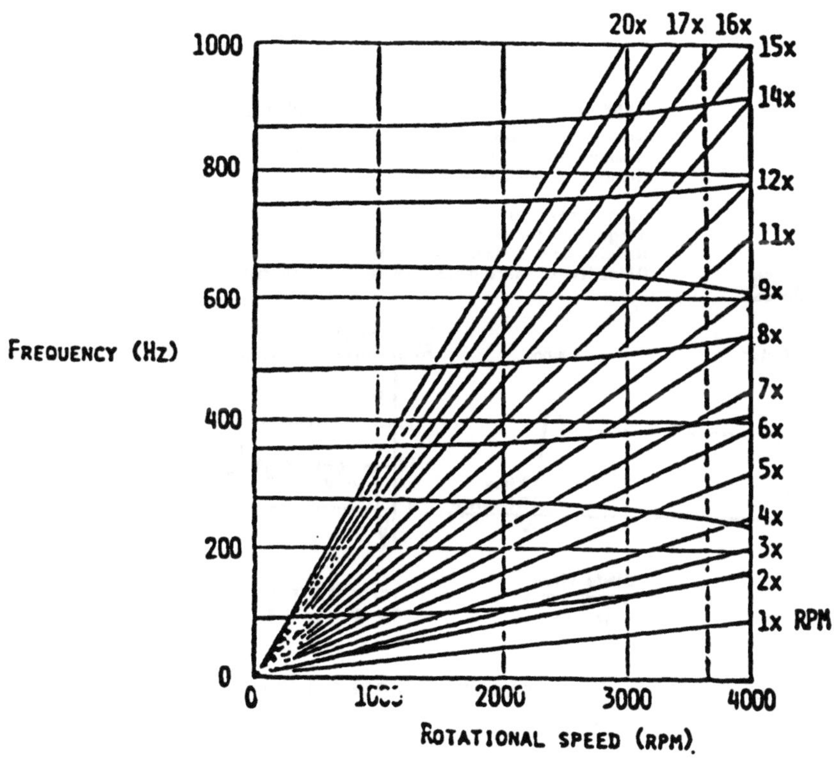

Figure 2. Campbell Diagram for LP Stage Blade Group

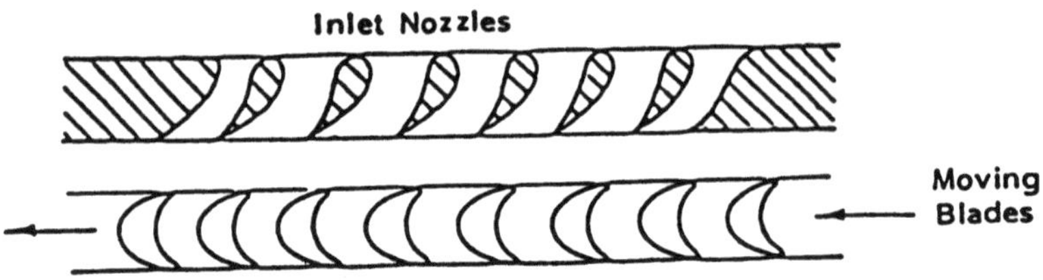

Figure 3. Typical Admission Arc in Partial Admission Stage

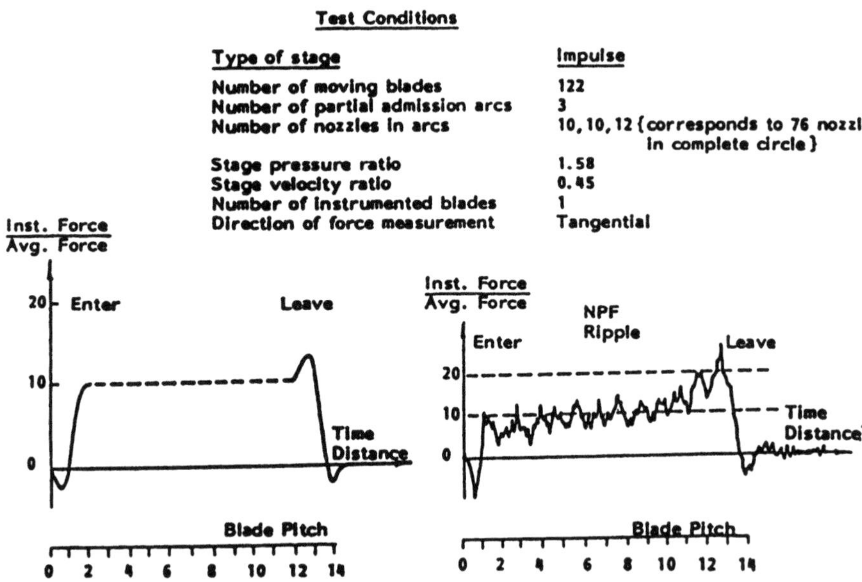

Figure 4. Comparison of Water Table Partial Arc Result with Typical Steam Turbine Result

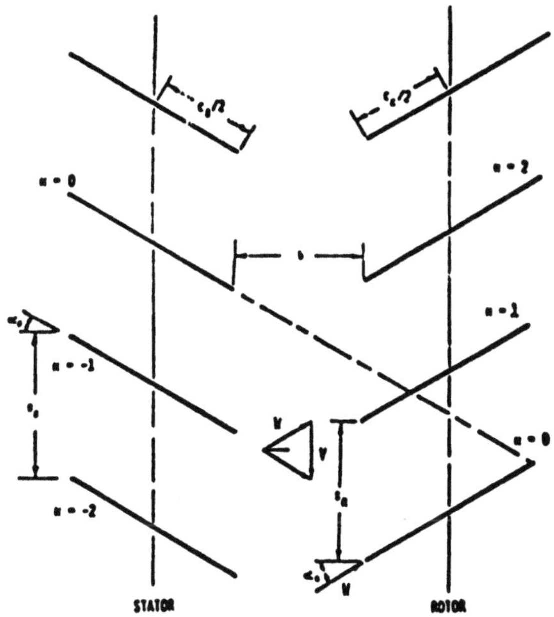

Figure 5. Kemp-Sears Cascade Configuration

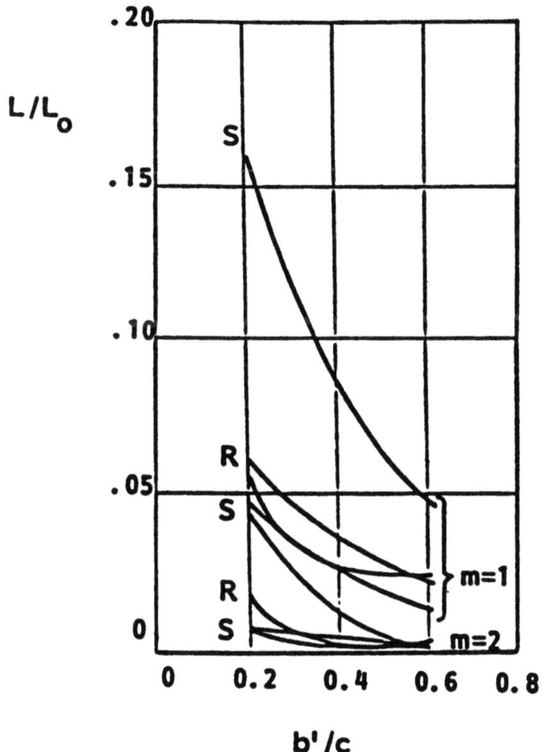

Ratio of Unsteady Lift to Steady Lift vs. Spacing Ratio (Stator Pitch Equals Rotor Pitch).

Figure 6. Results From Kemp-Sears Analysis

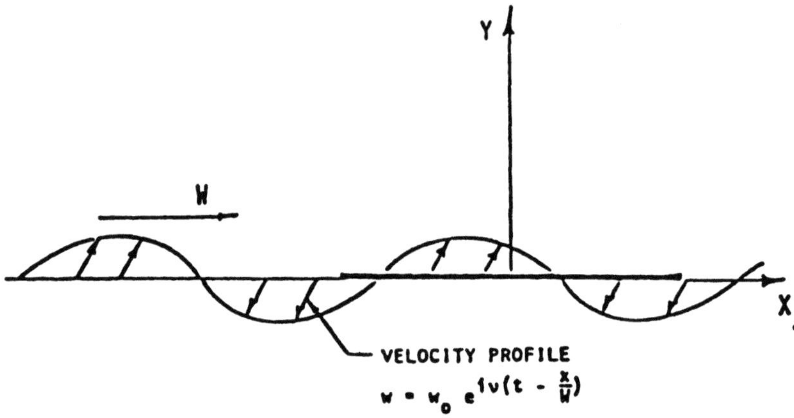

Figure 7. General Unsteady Gust Velocity Profile

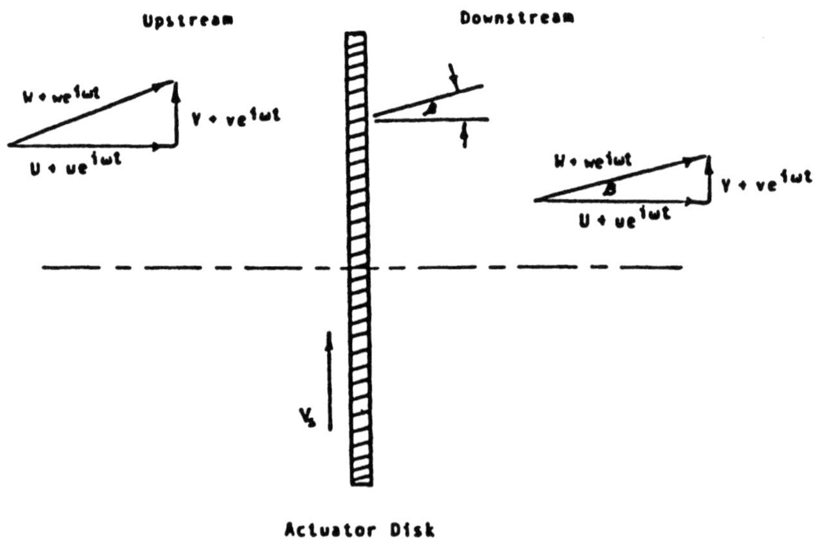

Figure 8. Actuator Disk and Velocity

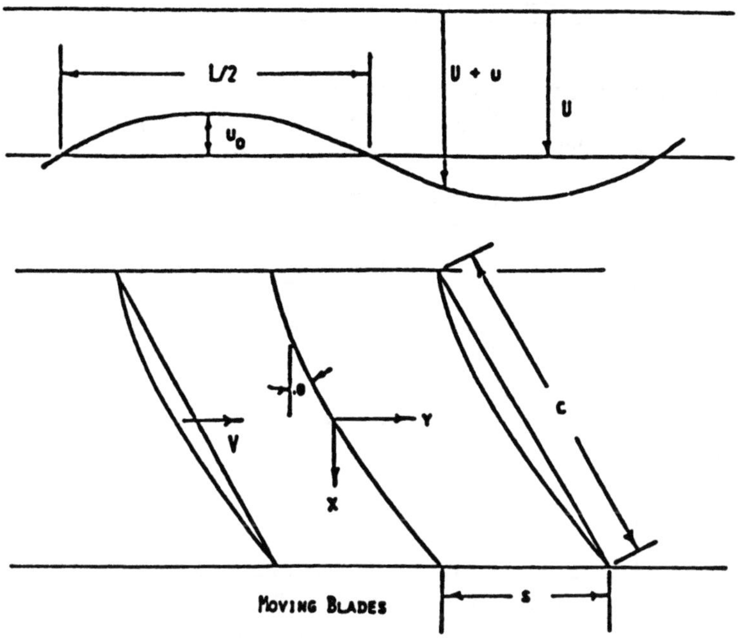

Figure 9. Cascade Notation of Henderson-Horlock

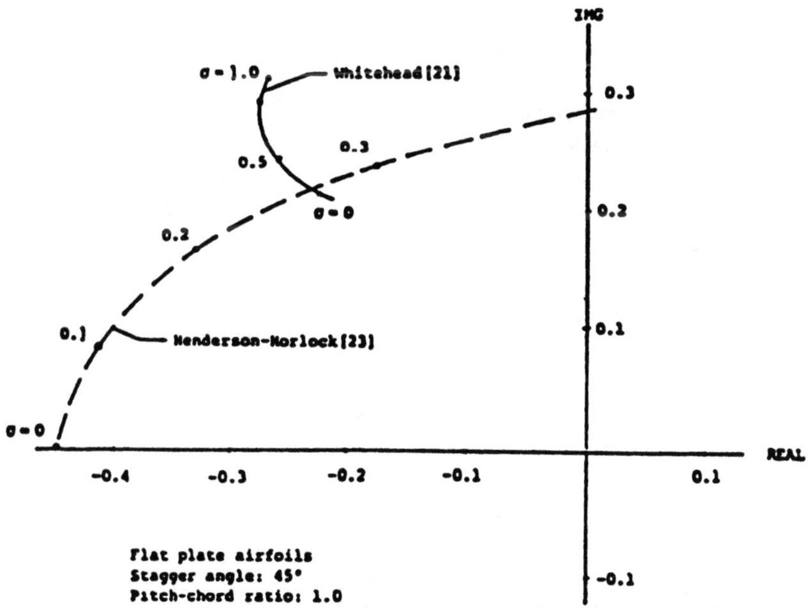

Figure 10. Real and Imaginary Parts of Unsteady Lift, Henderson-Horlock

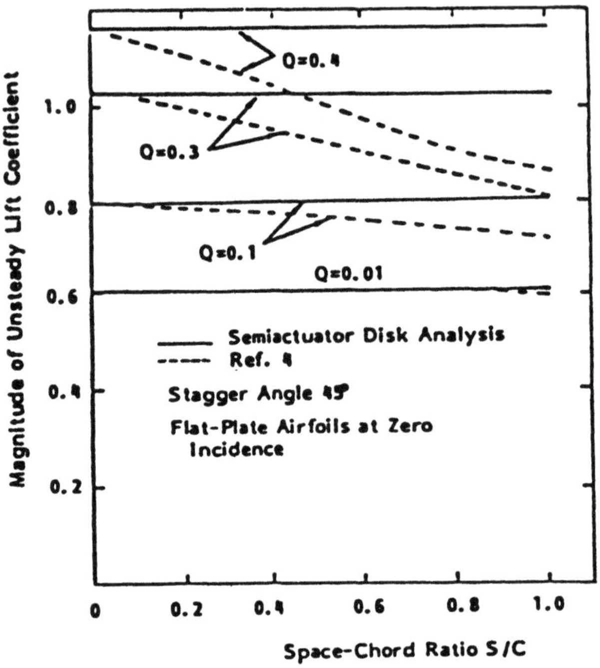

Figure 11. Comparison of Predicted Magnitude of Unsteady Lift Coefficient

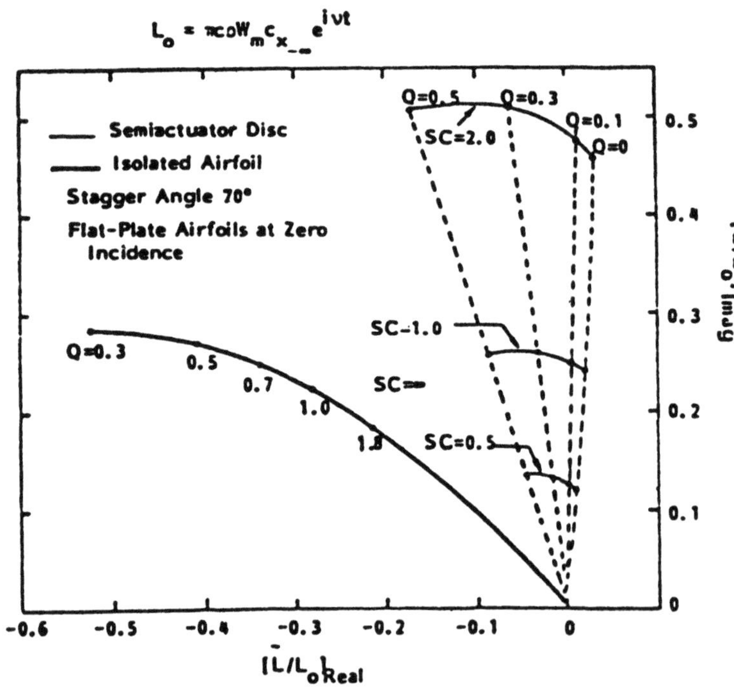

Figure 12. Comparison of Semi-Actuator Disk and Isolated Airfoil Theories

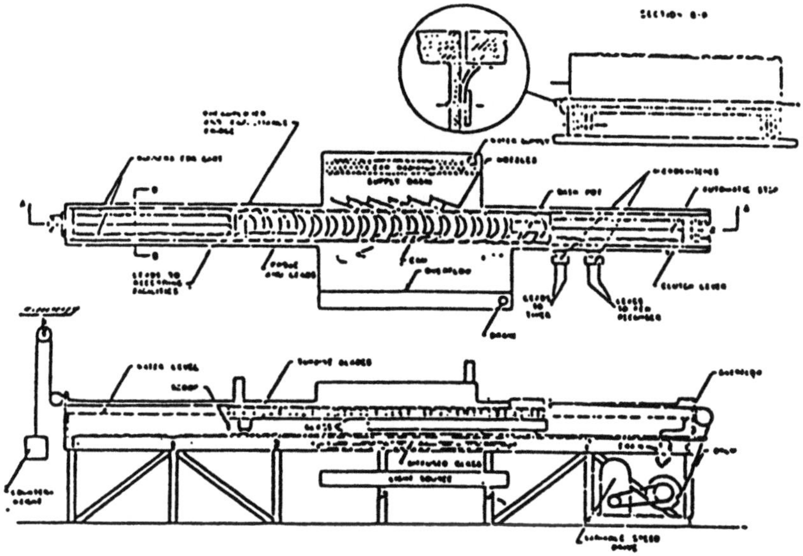

Figure 13. Partial Admission Test Rig. Heen and Mann

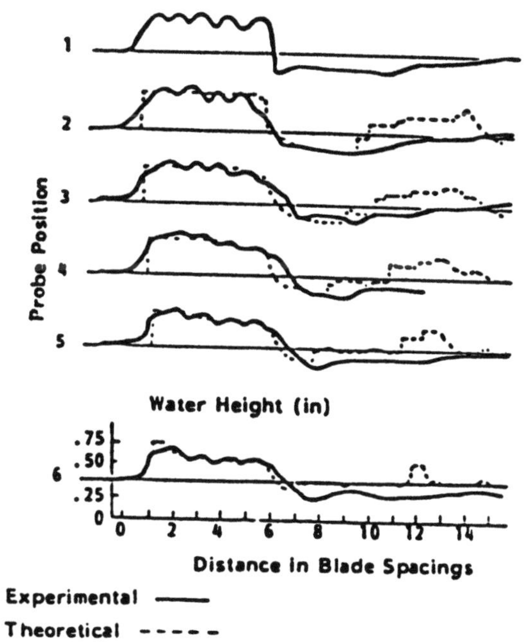

Figure 14. Water Depth vs. Distance for Several Locations in Moving Blade Passage from Heen and Mann. NOTE: Position 1 at Lead Edge, Position 7 at Trail Edge

Figure 15. STI Water Table Apparatus

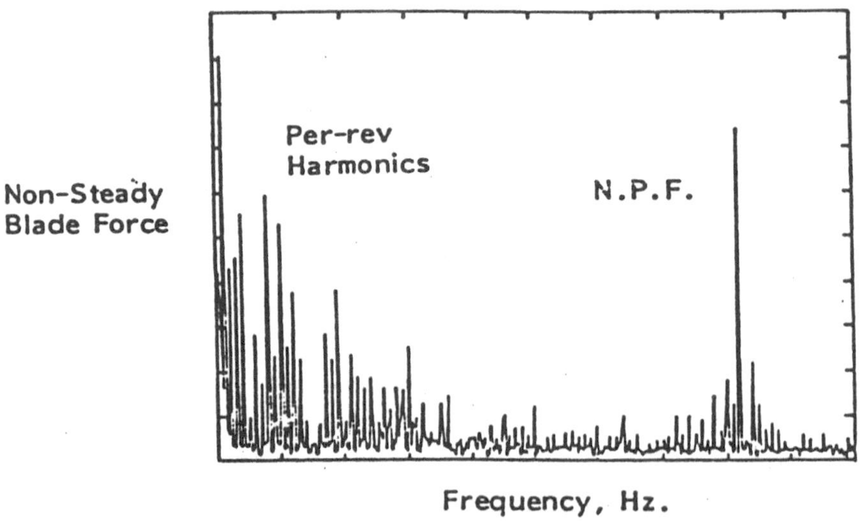

Figure 16. Tangential Load Magnitude vs. Frequency

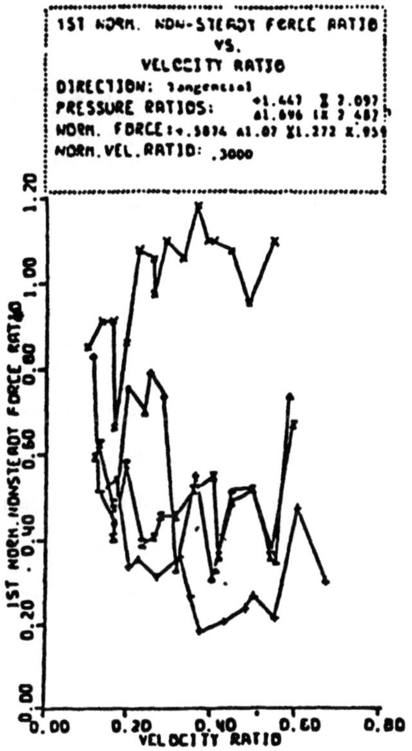

Figure 17. Normalized Tangential Force Ratio vs. Velocity Ratio

Table 1. Flow Path Excitation Sources and Harmonic Ranges

Rotor	Harmonic	Typical Sources
HP	High per-rev (40 x)	Nozzle tolerance harmonics Upstream wake degeneration Structural turbulence
	Nozzle passing frequency	Nozzle wakes
	2 x NPF 3 x NPF	Harmonics from nozzle wakes Harmonics from nozzle wakes
IP	Medium per-rev (20 x)	Nozzle tolerance harmonics Upstream wake degeneration Structural turbulence
	NPF	Nozzle wakes
LP	One per-rev	Relative displacement nozzle blades
	Two per-rev	Diaphragm joints
	Multiple per-rev	Structural supports in flow path
	Medium per-rev	Nozzle tolerance harmonics
	High per-rev	Nozzle tolerance harmonics Upstream wake degeneration Structural turbulence

Table 2. Analogous Quantities for Gas and Water Flows

Density ratio (ρ/ρ_o)	= Height ratio (h/h_o)
Temperature ratio (T/T_o)	= Height ratio (h/h_o)
Pressure ratio (p/p_o)	= (Height ratio)^2 $(h/h_o)^2$
Velocity of sound ($a = \sqrt{\gamma g R t}$)	= Gravity wave propagation velocity ($c = \sqrt{gh}$)
Mach number (M)	= Froude number (F_r)

CHAPTER 11.3

THE DIAGNOSIS AND CORRECTION OF STEAM TURBINE BLADE PROBLEMS

N.F. Rieger
Stress Technology Incorporated, Rochester, New York, USA

ABSTRACT

Several important types of turbine blading failure are discussed. A review of the basic causes of blading failures, e.g., fatigue, corrosion, stress corrosion cracking, erosion, etc., is given. Procedures for calculating stresses associated with such failures and the number of cycles to failure are reviewed. The familiar Goodman diagram procedure for cycles to crack initiation is included, plus a fracture mechanics approach to give the number of propagation cycles. Four case histories of turbine blade failure are presented in detail, with operating conditions, diagnostic procedures used to determine failure cause, and the remedies chosen to avoid further blading failures. Thirteen references to the subject literature are included.

11.3.1 Introduction

Service failures of turbine blades are infrequent but costly events. Electrical utility records show nearly 30 percent of all steam turbine forced outages are attributable to blade problems such as cracking, erosion, blade fracture, etc. The duration of blade related turbine outages may range from several days for a simple blade replacement in a small unit, to several months for a significant blading failure involving consequential damage in a large unit. Outage duration is obviously influenced by the availability of replacement parts, as well as repair time. Similar circumstances apply to industrial drive turbines and to marine propulsion turbines.

This chapter discusses several causes of steam turbine blading failures, identifying a number of important factors relating to these failure causes. Certain corrective measures which have been uses successfully in the past to overcome such problems are indicated. Knowledge of potential problem areas and of corrective measures is of value to designers and turbine operators attempting to avoid similar problems. Procedures for the analysis of stress related blading failures involving both high cycle and low cycle fatigue are described. These procedures permit quantitative assessments to be made of cases involving major stress related failure mechanisms such as fatigue, corrosion fatigue, and stress corrosion. Several case histories of blade failures are described, together with practical remedies which were used to overcome these failures.

Most blading problems present an unclear variety of evidence when the turbine is first opened. The first task when looking for the failure cause is to carefully record and evaluate the failure data and operating conditions which the blading has experienced. Identification of the failure cause is the first major step toward prescribing an effective solution. However, problem diagnosis may be a secondary objective in the period immediate following the failure. The turbine operator usually wants to 'get running' again, typically using some interim arrangement such as re-blading with available replacements or removal of the damaged row. Properly utilized, this situation offers a valuable opportunity to conduct a more thorough investigation and diagnosis of the problem as a basis for a more permanent fix to be installed at a future outage.

11.3.2 Blade Loading Conditions

The ability of a blade to support its applied loads depends on:

a) Strength of blade material in its environment.

b) Magnitude and distribution of steady mean stresses.

c) Magnitude and distribution of alternating stresses.

d) Loading history, including power and overspeed conditions.

Strength of the blade material is influenced by possible corrosive effects in the environment, by mean and alternating stress levels, by transient peak overload conditions, and by the number of applied load cycles which the most highly-stressed blade regions are capable of withstanding, including any residual effects of the blade forming process. Turbine blades may be subject to complex three-dimensional stress conditions at their attachments to the rim of the disk. Both the mean stresses and the alternating stresses have localized maximum values at such locations, i.e., at the cover-tenon junction, at the tiewire attachments, at certain locations in the airfoil section, at the airfoil-platform junction, and in the blade root/disk attachment region. At each location the mean stress value depends on centrifugal and steam loading conditions. Alternating stresses result from steam (and other) dynamic stimuli, from the modal response properties of the blade group, from the degree of modal damping involved, and from the extent of the dynamic coupling which occurs between the stimulus and the blade group modes. Loading history involves such factors as overspeed events, base load operation vs. peaking operation, and machine MWe load profile. It may also involve more subtle dynamic factors such as circumferential pressure distribution resulting from exhaust hood geometry, from reheat/extraction port locations, size, arrangements, from manifold strut arrangements, water ingestion, condenser vacuum rupture, and electrical line switching transient conditions.

11.3.3 Diagnosis of the Failure Mechanism

A variety of vibration sources may exist within a turbine stage, but in most instances the resulting blade vibration amplitudes and associated stress levels are small and insignificant. Occasional cases are observed where the vibration amplitudes of the blades have been shown to be large enough to cause blade failure by fatigue. More frequently, the blade vibration plus some other significant factor, e.g., corrosion, or residual stress, was also involved. Correct diagnosis of the failure mechanism is necessary before modifications can be proposed with confidence. The following questions may provide useful information when seeking the cause(s) of a given blade failure:

a) Does the evidence suggest that failure was due to (a) high cycle fatigue, (b) low cycle fatigue, (c) stress corrosion, (d) corrosion assisted fatigue, (e) erosion, (f) creep, (g) other sources, e.g., water ingestion?

b) What were the tangential, axial, and group natural frequencies of the blades under operating conditions? What were the associated mode shapes? Is the Campbell diagram available?

c) Does the Campbell diagram indicate the possibility of resonance between any blade group mode and any per-rev excitation harmonic (1x, 2x, etc.)? Which mode is most likely to become resonant considering statistical scatter of natural frequency values?

d) Was resonance possible between any blade group mode and any harmonic of nozzle-passing frequency (1 x NPF, 2 x NPF, etc.)?

e) Where were the failure initiation sites located? Is there evidence of local damage from corrosion, erosion, impact, or other initiating cause in that region?

f) What are the operating load and speed cycle history details for the machine since the original spin pit proving tests? How many overspeed governor trips have occurred? What speeds were reached in such cases? How long were the blades at overspeed, and could they have been resonant in this condition?

g) What was the chemical history of the steam operating conditions? Where was this sampled?

Other circumstances such as location of Wilson line in the machine, electrical network load variation details, and unit thermal cycling profile details may also be important. Some known features of several important blade failure mechanisms are discussed in the following section. It is evident that monitoring of speed and load to identify any transient conditions can provide important diagnostic information in such instances.

11.3.4 Types of Blading Problems

Fatigue:

Fatigue in turbine blades is broadly classified as either high cycle fatigue or low cycle fatigue. High cycle fatigue is often associated with a locally high mean stress level and moderate dynamic stresses. With high cycle fatigue, a large portion of the time to failure is taken up with the initiation of the fatigue crack. When a crack develops, the stresses at the crack front are much increased, and crack propagation usually takes place quite rapidly under the same alternating blade load conditions which caused the crack to initiate. Several known causes of steam turbine blade failure are listed in Table 1.

Low cycle fatigue is commonly associated with fewer load cycles applied through a much larger strain range than that which causes high cycle fatigue failure. A typical LCF failure cycle would involve strains from zero to a maximum such as the start-stop cycling of a blade or disk due to centrifugal force. For any turbine blade in which the local maximum stress exceeds the material yield point during such a load cycle, fewer of these load cycles will be needed to cause a crack to initiate (and subsequently to propagate), compared with the number of cycles and propagation rates observed in high cycle fatigue.

Many sources of harmonic excitation exist within a steam turbine stage. Steady harmonic excitations are continuously applied to the blades from sources such as nozzle wake excitations. Under resonant conditions, these excitations may cause large dynamic stresses to occur due to the low damping which is known to exist in most turbine blades. Transient blade excitations of large magnitude may also result from network electrical fault conditions at the generator, or from partial steam admission on startup, and so on. In many observed failure instances the blade vibration is observed to be intermittent (from the beach marks on the fracture surface). This tends to cause slower overall crack growth rates, but the growth increment is often larger where large dynamic stresses are involved.

High cycle fatigue failures can usually be recognized by visual inspection from the characteristic pattern of lines (called beach marks) which radiate from the crack initiation site. The surface may be uniformly polished from the rubbing of the crack surfaces against each other during vibration. The surface may also be tinted, depending on whether corrosion has resulted from the gas/steam environment. Frequently there is intermittent growth of the fatigue surface. This indicates intermittent growth of the fatigue crack, showing that crack-driving excitation was not constantly applied throughout the blade fatigue life. See Figure 1.

Many high cycle blade failures originate at some structural discontinuity or stress raiser. Such failures are frequently related to high local steady stresses, e.g., from centrifugal blade loading, as well as high dynamic stresses. With high steady stresses, more moderate vibratory stresses may cause a crack to initiate and propagate from the stress raiser. The same conditions can also cause an existing crack to propagate and grow until the component fails.

Low cycle turbine blade fatigue failures are frequently associated with corrosion or high temperature. The influence of these effects on fatigue is discussed later. Where cyclic stresses alone have led to low cycle fatigue failure, the progressive development of the crack can often be identified in electron microscope photographs, Figure 2.

Corrosion:

Corrosion assisted failures have occurred in the blade attachment region and in disk attachment, as well as in the vane, tiewire, tenon, and cover sections of the blade. Such failures typically occur at points of high operating stresses: the presence of dynamic stress is not required for stress-corrosion failure to occur. However, corrosion fatigue may occur where large dynamic stresses are applied with high steady stress in a suitable environment. This aspect is discussed in the next section.

Corrosive attack on blade and disk materials may arise from chemical impurities in the steam, such as sodium and potassium chlorides, sulfides, and carbonates. These substances usually exist in the steam in small quantities. Efforts are made to reduce chemical impurities by feedwater treatment. The effect of even very small quantities (parts per billion) may be concentrated by entrapment within grooves and cracks. Where such entrapment occurs at or near high stress regions, stress-corrosion may result. Evidence of concentrated corrosive attack may range from general degradation of the surface quality to corrosion failure of a component. See Figure 3.

Corrosion coupled with component stress and steam or water erosion may result in sufficient deterioration of the blade airfoil surface to affect the operating performance over a period of time. Corrosive buildup of deposits on blade surfaces can adversely affect stage operating efficiency by several percent. This has occurred in large utility steam turbines, process turbines, and geothermal steam turbines. Stress accelerated breakdown of surface quality is another important practical source of turbine blade degradation.

The possibility of stress-corrosion cracking and failure of a given component may be assessed by fracture mechanics procedures. Visual data which suggests this type of failure are the presence of white or gray chloride, sulphide, and/or carbonate deposit (nodules) forming a local coating over the surfaces of the moving and/or stationary blades. Signs of corrosive pits (small or large) may be evident especially near known regions of high stress, e.g., notches. Additional data can be obtained from microscopic examination of the same surfaces, Figure 3. This may reveal additional corrosive degradation of the surface, and a variety of small and large pits. Medium power microscope studies of sections through the surface may reveal that the progress of the crack has been aided by corrosive attack along the grain boundaries (intergranular cracking). The most informative source of such data is photographs from the scanning electron microscope which reveals the presence of corrosive attack as large nodules of corrosion products. See comments in case histories.

A recent paper by Jonas [1] discusses the general problem of corrosive attack on components from steam impurities. Table 2 herewith is taken from this paper. It identifies the locations, component materials, and associated chemical deposits observed at the sites of a variety of turbine plant problems. Three regions are specified as most susceptible to corrosion: (a) regions where metal or steam temperatures are around the melting points of corrodents, e.g., NaOH T_m = 604°F, (b) regions immediately ahead of, or at first condensation, e.g., LP turbine stage at Wilson point (pitting, stress-corrosion, and corrosion fatigue of blades and disks occurs most often in this region), and (c) superheated metal surfaces where impurities can concentrate by evaporation and drying. Long term (24,000 hours) tests on certain turbine steels at 150°F in a 28 percent NaOH solution have shown that stress corrosion cracking may occur at stresses as low as 30 percent of the material yield strength.

Recent utility turbine research programs have begun to develop comprehensive methods for chemical monitoring of steam turbine plants. A useful description of appropriate tests and water/steam monitoring requirements is given in the above paper by Jonas.

Corrosion Fatigue:

Corrosion assisted fatigue is probably the major source of steam turbine blade fatigue failures. Corrosion fatigue most frequently occurs in a corrosive environment where high steady stresses are applied together with high alternating stresses. Turbine blade vibration tests by many investigators have shown that even under conditions of nozzle resonance, blade types which have been known to fail during operation often do not develop dynamic stresses of sufficient magnitude to cause fatigue without the presence of corrosion. Such conditions suggest that some other factor must be involved, and these failures can frequently be explained where it can also be shown that significant corrosive attack must also occur along with fatigue conditions in the same operating environment. Laboratory test evidence has demonstrated that the material fatigue strength (endurance limit) may be immediately reduced by as much as seventy percent by a sufficiently aggressive chemical environment. On occasions, such chemical environments appear to have existed in, (a) certain marine turbine steam conditions, e.g., from sodium hydroxide in the make-up water, (b) from inadequate demineralizers in main utility stations, and (c) process steam turbines. Otherwise unexplainable blade fatigue failures can be accounted for quite readily where such circumstances can be shown to exist. Corrosive environments may accelerate both high cycle fatigue and low cycle fatigue if the operating conditions and chemical concentrating mechanism are right.

Corrosion fatigue in turbine blade steels has been studied in depth in recent years, and certain fracture mechanics data have been given, e.g., by Clark [2]. Figure 4 shows the rate or crack growth da/dN vs. stress intensity factor ΔK for a 304 stainless steel in a three percent caustic environment. Crack growth for the same steel in air is also shown for comparison. It is seen that the difference in crack growth rates for the same ΔK value, i.e., stress level, is approximately 3:1 for the caustic environment. this is another way of stating that the fatigue strength of the test components under the corrosive attack shown in this instance was only about one-third of the fatigue strength of the same components in air, i.e., seventy percent reduction as noted above.

An important initial source of corrosion fatigue may develop on site when rotors are left exposed to an aggressive environment without other protection for a considerable period prior to erection. Particularly damaging is the case where the protective coating has been removed from a rotor and blades, which are then left exposed to the moist, outdoor environment near a river or even the sea. The initial chloride pitting which may occur in such circumstances may later provide nesting sites for steam, etc., impurities in high stress regions which can accelerate the tendency toward blade fatigue.

Erosion:

Surface erosion can be a significant problem in all stages of a turbine. Surface erosion from hard particles (usually boiler exfoliation) can damage the HP and IP stage blading, and wet steam erosion can damage the leading edge of the long LP blades, usually from mid-height out to the cover. Cover damage from wet steam erosion can also be significant. Such erosion is caused by high velocity particles of steam condensate striking the blade lead edge over a period of time and eroding the material away. Certain wet region blades have been designed for many years with an erosion shield (stellite strip) which is bronze welded on to the blade lead edge to protect against wet steam erosion.

Signs of early erosion may often be observed on those blade leading edges which project noticeably out into the incoming steam, beyond the other blades in the same row. This condition may occur from minor misalignment on assembly, and this erosion is of no special significance unless it continues and presents an evident major damage problem. Replacement of damaged erosion shields is a straightforward procedure which can now be undertaken in the field, as well as in the manufacturer's shop. Suitably located LP moisture separators also help to decrease the rate of blading erosion.

Exfoliation of tube scale is another form of erosion which occurs in boiler tubes, superheater tubes, inlet steam pipes, and from condenser pipes. The scale develops from oxidation and corrosive attack from the feedwater condensate and from steam impurities. The scale is eroded away, and on passing the turbine, may damage the blading and may accumulate in the drains. Geothermal turbines are especially prone to scaling and exfoliation damage because of the high corrosive and impurity content of the inlet steam. Erosion products should be monitored as part of the turbine system chemical monitoring program.

11.3.5 Stress-Related Blade Group Cracking Theory

The following theoretical approach is general and may be used to develop an understanding of possible causes for specific cases of blade cracking which appear to involve fatigue, corrosion fatigue, or stress corrosion cracking. This method follows from the approach which was pioneered by Prohl in reference [3].

Consider a group of blades rotating in an axial flow turbine, Figure 6. The total stress at any location is due to two sources, the steady mean stress σ_m and the dynamic or alternating stress σ_a. At any instant during operating where a single harmonic response component predominates, the total stress is given by:

$$\sigma_t = \sigma_m + \sigma_a \cos \omega t$$

where ω is the circular frequency of the alternating stress. The mean stress results from the combined action of the centrifugal load due to turbine rotation, and from the steady blade bending load from the steam forces which drive the turbine.[1] These mean stress components combine to give the nominal steady extreme fiber stress, σ_{mo} at the blade (or root section):

$$\sigma_{mo} = \sigma_{co} + \sigma_{bo} = P/A + MC/I$$

where P is the centrifugal load from blade rotation acting at the section, A is the section area, M is the local bending moment due to the steam load on the blade, c is the extreme fiber distance from the neutral axis, I is the appropriate second moment of area of the blade section, and σ_{co} and σ_{bo} are the nominal centrifugal and bending stresses at the section (no stress concentration effects). The mean stress remains constant for a given blade arrangement at the specified speed and power output. Corresponding expressions may be written for steady stresses in the blade group cover and tenons.

Alternating stresses in the blade may arise from several causes, of which harmonic excitation from the nozzle wakes is widely recognized as one potentially significant contributor. The frequency of excitation f from the nozzle wakes is given by:

$$f = Nk = 2\pi\omega \text{ cycles/sec. (Hz)}$$

where N is the rotor speed rev/sec and k is the number of uniformly spaced nozzles[2] around the 360° circumference. Alternating stresses from nozzle excitation depend on several factors:

a) The magnitude of the steam exciting force or harmonic stimulus, expressed as a stimulus factor S.

b) The damping of the blade and its attachment expressed as a logarithmic decrement δ.

---

[1] Additional forces from torsion, centrifugal untwist, etc., may of course also apply.

[2] k may also be considered as the number of per-rev harmonic waves around the diaphragm circumference. This allows per-rev nozzle excitations to be considered.

c) The response factor K which is a measure of the ability of the blade group to accept energy input from the nozzle stimulus.

Blade harmonic stimulus is usually expressed as a proportion of the steady steam bending force F acting on the blade, i.e., $S = \Delta F/F$, where ΔF is the amplitude of the time-varying steam force. In practice, values of S may range from below 0.02 in smooth-running stages to above 0.20 in rough stages with off-optimum conditions. See references [4] and [5].

Damping in blade groups can arise from several sources such as rubbing friction in the attachment areas (root, cover), from material hysteresis and from gasdynamic effects on longer blades. The magnitude of blade group logarithmic damping value δ may vary substantially from one application to another, but the general range is from about 0.002 to 0.030 for conventional AISI 403, 12 chrome steam turbine blades, depending to some extent on blade geometry and the mode of vibration involved. See references [6] and [7].

The resonant response factor K depends on the excitation parameter $E = (nq/m)$, where n is the harmonic number (n = 1, first order: n = 2 second order, etc.), q is the number of nozzle inlets and m is the number of blades. The variation of the response factor K ranges from 1.0 under conditions where the blade group can receive strong energy input from the nozzle stimulus down to zero where the phasing of the stimulus is such that the blade vibrations are not readily excited.

As each turbine stage may develop many excitation harmonics, and each harmonic may act on several blade group modes, the influence of the excitation parameter on each blade group response in the frequency ranges of interest should be examined. A convenient procedure for determining the excitation harmonics and blade modes of interest was first given by Campbell [8] in which the natural frequencies of the blade modes are plotted as ordinate and the rotor speed is plotted as abscissa, see Figure 9. Radial lines from the origin corresponding to once-per-rev (1x), twice-per-rev (2x), etc., nozzle passing frequency (Nk), twice NPF (2nK), etc., are also plotted. Speed regions of intersection between blade frequencies and excitation harmonics are then noted, with particular reference to regions of sustained operation, e.g., operating speed range. The Campbell diagram shown in Figure 9 indicates the possibility of blade resonance in the axial torsional mode with the 2x NPF excitation, and also excitation of the second type tangential mode by NPF. See Case History 1 for details.

Resonant stresses are related to stimulus S, damping δ, and blade group dynamic response factor K, by the expression:

$$\sigma_{ao} = \frac{\pi}{\delta} KS \, \sigma_{bo}$$

where σ_{ao} is the nominal resonant alternating extreme fiber stress at the blade (or root) section, and the σ_{bo} is the nominal bending stress at the section, defined previously. It is evident that the practical combinations of blade damping, nozzle stimulus, and dynamic response factor may lead to resonant stresses σ_{ao} at certain blade cross sections which could approach or greatly exceed the nominal mean bending stress σ_{bo} at that section. It should further be noted that in practice, the resonant condition is often sharply defined. Sustained operation at the resonant peak condition is therefore unlikely to occur for long periods, though some lesser stress magnification should always be expected for operation in this region.

To determine whether the stress conditions at a given location could be responsible for blade cracking during operation, it is necessary to compare the local stresses with the appropriate strength criterion for the blade material at that section. To obtain an indication whether the nominal stresses σ_{ao} and σ_{bo} are likely to initiate high cycle fatigue cracking, a procedure due to Heywood [9], and adapted by Rieger and Nowak [10] for steam turbine blades using the Goodman diagram may be used as follows. The appropriate fatigue life envelope for unnotched specimens in the steam environment is modified in a specified manner to account for mean stress, local stress raisers, cycles to failure, and size effect. The new (notched, etc.) fatigue envelope then becomes the crack initiation criterion against which the nominal stresses σ_{mo} and σ_{ao} are compared. A point falling outside this region can be expected to initiate a crack in the number of cycles assumed in the calculation of the notched envelope, i.e., $N = 10^4 \ldots 10^5 \ldots 10^7$, etc. For high cycle fatigue, crack initiation commonly represents the larger portion of the fracture life, and the time to propagate the crack (which is not considered in this approach) represents the remainder of the component life.

Where the initial defect size is known from inspection or can be assumed, an alternate approach using fracture mechanics procedures may be used to estimate the number of load cycles required to propagate a crack, and cause failure of the defective component. Suitable materials data obtained from tests on fracture mechanics specimens within a similar environment and loading, in accordance with standard ASTM testing procedures, is required. The rate of crack propagation da/dN is related to material properties A, n., stress intensity range ΔK, and applied stress ratio by the expression:

$$\frac{da}{dN} = \frac{A\,[\Delta K]^{n}}{(1-R)^{0.5}}$$

ΔK	is the range of stress intensity factor = $c\Delta\sigma\ \pi a$
R	is the stress ratio, K_{min}/K_{max} or $\sigma_{min}/\sigma_{max}$
A, n	are material properties
C	is a geometric factor for the crack model involved
Δσ	is the stress range, $\sigma_{max} - \sigma_{min}$
a	is the crack length
N	is the number of stress cycles

For a crack in a notch-free region:

$$\sigma_{max} = \sigma_{mo} + \sigma_{ao}$$

$$\sigma_{min} = \sigma_{mo} - \sigma_{ao}$$

$$\Delta\sigma = 2\sigma_{ao}$$

$$R = \sigma_{min}/\sigma_{max}$$

For a crack in a sharply notched region:

$$\sigma_{max} = K_t^m\sigma_{mo} + K_t^a\sigma_{ao}$$

$$\sigma_{min} = K_t^m\sigma_{mo} - K_t^a\sigma_{ac}$$

$$\Delta\sigma = 2K_t^a \sigma_{ao}$$

$$R = \sigma_{min}/\sigma_{max}$$

σ_{max} and σ_{min} are the maximum and minimum values of the total stress at the location in question. A typical relation between da/dN and ΔK for a 4340 steel in a three percent caustic solution is shown in Figure 4. This chart does not include the effect of stress ratio R. In most cases the stress field in the body changes with distance into the body. This causes the stress at the crack tip to change and so influences the rate of crack propagation. Calculations which must include the effect of the above factors are most conveniently performed with a suitable fracture mechanics computer program, such as the CRACKS [11] or BIGIF [12]. The end result of such a calculation is a value for a specific number of cycles to propagate the crack to failure when Kmax = K_{IC} (fatigue) or K_{ISCC} (corrosion fatigue), or until a length is reached at which the crack stops propagating.

This fracture mechanics procedure is suitable for a time-to-failure analysis at any location in the blade group with a crack of known (or assumed) proportions, and for any material for which suitable fracture mechanics data is available for the blade operating environment. Where such input data is difficult to obtain or specify, the above procedure may be used to provide a bounding analysis suitable for determining the performance of a hypothetical crack under assumed minimum or maximum stress and material conditions based on experience.

11.3.6 Case Histories of Blade Failures

Case 1 Nozzle Resonance of HP Marine Turbine Blades [13]:

Both rotors were 55,000 SHP turbines operating at 3500 rpm. The blading of both HP rotors sustained damage. The complete blading of the ninth stage of the starboard turbine was missing and eight blades were missing from stage 10. On the port rotor, seven blades were missing from the ninth stage and there was some cracking in stages 8 and 10. The cracks occurred near the vane-platform junction where the vane overhung the platform. In most instances the cracks appeared to have propagated from beneath the overhung trailing edge, horizontally into the blade airfoil section, Figure 7 (a). One blade only was broken at mid-height, in the eighth stage of the port turbine.

A comprehensive investigation was made of the failure. Frequency calculations and vibration tests were performed on the original blades, and on the modified blades. Much evidence was found to show that the failures were due to high cycle fatigue from vibrations in the second type tangential (out-of-phase) mode, Figure 8 (a). The cracking pattern corresponded to the calculated distribution of modal amplitudes shown, both in magnitude and location. The Campbell diagram Figure 9 showed that the eighth, ninth, and tenth stages could resonate at propeller shaft speeds between 90 rpm and 174 rpm. The ship operating log, Figure 10, showed 2610 minutes of operation at 148 rpm. In this condition, the ninth stage could resonate in its second tangential mode.

Several remedial changes were included in the blading redesign. The overhung stress raiser was eliminated by smoothing the vane into the platform of the replacement blades, Figure 7 (b). Tiewires were added by brazing to eliminate the second tangential mode from the range of nozzle resonance, Figure 8 (b). Diaphragm changes (decreased number of nozzles) were also considered, but were not used because such changes were shown to be effective in removing the eighth stage alone from resonance. Also, larger nozzle passages frequently lead to higher exciting forces. Alternatively, an increased number of nozzles might have been considered.

Case 2 Nozzle Resonance in Ninth Stage of Process Steam Turbine:

A rash of shroud and vane cracking incidents had occurred in several process turbine drive units. Generally the shroud of a six-blade group had cracked, and also several vane sections near the platform in the eighth stage. The blade surface adjacent to the failure site was pitted, and white, solid deposits of unknown type (of contaminant) were attached to the blade surface. A diaphragm change from thirty-four inlet nozzles to forty-six inlet nozzles had increased the blade life somewhat (six weeks to forty weeks), but had not eliminated the problem.

Analysis showed that the second in-phase tangential mode of the blade group coincided with NPF at full operating speed using the thirty-four nozzle diaphragm. Further, the Campbell diagram, Figure 11, showed that although a change from thirty-four nozzle openings would cease to excite the second in-phase tangential group mode with the second NPF harmonic (2 x 234), the (1 x 46) NPF harmonic would then excite several second type tangential (out-of-phase) modes.

The fix proposed in this instance was a blade profile redesign which removed the blade groups from the troublesome resonance harmonics within the operating speed range identified above, without introducing other resonance problems. The blades were also detuned as shown in Figure 12, so that the excitation factor E would be 0.183 and the corresponding response factor K would then become zero. In the original design, E was 0.352 and the resonant response factor K had been 0.275. The new blades were therefore less responsive to nozzle stimulus.

Other factors requiring attention were the chemical content of the steam and the location of the Wilson line (dry/wet) in the turbine. Suitable de-mineralizers should have been prescribed and thoroughly maintained in view of the extreme, i.e., coated corrosion fatigue situation which existed. If the Wilson line corresponds to the ninth stage, care should be taken to shield the blade attachment regions in some manner. The chemical functioning of the turbine steam system should have been monitored following the re-installation to ensure that solids and impurities were within acceptable limits, see Jonas [1].

An apparent alternate fix would have been to detune the blades using a bronze welded tiewire, as was done in Case 1. This would suppress the out-of-phase tangential modes, and the forty-six nozzle diaphragm would not excite the second in-phase tangential mode, as noted. This was not done as the blade required re-design to remove the vane overhang.

Case 3 Stress Corrosion in a Fifth Stage 200 Mw Utility Turbine:

Catastrophic rupture occurred in the blade root section of thirteen fifth stage axial entry blades after eleven months of on-line operation, with considerable consequential damage to flow guides and to blades in adjacent stages. The failed row contained 324 moving blades, each about six inches average vane height, arranged in groups of five and six. Blade material was 403 stainless steel with nominal UTS 105 ksi, and yield stress 85 ksi in air. The adjacent inlet nozzle row had 240 nozzles. Extensive white-colored chemical deposits with average ph value of eleven, coated the general region (nozzles, moving rows), near the failed stage, Figure 13. Pitting in the remaining blades of the row ranged from slight to severe. 325 additional cracks were found (173 blade, 152 disk steeples) of varying sizes, mainly in the contact hook regions of the blade attachments. The location of the row coincided with the location of the Wilson line of the rotor at full power.

Metallographic examination showed branched intergranular cracking followed by transgranular cracking inward from the highly stressed blade root notch surface, Figure 14.

Scanning electron microscope studies showed pitting in the vicinity of the primary fractures and secondary intergranular fractures linking the corrosion pits. The fracture surfaces were relatively clean, indicating that little rubbing or polishing had taken place since cleavage occurred. This suggests that no predominant HCF or dynamic stress mechanism was involved. The high ph white coating was composed of NaOH, and Na_2CO_3. This indicates that the initial cracking had been assisted by corrosion in the highly stressed hook region of the blade root. Furthermore, the original material away from the corrosion sites still had the strength and impact properties required in the original material specifications. A typical SEM photograph of the fracture surface is shown in Figure 15.

Further investigation revealed that the boiler feedwater chemistry during operation had contained dissolved solids, iron, and sodium (77 ppb compared with 30 ppb specified) in excess of prescribed limits, despite the use of feedwater demineralizers. Stress corrosion of the blade root material under high stress conditions was the primary cause of this blade failure, based on (a) high sodium and other salt deposits, (b) wet/dry Wilson line at failure location, (c) widespread cracking in the vicinity of the high stress locations, (d) general pitting of adjacent surfaces, (e) absence of plastic flow and beach marks on the failure surface, (f) corrosion products seen in many SEM scans, and (g) corrosion fractography observed in sectioned failure regions. The principle remedy was improved steam quality and boiler feedwater chemistry by improved de-mineralizer control and monitoring.

Case 4 Fatigue Failure of Fourth Stage Marine LP Turbine Impulse Blades:

A single side entry fourth stage blade failed catastrophically in the root section at the first hook after thirteen months of service at sea. On inspection it was found that eleven additional fourth stage blades were also cracked in the same region, and that all cracked blades were end blades of seven-blade groups. In addition, a total of eight disk root sections were found to be cracked in the same stage. These failures were found by magnetic particle inspection. Hardness and chemistry checks of the blade and disk root sections were also made. Hardness was found to be in the RC 20-21 average, as required. Macro-etched sections of the cracked root were carefully inspected at 20x magnification.

The failure mechanism was found to be high cycle fatigue, as indicated by the beach mark progression of the crack front along several regions of the failed surface. The multiple crack origins indicate a wide distribution of the initial crack driving stress mechanism along the root hook notch. No oxides or corrosive deposits were observed on the fracture surface. Microexamination showed that the root cracks were relatively straight and transgranular without branching. This further suggests that pure fatigue was the cause of failure, though no electron microscope studies were made on this occasion.

Several possible causes of these fatigue failures were identified. First the tangential out-of-phase group modes were found to lie close to nozzle resonance, due to an inaccurate design estimate of blade root stiffness. Second, the root fillet radius of 0.031 inches was quite small, and this gave rise to magnified fillet steady and dynamic stresses. Third, variations in nozzle geometry from stage arrangement around the nozzle row were found to give 4:1 variations in the magnitude of nozzle stimulus. This raises the additional possibility of per-rev stimulus from this nozzle geometry variation, but this possibly was not considered further. Fourth, a small amount of stress corrosion may have occurred judging from minor discoloration observed on the crack surface. Fifth, a further source of significant excitation was thought to have occurred from four condensate extraction ports located around the circumference adjacent to the fourth stage moving blade row.

Design modifications were made as follows: (a) long-arc shrouding was introduced to suppress the troublesome resonant group modes, (b) the notch fillet radii were increased to 0.060 inches to reduce the fillet stresses, (c) the inlet nozzle geometry was made uniform around the diaphragm and the number of inlet nozzles was increased from 92 to 120, and (d) a flow-smoothing baffle was inserted to remove the flow disturbances created by the four extraction openings. No further failures have occurred since the introduction of these modifications.

11.3.7 Conclusions

- Turbine blade problems may result from design, manufacture, materials properties, steam/gas quality (erosion), steam/gas chemistry (corrosion, exfoliation), and abusive operation (water extraction, condenser flooding). Design practices, operating practices, and turbine plant specifications should address each of these potential problem areas.

- The primary diagnostic tools for analysis of turbine blade failure causes are:

a) Electron microscope to investigate failure mechanism and identify the role of corrosion in the failure.

b) Surface microscopy and section microscopy for defining the cracking mechanism, and for basic material quality assessment.

c) Water/steam chemistry records to determine role of corrosion.

d) Blade group and disk natural frequencies, modes and static/dynamic stress calculations, to determine role of operating stresses.

e) Fracture mechanics testing of failed component material to determine quality of material supplied, corrosion resistance, and crack propagation characteristics in operating environment.

- The failure surface and fracture sections should always be examined microscopically when the causes of a failure are being sought. Examination by Scanning Electron Microscope is a valuable additional aid for determining whether the failure is associated with corrosion, corrosion fatigue, or fatigue alone.

- Each stage of blading should be checked as to whether nozzle resonance and per-rev resonance may occur during operation. A Campbell diagram should be developed containing the first six modes of the blade group for each stage. The calculated modes and frequencies should be checked by vibration tests performed in the manufacturer's shop, on several blade groups around the circumference of each stage.

- High cycle fatigue is usually related to some resonant operating condition. HCF may be identified by the presence of polishing, beach marks, crack propagation increments under SEM, and final static rupture on the failure surface. Multiple staining lines indicate the occurrence of intermittent crack propagation from short periods of high dynamic stresses.

- Corrosion fatigue may be induced by concentration of steam impurities acting on high stress regions in the presence of dynamic stresses. The source of such corrodents may be in the steam itself, in the steam chemistry control apparatus, in the steam chemistry specifications, or in the original turbine erection environment.

- o Erosion of turbine blades and stage inlet guides may occur from boiler and tube exfoliation, and from wet steam impact. Such erosion may lead to performance degradation, and to degradation and failure of the working components.

11.3.8 References

1) Jonas, I., 'Turbine Steam Purity,' Combustion Magazine, p. 11, December 1978.

2) Clark, Jr., W. G., 'Evaluation of the Fatigue Crack Initiation Properties of Type 403 Stainless Steel in Air and Steam Environments,' ASTM STP 559, ASTM, pp. 205-224, 1974.

3) Prohl, M. A., 'A Method for Calculating Vibration Frequency and Stress of a Banded Group of Turbine Buckets,' Trans. ASME, pp. 169-180, January 1958.

4) Heyman, F. J., 'Turbine Vibrations Due to Nozzle Wakes,' ASME Publication Paper Number 68-WA/PWR-1, Presented December 1-5, 1968.

5) Rieger, N. F., Wicks, A. L., 'Non-Steady Force Measurements and Observations of Flow in Three Turbine Stage Geometries,' Journal of Engineering Power, Vol. 100, p. 525, October 1978.

6) Lazan, B. J., Damping of Materials and Members in Structural Mechanics, Pergammon Press, Incorporated, New York, 1968.

7) Wagner, J. T., 'Blade Damping Tests,' Westinghouse Engineering Report EC-401, NOBS N00024-67-C-5494, May 1969.

8) Campbell, W., 'Tangential Vibration of Steam Turbine Buckets,' Trans. ASME, pp. 643-671, 1925.

9) Heywood, Designing Against Fatigue of Materials, Reinhold Publishers, London, England, 1962.

10) Rieger, N. F., Nowak, W. J., 'Analysis of Fatigue Stresses in Turbine Blade Groups,' Workshop on Improved Turbine Availability, Electric Power Research Institute, Palo Alto, CA, January 1977.

11) CRACKS: 'A FORTRAN IV Digital Computer Program for Crack Propagation Analysis,' USAF Flight Dynamics Laboratory, Wright-Patterson AFB, Dayton, OH, March 1970.

12) BIGIF: 'Fracture Mechanics Code for Structures,' Program Users Manual, Failure Analysis Associates, Palo Alto, CA, December 1978.

13) Fleeting, R., Coats, R., 'Blade Failures in the HP Turbines of R.M.S. Queen Elizabeth 2 and Their Rectification,' Trans. Marine Engineers, London, England, October 28, 1969.

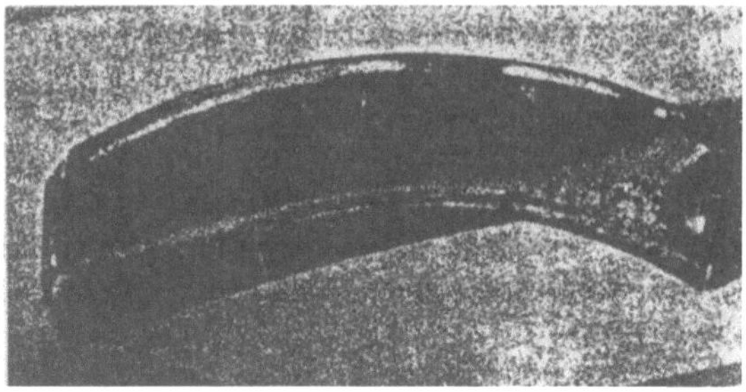

Figure 1 Fatigue Surface with Staining Pattern

Figure 2 Electron Micrograph Showing Fatigue Striations in 4340 Steel

Figure 3 Corrosion Pits on Root Hook Near Fracture Initiation Site

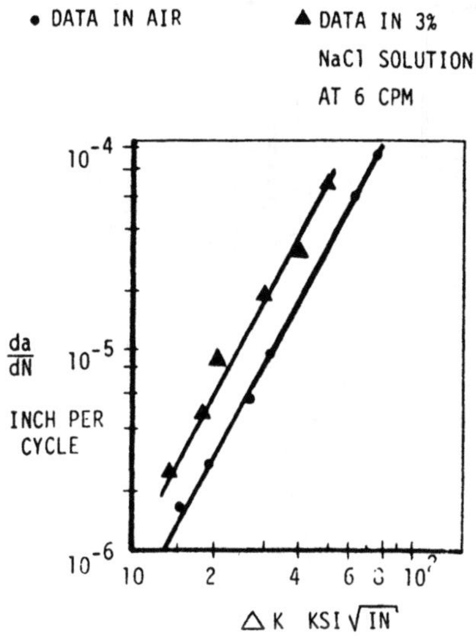

Figure 4 Crack Growth Data for 304 Steel in Air and Three Per Cent Solution

Figure 5 Medium Pressure Blade Profiles in Axial Flow Turbine Stage

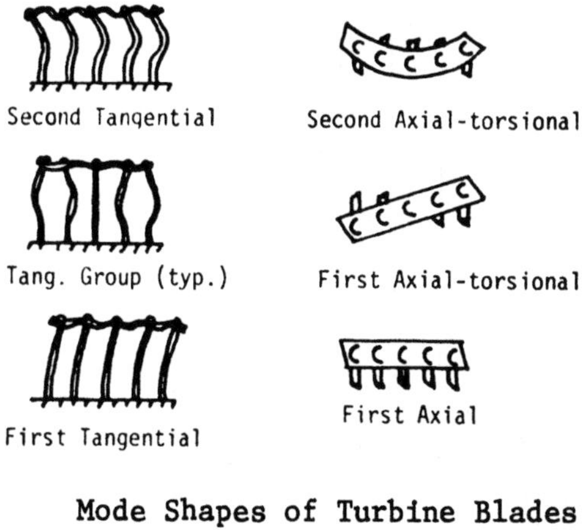

Figure 6 Mode Shapes of Turbine Blades in Groups

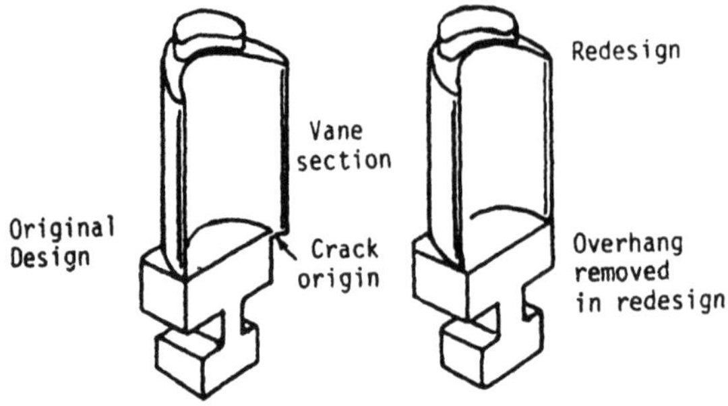

FIGURE 7. DETAIL OF FAILED BLADE AND MODIFIED REPLACEMENT BLADE, VANE OVERHANG REGION.

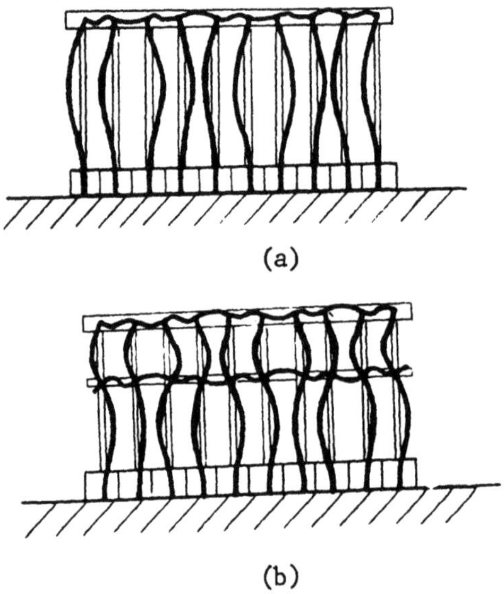

Figure 8 Second Type Tangential Group Mode (a)
Second Type Tangential with Tiewire (b)

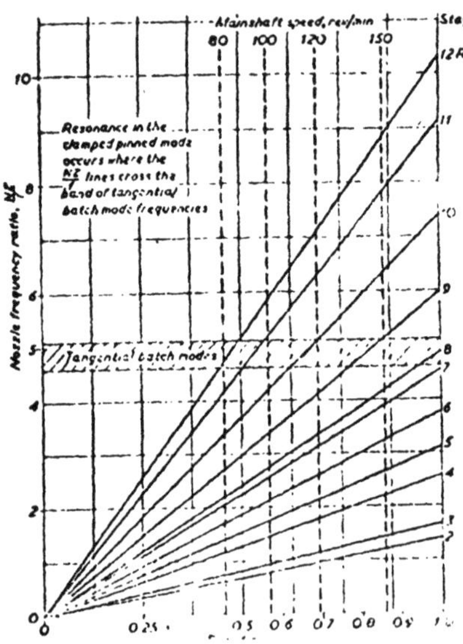

Figure 9 Campbell Diagram for Fatigued Marine Turbine Blade

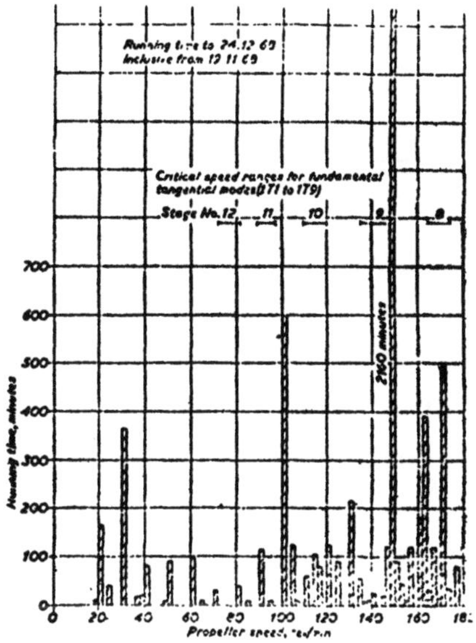

Figure 10 Turbine Log. Times at Various Speeds

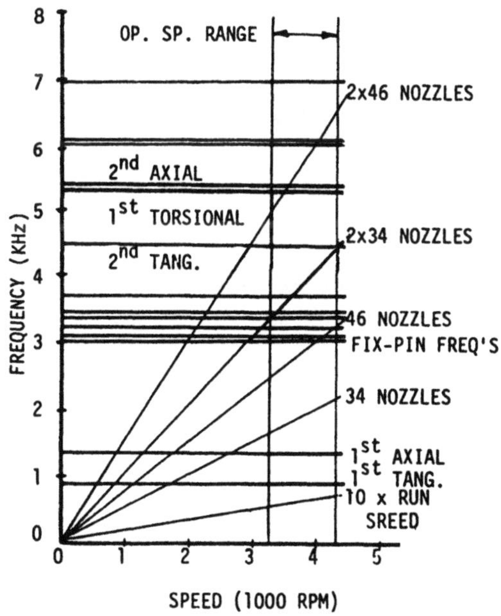

Figure 11 (a) Campbell Diagram for Failed Blades

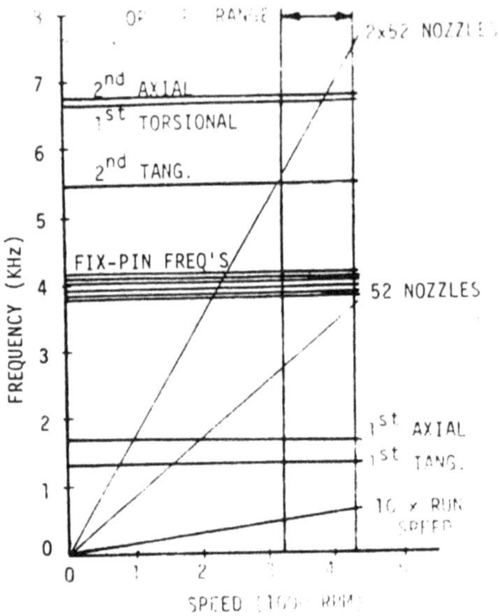

Figure 11 (b) Campbell Diagram for Replacement Blades

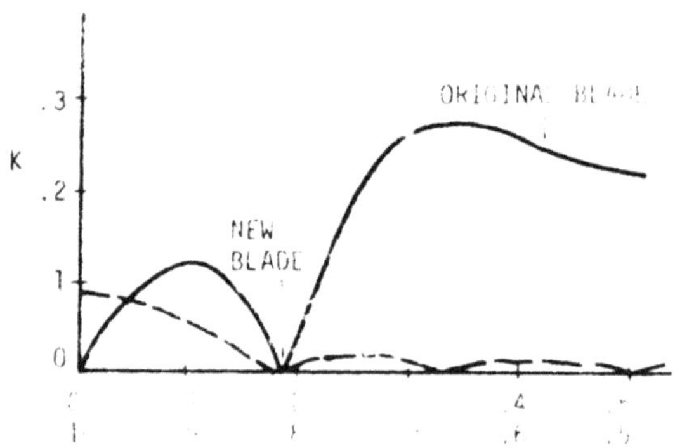

Figure 12 Resonant Response Factor Re-Design

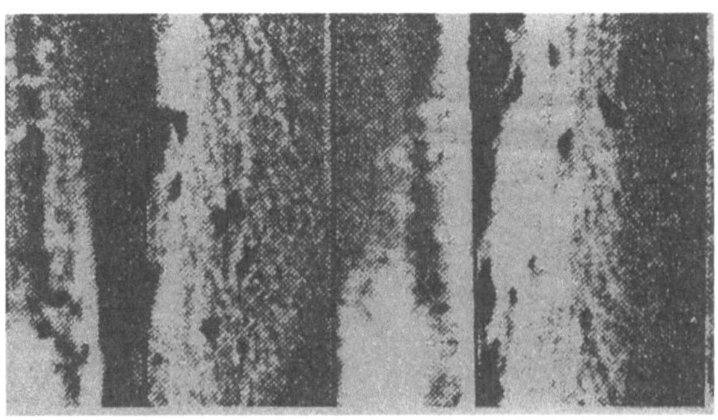

Figure 13 Blade Vane Sections Showing White Chemical Coatings

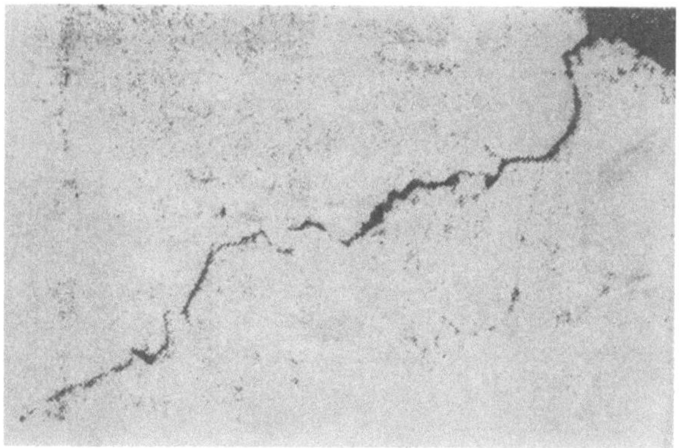

Figure 14 Crack Origin and Propagation From Stress Raiser into Component

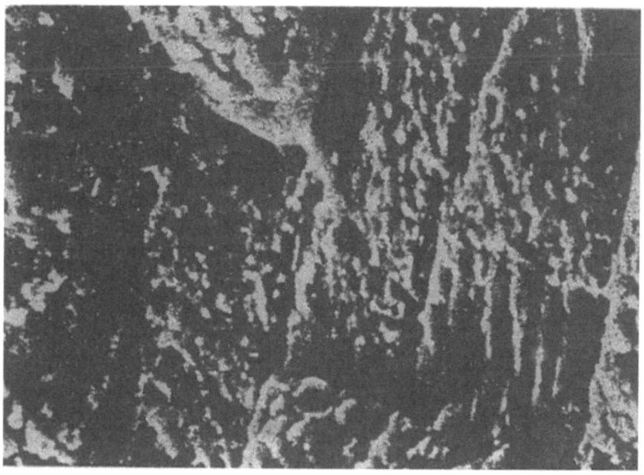

Figure 15 Electron Micrograph Showing NaOH Nodules on Surface of Crack

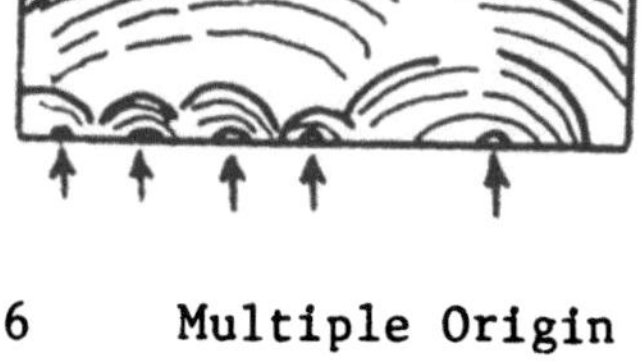

Figure 16 Multiple Origin Cracking

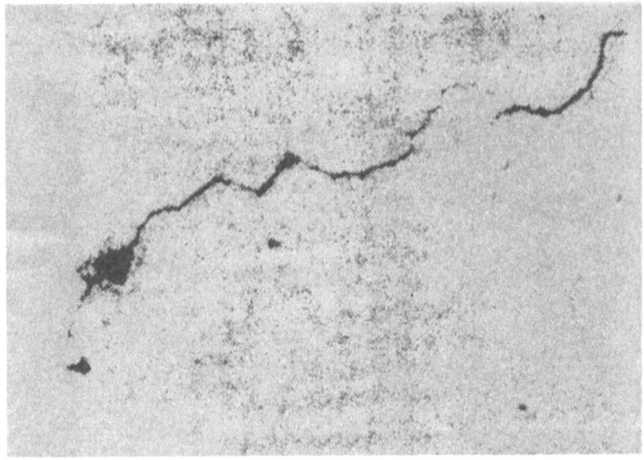

Figure 17 Crack Path - Straight, Intergranular

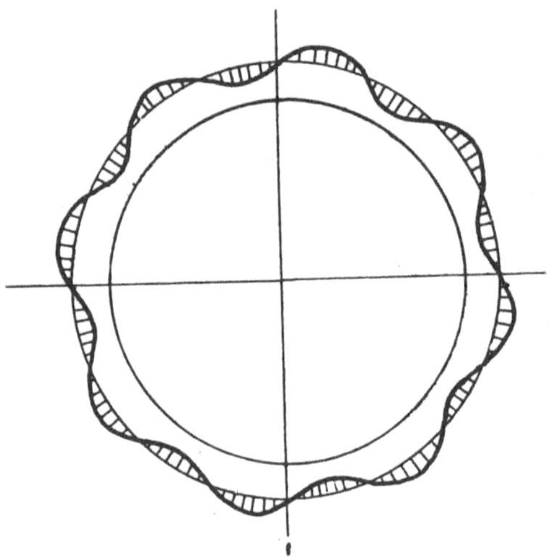

(a) Eight-Per-Rev Harmonic Excitation Around Nozzle Circumference

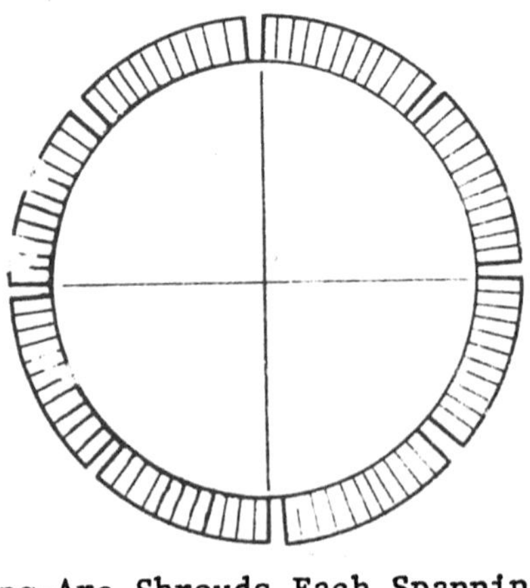

(b) Long-Arc Shrouds Each Spanning One Excitation Wave at all Positions

Figure 18

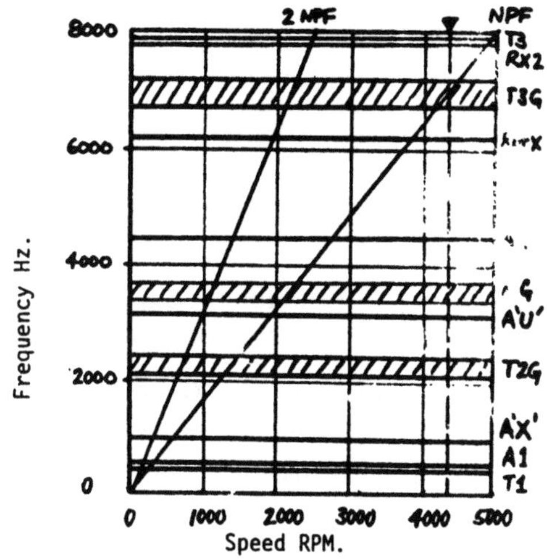

Figure 19 Campbell Diagram, 4th Stage, 96 Nozzle Diaphragm - Failed Row

Table 1 Possible Causes of Turbine Blade Failures

FAILURE MECHANISM	SOURCE	LOCATION
Fatigue	Unsymmetrical stage flow Nozzle resonance Partial admission Torsional transients Negative sequence currents Excessive condenser pressure	General
Corrosion	Excessive corrosion agents in steam/ feedwater	General
	Concentration mechanism	Wilson Line
Corrosion Fatigue	Corrosion plus vibration source	General
Erosion	Wet Steam	LP stages Erosion shield, cover Wilson Line
Ash Deposit	Combustion residue in gas stream	Blade lead edge
	Boiler, pipewall exfoliation	HP stages
Water Ingestion	Moisture separators Condenser	Adjacent LP stages

Table 2 Industry Experience Stress Corrosion and Corrosion Fatigue

PART	MATERIAL	Y.S., KSI	ENVIRONMENT
Pipe expansion joints	Inconel 600	35	Melted caustic
	304SS	30	611 to 710°C
HP turbine bolts	Re 26	—	High sodium
	Pyromet 860	—	in deposits
	Steel	—	—
	Incoloy	—	NaOH-NaCl in deposits
LP stationary blades	Welded 304	—	Na_1K_1Cl identified in cracks
LP stationary blades	304	30	Caustic and chlorides
IP rotating blades	Re 26	—	Caustic in steam and deposits
LP rotor	1Cr1Mo 1/4 V	87	Caustic carry-over, 140°F 3 psi at saturation
Wheel dovetails in boiler feed pump turbines	NiCrMoV	97	Caustic in deposits
Shrunk on wheels	1Cr 1/2 Mo	123	Caustic wash
Shrunk on discs	3Cr 1/2 Mo	117	Up to 215 ppm of sodium hydroxide in boilers
Small turbine rotor discs	CrMoV	126, 133	Caustic and chloride in deposits
HP inner cylinder horizontal joints	—	—	—
HP rotor dovetails	CrMoV	—	Hydrogen sulfide (229 ppb in condensate) from sodium sulfite treatment
LP rotor	NiMoV	—	
HP discs	NiCrMo	117	Caustic from demineralizer in deposit
HP blade pins	Re 26	—	High caustic in feedwater, steam and deposits source demineralizers
HP seal springs	Inconel 718	—	
HP bolts	AISI 4130	—	
LP discs	NiCrMoV	—	
HP inner cylinder, horizontal joints	CrMoV	—	NaOH, NaCl in deposits Source: demineralizers
Two LP discs (ESCOM)	2% NiCrMoV	114	—
Four LP discs (SEVC)	2% NiCrMoV	114	—
Two LP discs	NiCrMoV	130	Caustic suspected

Table 2 (cont') Industry Experience Stress Corrosion and Corrosion Fatigue

PART	MATERIAL	Y.S.,KSI	ENVIRONMENT
Last minus one rotating blades (fossil turbine)	12Cr, 17-4 PH	--	Chlorides and sulfates in deposits
Single cylinder of turbine	Cast iron	—	High oxygen and a small sulphuric acid from hydrolyzed Na_2SO_4 plus oxygen
LP blades	12Cr	--	Mixture of organic acids and acetic acid
First row LP	12Cr	—	
LP discs and blade fastenings	Low alloy steel	—	
LP blades and shrouds (magnox reactor unit)	Stainless Steel	--	Organic acids (acetic, propionic, butyric) from surface cooling water organic compounds (humic acids)
Last row LP rotating blades	12Cr hardened	--	Inorganic acids in steam, ferrous chloride and sulfate in deposits
Stationary blades		--	Hydrochloric acid in steam (seawater cooling + powder)

CHAPTER 11.4

AN IMPROVED PROCEDURE FOR COMPONENT LIFE ESTIMATION WITH APPLICATIONS

N.F. Rieger
Stress Technology Incorporated, Rochester, New York, USA

ABSTRACT

A general procedure is described for the calculation of fatigue initiation life. This procedure was developed for the investigation of component fatigue failure, but it is also suitable for the design of new components. The procedure incorporates the effects of steady stresses, dynamic stresses from excitation spectra, and material cyclic properties. The Rainflow Cycle Counting procedure is used to incorporate complex response waveforms, and the Local Strain Approach is used to determine the amount of fatigue damage from each harmonic component. The resulting cumulative damage is evaluated using Miner's law, to provide an estimate of component life under given stress and thermal load conditions. Several examples which describe the application of this procedure are given.

11.4.1 Introduction

This paper describes a comprehensive procedure for estimation of component fatigue initiation life. By initiation life is meant the period between the first loading of the new component through the initiation of a fatigue crack at the location under investigation. This period is here referred to as crack 'initiation life,' and it is distinguished from crack 'propagation life,' in which the crack propagates until component failure occurs by rupture. The purpose of the paper is to explain this initiation life procedure in detail, and to show that all major factors which influence fatigue crack initiation are included in the calculation sequence.

This procedure was originally designed to meet the needs of component failure investigations, through a comprehensive, computerized procedure which gave improved estimates of component initiation life under practical operating circumstances. During such studies, information on material properties can often be obtained only approximately from records, and possibly by the testing of coupons from failed components. During failure investigations details of relevant operating circumstances are usually obtained from operating records and from discussions with plant engineers. This life analysis approach therefore does not contain any in-built factor of safety, because the requirement is for an accurate estimate of the life of the failed component. The same procedure can be used under design conditions to provide 'true' life values, within current limits for such estimations. This procedure can then be used to design a life factor-of-safety into the component, through design adjustments which give lifetimes suitably beyond the required service life. Any difference between required life and anticipated service life then becomes the factor-of-safety of the design.

It may be argued that the well-known Goodman diagram procedure leads to similar results, and is therefore adequate for design purposes. The conventional Goodman diagram is however, limited to constant life characteristics, and to single component harmonic stresses. The method proposed in this paper is able to address the problems of life estimation in much greater detail, because it includes the following factors into a single life result:

a) Multi-component vibratory stresses are included directly using Rainflow cycle counting to prepare the cyclic data for Miner's law, or other multi-component cumulative damage criterion.

b) The proposed procedure is based on the Local Strain method. Strain life data can be obtained for actual material samples under precise strain loading, and temperature and chemical environment control. Life tests can then be run to many times the anticipated lifetime.

c) Strain-life curves are developed using test specimens and loads which simulate actual stress levels, R-ratios, and temperatures experienced by the component during service life. These strain life results replace the conventional S-N curves, and they can include creep effects if needed.

d) Goodman envelopes are often not available for fatigue lives of less than 10^7 cycles. Furthermore, such data is invariably stress-based rather than strain-based, and may involve wide scatter without strain-controlled testing. Stress-based procedures also lead to complications when the calculated elastic stresses exceed the material yield point. Goodman diagram data for high temperatures is only available in rare instances.

The fatigue life procedure described in this chapter is presented with definitions of the major terms involved. For convenience it is assumed that results for both steady stresses and dynamic stresses are available in sufficient detail from finite element calculations, in order to commence the initiation life procedure. Such stress calculations should address all major loading aspects, including component validation tests, anticipated service load (and overload), and also temperature influences where these affect material properties and stresses. Elastic stresses are adequate for this purpose, as will be described. Several examples from practice in which this procedure was used are described.

11.4.2 Notation

A Cross-section area under uniaxial load

A_o Cross-section area at zero load

b Plastic strain life exponent

c Elastic strain life exponent

D Damage

D_j Damage under given loads

E Modulus of elasticity

K' Cyclic strength coefficient $\sigma_f'/\varepsilon_f'^{n'}$

K_t Theoretical stress concentration factor

K_σ Stress concentration factor

K_ε Strain concentration factor

N_f Cycles to crack initiation

n_i Number of cycles at strain level

N_i Cycles to cracking at strain level

$2N_f$ Stress reversals to crack initiation

n' Cyclic strain hardening coefficient b/c

P Applied steady load

q Notch sensitivity factor

R Ratio $\sigma_{min}/\sigma_{max}$

$\Delta\sigma$ Stress range

$\Delta\varepsilon$ Strain range

σ_f True fracture ductility

σ_f' Fatigue ductility coefficient

σ_e Elastic stress: from linear F.E. calculations

σ_f True fracture strength

σ_m Mean stress

σ_T Total stress

ε_f' Fatigue strength coefficient

ε_m Mean strain

ε_T Total strain

11.4.3 Definitions

Engineering Stress

$$\sigma_o = P/A_o$$

True Stress

$$\sigma = P/A = \sigma_o(1 + \varepsilon_o)$$

Engineering Strain

$$\varepsilon_o = \sigma_o/E = \Delta 1/1$$

True Strain

$$\varepsilon_o = \sigma_o/E = 1n\ (A_o/A) = 1n\ (1 + \varepsilon_o)$$

True Fracture Strength

$\sigma_f = P_f/A$ (plus correction for triaxiality)

True Fracture Ductility

$\varepsilon_f = \ln (A_o/A_f)$

Half Cyclic Strain Range

$1/2\Delta\varepsilon = 1/2\Delta\varepsilon_p + 1/2\Delta\varepsilon_e$

Half Plastic Strain

$\Delta\varepsilon_p = (\Delta\sigma/2K')^{1/n'}$

Half Elastic Strain

$\Delta\varepsilon_e = (\Delta\sigma/2E)$

Half Elastic Cyclic Stress

$1/2\Delta\sigma_e = \sigma_f' \ (2N_f)^b$

Half Elastic Strain

$1/2\Delta\varepsilon_e = 1/2(\Delta\sigma_e/E) = (\sigma_f'/E)(2N_f')^b$

Half plastic Strain

$1/2\Delta\varepsilon_p = \varepsilon_f' \ (2N_f)^c$

R-ratio

$R = \sigma_{min}/\sigma_{max}$

Notch Sensitivity

$q = (K_f - 1)/(K_t - 1)$

Fatigue Factor

$$K_f = \{1 + (K_t - 1)/(1 + a/r)\}$$

a = constant, r = radius

Neuber's Rule

$$K_t^2 = K_\sigma K_\varepsilon \quad (\sigma_e^2 = \sigma E\varepsilon)$$

11.4.4 Previous Work

The procedure described herein is original, though portions of it have been described previously by several authors [1] [2]. Development of the Local Strain method is primarily due to Morrow, Socie, and Dowling [3] [4] [5], though many other studies and developments have been made concerning this procedure in recent years. The Rainflow cycle counting procedure has been described and discussed by Dowling [6]. A recent assessment of the current accuracy of the procedure is given in [7]. Valuable studies of cumulative damage have recently been reported by Conway, Stentz and Berling [8], and by Battacharya [9].

11.4.5 Factors Affecting Component Life

Steady stress increases the local strain level at the location under study, and load cycling (dynamic stress) occurs above and beyond this strain level. Stress cycling about the steady stress value within the elastic region dissipates relatively little energy, and leads to relatively little cyclic damage. Stress cycling about the steady stress value in the plastic region leads to greater energy dissipation, and to greater cyclic damage and loss of fatigue life.

Dynamic loading may be classified as (a) simple harmonic, (b) general periodic, e.g., saw-tooth waveform, (c) transient or impactive, (d) spectral or random. General dynamic load spectra may contain any or all of the above types of loading. In practice, component loadings obtained from testing are commonly presented as frequency spectra containing multiple harmonic components.

Dynamic strain causes the cyclic damage which can lead to crack initiation. Cyclic strains may result from low cycle loading, e.g., machine start-stop sequence, and from high cycle loading, e.g., forced vibration. Cyclic strain damage can often be observed from electron micrographs as a pattern of ridges in the material fracture surface. Dynamic strains are caused by the effect of the loading spectrum on the dynamic properties of the structure. Resonance may then cause certain harmonic components to dominate the dynamic strain spectrum.

Material strength determines the number of strain cycles a component can withstand at a given location at a given frequency, under the applied preload steady stress. Charts of material strength data are prepared from fatigue tests on standard samples of the material having similar chemical composition, material properties, and heat treatment as those of the prototype. A recent development is the strain life curve, as shown in Figure 1. This approach is used in the present initiation life procedure.

Component life history. The loading to which a component is exposed will strongly influence both the steady preload strain level and the dynamic strains which occur at a given location. The time to which a component is exposed to such loadings influences the fatigue initiation life of the component. Many comprehensive component load histories involve repetitions of parts of the overall loading sequence, as shown in Figure 2. Important aspects of any load history are:

a) Initial startup cycle (to 100% speed).

b) Transient excitation of structural modes during runup and shutdown.

c) Initial overspeed test cycle (to 120% speed or more).

d) Several balancing speed cycles in spin pit (to 100% speed).

e) Thermal tests at speed. Various start-stop cycles (100% speed).

f) Initial startup cycles in plant (to 100% speed).

g) Overspeed trip settings in plant (typically 110% and 112% speed).

h) On-stream or on-line startup (to 100% speed, several times).

i) On-load (to 100% speed).

j) Process establishment. (Several 0-100% cycles, thermal transients).

k) Process dynamic loading (gas pulsations, torsional transients, etc.).

l) Process load variations (load, speed, temperature, flow, etc.).

m) Planned outages (100%-zero-100% speed; thermal transients).

n) Unscheduled outages. Process problems, failures, etc.

Each of these factors must be incorporated into the load history of the component, as applicable. Previous load history details are often assembled by estimating probable cycles from typical procedures, records, and operator recollections. Anticipated loadings will include items from the above list, and are arranged as typical load-blocks to be applied to the component: see Figure 2. Note that process thermal history and stresses may be an important aspect of a given history.

Cumulative damage. The applied load history contains both low cycle and high cycle load components. Low cycle loading typically involves the largest strain ranges, and provides the pre-strain about which the dynamic cycling occurs. High cycle loading typically contains multiple frequencies, each with different load magnitudes. This results in many more load cycling harmonic components, but these components usually have considerably smaller magnitudes than the low cycle components. All loadings together result in cumulative damage to the component, for which the effects can be assessed using the Rainflow cycle counting method, as described later under Cycle Counting Procedure. The influence of load cycling on the stress-strain curve is illustrated in Figure 3. Each time a cycle is completed in the manner shown, it counts as one load cycle having a specific strain range and frequency. A computer program keeps account of the number of load cycles which are completed at a given strain level, in the procedure described herein.

11.4.6 Calculation Procedure for Initiation Life

Prerequisites. The following input data are needed for conditions at the location being considered, to perform the initiation life calculation.

a) Steady stress value corresponding to each load level in the load history. Precise data from finite element calculations is preferred.

b) Dynamic stress value corresponding to each load condition in the load history. Finite element data is preferred.

c) Material properties for the component material obtained from strain-life tests on samples, in the operating environment, at each operating temperature in the load history. A cyclic stress-strain curve for the material is also required.

d) Load history details for number of cycles during each of the life history conditions (a) through (n) are listed previously.

Examples of such data are given in the case histories given later in this chapter.

Cyclic stress-strain curve. Where a material is loaded cyclicly, the stress-strain response does not follow the well-known pull-test or monotonic load deflection curve. The needed cyclic stress-strain data is obtained from the locus of the hysteresis loop peaks, as shown in Figure 4. This data is obtained from laboratory tests on material samples under simulated operational conditions (preload, temperature, environment). The cyclic stress vs. strain curve is then curve-fitted in the computer for life calculations, using the following relations:

$$\frac{\Delta\varepsilon}{2} = \frac{\Delta\sigma}{2E} + \left(\frac{\Delta\sigma}{2K'}\right)^{1/n'} \qquad (1)$$

where $K' = \sigma_f'/\varepsilon_f'^{\,n'}$

and $n' = b/c$

Figure 3 shows how preload and load cycles influence component strain cycles, in the elastic and plastic regions of the stress-strain curve.

Stress history. The required steady stress distribution is calculated using some linear elastic procedure, such as the finite element method. This provides high computational precision in the stress results, and it allows stress conditions for other operating conditions (such as lower speeds or higher temperatures) to be found by proportioning the results obtained from a single calculation, in accordance with the known structural or thermal laws.

Dynamic stresses are found in a similar manner, taking into account the relation between the exciting frequencies in the load spectrum and the natural frequencies in the structural response. Changes in vibration amplitude are again proportional to the applied forcing, as mentioned above. Changes in dynamic stress magnitude due to changes in exciting frequencies, e.g., due to speed changes, or in the process gas excitations, can be addressed by use of equation 2, as follows:

$$\frac{\sigma_d}{\sigma_n} = \frac{2\xi(\omega/\omega_n)}{[\{1 - (\omega/\omega_n)^2\}^2 + \{2\xi\omega/\omega_n\}^2]^{1/2}} \tag{2}$$

where σ_d is the dynamic (modal) harmonic stress, σ_n is the resonant stress for the mode, ξ is the modal damping ratio, ω is the excitation frequency and ω_n is the natural frequency of the mode.

The use of this expression can be demonstrated as follows. The excitation spectrum shown in Figure 5 shows typical process conditions corresponding to the dynamic portion of the load history in Figure 2. If the exciting frequencies and harmonic forcing are then changed from Condition A to Condition B, the dynamic response of the structure in its various modes of vibration will vary in accordance with equation 2, as shown in Figure 6.

Strain life behavior. The strain life of the material as determined by test specimens under controlled conditions is governed by equation 3, as follows:

$$\frac{\Delta\varepsilon}{2} = \frac{\sigma_f'}{E}(2N_f)^b + \varepsilon_f'(2N_f)^c \tag{3}$$

This forms the basis for the fatigue initiation life calculations. The specimen tests provide data which allows (a) the fatigue ductility coefficient ε_f', (b) the fatigue strength coefficient σ_f', (c) the fatigue strength exponent b, and (d) the fatigue ductility c, to be determined. This then allows the strain life curve for the material, equation 3 to be calibrated. The resulting analytical expression is then used to calculate the true strain at stress levels resulting from the various aspects of the finite element calculations.

The influence of mean stress on the strain-life behavior may be considered in several ways. First, it may be considered directly during laboratory testing, by performing tests at different R-ratios, so that the resulting curve includes the influence of mean stress directly. Secondly, the mean stress effect may be included directly in the strain-life expression, using the form shown in equation 3a:

$$\frac{\Delta\varepsilon}{2} = \left(\frac{\sigma_f - \sigma_m}{E}\right) (2N_f)^b + \varepsilon_f' \, (2N_f)^c \qquad (3a)$$

A third method is to use the Smith-Watson-Topper hypothesis, which is described by equation 4:

$$\sigma_{max} \, \frac{\Delta\varepsilon}{2} = \frac{\sigma_f'^2}{E} (2N_f)^{2b} + \sigma_f' \, \varepsilon_f' \, (2N_f)^{b+c} \qquad (4)$$

The second of these procedures, i.e, modified strain-life equation, is used in the case histories which follow.

The ability to include the effects of calculated stresses at levels which exceed the dynamic yield stress of the material is another important aspect of the procedure presented here. This step is accomplished using Neuber's rule, as follows:

$$\sigma_e^2 = \sigma E \varepsilon \qquad (5)$$

This expression allows calculated linear stresses to be realistically converted to dynamic stresses and strains, using the dynamic stress-strain curve which is based on data obtained from the cyclic strain tests. This data is next converted into an analytical form using equation 3.

Equations 3 and 3a give the relation between dynamic stress and strain throughout the component strain range tested. When used with Neuber's rule, this convenient procedure allows the linear stresses obtained by the finite element or other method to be converted by the computer into practical elastic-plastic strains. These operations are included in the computer program developed for life calculations.

Cycle counting procedure. Under spectral loading the dynamic strain conditions at critical locations in a component may have very complex waveforms, in keeping with the dynamic spectrum at that location. Several procedures exist to deal with this condition, of which the Rainflow cycle counting procedure is well known. Simply stated, this procedure consists of dividing the complex waveform into a sequence of simple cycles, and then counting the number of stress cycles within a given strain range. The resulting number is then compared with the tested fatigue life of the material at this strain level.

The complex waveform shown in Figure 7 can be sub-divided into constituent cycles (or strain reversals) by drawing from each significant peak a line in the direction of increasing time until this line again intersects the complex waveform, as shown. Each time this occurs, one cycle of damage has been completed at a corresponding strain level. Cycles within cycles may also occur, and the corresponding cyclic damage is shown on the dynamic stress-strain curve in Figure 3. Computer sub-routines are available to accomplish this cycle counting process automatically. When the process is completed, for a given load history segment (corresponding to a particular operating condition) the number of cycles occurring at a given strain level is identified. A table of such cyclic damage constitutes the output from the load-cycle counting sub-routine.

Cumulative damage. The best known cumulative damage assessment procedure is Miner's law, which states:

$$\sum \frac{n_i}{N_i} = 1 \tag{6}$$

This criterion is used in the procedure described in this paper. The number of cycles n_i occurring at a given strain level is obtained from the Rainflow cycle counting procedure described above. The number of cycles to failure N_i based on test sample data, adjusted for mean stress effects, is then obtained from the strain-life equation. Particular values are then divided into corresponding n_i values to form the damage portion n_i/N_i. This procedure is repeated for each of the other strain levels to form the remaining cycle ratios for the forcing response. These cycle ratios are then summed, and this sum is compared with the criterion value 1.0, in accordance with Miner's law.

For practical loading conditions, the load segment time is often relatively short, and represents only a portion of the anticipated life of the component. Each segment of the load history may therefore be interpreted as contributing a portion of the total damage which the component can sustain: see Figure 2. If the loading period j causes damage D_j, then the total damage $D = \Sigma D_i$ for the period of operation may be obtained from the summation of the individual damage D_i components. Furthermore, the damage which occurs in a specified time period D_t, representing the total number of dynamic periods completed within the time period t may be used to calibrate the operating life in years. For example, if $D_t = 0.25$ damage units acquired in one year, then $D = D_t = 1.0$ damage will be acquired in a total operating life of four years, under the same circumstances, in keeping with Miner's law. Component failure could then be anticipated in 4 years.

With a computer program it is apparent that the above process can take into account any number of cycles within a given time frame, at any stress level. This damage can then be compared with Miner's criterion to determine whether component failure is likely to result within the specified period. Alternatively, once the amount of damage, however complicated, acquired within a specified time is known, then the reciprocal of this value will give the anticipated life of the component under similar circumstances. In this manner, component life can be assessed, in a manner which includes wide variations in load conditions and complex operating conditions.

11.4.7 Case History 1: Shaft Keyway Fatigue Life

Problem: Consider a steel shaft made from AISI 4340 steel in a compressor drive system, with a possible fatigue failure point at a keyway. The material properties for the shaft material were obtained from fatigue tests, and are listed in Table 1. The loading history on the keyway, Figure 8, shows both high and low stress cycles. Each startup includes a 20% overload peak stress (point d) resulting from motor overshoot. During normal operation, the shaft vibrates torsionally at 1,000 Hz due to tooth-mesh excitation from the drive gears. The resulting alternating torque varies by 6.7% about the mean torque. The maximum elastic mean stress due to the mean torque is 150 ksi at the keyway stress calculation.

The strain history of the shaft is shown in Figure 3. The compressor driver operates 8 hours per day for 260 days per year and has only one startup/shutdown cycle per year (points c through g). The shaft is subjected to 2.88×10^7 vibratory stress cycles each day (points e through f). The loading history also includes the manufacturer's overload test (points a through c) to about 133% of the required operating load. The elastic stress due to this overload is 200 ksi. The local strain method is used to predict the fatigue life of the shaft.

The local stresses and strains are first computed from the loading history (keyway elastic stress σ_e vs. time) using equations 1 and 3. Typical results are summarized in Table 2 as local stress vs. strain data. For instance, a peak stress σ_e, of 200 ksi (point b) during the manufacturer's overload test results in local plastic deformation at the keyway, and a local stress σ of 123.7 ksi. This value of σ is computed from Neuber's rule, equations 1 and 5, using Newton-Raphson iteration, along with the local strain ε. These computations may be summarized as follows:

$$\sigma\varepsilon = \frac{200^2}{29{,}500} = 1.356$$

$$\sigma\varepsilon = \sigma\left(\frac{\sigma}{E}\right) + \sigma\left(\frac{\sigma}{K'}\right)^{1/n'}$$

$$1.356 = \sigma\left(\frac{\sigma}{29{,}500}\right) + \sigma\left(\frac{\sigma}{180}\right)^{1/0.075}$$

$$\sigma = 123.7 \text{ ksi}$$

$$\sigma\varepsilon = 1.356$$

$$\varepsilon = \frac{1.356}{\sigma} = \frac{1.356}{123.7} = 0.011 \text{ in./in.}$$

Because the shaft initially consists of material with zero damage, the initial overloading by the manufacturer occurs as a monotonic stress vs. strain curve. However, the subsequent duty cycles occur as cyclic hysteresis loops as discussed previously. Note that the overload test results in a residual compressive stress of 69 ksi at point c, Figure 3.

Both the loading history and the local stress vs. local-strain diagram show both low and high cycle loads. Low cycle loading is represented by the hysteresis loop, points c through g of Figure 3, whereas the high cycle loading consists of fluctuating stress levels (between points e and f). Although not detailed here, the total strain amplitude ε_T and the mean stress σ_m are respectively computed for the low cycle loading to be 0.004520 in./in. and 18.3 ksi, and for the high cycle loading to be 0.00539 in./in. and 71.1 ksi.

The fatigue damage due to both high and low cycle loading is determined by solving equation 3 numerically. The results of this computation are listed in Table 3, for fatigue damage data. Table 3 also lists the damage per year which then can be summed to determine the total fatigue damage which occurs each year due to both types of loading. Thus:

$$\frac{1}{N_T} = \frac{n_1}{N_{f(1)}} + \frac{n_2}{N_{f(2)}}$$

$$= \frac{260}{4.153 \times 10^{-4}} + \frac{7.488 \times 10^{9}}{1.400 \times 10^{-13}}$$

$$= 0.1090$$

$$N_T = 9.17 \text{ yr}$$

Finally, the fatigue life is predicted as the inverse of the total damage per year, or N_T = 1/0.1090 = 9.17 yr. This fatigue life analysis shows that low cycle loading due to machine startup and shutdown was the primary factor leading to fatigue damage at the keyway.

11.4.8 Case History 2: Six Blade Group

Blade group model. Figure 9 shows a 3D finite element model of a blade group consisting of six blades, a cover band, tenons, vanes, platforms, and blade-root/disk-root sections. Suitable boundary conditions were applied to the cover section, and to the disk root section, to ensure that the model would simulate the blade group response under steady and dynamic loading. The model was then subjected to 3600 rpm rotational loading, and to a 10 psi steam pressure drop applied to the vane upstream surfaces, to the cover underside, and to the platform upstream surface. The platform nodal displacements were obtained from this calculation for use in further detailed 2D stress calculations.

The 2D gridwork shown in Figure 10 was used to obtain stresses in the attachment region in finer detail. Very fine elements were used in notch radius regions where higher stresses are normally encountered. Twelve gap elements were used to transfer the hook loadings from the blade root to the disk, to account for the indeterminate support conditions, and to study tolerance effects on attachment stresses. Details of maximum stress found for both loading cases are given in Table 4. Figure 11 shows the deformed shape and stress distribution for the root section with zero tolerance mismatch when subjected to steady load.

Steady stresses. The maximum elastic stresses found in the root attachment occurred in the upper hook radius region, adjacent to the contacting surfaces. Maximum elastic stress values found in the blade attachment are listed in Table 4. These values were used to define the plastic strain using Neuber's rule as described previously.

The overall effect of the machining tolerances was to increase the upper blade hook maximum stress by 10%. Under the same conditions, the maximum stress in the middle blade hook and lower blade hook was reduced by 20%. The maximum stresses in all hook-notch regions were greater than the material yield stress, indicating that local plastic flow will occur in those locations. The work hardened regions are quite localized and are not expected to influence the results greatly. The hand-calculated centrifugal force for a single blade was 68,555 lbs. at 3600 rpm. This force was checked against the maximum total force in the gap elements, and the validity of the calculation was confirmed.

Dynamic stresses. Resonant stresses resulting from harmonic forcing were calculated for the first 20 modes. A constant harmonic forcing of 10% of the steam-bending loads in the axial and tangential directions was applied, with appropriate load phasing between the blades. The 2D root section model was used to calculate accurate stress details for five modes which showed significant vibratory displacements. The other modes were not included because, even at resonance they showed very little response. These results are plotted in the dynamic stress spectrum in Figure 12. Table 5 lists the natural frequencies, and Table 6 gives dynamic stress results for these five modes.

Dynamic stress was calculated for the first 80 per-rev excitation harmonics, and the rms total dynamic stress in each mode was obtained as the square root of the sum of the squares from these stresses, to account for phasing differences. Figure 6 illustrates the procedure for detuning the stress from two resonant modal responses. The excitation spectra were based on water table test results for similar stages. Forcing frequencies above 5000 Hz gave negligible response amplitudes. Damping was assumed to vary from log. dec. $\delta = 0.015$ at 1000 Hz to $\delta = 0.025$ at 4000 Hz.

The calculated results showed that the R1 mode and T22 mode were responsible for the highest dynamic stress in the upper hook-notch region. The calculated rms total dynamic stress was 3335 psi. The resonant stresses in mode R1 and mode T22 were 5652 psi and 5387 psi respectively.

Fatigue analysis. The blade group had operated cyclicly, and had experienced frequent starts and stops. Operating records of startup, overspeed, trip settings, load rejection outages and service hours were available for load history input preparation. An average number of 170 start-stop cycles and one 10% overspeed test per year was used for nominal operation, i.e., one load block. The fatigue life evaluation was based on the startup overspeed block and the nominal operating years as succeeding load blocks. Figure 13 shows the type of load-life history used for the fatigue calculations.

A material test program was conducted to obtain fatigue data at the blade operational temperature of 635°F. This program has been described by Morrow [10]. Fatigue life due to start-stop cycles was obtained by testing specimens cut from blade material. The cyclic strain-life curve obtained for 403 stainless steel at 635°F is shown in Figure 14. The material data is summarized in Table 7. These data provided the necessary information for the fatigue damage calculation.

Low cycle fatigue. Details of unit service hours showed that a typical operating year has one 10% overspeed run, and 170 normal start-stop cycles. The elastic stresses in the leading blade upper hook notch region during 3600 rpm operation were:

Case 1: Perfect contact all hooks at zero speed-138,800 psi

Case 2: Contact upper hook only at zero speed-140,240 psi

Case 2 corresponds to a maximum stress of 168,288 psi during the 10% overspeed test, and a maximum stress of 201,946 psi during the initial 20% overspeed tests. Elasto-plastic stress and strains were obtained from these elastic stresses using Neuber's rule.

Figure 13 shows the nominal stress history including the 20% overspeed spin pit test by the manufacturer, the 10 % overspeed which is done once a year to set the governor, the start-stop cycles at normal speed, and the vibratory stress cycles. The corresponding stress-strain path for the upper hook notch material follows the load history in Figure 13. The material was loaded from its initial stage at point 0 to the maximum stress of 93 ksi at point 1 during spin pit test. After the 20% overspeed test, the residual stress in this region was -57.5 ksi in compression. Point 2 is the material in the stage of residual stress and strain. During the 10% overspeed test, the material was brought to point 3.

The stress due to this overspeed was 66.1 ksi. During steady load operation the stress-strain condition was at point 4. The steady stress due to normal speed operation was 55.9 ksi. When the vibratory stress reached the positive peak the total stress went to point 5. When the stress reached a minimum value, the stress dropped to point 6. The stress-strain condition went back to point 2 when the unit was brought to zero rpm.

Table 8 summarizes the strain amplitudes and fatigue damage rate for the 10% overspeed cycle and the normal speed cycles. The total damage in one operating year is:

$$D_j = (0.269 \times 10^{-3}) + (170 \times 0.108 \times 10^{-3}) = 0.1865 \times 10^{-1}$$

$$Y_f = \frac{1}{D_j} = 53.6 \text{ years}$$

This means that the blade would have a nominal life of 54 years in low cycle fatigue. In a practical sense this figure is regarded as statistically marginal.

<u>High cycle fatigue.</u> Per-rev excitations stimulate the blade groups to vibrate in different modes. The exact forcing function for the blade group was not known at the time these calculations were made. High cycle fatigue analysis was based on calculated vibratory stresses obtained by the procedure described previously, including the mean stress effect, equation 3a.

The magnitude of the mean stress σ_m for high cycle loading calculation is obtained using Neuber's rule. The elasto-plastic stress due to steady forces at 3600 rpm was 55.9 ksi. The mean stress for vibration in this condition becomes 53 ksi. High cycle vibratory strains were found to be predominantly elastic in nature. A Rainflow cycle counting technique is used for the analysis. Damage was calculated and accumulated linearly throughout the entire loading block using Miner's law.

<u>Cumulative damage.</u> Fatigue damage calculations were made again assuming that resonance occurred in five significant modes. Table 8 shows the results of resonant stresses and fatigue damage for each of the five modes in one year. Resonance in the T22 mode was found to cause the most damage. The resonance of R1 group rotational mode was also dangerous for the leading blade in the group if the blade group experienced the loading in Load Case 1. The total fatigue damage was computed by Miner's law, as follows:

$$D_j = D_H + D_L$$

where D_H is fraction of the fatigue life used per year due to high cycle strains, D_L is fraction of the fatigue life used per year due to start-stop cycles, and D_j is total fraction of the fatigue life used per year.

The results indicated that if resonance occurred in the R1 mode, the leading blade life would be 56.7 months. If resonance occurred at T22 mode, the same blade would have a life of about 19.6 months.

A fatigue life of 19.6 months in the T22 mode appears to be inadequate for blades operating under the above circumstances. The R1 mode life of 56.7 months also appears to be statistically marginal. These estimated life values are both consistent with the actual blade life observed in practice.

11.4.9 Conclusions

- A comprehensive procedure has been described and demonstrated for the calculation of fatigue initiation life.
- The procedure requires accurate input data on steady stresses, dynamic stresses, material cyclic properties, and component load history.
- The resulting fatigue initiation life value has state-of-the-art accuracy for the various technologies involved.
- The procedure may be used for new component design evaluation. Component life vs. required life then represents the factor-of-safety of the component.

11.4.10 References

1) Steele, J. M., Lam, T. C., 'Improving the Accuracy of Fatigue Analysis,' Machine Design Magazine, Penton Publishers, Cleveland, OH, August 1, 1983.

2) Rieger, N. F., 'Blade Fatigue,' Invited Address, Rotordynamics Session Proceedings, Sixth IFToMM Congress, Theory of Machines and Mechanisms, New Delhi, India, December 1983.

3) Socie, D. F., Morrow, J. D., 'Review of Contemporary Approaches to Fatigue Damage Analysis,' Chapter 8 - Risk

and Failure Analysis for Improved Performance and Reliability, J. J. Burke and V. Weiss Eds., Plenum Publishing Corporation, 1980.

4) Socie, D. F., 'Fatigue Life Prediction Using Local Stress-Strain Concept,' Paper Presented at 1975 SESA Spring Meeting, Chicago, IL, May 1975.

5) Dowling, N. E., Brose, W. R., and Wilson, W. K., 'Notched Member Fatigue Life Predictions by the Local Strain Approach,' Fatigue Under Complex Loading: Analyses and Experiments, SAE, Vol. 6, 1977.

6) Dowling, N. E., 'Fatigue Failure Predictions for Complicated Stress-Strain Histories,' Jnl. of Materials, JMSLA, Vol. 7, No. 1, March 1972, pp. 71-87.

7) Rieger, N. F., 'Factors Affecting the Fatigue Life of Turbine Blades and an Assessment of Their Accuracy,' Proceedings, 55th Shock and Vibration Symposium, Dayton, OH, October 1984.

8) Conway, J. B., Stentz, R. H., Berling, J. T., 'Cumulative Damage Concepts,' Chapter 5, Fatigue, Tensile, and Relaxation Behavior of Stainless Steel, TID-26135 U.S. Atomic Energy Commission, Technical Information Center, Oak Ridge, TN, 1974. (Also Mar-Test, Inc., Cincinnati, OH).

9) Battacharya, A., 'Cumulative Damage of Carbon Steel Specimens in Tension-Impact Fatigue,' Ph.D. Thesis, Banaras Hindu University, Varanasi, India, 1984.

10) Morrow, J. D., 'Laboratory Simulation of the Low Cycle Fatigue Behavior of the Hook Region of a Steam Turbine Blade Subjected to Start Stop Cycles,' ASME Proceedings, Fourth National Congress on Pressure Vessel and Piping Technology, Portland, OR, June 1983.

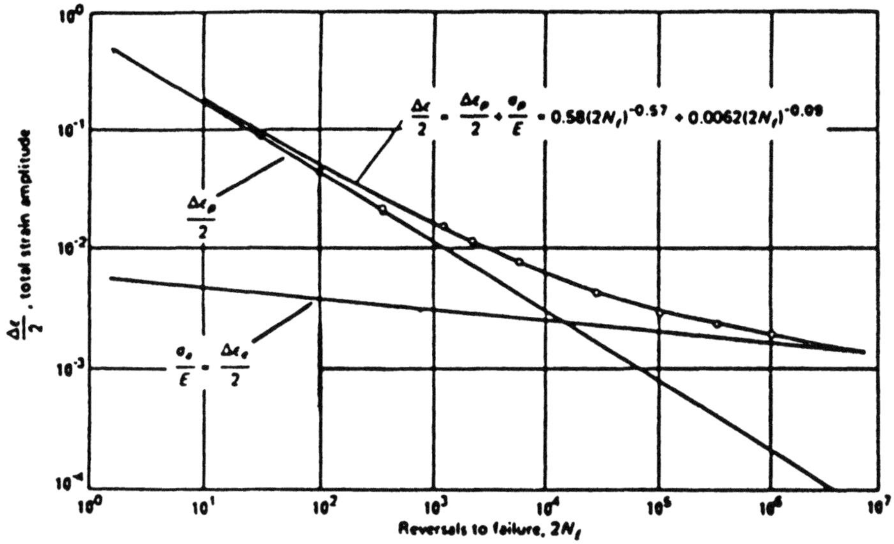

Figure 1 Total Strain vs. Fatigue Life for Annealed AISI 4340

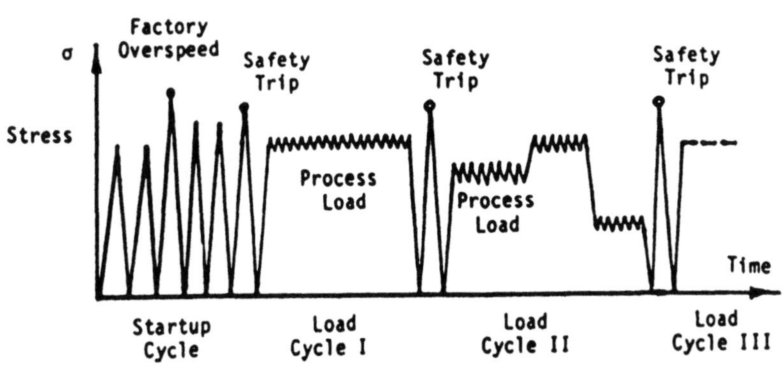

Figure 2 Typical Load History - Start and Process Cycles

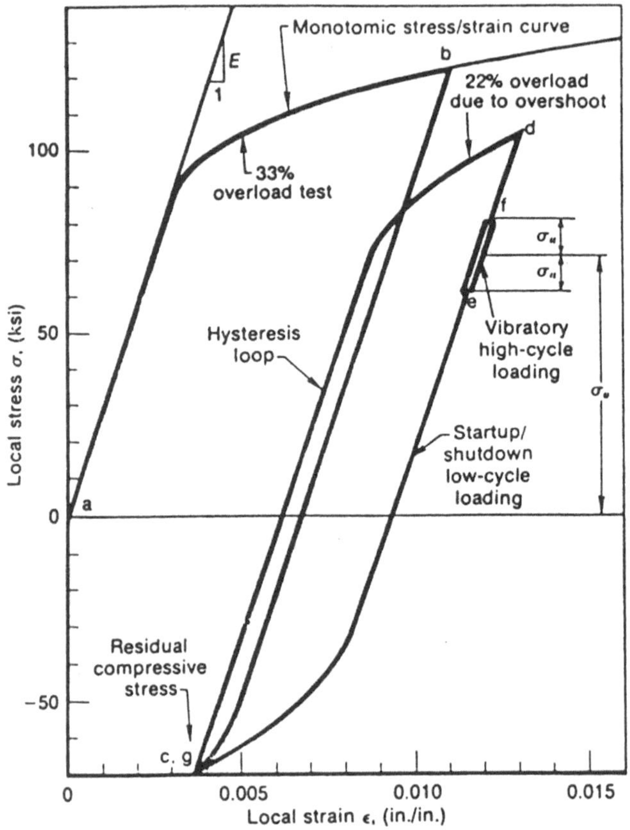

Figure 3 Stress-Strain Loading History

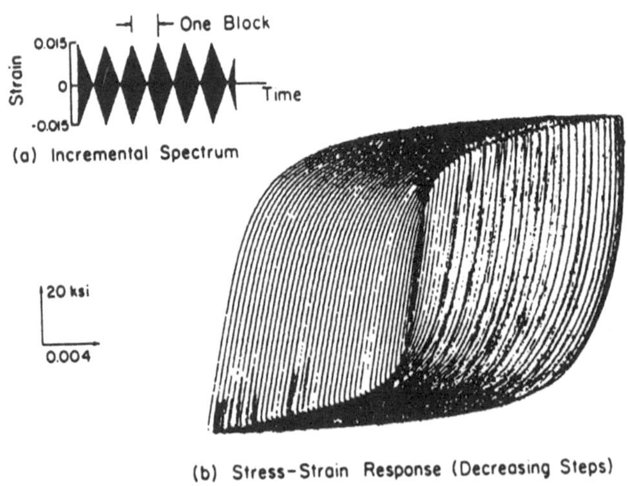

Figure 4 Dynamic Stress-Strain Curve Developed from Hysteresis Loops

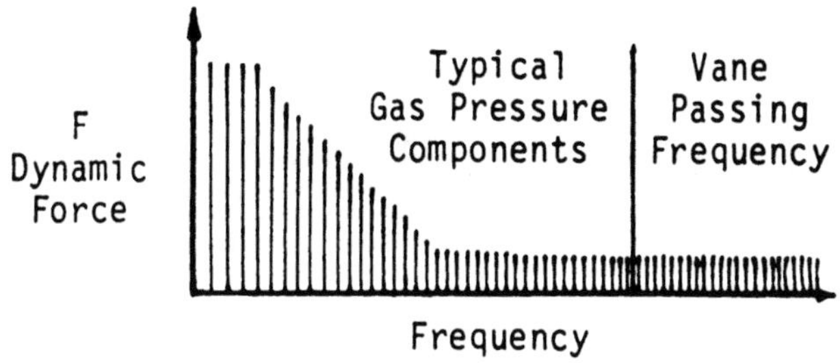

Figure 5 Excitation Spectrum

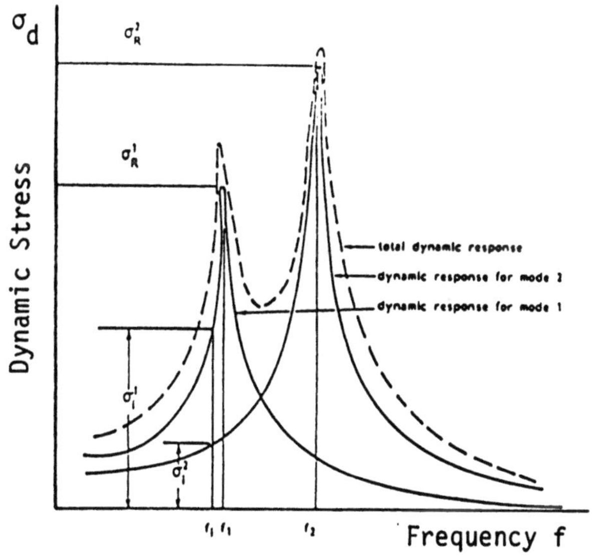

Figure 6 Stress Detuning Procedure

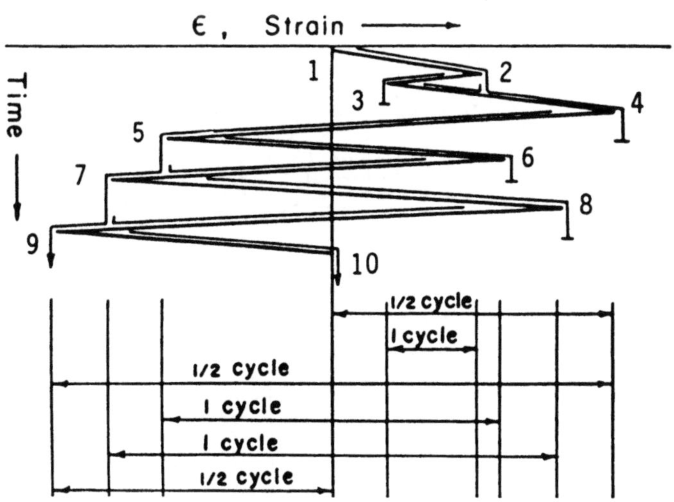

Figure 7 Rainflow Cycle Counting Procedure

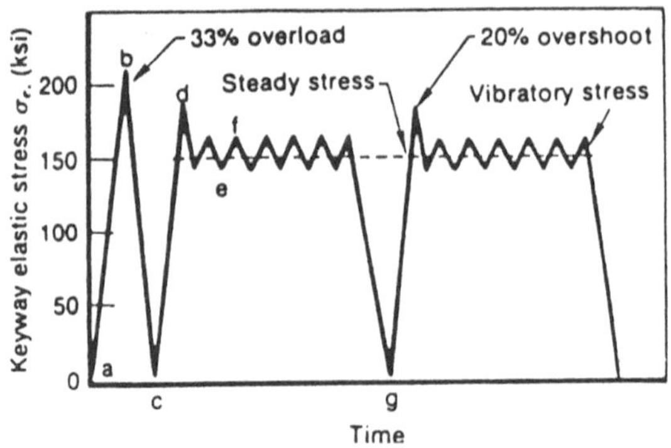

Figure 8 Loading History for Keyway

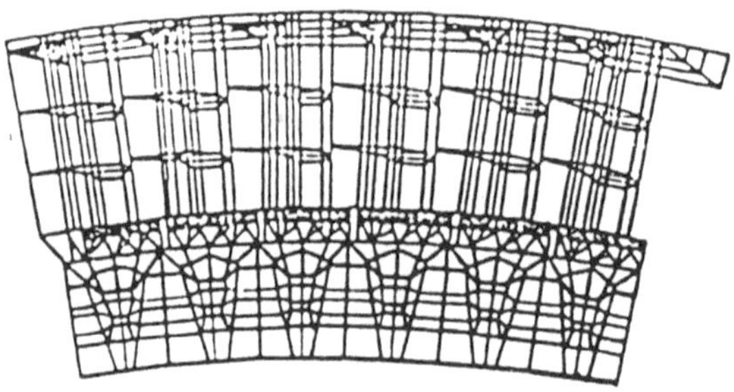

Figure 9 3D Blade Group Model

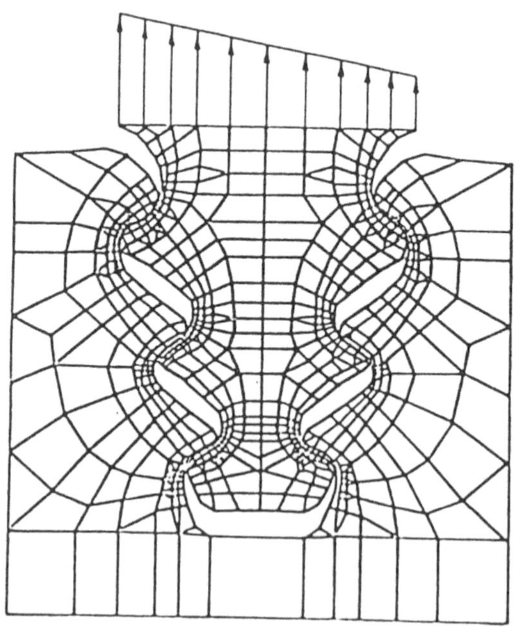

Figure 10 2D Root Attachment Model

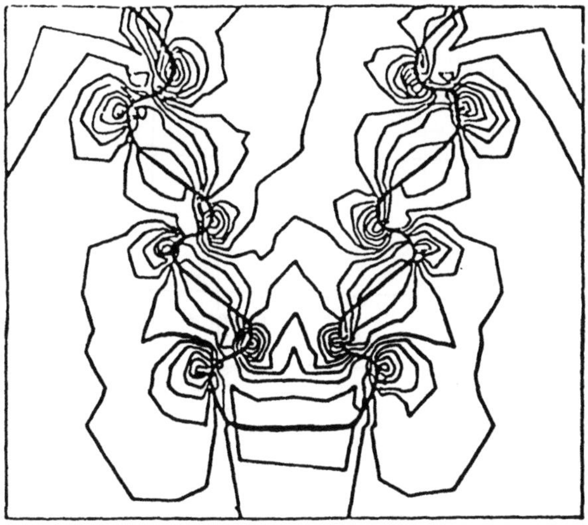

Figure 11 Stress Distribution at Attachment Area with Zero Tolerance Mismatch

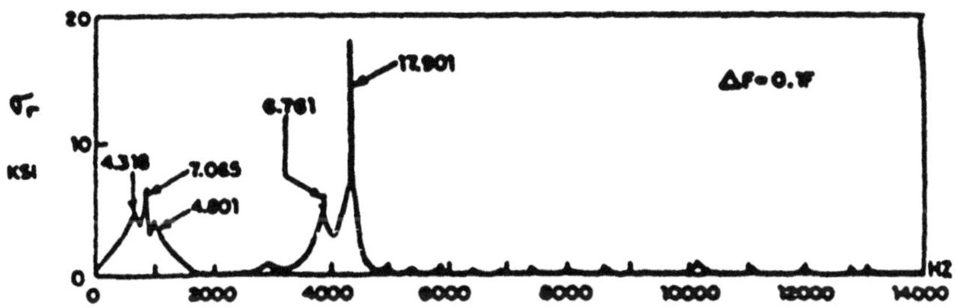

Figure 12 Response Spectrum: Stress Amplitude vs. Frequency

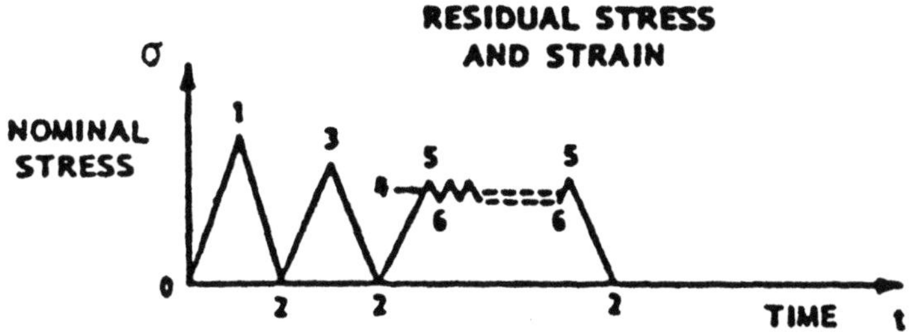

Figure 13 Applied Load History for Turbine Blade

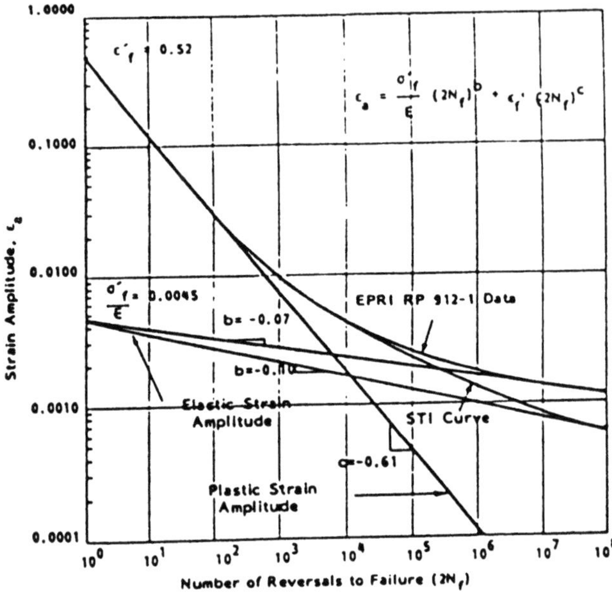

Figure 14 Strain Life vs. Number of Cycles for ASM Stainless Steel

Table 1 Material Properties for AISI 4340 Steel

Modulus of elasticity, E	**29,500 ksi**
Cyclic strength coefficient, K'	**167 ksi**
Cyclic strain hardening exponent, n'	**0.100**
Fatigue ductility coefficient, ε_f'	**1.060**
Fatigue strength coefficient, σ_f'	**168 ksi**
Fatigue strength exponent, b	**-0.075**
Fatigue ductility exponent, c	**-0.75**

Table 2 Computed Local Stresses and Strains

Point on Load History Diagram	Local Stress, σ (ksi)	Local Strain ε(in./in.)
a	0.0	0.0
b	123.74	0.01096
c,g	-69.18	0.00393
d	105.6	0.01297
e	61.1	0.01229
f	81.1	0.01161

Table 3 Computed Fatigue Damage Data

Type of Loading	Total Strain Amplitude, ε_a (in./in.)	Mean Stress, σ_o (ksi)	Fatigue Damage/Cycle, $1/N_{f(i)}$	Number of Cycles/Yr. n_i	Damage/Yr. $n_i/N_{f(i)}$
Low Cycle	0.004520	18.3	4.153×10^{-4}	260	0.10798
High Cycle	0.000339	71.1	1.400×10^{-13}	7.488×10^{9}	1.0484×10^{-3}

Table 4 Maximum Stresses in Blade/Disk Attachment Area

MAXIMUM STRESSES IN BLADE AND DISK ATTACHMENTS

Location	Case	Blade	Disk
Upper hook	Case 1 Case 2	138800 psi 140240 psi	149470 psi 156890 psi
Middle hook	Case 1 Case 2	144730 psi 112250 psi	161280 psi 132640 psi
Lower hook	Case 1 Case 2	126340 psi 103770 psi	158490 psi 133340 psi

Table 5 Blade Group Natural Frequencies

MODEL	MODE	STI	NASTRAN
6 BLADES	A1	863	759
	R1	990	888
	T1	1052	919
	U	1644	---
	S	3119	---
12 BLADES	A1	850	763
	R1	901	816
	T1	1002	922
	U	1044	931
	S	1320	1249

Table 6 Detuned Dynamic Stresses in the Leading Blade Upper Hook Notch Region

DYNAMIC STRESSES PSI

	A1	R1	T1	A3	T22	TOTAL, rms
CASE 1 SPECTRUM	503	857	613	1980	2417	3335
CASE 2 SPECTRUM	195	335	242	1754	3866	4270

Table 7 Material Properties of ASM 403 Stainless Steel

	Test Data Morrow [10] 635°F
Elastic Modulus, E, ksi	28,000
Fatigue Strength Coefficient Exponent, σ'_f, ksi	131.0
Fatigue Strength Exponent, b	-.083
Fatigue Ductility Coefficient ϵ'_f	0.381
Fatigue Ductility Exponent, c	-.58
Cyclic Strength Coefficient, K', ksi	151.0
Cyclic Strain Hardening Exponent n'	0.143

Table 8 Fatigue Damage Evaluation for Six-Blade Group

Mode		A1	R1	T1	A3	T22
Natural Frequencies (Hz)		863	990	1052	3849	4481
Load Case I	Stress Amplitude (ksi)	3.45	5.65	3.70	2.37	5.39
	High Cycle Fatigue Damage $1/Y_f$	1.21×10^{-3}	1.93×10^{-1}	2.97×10^{-3}	1.26×10^{-4}	5.45×10^{-1}
	Years to Failure Y_f	825	5.19	337	7.91×10^{3}	1.84
Load Case II	Stress Amplitude (ksi)	1.29	2.12	1.44	2.03	5.44
	High Cycle Fatigue Damage $1/Y_f$	6.47×10^{-8}	1.06×10^{-5}	2.37×10^{-7}	2.69×10^{-5}	5.97×10^{-1}
	Years to Failure Y_f	1.55×10^{7}	9.38×10^{4}	4.22×10^{6}	3.72×10^{3}	1.67

CHAPTER 11.5

DAMPING PROPERTIES OF STEAM TURBINE BLADES

N.F. Rieger
Stress Technology Incorporated, Rochester, New York, USA

ABSTRACT

This section describes a program of tests to examine the damping properties of several types of steam turbine blades. These studies were conducted in a non-rotating damping test rig which used real blades mounted in corresponding real disk root attachments. Simulated centrifugal loading was applied to the blades as a lengthwise axial pull. The rig and the test program are described. The damping properties of the blades studied were found to conform primarily to effects associated with material hysteresis damping. The influence of attachment friction appears to be negligible for test loads and vibratory amplitudes encountered in practice. A small level of fluid damping from the gas stream appears to be possible where the blades are subject to working pressure loads.

11.5.1 Introduction

This section describes a series of experiments which were undertaken to determine the nature and magnitude of damping in steam turbine blades. The test program included high pressure blades, intermediate pressure blades, and low pressure blades. The program was undertaken using a special type of test rig, Figure 1, which simulated the blade loading under rotating conditions, by the application of a known preload. A statistically significant number of identical blades were involved in each test. This gave some indication of the amount of test scatter which could result in the damping data under controlled test conditions from differences between blades and their assembly conditions within their root attachments. The tests examined the influence of blade type, attachment root type, surrounding environment (air, steam, vacuum), axial load, mode of vibration, and other factors.

The objectives of this study were:

a) To study the damping properties of several types of steam turbine blades in actual root attachments under simulated operating conditions.

b) To examine the nature of the damping which is evident in the test results.

c) To evaluate the magnitude of the damping from the decay traces for a variety of blade types, root attachment designs, blade lengths, and operating conditions.

d) To determine the relative contributions of the various damping mechanisms to the total damping under these conditions.

e) To evaluate the significance of applied root load, vibration amplitude, root type, blade length, and gas environment as possible contributing parameters in each case.

Details of the test rig used, and of the test conditions and results achieved are given in the following sections.

11.5.2 Previous Blade Damping Studies

There is surprisingly little information in the open literature on the damping properties of steam turbine blades. Blade manufacturers have developed extensive test data relating to materials performance, and to blades and blade group performance from vibration tests.

Grady [1] tested several dummy intermediate pressure blades in disk attachments using a pull-test machine, and used a dynamic shaker to excite the blades to specified force levels. Attachment pre-stress conditions were changed by varying the root interference fits. It was shown that an optimum root flexibility condition could be developed which would result in dynamic stresses about 1/3 to 1/2 of those in a rigid (welded) joint. Where root stiffness is high, less damping is developed. Considerable variability existed in this optimum root flexibility condition, which was found to depend on blade assembly procedures and on blade vibration history.

Grady [2] performed tuning fork vibration tests in air to measure blade material damping under simulated centrifugal load conditions in a tensile test machine. Data was obtained at zero centrifugal load through a range of dynamic stress to

30,000 psi. The results showed that the material log. dec. values were highest with high dynamic stresses. A maximum test log. dec. value of 0.033 was obtained. Damping decreased with increase in centrifugal load, and increased with frequency of excitation.

Wagner [3] conducted a program of damping tests on rotating steam turbine blade groups in a test turbine. The magnitude of blade damping, the influence of root details and of blade sizes on damping, and the effectiveness of certain damping devices were studied. The test blade groups were rapidly rotated through axially directed water jets which impulsively excited them into vibratory motion. The blade response was obtained with weldable strain gages and a slip ring assembly. Substantial scatter was observed in the log. dec. results. The damping for impulse blade groups was found to be greater than for long reaction type blades. Slightly higher damping with higher stress values was observed. In one test the damping remained substantially constant over 84,000 cycles at a maximum stress value of 30,000 psi. Undershroud welding caused an insignificant reduction in damping.

Rieger and Beck [4] found similar results to these of Grady and Wagner. Their tests were performed on blade pairs inserted in disk segments, with an axial pre-load applied during testing. Log. dec. values increased in a linear manner with increase in vibratory amplitude, and decreased with increase in root load. Values of log. dec. within the same general range as those obtained by Grady and Wagner were obtained (δ = 0.01 to 0.04) for similar blades, at similar vibration amplitudes. Rieger and Beck observed higher log. dec. values in axial vibration (δ = 0.02 to 0.08) than in tangential modes.

Brown [5] tested a blade row in a rotating test apparatus to determine the blade row damping values. A steam jet was used for excitation, and both impulse and harmonic excitations were used to excite the blades. The damping results were presented as a ratio between the log. dec. decay test results and the resonance bandwidth test results. No absolute data are given. Brown reported difficulty extracting the damping results due to beating, multiple resonances, etc. Impulse excitation was observed to initiate several modes simultaneously.

Gotoda [6] studied the influence of damping on steam turbine blade vibrations. He identified damping contributions from the blade material, from structural effects, and from aerodynamic damping. Gotoda used the resonance decay method to obtain data, and showed that the log. dec. increased in proportion to dynamic stress. Tests were next conducted on tangential entry tee head steam turbine blades, tightly fitted as single blades into the disk root groove. Higher damping values were obtained in this case than in the tuning fork tests. Damping was again found to vary linearly with dynamic stress, and the increase in such values over the material test results was attributed to root friction. The scatter in results was attributed to individual differences between blades. At high stress this scatter was quite large, and the log. dec. vs. dynamic stress relationship was no longer linear, above σ = 12.0 Kg/mm^2. Calculations by Gotoda demonstrated that the value of stress index was 3, for the blade material used. Further data was given for the variation of log. dec. with stress for simple blades, blade groups, and tuning forks. The form of the damping vs. stress relation was linear for the blades tested, except as noted above.

An early study of enhanced damping in hollow blades was made by DiTaranto [7] who tested blades containing a core of 0.005 inch diameter radial wires. The damping in a stationary blade with wires pre-stressed to 7,000 psi was an order of magnitude higher than that of the stationary hollow blade without wires for equal excitations. The damping in a blade rotating at 8,000 rpm was further found to be twice that of the stationary blade with wires stressed to 7,000 psi. DiTaranto discussed the following methods for enhanced damping: (a) insertion of particles and/or wires of different sizes into hollow blades, (b) use of two flat cantilevered plates under normal force, rubbing together as they vibrated, (c) root damping of the vibrating blade, with normal force being the critical factor. Too great a force was found to prevent relative blade root motion: (d) tuned dampers, which were effective in narrow frequency bands.

Srinivasan, Cutts, and Sridhar [8] studied compressor fan blade damping. Root damping was found to be negligible at operating speeds for the dovetail root attachment structure. The material damping in the titanium alloy blades used was also found to be very small. The shroud interface contacts were found to be the most useful source of damping in such fan blade structures, but the authors indicate that damping from shroud or platform rubbing may be difficult to define for consistent results. Material damping log. dec. values for titanium blades were found to range from 0.0006 to 0.0015.

Jones and Muszynska [9] investigated the influence of slip between components in the blade platform region. The purpose of this study was to determine the extent to which damping in a blade geometry could be optimized by allowing contact between the blade platform and the disk to occur at high speed. Approximate non-linear equations of motion were derived and solved using a harmonic balance procedure. It was found that high levels of slip damping could be achieved if the relative component stiffnesses were properly selected. This paper does not contain any significant experimental results, but testing is proposed together with a spin pit investigation, to validate the influence of practical platform dampers.

Muszynska, Jones, Lagnese, and Whitford [10] analyzed the non-linear response of a set of compressor blades mounted on a rigid disk and interconnected with a dry friction coupling near each blade platform. Each blade consists of a discrete two-mass system connected to adjacent blades by a flexible link, with a dry friction contact. Effects from blade-to-blade dry friction contacts are also included. A single blade was initially investigated experimentally to obtain modal values for the blade models. Several system effects were examined, including blade mistuning and influences from the magnitude and distribution of exciting forces and phase differences between the exciting forces on adjacent blades. Conclusions concerning the influence of dry friction in reducing blade response amplitudes and on the influence of mistuning on the resulting blade amplitudes are given. The fundamental modes of the blades were accurately modeled and their parameters were precisely identified. An optimum value of the friction parameter was found to exist for which response amplitudes are at a minimum for all blades. The effect of blade mistuning was to introduce blade-to-blade coupling through the stiffness elements connecting the blades. This led to typical mistuned behavior in which several response peaks were observed, scattered on both sides of the tuned system resonant frequency.

Attachment Slip Studies:

Hanson [11] described a rig for testing gas turbine and compressor blade damping properties, which consisted of a disk carrying a single blade which could be rotated at speeds up to 15,000 rpm. The blade could be excited by an air jet or by the impact of a small ball. Blade decay rates following impact were observed. Data is given on root friction damping coefficient vs. rpm, and on material damping vs. rpm, for several types of blade roots. For these blades fir tree roots were found to give the highest rates of amplitude decay. Pin root and wedge root blades gave smaller damping values. In all instances the damping increased as rpm increased.

Hanson, Meyer, and Manson [12] investigated a proposed compressor blade friction damping device. Using the same rotating blade apparatus, they demonstrated a 3:1 decrease in blade vibration amplitudes when this device was used on compressor blades. Similar platform friction devices were recently developed and tested during studies of blade vibration associated with the space shuttle turbopump engines.

Beards [13] evaluated the likely effectiveness of root slip damping for suppressing vibrations of compressor blades based on an examination of the results obtained by Hanson [11], Hanson, Meyer, and Manson [12], and Goodman and Klumpp [14].

Jones and Muszynska [15] developed a simple two-mass analytical model to represent the vibrational behavior of a jet engine compressor blade in its fundamental mode, allowing for slip at the blade attachment interface. This work is an attempt to develop a theoretical basis for earlier work by Hanson, Meyer, and Manson [12], and others. Vibration tests were carried out to experimentally verify the analysis using a particular blade geometry. A blade with a simple root geometry was inserted in a heavy fixture containing a matching axial entry tee head root. The blade was loaded with radial forces to simulate the centrifugal load on the root during operation. Tests were conducted to determine the response amplitude at various frequencies corresponding to several centrifugal loads. The non-linear equations of motion with slip at the root were developed and solved by the method of matched parameters. Correlation between analysis and experiment showed that it is possible to model the blade response using a simple two-degree system with slip at the root level. The response curves obtained follow the response for a linear model until slip occurs at the loaded interface. Beyond this point the blade vibration amplitude ceased to grow, and remained constant over a substantial range of frequency.

Muszynska and Jones [16] investigated the dynamic response of blades on a rigid disk allowing for interface slip at the attachment interface and blade hysteretic damping. Two types of mass model of a single blade in a disk were used. The first model allowed for root slip and blade hysteresis, and the second model allowed a damper concept which incorporated platform slip to be studied. Results were compared with experimental response data for a single blade in a test fixture under low level harmonic excitation. It was demonstrated that the contribution of root slippage to blade damping falls off with increasing rotational speed, but this was compensated with a suitably sized damper device in the second model.

Material Damping:

Material damping has been studied by many investigators. This literature has been discussed comprehensively by Lazan [17]. Rowett [18] investigated the torsional damping properties of certain grades of steel shafting, and appears to have first suggested the stress damping law $D = J\sigma^n$. Kimball [19] obtained log. dec. damping data for several grades of carbon steels and alloy steels. Lazan studied the general nature of material damping, and presented results for almost 2000 materials in his treatise on this subject. Lazan's results are presented in terms of the above material damping law in chart form and in a data glossary, in which specific damping energy is related to dynamic stress. An application of Lazan's procedure to first mode vibration of a turbine blade is given.

Lazan's results again show that log. dec. values will increase with dynamic stress, i.e. with vibration amplitude, where material damping is the dominant mechanism. Similar results were obtained by Robertson and Yorgiadis [20] in an early study of energy dissipation in several metals under torsional and axial loadings.

11.5.3 Apparatus

Blade test rigs of the type shown in Figure 1 were used in this program. The principle of the rig is illustrated in Figure 2. Several rigs of different sizes were built to accommodate different blade lengths. Damping properties of steam turbine blades in disk segments with actual root attachments were measured. Vibration decay tests were conducted using blades which had been welded together in pairs at their tips: see Figure 3. The purpose of this blade pair arrangement was to carry the lengthwise tensile pre-load which was intentionally developed using relative thermal shrinkage (initially) and hydraulic load cells (later) between the blade pairs and the rig frame to simulate the blade centrifugal loading which is applied to the root attachments during operation. Damping data was obtained from the pre-loaded blade pairs tested in their lowest mode of vibration, in both the tangential and the axial directions. Details of vane lengths, root attachment types, location in rotor, etc., involved in the test program are given in Table 1.

The primary requirements for the blade test apparatus were: (a) to provide representative support conditions for the blade and root attachment structure during testing, (b) to apply a lengthwise load to the blade pairs during testing in a manner which simulates applied centrifugal loading on the blade, and (c) to isolate the blade pair from unwanted rig or apparatus natural frequencies. It was also required that the natural frequencies and mode shapes of the pre-loaded blade pairs should be reasonably close to those which occur in the grouped blades during operation in the turbine. This requirement was achieved by making a natural frequency calculation of the original blade, and by then adding a suitably proportioned flexure link between the blades, Figure 3. The flexure link transmits the end-applied blade tensile load. This arrangement typically allowed the tangential and axial natural frequencies to be approximated to within 15 percent. Some inaccuracies remain between the relative mode shapes of these two conditions in the blade tip flexure link region, but calculations showed that the mode shape details in the attachment region and in the lower portion of the vane are in good agreement.

A requirement of each test rig was that the natural frequencies and mode shapes of the rig should not influence, or interact significantly, with any mode of the blade pair. Prior to rig construction, a detailed finite element calculation was made to identify the first ten or more natural frequencies of the test rig, both with the blade pair inserted, and without blade pair inserted. Each rig was also tested by rapping, to locate actual natural frequencies. The test apparatus was finally modified as needed so that the required blade natural frequencies occurred in dynamically inert frequency regions ('dead zones') of the test apparatus.

The environmental chamber shown in Figure 4 was built to investigate the influence of the surrounding gas environment on the blade damping properties. Gasdynamic damping effects from surrounding air, from steam (at various pressures and temperatures), and in vacuum were tested using this chamber. The chamber completely enclosed the blade pair and disk attachment segments. The design and construction of the environmental chamber used for evaluation of the 25 inch vane, low pressure, curved axial entry blades involved the following:

a) A simulated low pressure stage environment between 0.8 and 1.5 inches of Hg and 100°F.

b) A steel tube which surrounded the blade pair and attachment, using a pair of flexible bellows at either end to accommodate rig thermal and other expansions, from the steam tests and applied pre-load.

The blade pairs were mounted in the rig in the usual manner and during testing were pre-loaded using hydraulic load cells. All blade and attachment components were contained within the environmental chamber during the test program.

The test arrangement consisted of five openings in the environmental test chamber: two for strikers to excite the blades, two for the transducers, and one for the vacuum draw-off connections. The location of the transducers and the strikers within the environmental test chamber is shown in Figure 4.

The amount of test data obtained from this program was substantial. Between ten and twenty-five blades of each type were tested, over a range of pre-load, initiating amplitude, environment, etc., conditions, and each test was repeated five to ten times, to ensure that representative statistical results had been obtained. In anticipation, a computer aided data acquisition system was developed and used to reduce the test data. This system is shown in Figure 5. A calibrated decay signal from the blade transducer was captured on a spectrum analyzer, visually reviewed, and then processed. Accepted traces were then transmitted by a data link to the memory of a minicomputer. This data was then processed and printed as a tabulation of log. dec. vs. decreasing vibration amplitude, with test details.

11.5.4 Experimental Damping Results

Seven blade configurations were tested to determine their damping properties. These included two high pressure blade designs, three intermediate blade designs and two low pressure blade designs. Statistical analyses were performed on the damping results for selected tests. The influence of root coatings, of root modifications, and of several operating environments on damping was also investigated, for low pressure blades. The results obtained are described in this section, and their significance is discussed in the following sections.

High Pressure Blade Tests:

5 cm Vane, Straddle Mounted, Tangential Entry. Figure 6:

Sixteen blade pairs were tested in air for damping in the tangential and axial directions, each under four centrifugal load conditions (1000, 2000, 3000 and 5000 lb.). The results, Figures 7 and 8, show that the log. dec. values increase with blade tip displacement in both axial and tangential modes. Damping is approximately linearly related to the vibration amplitude. Increasing the centrifugal load progressively decreased the damping value toward a lower limit value. This result has been further confirmed in subsequent tests.

10 cm Vane, Ball and Shank, Axial Entry Blade. Figure 9:

Damping values were obtained from the tests conducted in air on two sets of eleven pairs of ball and shank blades, in the tangential and axial modes. Both long shank and short shank blades with 4 inch vanes were tested. Typical results for log. dec. vs. amplitude are shown in Figure 10. Damping increased with increased blade tip displacement in the tangential mode, for both long and short shank blades. For long shank blades, the axial mode damping increased only slightly with increased load. For short shank blades, axial damping increased by an order to magnitude over axial damping of the long shank blades.

Comments on High Pressure Blade Results:

High pressure blade damping appears to be related to vibration amplitude in a linear manner, and inversely related to centrifugal load.

Intermediate Pressure Blade Tests:

15 cm Vane Fir Tree Root, Axial Entry. Figure 3:

Damping data was obtained for twenty-three blade pairs tested in air. Results for log.dec. values are shown in Figure 11 for the tangential and axial directions, as a function of blade tip displacement and centrifugal load. Damping increased linearly with increased tip displacement, and decreased with increased centrifugal load.

23 cm Vane, Fir Tree Root, Axial Entry Blade:

Twelve blade pairs were tested in air. Tests were conducted in both the tangential and axial modes. Damping values increased with blade tip displacement, and decreased with increased centrifugal load.

Comments on Intermediate Pressure Blade Results:

Damping results for IP blades with fir tree roots again indicate the trends observed with the high pressure blades, i.e., damping increased with tip displacement, and decreased with increase in centrifugal load toward a limit value.

Low Pressure Blades:

70 cm Vane, Fir Tree Root, Curved Axial Entry Blade. Figure 12:

Twenty-two pairs of blades were tested in their disk root sections, for seven centrifugal load conditions. Trends for both tangential and axial modes are shown in Figure 13. The curve showed a small negative hook followed by a shallow curve of positive slope. The spread of damping values for centrifugal loads of 5000, 9000, and 13,4000 lbs. were statistically analyzed.

Comments on Low Pressure Blade Results:

Tests on the curved axial entry blades showed some interesting features. Initially, the blade damping decreased to a minimum value, from which it increased with centrifugal load. These blades were much longer, the attachment was curved, and it was oriented across the disk rim at an angle. These results appear to be some combination of effects seen in tangential and axial modes of previous tests. The two-part curve may be an expanded version of the characteristic seen for stiffer blades, e.g., Figure 11. Damping increases with larger tip amplitudes of vibration. The LP vane section is generally more flexible than other blade types, and so the damping which results from flexure occurs mostly toward the vane tip.

Environmental Effects: Low Pressure Blades. Figure 14:

Ten pairs of 25 inch low pressure blades were tested in air, in vacuum, and in steam. Statistical data was compiled from the multiple testing performed on each pair. Negligible differences were detected between the mean damping values obtained for the three environments.

Root Coating Effects. Figure 15:

The results from the root coating test on 25 inch low pressure blades indicate that attachment coatings consisting of thin films of structural adhesives on the hook faces have only a small effect on the overall blade damping in the normal operating range.

Influence of Root Modifications. Figure 16:

Six 25 inch vane low pressure blades with the following relieved hook conditions were tested in their disk attachments. Two sets of conditions applied: (a) the two upper hooks were relieved (reduced contact loads) and the four lower hooks were therefore equally loaded, and (b) the two middle hooks were relieved, and the upper and lower hooks were equally loaded. Little difference was found between results obtained from these two test conditions. Under circumstances where the relieved upper hooks could slide, case (b), the damping values were higher. This condition is not expected to occur under normal conditions.

11.5.5 Discussion of Test Results

The blade types and test conditions involved in this program are listed in Tables 1 and 2. The high pressure blades showed tangential log. dec. values ranging between 0.012 and 0.18, and axial log. dec. values between 0.024 and 0.21.

Corresponding values for the IP blades range from 0.008 to 0.09 in the tangential direction, and from 0.015 to 0.15 in the axial direction. Log. dec. values for the low pressure blades range from 0.005 to 0.030 in the tangential direction, and from 0.005 to 0.035 in the axial direction. The range of variation of log. dec. in each instance is due to variation of dynamic amplitude, i.e., to dynamic stress range values, and to test conditions, in particular the applied pre-load values.

Damping data for short high pressure blades are shown in Figures 7 and 8, for a tangential entry root, and in Figure 10 for a ball type axial entry root. The general form of the curve in each instance is similar. For tangential entry blades, log. dec. increases with blade tip displacement in both the tangential and axial directions. Centrifugal load tends to decrease the damping for higher centrifugal load values. This trend is more pronounced for tangential vibrations than for axial vibrations. The ball type roots repeat these trends, but for the short shank blades the axial vibration damping is almost constant with blade tip displacement and centrifugal force.

Results for 6 inch blades with axial entry fir tree roots are shown in Figure 11, and these again show an increase in log. dec. with blade tip displacement. Data is given for tangential vibrations and for axial vibrations, in air. The damping relationship is seen to be almost linear in these instances, though the vibration test amplitudes were relatively small. Later tests established that damping continues to increase with higher vibration amplitudes. Increased centrifugal load causes the log. dec. values to decrease, as was found by Lazan [17] for damping which is related to material hysteresis.

Results from tests on 25 inch blades with curved axial entry attachments under a variety of conditions are shown in Figure 13. Log. dec. vs. blade tip displacement is shown for several pre-load values, for axial vibrations, and for tangential vibrations. In this instance, a higher mean stress appears to increase the blade damping, at loading values beyond a certain lower limit. The reason for the difference between these results and, for example, those in Figures 7 and 11 is thought to lie in the interaction between the local pre-load and the dynamic stress. This causes local plastic deformations to occur at higher pre-loads, which would give higher material damping values. Higher amplitudes of vibration were achieved for the tangential vibrations than for the axial vibrations.

Figure 14 shows the influence of gaseous environment on the test results, for tests involving new roots (assembled for the first time), and old roots (which had been used in previous tests). Generally speaking there is little difference between the values obtained under vacuum and in air. Damping data with used roots in air gave somewhat lower values than air tests with new roots. These tests were conducted using a 25 inch blade loaded lengthwise with a force of 13,400 lbs. which is smaller than typical operating attachment loads carried by such blades.

Figure 15 shows the influence of root coatings on log. dec. for various amplitudes of vibration. Results for uncoated roots were compared with those obtained from roots coated with epoxy, and with loctite. It was found that the epoxy coatings would smear throughout the attachment upon insertion, and were only present over portions of the root surface when the blades were removed. The influence of loctite appeared to decrease the overall damping by a small amount, probably because of its effectiveness in securing the blade into the disk attachment.

Figure 16 demonstrates the effect of severe mal-distribution of root tolerances in LP blade attachments. For similar applied loads, relief of the middle hook contact was found to increase the damping more than for relief of the upper hook. This is because the middle hook relief applied a greater load to the upper hook, where damping and movement is most likely to occur. The test data is consistent for loads between 5,000 lbs. and 13,400 lbs. These results may be compared with similar data for unmodified roots in Figures 13 and 14.

Variation of log. dec. with centrifugal load increase is shown in Figure 17. The vertical bars show in the data scatter limits for one standard deviation. This data applies to vibration amplitudes up to 3.7×10^{-3} inches in the tangential direction, and to 0.7×10^{-3} inches in the axial direction. It is seen that the log. dec. is almost constant for centrifugal load variation in the axial direction, but it decreases with load increase for vibrations in the tangential direction. These results are again in keeping with results obtained by Lazan [17] for material hysteresis.

Figure 18 shows additional damping data for a 30 inch LP blade, where log. dec. is plotted against vibratory stress. The blade pairs used had curved axial entry roots. Dynamic stress values were obtained with a strain gage attached on the vane near the root. It is seen that log. dec. decreases in the lower stress region, and then increases as the dynamic stress increases further. This is again due to the complex interaction between the dynamic stress, the mean stress under test conditions, and the material properties, as mentioned above.

11.5.6 Conclusions

- o Blade damping increased with increase of vibration amplitude in most instances, for both tangential vibrations and axial vibrations. This trend was most evident for HP and IP blades.

- o Damping in low pressure blades increased somewhat with increase in vibration amplitude, but at a lower rate than in HP and IP blades. Average log. dec. values were moderately constant with vibration amplitude.

- o Damping values decrease toward a constant limiting value with increased attachment centrifugal load effect. The limit appears to be the rigid attachment condition.

- o No significant changes in damping values were observed with LP blades where root coatings were applied.

- o No change in damping was found from tests involving the presence of stationary air, steam, or vacuum surrounding the blades. Gasdynamic damping evidently derives from doing work on the local pressure distribution. This only becomes significant with a moving gas stream under operating conditions.

11.5.7 References

1) Grady, R. F., 'Investigation of Dovetail Damping Contribution of Propulsion Steam Turbine Buckets,' General Electric Company Report NOB594390, Submitted to Bureau of Ships, Department of the Navy, November 1967.

2) Grady, R. F., 'Investigation of Material Damping Properties of Propulsion Turbine Blade Material,' General Electric Company Report NOB5-94390, Submitted to Bureau of Ships, Department of Navy, December 1967.

3) Wagner, J. T., 'Blade Damping Tests,' Westinghouse Engineering Report EC-401, NOBSN00024-67-C-5494, May 1969.

4) Rieger, N. F., Beck, C. M., 'Damping Tests on Steam Turbine Blades,'EPRI Project RP-1185-1, Palo Alto, California, 1980.

5) Brown, W. G., 'Determination of Damping Value.for Turbine Blades,' ASME Design Engineering Technical Conference, 81-DET-131, Hartford, 1981.

6) Gotoda, Hidemi, 'An Analysis on Resonant Stresses in Steam Turbine Blades,' Technical Designing Department, Kawasaki Heavy Industries, Ltd., Kobe, Japan, 1974.

7) DiTaranto, R. A., 'Blade Vibration Damping Device,' Journal of Applied Mechanics, Trans. ASME, Vol. 80, pp. 21-27, 1958.

8) Srinivasan, A. V., Cutts, D. G., Sridhar, S., 'Turbojet Engine Blade Damping,' United Technologies Report No. R81-91441031, Submitted to NASA (Report No. CR-165406). July 1981.

9) Jones, D.I.G., Muszynska, A. 'Design of Turbine Blades for Effective [Slip Damping,' The Shock and Vibration Bulletin, Part 2, Washington, D.C., September 1979.

10) Muszynska, A., Jones, D.I.G., Lagnese, T., and Whitford, L., 'On Non-linear Response of Multiple Blade Systems,' Shock and Vibration Symposium, San Diego, California, October 1980.

11) Hanson, M. P., 'A Vibration Damper for Axial Flow Compressor Blading,' Proceedings Society Experimental Stress Analysis XIV, pp. 155-162, 1955.

12) Hanson, M. P., Meyer, A. J., and Manson, S. S. 'A Method for Evaluating Loose Blade Mountings as a Means of Suppressing Turbine and Compressor Blade Vibrations,' Proceedings Society Experimental Stress Analysis, 10 (2) pp. 103-116, 1953.

13) Beards, J. F., 'Damping in Structural Joints,' Shock Vibration Digest, (8). pp. 35-41, 1979.

14) Goodman, L. E., Klumpp, J. H., 'Analysis of Slip Damping with Reference to Turbine Blade Vibration,' Journal of Applied Mechanics, Trans. ASME, 23, (3), pp. 421-429, September 1956.

15) Jones, D.I.G., Muszynska, A., 'Vibrations of a Compressor Blade with Slip at the Root,' The Shock and Vibration Bulletin, Washington, D.C., September 1978.

16) Muszynska, A., Jones, D.I.G., 'On Discrete Modellisation of Response of Blades with Slip and Hysteretic Damping,' Proc. of the Fifth World Congress on Theory of Machines and Mechanisms, 1979.

17) Lazan, B. J., 'Damping of Materials and Members in Structural Mechanics,' Pergammon Press, Inc., New York, 1968.

18) Rowett, F. E., 'Elastic Hysteresis in Steel,' Proceedings, Royal Society of London, Vol. 89, London, 1914.

19) Kimball, A. L., 'Vibration Problems Part 5: Friction and Damping in Vibrations,' Trans. ASME, 63, pp. A-135 A-140, 1941.

20) Robertson, J. M., Yorgiadis, A. L., 'Internal Friction in Engineering Materials,' Journal of Applied Mechanics, Vol. 13, No. 3, p. A13182, September 1946.

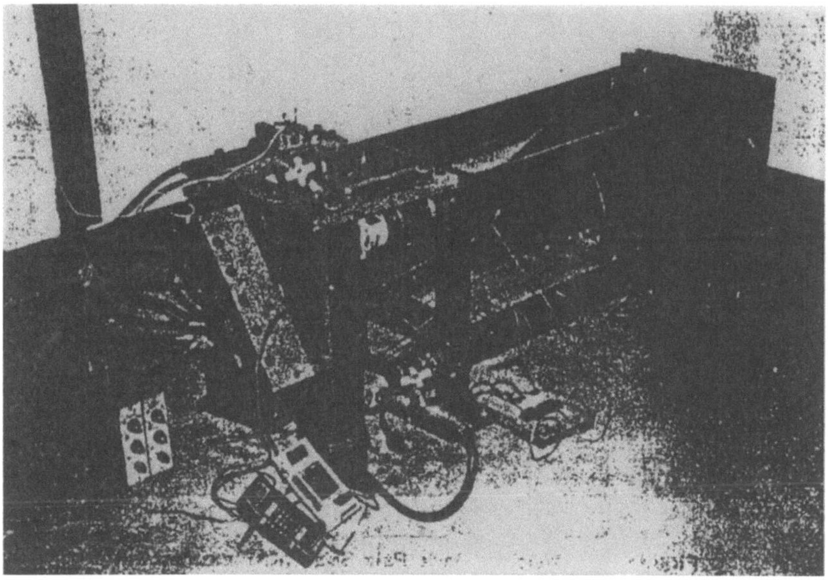

Figure 1 Blade Damping Test Apparatus

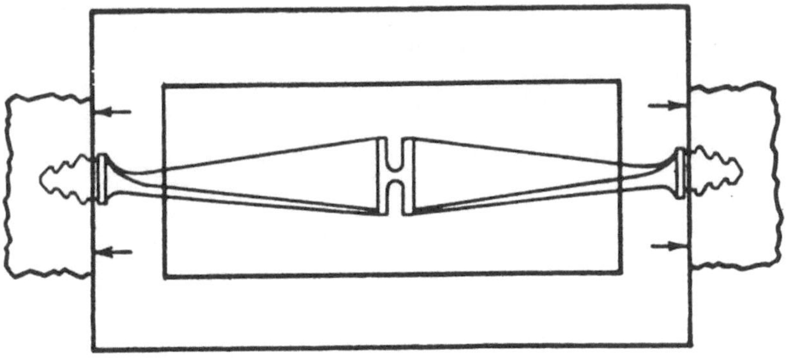

Figure 2 Principle of Test Rig

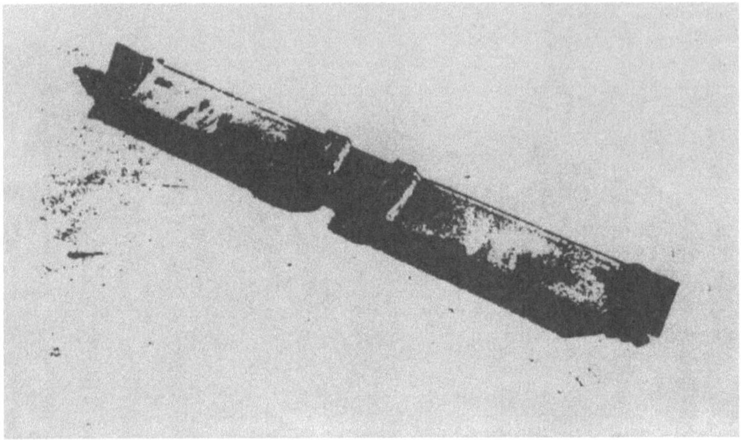

Figure 3 Welded Blade Pair Showing Flexure Link

Figure 4 Test Rig with Environmental Chamber

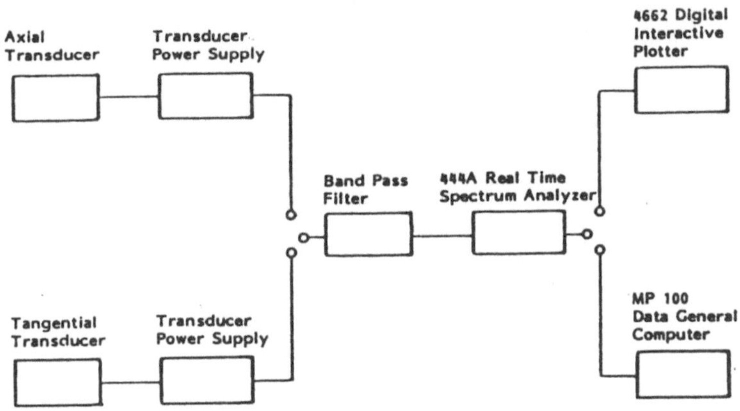

Figure 5 Computer Aided Data Acquisition System

Figure 6 2" Straddle Mounted Tangential Entry High Pressure Blade Pair

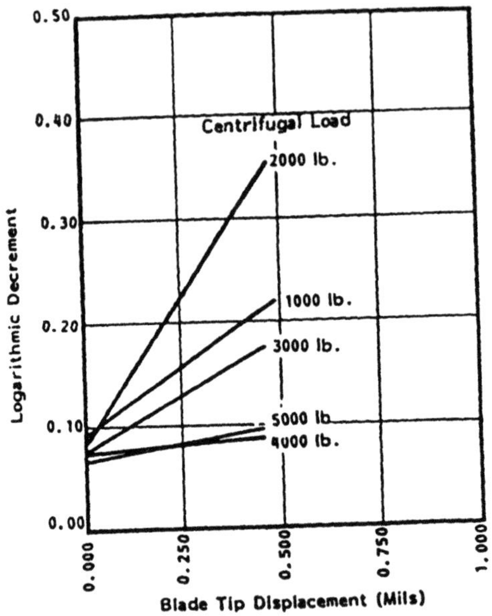

Figure 7 Logarithmic Decrement vs. Tip Displacements and Centrifugal Load. Tangential Mode 2" Straddle Mount. Tangential Entry High Pressure Blade

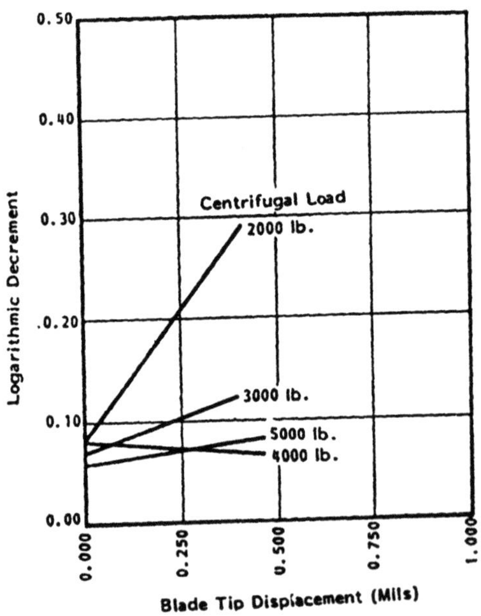

Figure 8 Logarithmic Decrement vs. Tip Displacements and Centrifugal Load (Axial). 2" Straddle Mount. Tangential Entry High Pressure Blade

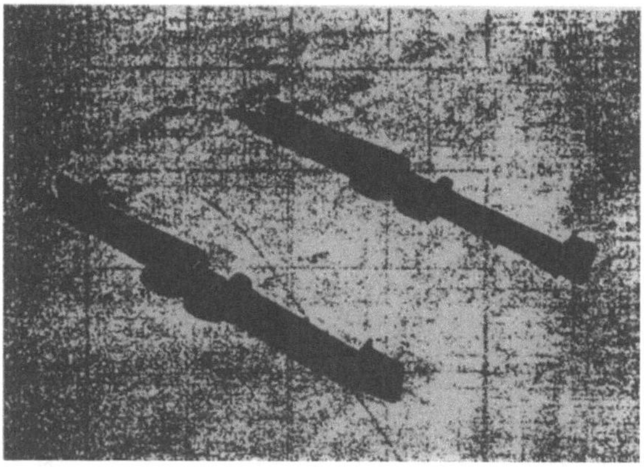

Figure 9 4" Ball and Shank - Axial Entry High Pressure Blade Pairs

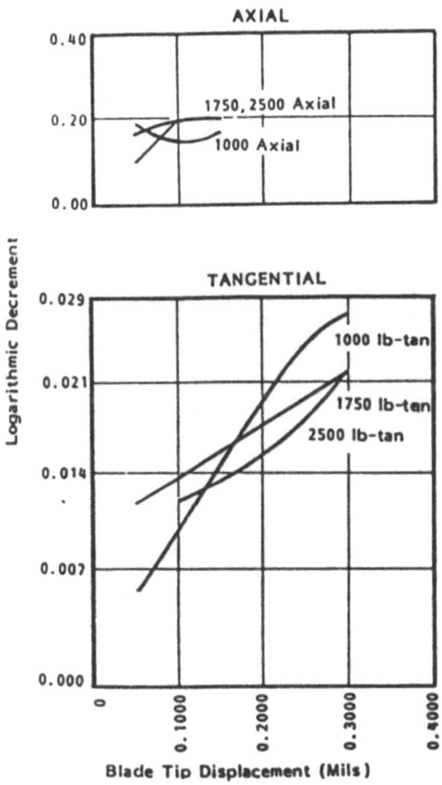

Figure 10 Logarithmic Decrement vs. Blade Tip Displacement and Centrifugal Load - Short Shank 4" High Pressure Blade Pairs

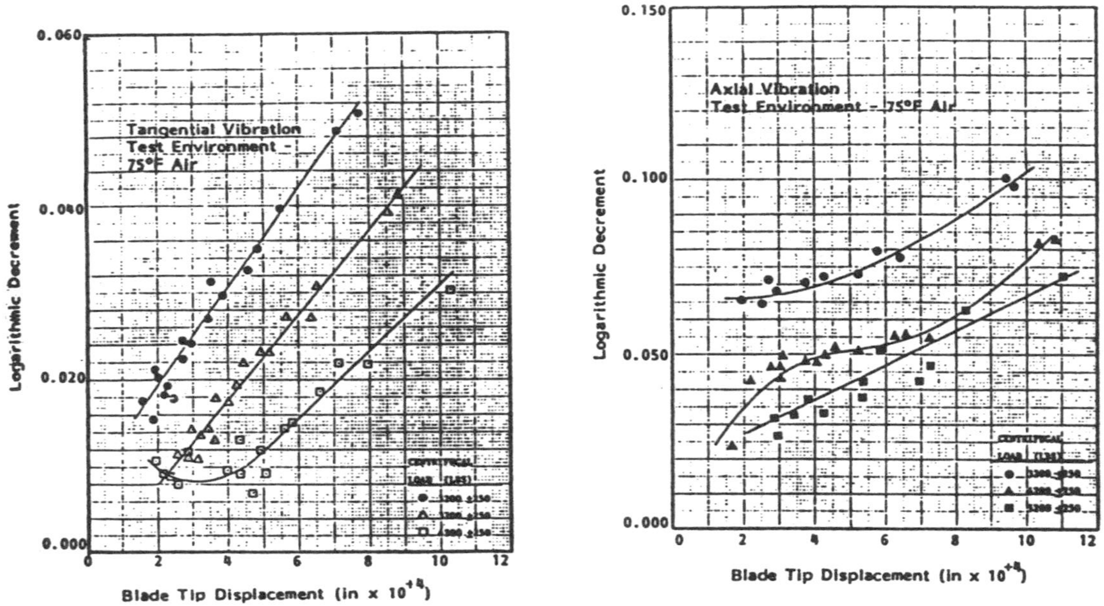

Figure 11 Logarithmic Decrement vs. Blade Tip Displacement and Centrifugal Load - 6" Axial Entry Fir Tree Root, Intermediate Pressure Blade

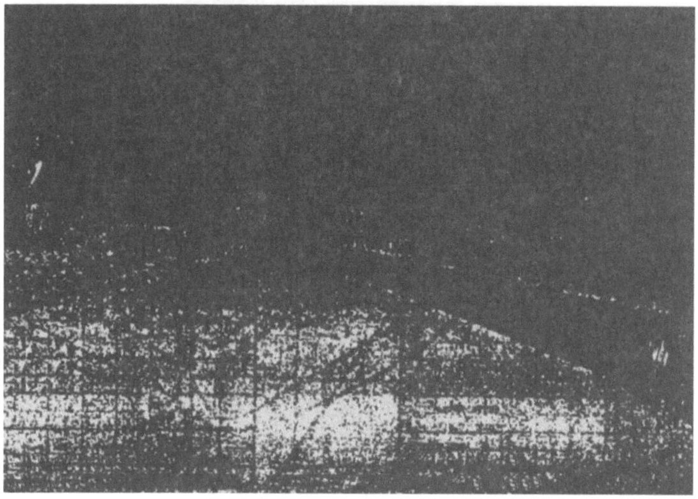

Figure 12 25" Curved Axial Entry Fir Tree Root - Low Pressure Blade Pair

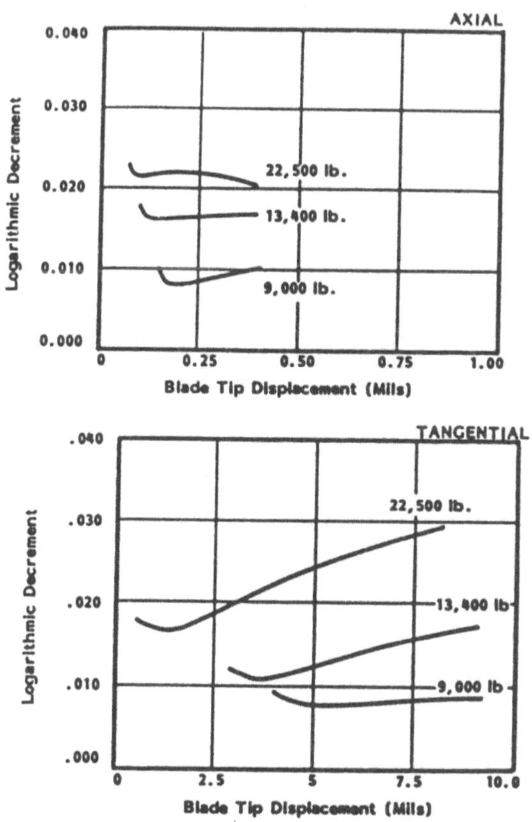

Figure 13 Logarithmic Decrement vs. Blade Tip Displacement and Centrifugal Load - 25" Fir Tree Root - Curved Axial Entry - Low Pressure Blade

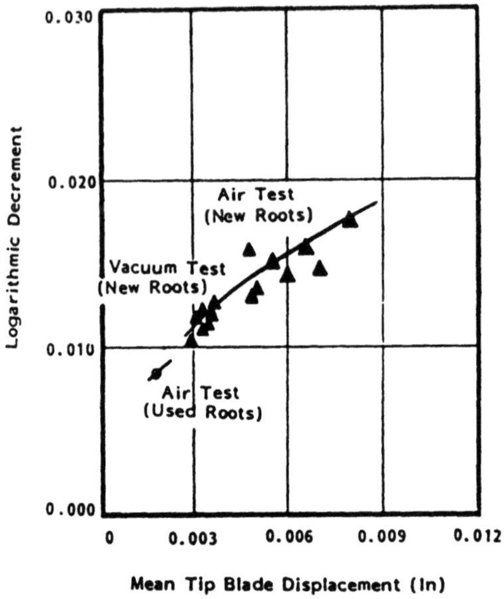

Figure 14 Environmental Effects - Logarithmic Decrement vs. Tip Displacement - Tangential Air/Vacuum/Steam - 9000 lbs.- 25" Curved Axial Entry - Fir Tree LP Blade

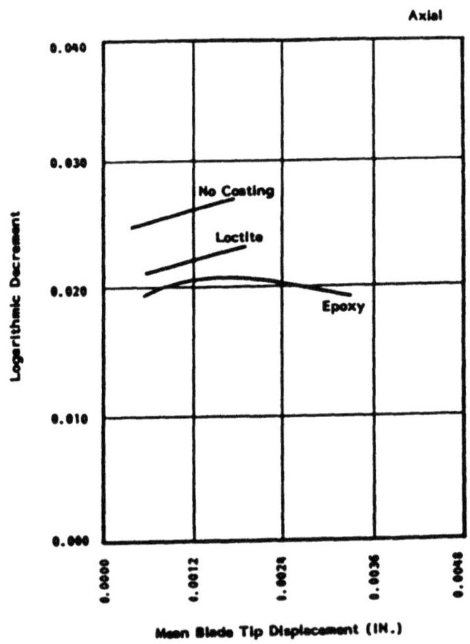

Figure 15 Root Coatings Comparison - Logarithmic Decrement vs. Tip Displacement - Axial - 13,400 lbs. - 25" Curved Axial Entry - Fir Tree LP Blade

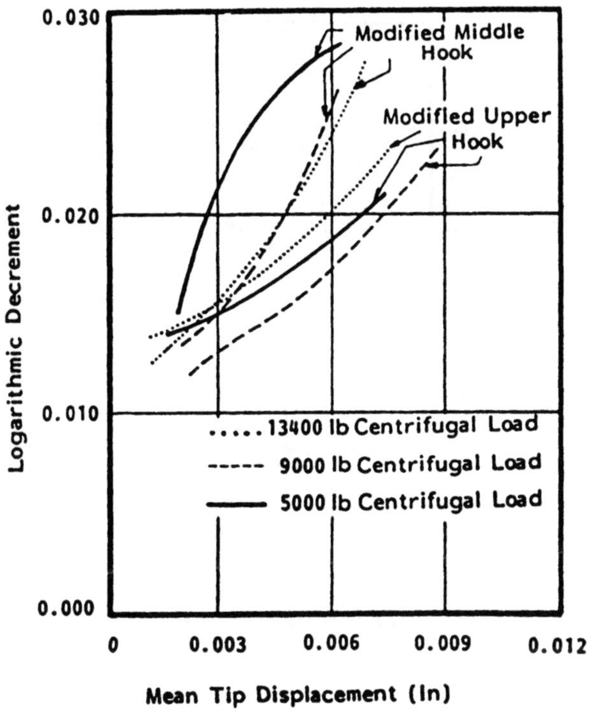

Figure 16 Modified Root Load Study - Logarithmic Decrement vs. Tip Displacement Tangential 25" Curved Axial Entry - Fir Tree Root - Low Pressure Blade

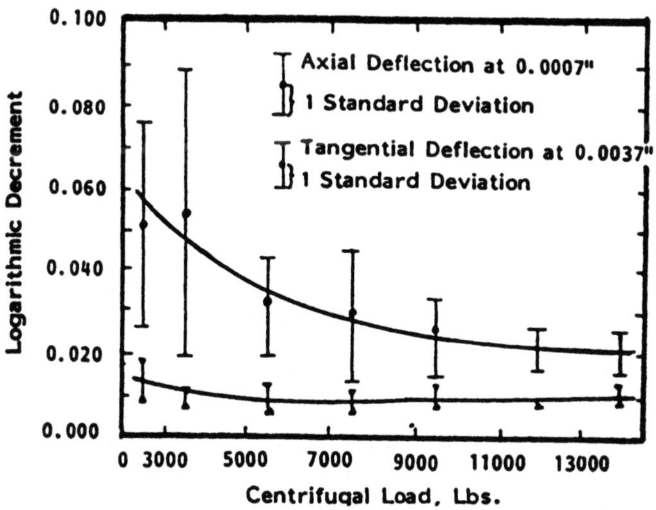

Figure 17 Variation of Damping with Preload Showing Limits of the Scatter. 25" LP Curved Axial Fir Tree Entry Blade Pair. Specified Vibratory Amplitudes.

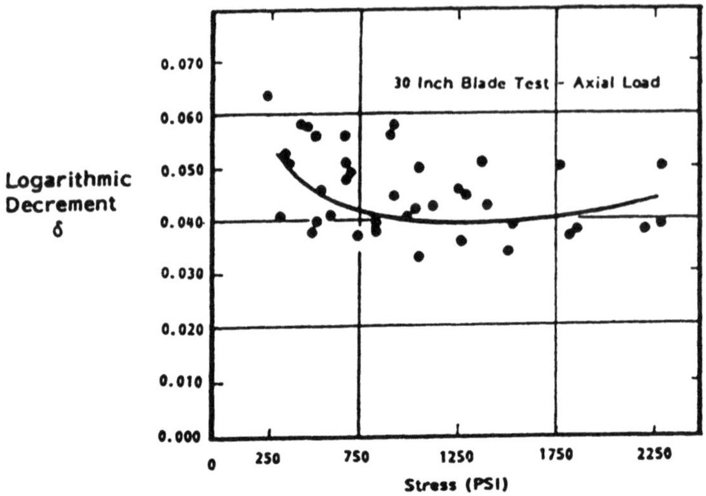

Figure 18 Variation of Damping with Vibratory Stress. 30" LP Blade Pair - Axial Vibration - Curved Axial Entry Root - Preload 22,500 lbs.

Table 1 Details of Turbine Blades and Root Attachments Tested

ROOT TYPE	VANE LENGTH	ENTRY	STAGE
Straddle Mount	2.00	Tangential	First
Ball and Shank	3.50	Axial	L-3
Fir Tree	6.00	Axial	L-5
Fir Tree	9.50	Axial	L-1
Fir Tree	25.00	Curved Axial	Last

Table 2 Range of Logarithmic Decrement Values

Stage	Vane Size	Root Type	Range of Axial Log. Dec.	Range of Tangential Log. Dec.	Comment
HP	2"	Tangential straddle mount	0.05–0.12	0.07–0.18	Air environment
HP	4"	Ball & shank short	0.12–0.21	0.012–0.021	Air environment
HP	4"	Ball & shank long	0.024–0.041	0.012–0.024	Air environment
IP	6"	Axial fir tree	0.015–0.035	0.008–0.055	Air environment
IP	9.5"	Axial fir tree	0.015–0.15	0.015–0.090	Air environment
LP	25"	Curved axial fir tree	0.005–0.035	0.008–0.030	Air environment
LP	25"	Cuved axial entry	0.010–0.020	0.008–0.010	Air/Vaccum/Steam
LP	25"	Curved axial entry	0.020–0.024	0.005–0.010	Loctite & epoxy coatings
LP	25"	Curved axial entry	0.020–0.025	0.020–0.030	Adjacent root study
LP	25"	Curved axial	0.025–0.030	0.015–0.028	Modified root study

CHAPTER 12.1

MAGNETIC BEARINGS

G. Schweitzer
Institute for Mechanics, ETH Zurich, Switzerland

ABSTRACT

Magnetic bearings have some distinct advantages. They do not generate wear and they do not need lubrication. These features make them attractive for vacuum applications. And their dynamic behaviour can be adjusted in a wide range, which allows active vibration damping and control.

This chapter presents the state of the art for the design of an electromagnetic bearing system. It introduces first the main elements and then discusses control and system aspects. Models for describing an elastic rotor and its active vibration control are included. The characteristics and the losses of such a suspension system are detailed.

Several applications are demonstrated, and future trends are indicated.

CONTENTS

1. INTRODUCTION

Magnetic bearings can support a rotor in such a way that it levitates freely without any contact. Furthermore the dynamics of this suspension can be easily adjusted in a wide range for various applications. These two main properties already make the magnetic bearing a very attractive device for solving the classical bearing problem. On the other side the magnetic bearing is complex, expensive, usually not readily available from the shelf, and up to now only used for some advanced machinery. In the following the state of the art is presented so that future trends can be derived.

Magnetic forces are generated either by permanent magnets, electrodynamically or electromagnetically. In the constant field of permanent magnets, however, a ferromagnetic body cannot hover in a stable way /BR 39/, and electrodynamic forces are usually too small or still too difficult to generate to be of actual technical interest. They are used where small forces are sufficient, for example in space applications for the support of flywheels or for a micro-g-platform in a near zero-g-environment. Or they are used where the high currents necessary for large forces are generated by means of cryogenics as in prototypes for an electrodynamically levitated high-speed vehicle.

It is the electromagnetic force that is used most efficiently. For the levitation of guided vehicles a technology of its own has developed /GO 84/, that basically of course has some connection to magnetic bearings for rotors, too. Rotors have been supported magnetically at first for physical experiments. Spectacular 2.10E7 rpm have been reached while testing the strength of small steel balls under a centrifugal field of 1.0E8 g /BY 46/. Since then the electromagnetic rotor bearing has been applied to solve a number of different technical problems, and therefore the construction and the properties of the bearings differ remarkably. Numerous patents in this field have been issued or are pending. And a few types of bearings are already available commercially for use in machine tools and in turbomachinery /HB 85/. More details on applications will be given in chapter 6. Now at first the functional principle, some design goals, and the elements of the bearing system will be discussed.

2. FUNCTIONAL PRINCIPLE

Fig. 1 shows a hovering shaft S. Any deviation from this reference position is measured by a sensor, the measured signal is transformed into a control signal according to a suitable control law. The control signal is amplified and fed as a control current to the coils of the electromagnet, the actuator within this control loop. When the shaft for example starts to fall down it produces a measuring signal which leads to an increase in the

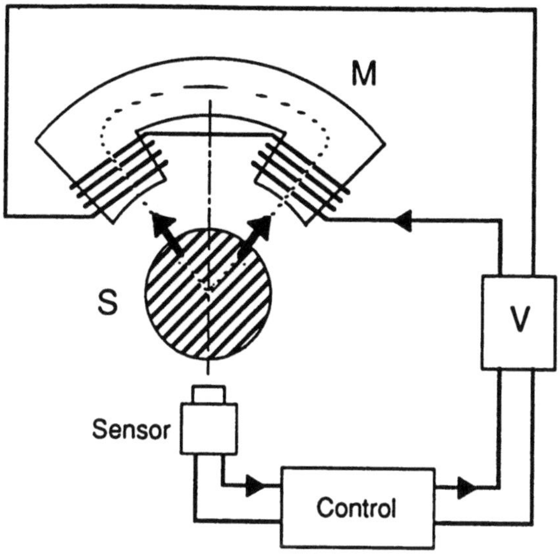

Fig. 1: Magnetic suspension

control current and thus the increasing magnetic force attracts the shaft again. Without the feedback the shaft would either fall down or be attracted by the magnet. The control law has to take care of the stability and of the dynamic properties of the hovering state. The magnetic forces can be made to be a function of the rotor motion in such a way that the actuator usually has spring and damping characteristics which suitably depend on the excitation frequency.

For fully suspending a complete rotor the simple loop explained above will not be sufficient. Fig. 2 shows part of the radial suspension of a rigid rotor with four degrees of freedom. For each degree of freedom a magnetic actuator has to be controlled individually. The control signals, however, depend on one another, i.e. each bearing force will depend on all sensor signals, leading to a typical multivariable control. The axial suspension of the rotor is not shown here. Its control is decoupled from the radial one and can be dealt with separately.

Figure 3 shows the hardware setup for such a bearing system. The rotor has a length of about 1 m, a diameter of 120 mm and a mass of 12 kg. The air gap of 10 mm is extremely large for technical purposes. The device was used for the exposition "Phänomena" in Zurich as a demonstration object. For measuring the rotor displacement within the large gap optical CCD-sensors are applied. Their signals are directly processed by a microprocessor and fed to switched power amplifieres. The block in between the two bearings is a simple asynchronous motor drive.

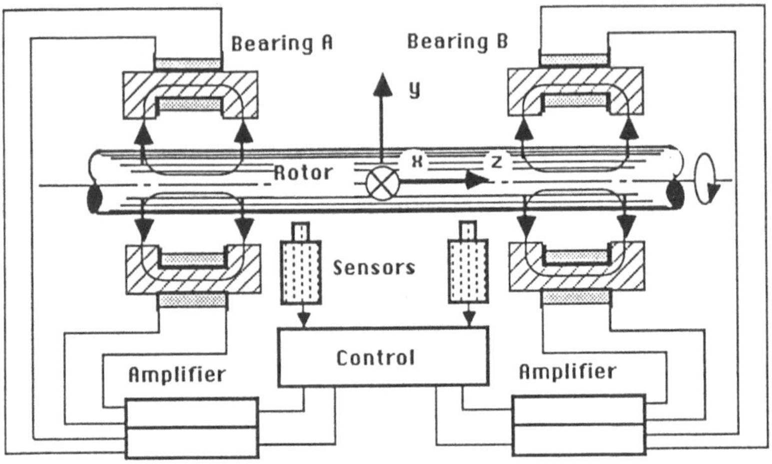

Fig. 2: Block-diagram for the radial suspension of a rotor in one plane

Fig. 3: Magnetic bearing system, front view with control unit, power supply and drive /TM 84/

3. DESIGN GOALS

Primary design goals have been to support a rigid rotor, because the rigidity of the rotor facilitates the control design essentially. Magnetic bearings are quite capable of supporting even a heavy rotor. The considerable freedom, however, in assigning dynamic characteristics to the actuator can be used not only for supporting a rigid rotor or even an elastic rotor but for its vibration control as well. Some efforts have already been undertaken to control various kinds of vibrations. Pietruszka and Wagner /PW 82/ show how an unbalanced rigid rotor can be made to rotate about its principal axis of inertia, thus avoiding vibrational unbalance forces on the bearing foundations. Gondhalekar et al. /GH 84,SA 84/ investigated vibrational control problems of an elastic rotor. Even the active damping of self-excited vibration caused by internal friction and of parametric vibration due to rotor asymmetry have been looked into.

This survey, however, will restrict itself to the basic features of the magnetic bearings and to their current main applications.

4. ELEMENTS OF THE MAGNETIC BEARING SYSTEM

The magnetic suspension of a rotor in reality is a well planned interconnection of various elements, a system. Its design is based on three steps, characterizing the different levels of expertise and teamwork necessary for a good overall solution. On the first level the elements of the bearing system have to be dealt with, the rotor, the sensors, the controller, the amplifiers and the magnetic actuators. Each of it has to be specified with respect to its tasks within the control loop. Their main characteristics will be presented in this chapter, emphasizing the most prominent element, the magnetic actuator. On the second level aspects of the control loop have to be discussed, and the control law has to be chosen according to the design goals and the desired overall characteristics. These design goals most often are determined by the specific application, and this has to be taken account of on the third level, where an efficient team work between different specialists will be essential. These steps, which of course have to be iterated, will be outlined in the next chapters.

4.1 Model of the rotor

The rotor obviously is the central part of the suspension system, and it has to be modeled mathematically as the plant within the control loop. At first let us describe a rigid rotor and then we will model an elastic rotor, which will be somewhat more complicated.

Model for rigid rotor: The rotor with a vertical axis (fig. 4) shall be kept in this reference position by two radial bearings and of course an axial bearing. The axial suspension with one degree of freedom is decoupled from the radial one and will not be dealt with here. The radial displacements are expressed either by the "analytical" coordinates for the translation of the center of gravity x,y and the inclination of the rotor α,β, or by the

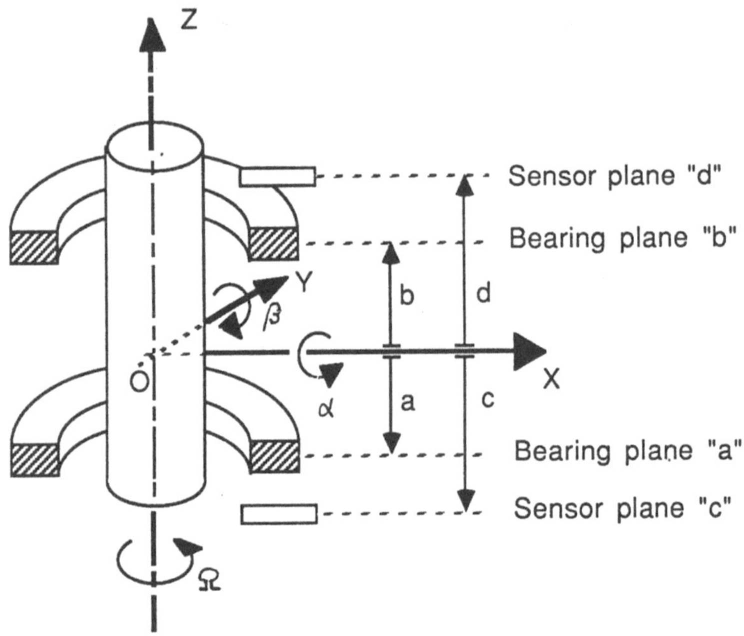

Fig. 4: Rigid rotor model /BL 84/

"technical" coordinates, the displacements x_c, x_d, y_c, y_d as measured by the sensors. The geometry is given in fig. 4, the mass of the rotor is m, the rotor is symmetric, any unbalance is small. Then the linearized equation for the radial motion $\underline{z}$ is obtained as

$$\underline{M}\,\ddot{\underline{z}} + \underline{P}\,\dot{\underline{z}} + \underline{S}\,\underline{z} = \underline{B}_f\,\underline{u}_f + \underline{V}_f\underline{s}\ , \quad \underline{z} = \left[\beta, x, -\alpha, y\right]^T \tag{4.1}$$

The vector $\underline{u}_f$ contains the bearing forces

$$\underline{u}_f = \left[f_{ax}, f_{bx}, f_{ay}, f_{by}\right]^T \tag{4.2}$$

The harmonic disturbances by unbalance are introduced by the vector

$$\underline{s} = \left[\sin\Omega t,\ \cos\Omega t\right]^T \tag{4.3}$$

The structural matrices for mass, gyroscopic effects and stiffness are

$$\underline{M} = \mathrm{diag}\left[I_x\ m\ I_x\ m\right]\ , \quad \underline{P} = I_z\Omega\left[\begin{array}{cc|cc} 0 & 0 & 1 & 0 \\ 0 & 0 & 0 & 0 \\ \hline -1 & 0 & 0 & 0 \\ 0 & 0 & 0 & 0 \end{array}\right]\ , \quad \underline{S} = \underline{0} \tag{4.4}$$

The gyroscopic matrix $\underline{P}$ is proportional to the rotational speed Ω and the axial moment of inertia I_z. It is the only matrix coupling the two lateral directions (x, β) and (y, α) with each other. The matrix $\underline{S}$ describes all position dependent forces except the bearing forces, and it contains pendulous forces and small coupling forces to the axial suspension which may be neglected here. Matrix $\underline{B}_f$ introducing the bearing forces consists mainly of the sign-valued coordinates a and b for the bearing locations. And the matrix $\underline{V}_f$ for the unbalance effects contains the products of inertia I_{xz}, I_{yz} and the excentricity e of the centre of gravity C from the (magnetic) rotor axis.

$$\underline{B}_f = \left[\begin{array}{cc|cc} a & b & 0 & 0 \\ 1 & 1 & 0 & 0 \\ \hline 0 & 0 & a & b \\ 0 & 0 & 1 & 1 \end{array}\right] \qquad \underline{V}_f = \Omega^2 \begin{bmatrix} -I_{yz} & I_{xz} \\ 0 & me \\ & \\ I_{xz} & I_{yz} \\ me & 0 \end{bmatrix} \tag{4.5}$$

The rotor displacements $\underline{z}$ are expressed in terms of the measured displacement signals $\underline{z}_s$ from the sensors by

$$\underline{z}_s = \left[x_c \; x_d \; y_c \; y_d \right]^T = \underline{T}\,\underline{z} \;, \quad \underline{T} = \left[\begin{array}{cc|cc} c & 1 & 0 & 0 \\ d & 1 & 0 & 0 \\ \hline 0 & 0 & c & 1 \\ 0 & 0 & d & 1 \end{array}\right] \tag{4.6}$$

This can be substituted into the equations of motion in order to express them in the measured variables $\underline{z}_s$ as well. Furthermore for control purposes it is appropriate to use the state space representation

$$\dot{\underline{x}} = \underline{A}\,\underline{x} + \underline{B}\,\underline{u}_f + \underline{V}\,\underline{s} \;, \quad \underline{x} = \left[\underline{z}_s^T \; \dot{\underline{z}}_s^T \right]^T \tag{4.7}$$

$$\underline{A} = \begin{bmatrix} \underline{0} & \underline{E} \\ \underline{0} & -\underline{T}\,\underline{M}^{-1}\,\underline{P}\,\underline{T}^{-1} \end{bmatrix}, \quad \underline{B} = \begin{bmatrix} \underline{0} \\ \underline{T}\,\underline{M}^{-1}\,\underline{B}_f \end{bmatrix}, \quad \underline{V} = \begin{bmatrix} \underline{0} \\ \underline{T}\,\underline{M}^{-1}\,\underline{V}_f \end{bmatrix}$$

In this representation the input of the rotor, the controlled bearing forces $\underline{u}_f$ are connected to the output, the measured displacement signals $\underline{z}_s$. Using control techniques the input has now to be generated such that the output behaves as desired, for example by letting the control forces act as damping and restoring forces.

Model for an elastic rotor: In principle the objective is the same as for the rigid rotor. We will establish a relation between the input, the control forces, and the output, the motion at distinct locations of the rotor. The model has to be of low order, so that it can be incorporated into the control loop more easily. For the derivation of such a model /BU 85/ let us start with a finite element model of the elastic rotor

$$\underline{A}_2\,\ddot{\underline{q}} + \underline{A}_1\,\dot{\underline{q}} + \underline{A}_o\,\underline{q} = \underline{p} \tag{4.8}$$

where $\underline{q}$ is a n x 1 vector of generalized displacements, $\underline{p}$ is the vector of generalized forces, and the $\underline{A}_i$ are structural matrices, characterizing inertia and elasticity as well as gyroscopic and nonconservative properties. Equ. (4.8) may be transformed into the complex frequency domain and expressed by

$$\left[s^2\,\underline{A}_2 + s\,\underline{A}_1 + \underline{A}_o\right]\underline{Q}(s) = \underline{P}(s) \tag{4.9}$$

The matrix in brackets is termed dynamical stiffness matrix and its inverse $\underline{H}(s)$ is the dynamical flexibility matrix

$$\underline{H}(s) = \left[s^2\,\underline{A}_2 + s\,\underline{A}_1 + \underline{A}_o\right]^{-1} \tag{4.10}$$

The elements of $\underline{H}(s)$, the dynamical flexibility transfer functions, are well-known in measurement and modal analysis techniques. A general way of reducing the large set of finite element equations is to truncate the modal representation of equ. (4.8). In order to find that representation we have to solve the eigenvalue problem of equ. (4.8). For the sake of simplicity - all the subsequent derivations can be extended to the general system (4.8) as well /BU 85/ - let us assume a simple elastic structure

$$\underline{M}_o\,\ddot{\underline{q}} + \underline{K}_o\,\underline{q} = \underline{p} \tag{4.11}$$

From the solution of the eigenvalue problem we obtain a set of eigenvalues $\omega_1, \dots \omega_n$ and the corresponding real normalized eigenvectors $\underline{u}_k$ with the modal matrix $\underline{U} = [\underline{u}_1, \dots \underline{u}_k, \dots \underline{u}_n]$.

Then the modal expansion of $\underline{q}$

$$\underline{q} = \sum_k \underline{u}_k\, z_k = \underline{U}\,\underline{z}\ , \qquad k = 1,\dots,n \tag{4.12}$$

leads to the modal representation of equ. (4.11)

$$\underline{M}\,\ddot{\underline{z}} + \underline{K}\,\underline{z} = \underline{U}^T\,\underline{p} \tag{4.13}$$

with

$$\underline{M} = \underline{U}^T\,\underline{M}_o\,\underline{U} = \mathrm{diag}\,(m_k)$$

$$\underline{K} = \underline{U}^T\,\underline{K}_o\,\underline{U} = \mathrm{diag}\,(k_k)\ , \quad \omega_k^2 = k_k/m_k$$

This modal representation is a time domain representation of the partial fraction expansion of the corresponding flexibility matrix $\underline{H}(s)$. Translating equations (4.12, 4.13) into the s-domain, yields

$$\underline{H}(s) = \underline{U}(\underline{M}\, s^2 + \underline{K})^{-1}\, \underline{U}^T = \sum_{k=1}^{n} \frac{\underline{u}_k\, \underline{u}_k^T}{m^k (s^2 + \omega_k^2)} \tag{4.14}$$

For the design of the controller, a model of the flexible structure, that relates the displacements and velocities at the location of the sensors to the excitation forces of the actuators, is required. By analyzing the closed loop system with respect to the actuator and sensor coupling points, it can be checked whether the design objectives have been reached, i.e. whether the system is stable and shows up with good damping performance. Also, test measurements for the closed loop system are easily carried out by exciting the structure with the available actuators. Hence, for both control design and closed loop analysis, we may confine the selection of coordinates to the coupling points of the actuators and sensors with the structure. A modal representation of the dynamical flexibility matrix of the elastic structure, with respect to the cited coordinates, provides the desired description, as will be shown subsequently.

The n_A actuator forces form the vector $\underline{w}$, the n_D measured displacements are given by $\underline{y}_D$, and the n_V measured velocities by $\underline{y}_V$. These vectors are related by appropriate incidence matrices to the nodal forces and displacements by

$$\underline{w} = \underline{I}_A^T\, \underline{p}\;,\quad \underline{y}_D = \underline{I}_D\, \underline{q}\;,\quad \underline{y}_V = \underline{I}_V\, \underline{q}\;,\quad \underline{y}^T = \left[\underline{y}_D^T, \underline{y}_V^T\right] \tag{4.15}$$

The elements of the incidence matrices are zero or one, when discrete actuators and sensors are located at nodal points. We assume that we can truncate the number of modes retaining only m essential modes, so that instead of equ. (4.12) we now have

$$\underline{q} \approx \sum_i \underline{u}_i\, z_i = \underline{U}_m\, \underline{z}_m \;,\quad i = 1,\ldots m\;;\quad m < n \tag{4.16}$$

with the truncated n x m modal matrix $\underline{U}_m$ and the truncated modal vector $\underline{z}_m$. This leads to the truncated set of modal equations

$$\underline{M}_m\, \ddot{\underline{z}}_m + \underline{K}_m\, \underline{z}_m = \underline{U}_m^T\, \underline{p} \tag{4.17}$$

With the subset of retained coordinates (4.15) we have in the s-domain a transfer maxtrix representation

$$\underline{Y}(s) = \underline{H}_G \, \underline{W}(s) \quad \underline{H}_G = \begin{bmatrix} \underline{T}_D \, \underline{U}_m (s^2 \, \underline{M}_m + \underline{K}_m)^{-1} \, \underline{U}_m^T \, \underline{T}_A \\ s \, \underline{T}_V \, \underline{U}_m (s^2 \, \underline{M}_m + \underline{K}_m)^{-1} \, \underline{U}_m^T \, \underline{T}_A \end{bmatrix} \tag{4.18}$$

Here the transfer matrix $\underline{H}_G$ is a generalized dynamical flexibility matrix, since it relates a force excitation (actuator forces) to nodal displacements and velocities (sensor signals).

By introducing the state vector

$$\underline{x}_e^T = \left[\underline{z}_m^T , \dot{\underline{z}}_m^T \right]$$

an equivalent state space representation is obtained

$$\dot{\underline{x}}_e = \underline{A}_e \, \underline{x}_e + \underline{B}_e \, \underline{w} \; , \quad \underline{y}_e = \underline{C}_e \, \underline{x}_e \tag{4.19}$$

$$\underline{A}_e = \begin{bmatrix} \underline{0} & \underline{E} \\ -\underline{M}_m^{-1} \, \underline{K}_m & \underline{0} \end{bmatrix}, \quad \underline{B}_e = \begin{bmatrix} \underline{0} \\ \underline{U}_m^T \, \underline{T}_A \end{bmatrix}, \quad \underline{C}_e = \begin{bmatrix} \underline{T}_D \, \underline{U}_m & \underline{0} \\ \underline{0} & \underline{T}_V \, \underline{U}_m \end{bmatrix}$$

All of the representations (equs. 4.17, 4.18, 4.19) will be suitable for the control design. The method can be extended to include other structural elements as well, i.e. foundation dynamics, by applying the building block approach. The rotor dynamics program MADYN /MA 82/ has been readily adopted to model magnetically controlled elastic rotors. Problems left over and to be discussed later are that of truncation and the control design itself.

4.2 Sensor, controller, amplifier

The sensors have to measure the rotor motion. The quality of this measurement is essential for the quality of the suspension. The main requirement is that there is no drift of the zero-output. The resolution is directly connected to the positioning accuracy of the rotor, i.e. when the rotor has to be positioned within 1 micron the sensor output has to be at least as sensitive as that. The frequency response of the sensors is part of the dynamic properties of the suspension. The sensors should be insensitive to magnetic fields and to temperature variations. Good results have been obtained with inductive sensors, eddy-current pick-offs and with optical sensors. In the latter case CCD-arrays with subsequent direct digital control have been used.

The controller has to process the measured signals. It may consist of filters, of the controller in the strict sense, and of a linearization unit. It may be built as an analogue network, or it is a microprocessor where the different tasks are taken over by algorithms. Section 5.1 will show more details on that. The control law itself has to be chosen according to the requirements of the specific application, leading to performance values, which are discussed later on.

The power amplifiers have to amplify the control signals and feed them to the bearing coils. The input of an amplifier, the control voltage, can generate an output voltage or an output current, and consequently the input/output- characteristics of the bearing have to be described accordingly. The bearing coils represent an essentially inductive load. Therefore the capability of the amplifier to drive a high-frequency current through this load mainly determines the frequency response, i.e. the dynamics, of the whole suspension system. Furthermore we distinguish between analogue DC-amplifiers and switched amplifiers. The first ones are usually applied when the load is small (less than 0.5 kVA) and when the amplifier losses are not important. Switched amplifiers work at switching frequencies outside the audio range. They do not yet seem to be commercially and easily available. In the following the properties of two amplifiers for two different bearings are listed:

Bearing load	180 N	160 N
Bearing diameter	80 mm	120 mm
Bearing gap	0.7 mm	10 mm
Type of amplifier	DC	switched with 28 kHz
Ampl. voltage/current/losses	±50/1A/100W	±300V/2.6A/30W

4.3 Magnetic actuator

The electromagnetic actuator is the element within the control loop, that transforms a voltage or a current input into the most desired output: the bearing force. Fig. 5 shows the magnetic

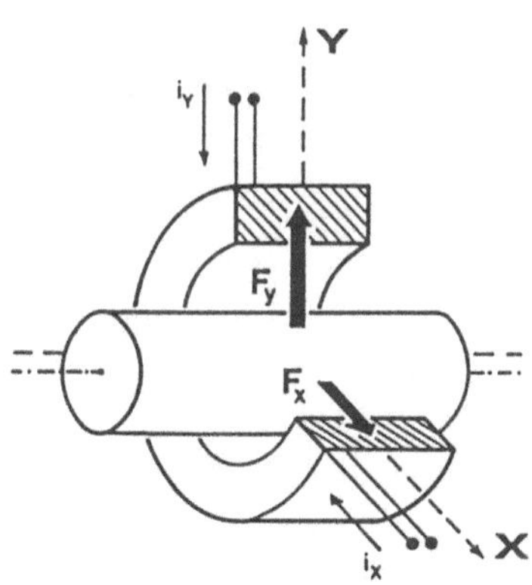

Fig. 5: Radial magnetic actuator as input/output element

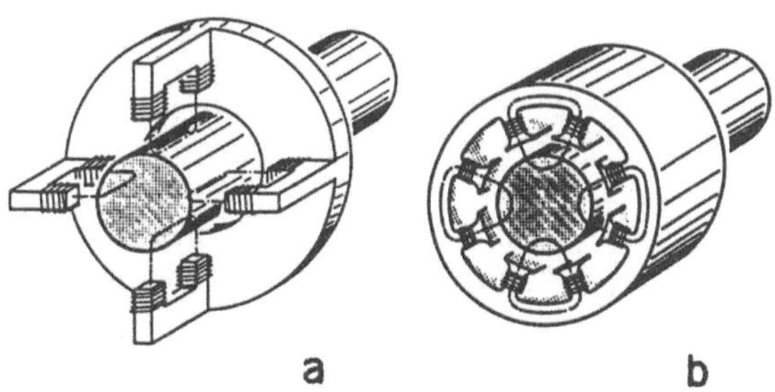

Fig. 6: Two different configurations for a radial bearing

actuator as such an input/output element, whereas fig. 6 already shows constructional details. The two typical constructions for a radial actuator differ mainly in the path for the magnetic flux with respect to the rotor axis, and as a consequence thereof the magnetic losses are different. The characteristics of the actuator have to be calculated from geometrical and electrical data, or they have to be measured.

For calculating the bearing forces the radial bearing of fig. 6b is assumed to consist of U-shaped elements as shown in fig. 7.

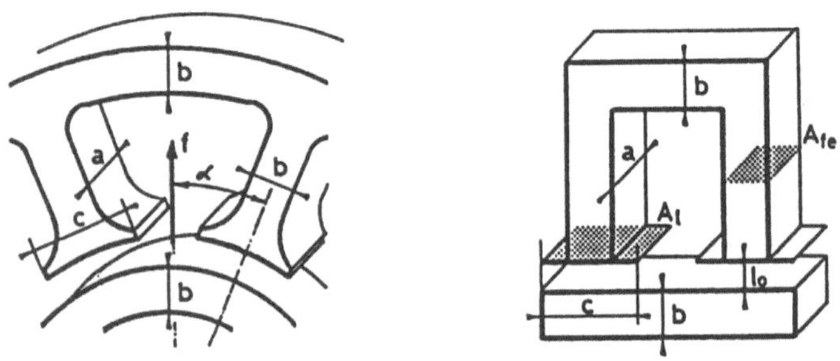

Fig. 7: Single U-shaped electromagnet as part of a radial bearing a) real geometry b) simplified geometry

Each of them exerts forces in the positive or negative X- or Y-direction. The force of each magnet is controlled by a control current. The force resulting from a certain current depends on the geometry of the magnet and on the used ferromagnetic material. The nonlinear behaviour of the ferromagnetic material makes the analytical calculation of the forces difficult. Therefore, the U-shaped magnets are modeled with a simplified geometry and then the forces are calculated numerically. Figure 7 shows the relation between the real iron path and the simplified one. The cross-sectional area Afe = a x b of the ferrous core is constant along the whole length lfe of the iron path. The width c of the pole shoes is used only for the calculation of the cross-sectional area Al of the airgap lo. With the simplified geometry, the measured magnetisation-curve B = B(H) can be modified to a socalled sheared magnetisation-curve which relates the flux Φ to the ampere-turns Θ. This relation is given by the two equations

$$\Phi = B\, A_{fe} \tag{4.20}$$

$$\Theta = H\, l_{fe} + B\, \frac{2A_{fe}\, l_o}{A_l\, \mu_o} \tag{4.21}$$

where Θ is the product of the current i and the number of windings n

$$\Theta = n\ i \qquad (4.22)$$

With equations (4.20) and (4.21) we can find the flux Φ for a given ampere-turn Θ. From the flux Φ we can calculate the force f

$$f = \phi^2 \cos \alpha / \mu_o\ A_1 \qquad (4.23)$$

The four equations connect the input current i, the airgap lo, and the force f on the rotor with the electrical and geometrical characteristics of the bearing. The variables can be time-varying as well. As long as no saturation occurs within the iron the force will be proportional to the square of the current i (fig. 8a) and it will be inversely proportional to the square of the airgap lo (fig. 8b). The deviation between calculated and measured values is mainly due to stray effects which have not been considered in the simplified mathematical model. The influence of stray fields, however, is small as has been shown by numerical flux-line investigations /TR 85/.

The experimental results on the forces were obtained by using a piezo-electric dynamometer for static loads and for load variations up to 400 Hz. Above that until 1400 Hz indirect measurements have been used by measuring the acceleration of a test rotor in order to identify the magnetic forces exciting the rotor motion. The amplitude and phase response for the force/current factor of fig. 9 is flat within the measuring range

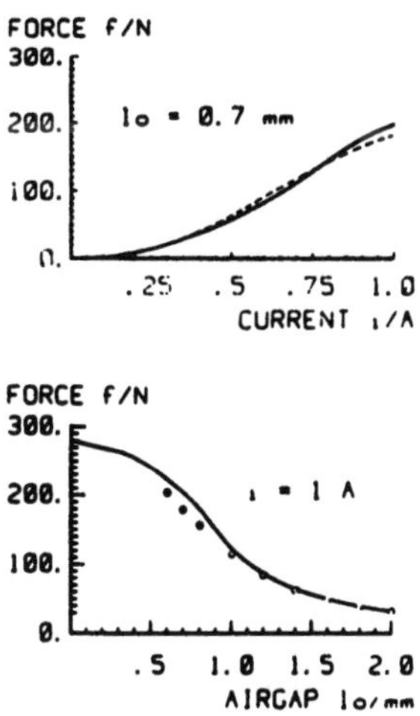

Fig. 8: Calculated (——) and measured (---,o) bearing force of a radial bearing (diameter = 80 mm, a = 40 mm, n = 720)

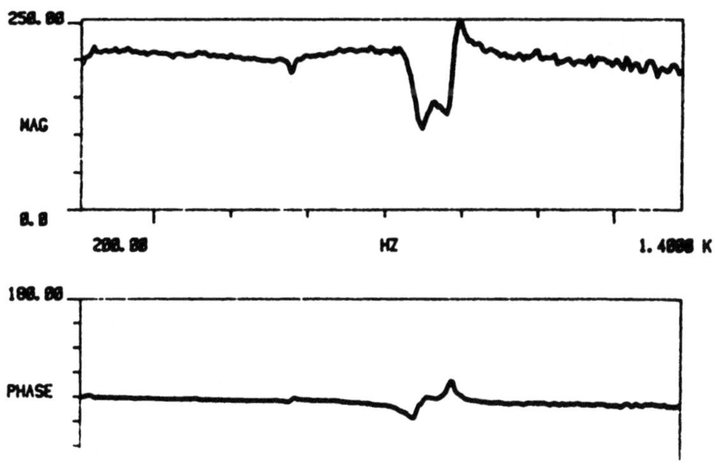

Fig. 9: Force/current factor depending on the excitation frequency

up to 1400 Hz and has to be corrected only at about 900 Hz where the test rotor had a bending resonance /TR 85/. That means that for this specific bearing the ratio of force to current for small signals is constant at least up to 1400 Hz.

The maximally permissible current depends on the copper losses in the windings of the coil and its cooling capacity.The simplest model for the cooling capacity is given by

$$P_{max} = A\ h\ T \tag{4.24}$$

where Pmax is the cooling power for the temperature-difference T, A is the surface of the magnetic bearing and h is a cooling factor, depending on the cooling medium. For a given application one can of course use much more sophisticated models. In most cases, the eddy-current losses and the hysteresis-losses in the stator are negligible compared to the copper losses. Therefore we can calculate the maximum ampere-turns for one magnet by

$$\Theta_{max} = n\ i_{max} = \sqrt{\frac{P_{max}}{4}\ \frac{A_w\ f_a}{\rho\ l_w}} \tag{4.25}$$

where Aw is the cross-sectional area of the coil, fa is the copper filling factor which gives together with Aw the copper's cross-section, ρ is the specific resistance of copper and lw is the length of one winding of the coil. In order to calculate the maximum forces for a magnetic bearing we proceed in the following steps:

1. Calculation of the cooling power, i.e. the maximum copper loss power , for a given geometry and cooling medium.

2. Calculation of maximal ampere-turns Θmax for the given geometry, where for example the space for the coils Aw and the length of one winding lw may be given.

3. With the equations (4.20) and (4.21) and the relation B = B(H) for the magnetisation-curve we get the flux Φ belonging to the given Θ .

4. Calculation of the bearing force with (4.23).

Of course this procedure may have to be iterated in order to obtain an optimal configuration.

A bearing may be termed optimal, for example, when for a given bearing space the magnetic force has a maximal value. Assuming the geometry to be fixed, except for one free parameter - the core width b - the above design rules are iteratively applied, leading to the results shown in fig. 10. For each air-gap lo there is an optimal core width b where the force f is maximal. The cooling power has been kept constant in this example.

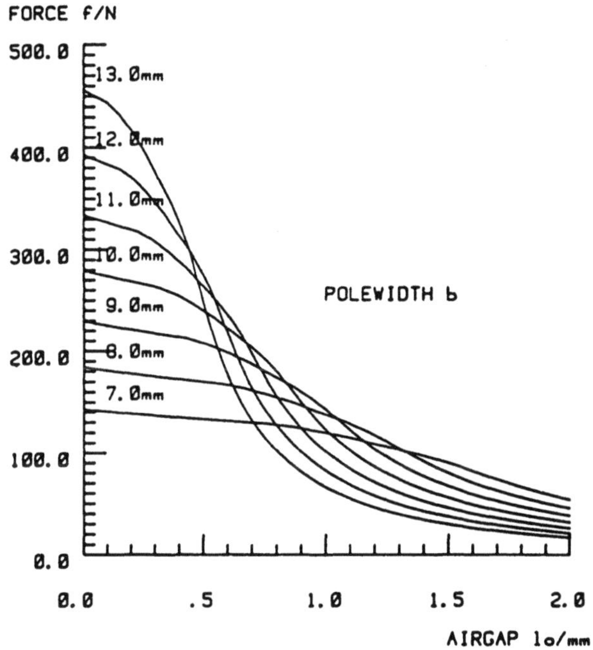

Fig. 10: Optimization of the force, depending on the airgap lo and the core width b

The force/current relation of fig. 8 is nonlinear. For the use of the bearing magnet within the control loop, however, a linear current-force characteristic is most desired. This can be achieved in two different ways. The nonlinear characteristic of fig. 8 can be linearized by a linearization network, or by a suitable algorithm in a microprocessor. Another very convenient method uses differential winding of the coils and premagnetization /SL 76/. In this case the magnetic force is given by

$$F(x,i) = k_x \, x + k_i \, i \tag{4.26}$$

where the force/displacement factor k_x and the force/current factor k_i depend on the premagnetization, the nominal air-gap and the geometry of the actuator. For small displacements and small control currents linear characteristics as in fig. 11 are obtained. A disadvantage of this linearization is that the premagnetizing current causes higher copper losses; the advantage is that this bearing is especially well applicable to control purposes.

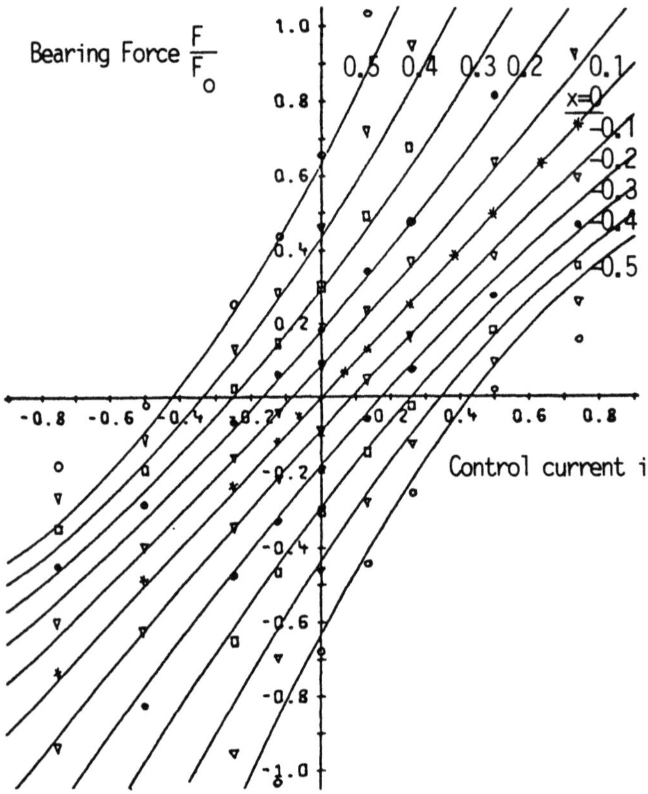

Fig. 11: Theoretical (——) and measured (∘ ▿ ▫ • *) bearing force characteristics for a premagnetized bearing (x ≙ relative displacement)

The load capacity of the bearing depends on its size and geometry, its magnetic material, cooling power, and on the control current. The specific load capacity amounts to 50 N/cm^2 for conventional transformer Si-alloys under a magnetic field of 1.5T. It can be increased by 50 % by using special Co-alloys. By increasing the bearing size obviously high loads can be supported. The potential of the actuator of course can only be used when it is suitably controlled and incorporated in the total rotor-bearing-system. This will be shown more explicitly in the following chapters.

5. SYSTEM ASPECTS

5.1 Control of rigid and elastic rotors

Any control requires information about the motion of the rotor. Here the rotor position is measured by distance sensors. The signals for the displacement speed can be produced in three different ways: by electronically differentiating the displacement signal, by directly measuring the velocity with suitable sensors, or by using a reduced Luenberger observer, an algorithm for estimating not measured data from measured ones. The various control schemes differ in how the available signals are used to derive control signals. In the following the control for a rigid rotor and for an elastic rotor will be outlined.

The control of a rigid rotor is based on linear multivariable control theory. A robust feedback, in the sense of low sensitivity to parameter changes, can be computed by minimizing the performance index J

$$J = \int_0^\infty (\underline{x}^T \underline{Q}\, \underline{x} + \underline{u}_f^T \underline{R}\, \underline{u}_f)\, dt \qquad (5.1)$$

where $\underline{x}$ and $\underline{u}_f$ are motion and control variables (eq. 4.7). When all the state variables $\underline{x}$, displacements $\underline{z}_s$ and their time derivatives $\dot{\underline{z}}_s$ are used, we have a complete (or central) state feedback. Then the control $\underline{u}_f$ depends linearly on the state

$$\underline{u}_f = - \underline{K}\, \underline{x} \qquad (5.2)$$

where $\underline{K}$ is the gain matrix. In our case (fig. 12a), with 8 state variables and 4 control variables for the two radial bearings, the gain matrix has 32 coefficients. Obviously this large number of coefficients makes any realization of the controller somewhat cumbersome.

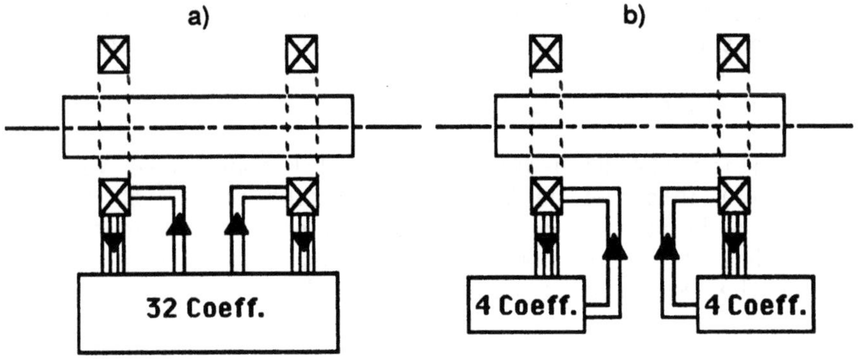

Fig. 12: Decentralized vs. central feedback structure

Therefore a decentralized control has been designed /BL 84/ where the control signals for each bearing only depend on the motion of the rotor at or near this bearing (fig. 12b). Thus the central control is broken down into smaller subsystems. Their coefficients are still determined by minimizing a quadratic error integral as in eq. (5.1), and the control retains to some extent the robustness properties of the complete state feedback. Now due to the simplified structure of the controller it is feasible to use direct digital control with a microprocesor. Cycle times of less than 1 ms have been achieved /TM 84/. The application of a microprocessor has the additional advantage that it can perform operational tasks and redundancy management at the same time much more versatile than an analogue network.

In any case, whether we use a central or a decentralized feedback, the dynamic behaviour of the rotor is determined by choosing the weighting matrices $\underline{Q}$ and $\underline{R}$ in eq. (5.1). The matrix $\underline{R}$ influences the amount of control power to be used, and the elements of $\underline{Q}$ weigh the displacements and velocities of the rotor. Another possibility to influence the dynamics is to assign the poles of the system behaviour, i.e. to assign natural frequencies and damping. Theoretically, for complete state feedback and unrestricted control forces even arbitrary dynamics can be assigned. Other methods, too, for designing a control loop, including integral feedback and disturbance compensation, have been applied successfully. Practical values that have been attained for system characteristics will be discussed in section 5.3.

The control of the elastic rotor can be accomplished by the same methods as for the rigid rotor. There are pragmatic solutions to the control design, a systematic approach, however, is still under investigation. The problems arise from the following reasons.

The control design has to be based on a reduced-order model of the rotor (equ. 4.17), containing only the most important modes of the elastic rotor. The question is whether the real rotor will indeed be controlled correctly by such a simplified controller. Actually there may be detrimental effects, and they will be explained subsequently by using the elastic rotor model.

Let us partition the high-dimensional modal coordinate vector $\underline{z}$ of equ. (4.13) into two parts. The important part with the low dimension n_m will be the one to be controlled, the other residual part of dimension n_r is the one to be neglected. Then equation (4.13) and the measurement equations (4.15) can be arranged in the following form

$$\begin{bmatrix} \underline{M}_m & \underline{0} \\ \underline{0} & \underline{M}_r \end{bmatrix} \begin{bmatrix} \ddot{\underline{z}}_m \\ \ddot{\underline{z}}_r \end{bmatrix} + \begin{bmatrix} \underline{K}_m & \underline{0} \\ \underline{0} & \underline{K}_r \end{bmatrix} \begin{bmatrix} \underline{z}_m \\ \underline{z}_r \end{bmatrix} = \begin{bmatrix} \underline{U}_m^T \\ \underline{U}_r^T \end{bmatrix} \underline{p} \tag{5.3}$$

$$\underline{y}_D = \underline{T}_D \underline{U}_m \underline{z}_m + \underline{T}_D \underline{U}_r \underline{z}_r , \quad \underline{y}_V = \underline{T}_V \underline{U}_m \dot{\underline{z}}_m + \underline{T}_V \underline{U}_r \dot{\underline{z}}_r$$

For design purposes it will be assumed that the controller is based on the reduced-order model only, being the m-subset of equations (5.3):

$$\underline{M}_m \ddot{\underline{z}}_m + \underline{K}_m \underline{z}_m = \underline{U}_m^T \underline{p} = \underline{U}_m^T \underline{T}_A \underline{w}$$

$$\underline{y}_{Dm} = \underline{T}_D \underline{U}_m \underline{z} \quad , \quad \underline{y}_{Vm} = \underline{T}_V \underline{U}_m \underline{z}_m \tag{5.4}$$

The relation between the control vector $\underline{w}$ and the measurement follows from the chosen law, which preferably will be a linear one and can be determined in the usual ways. Such a control design will result in the desired and "good" control for the reduced-order model (5.4), but when it is applied to the real full-order-system (5.3) the system qualities can alter essentially (fig. 13). In reality the measurements $\underline{y}$ do not only consist of the modeled part $\underline{z}_m$ but they also depend on the residual vector $\underline{z}_r$, causing the socalled "observation spillover".

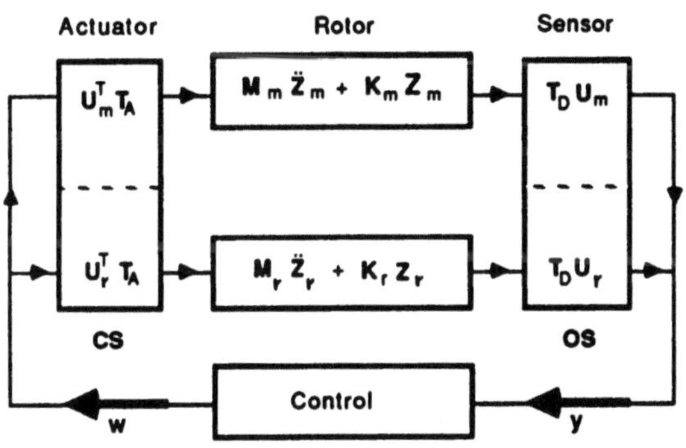

Fig. 13: Reduced order control applied to the real full-order system, demonstrating the observation spill-over (OS) and the control spill-over (CS)

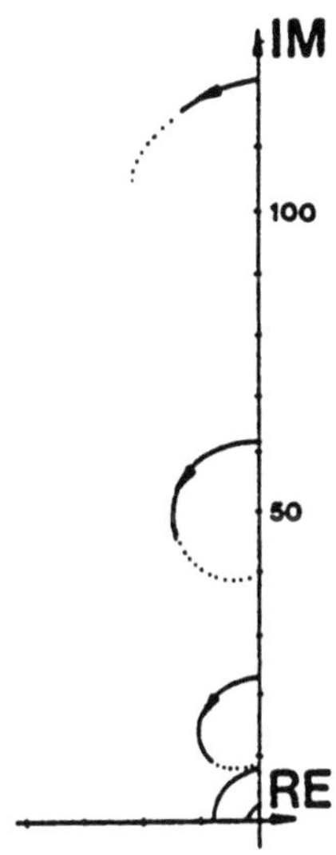

Fig. 14: Typical full-order-closed-loop eigenvalue curves for an elastic rotor with low stiffness magnetic bearings and increasing damping D assigned to the modes

Furthermore the control vector $\underline{w}$ does not only act on the modeled part but on the real full-order system itself which obviously contains the unmodeled part as well. This influence is called "control spillover". These spillover terms can change and deteriorate the behaviour of the real system and even destabilize it.

The objective of the control design now is to derive the control on the basis of the reduced-order system without exact knowledge about the residual, unmodelled part.

It is advisable to use a direct output feedback, i.e. the control forces depend linearly on the displacement and velocity measurements

$$\underline{w} = - \underline{G}_D \underline{y}_D - \underline{G}_V \underline{y}_V \tag{5.5}$$

with the yet unkown gain matrices $\underline{G}_i$. Introducing this control law and the measurement equations (4.15) into the full-order system equations (4.13) leads to the full-order closed-loop system

$$\underline{M} \ddot{\underline{z}} + \underline{U}^T \underline{T}_A \underline{G}_V \underline{T}_V \underline{U} \dot{\underline{z}} + (\underline{K} + \underline{U}^T \underline{T}_A \underline{G}_D \underline{T}_D \underline{U}) \underline{z} = \underline{0} \tag{5.6}$$

In order to avoid spillover effects the gain matrices $\underline{G}_i$ have to be determined such that the solutions of (5.6) are stable and show some desired and specified behaviour. In /SA 84/ the gain matrices $\underline{G}_i$ are derived so that the resulting reduced-order control shows robustness qualities with respect to parameter errors and truncation effects and leads to a stable full-order closed-loop system of even infinite dimension, if

- the rotor-bearing system is not unstable from the beginning (rigid body modes allowed),
- the reduced-order model contains at least all rigid-body modes,
- the sensors and actuators are collocated and their number is at least equal to the number of rigid-body modes.

This control approach includes the technically most interesting case where a real flexible rotor is suspended by actuators which at the same time have to control its elastic vibrations. As an example a flexible homogeneous beam supported at its ends will be considered. The two bearings will have to control the two rigid body modes, and as a design goal, two elastic modes should be simultaneously damped with these bearings, too. The gain matrices for the direct output control are determined such that a desired bearing stiffness is obtained, characterized by the natural frequencies of the two rigid body modes. Then the damping D assigned to the two rigid body modes and the two elastic modes is varied between 0 and 1, and the resulting root locus curves are represented in fig. 14. When we increase the damping the eigenvalues of the originally elastic modes are going back to the imaginary axis along semi-circles, ending there at the modal frequencies of a "fixed-fixed" beam.

5.2 Characteristics of the magnetic bearing system

Typical specification terms for rotor bearings are the maximal load capacity, the specific load capacity, stiffness, damping, frequency response, maximal angular velocity, and losses.

The maximal load capacity, as derived in chapter 4.3, depends on the bearing size, the magnetic material and the control current, and there are real bearings where it amounts up to 50 kN. When the actual load is larger than the allowable maximal load the rotor can not hover any more and touches upon the bearing.

The specific load capacity relates the maximum load capacity to the cross sectional area of the bearing hole. Characteristic values for usual Fe/Si-alloys with a saturation of 1.5 Tesla are 50 N/cm^2. With special Fe/Co alloys up to 80 N/cm^2 can be obtained.

The stiffness of the bearing is typically frequency dependent. For static loads - the frequency of the loading is zero - an integrating feedback of the displacement signals leads to a theoretically infinite stiffness. For load frequencies below the cut-off frequency of the bearing control, which usually is between 100 Hz and 1500 Hz, the stiffnesss characteristics can be shaped according to requirements. This property can be used to counteract disturbance forces. For load frequencies above the cut-off frequency the stiffness gradually decreases and finally becomes very small. Very high frequency forces are thus not actively transmitted by the bearing, which makes it quite useful for vibration isolation.

The damping that can be obtained by the bearing depends on the control laws and on the available force. This available force is given by the maximal load capacity, and it can be used as a restoring force, leading to the bearing stiffness, or as a damping force. The frequency dependence typically is the same as that for the stiffness. Of course any combination of damping and stiffness can simultaneously be generated as long as the total force does not exceed the load capacity.

The frequency response characterizes the dynamic behaviour of the bearing. The actuator itself corresponds to a series of R-L-elements. And a harmonic voltage input, with the highest amplifier voltage-amplitude possible, generates a current through the coils with a frequency response as in fig. 15. The cut-off frequency f_g corresponds to a time-constant of $T=1/f_g$ for a step-input, which can be easily measured. Most often, however, the power-amplifier is internally controlled as a voltage-current converter. This reduces the order of the overall-control-loop and reduces the time constant of coil and amplifier by the factor of the open-loop-gain of the amplifier. Thus the current of the bearing - and consequently the magnetic force - follows the input voltage practically without delay. Of course this behaviour is only valid within the current-frequency response of fig.15, i.e., the transfer function for amplifier and bearing refers to small

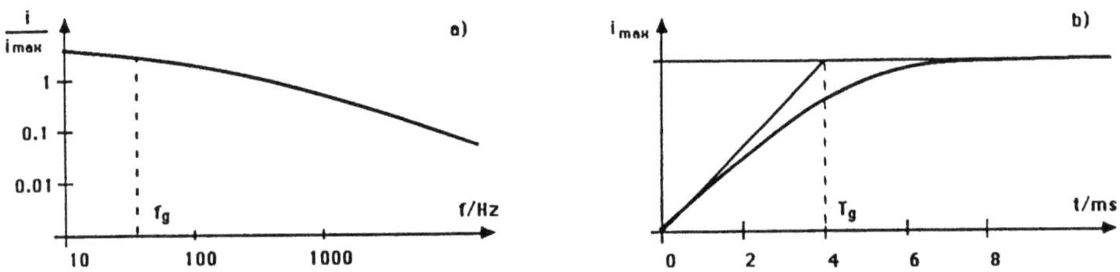

Fig. 15: a) Frequency response of the current through a bearing coil, with full amplifier voltage applied
b) step response

signals within the limits of the frequency response of fig.15. And within these limits the transfer function of the closed loop itself, determining the dynamics of the rotor suspension, can be shaped almost arbitrarily by a suitable control design.

Further dynamic characteristics and their relation to the size of the power amplifier are derived in /TR 85/.

The maximal rotor velocity depends on two factors. One is the strength of material of the rotor. Because of the laminated ferromagnetic sheets on the rotor circumferential velocities of more than 200 m/s are difficult to obtain. The second factor is the power of the rotor drive necessary to overcome rotor braking torques. As shown in the next section these losses are small compared to conventionally supported rotors, and they have to be considered only at high speeds.

The losses of the magnetic suspension mainly occur in the rotor, the actuator and the power amplifier /TR 85/. The rotor losses are the equivalent to the friction in hydrodynamic- or rollerbearings as these losses have to be overcome by the rotor drive, too. The magnetic losses in the rotor are caused by the modulation of the magnetic flux in the ferromagnetic part of the turning rotor, when it passes the poles of the stator magnets. This modulation is much lower in the bearing of fig.6a than in the bearing of fig.6b, because the rotor parts pass only poles with the same polarity. These magnetic losses are caused by hysteresis and by eddy currents. Hysteresis losses depend on the material, used for the ferromagnetic part of the rotor in the bearing area. They are proportional to the rotation frequency Ω and the square of the flux density B:

$$P_{1h} = kh * \Omega * B^2 \tag{5.7}$$

with kh being a material constant. The eddy current losses also depend on the material used. The ferromagnetic part of the rotor should be laminated and built from thin discs or rings to reduce the eddy currents. The losses are proportional to the square of the rotation frequency, and the square of the thickness s of the ferromagnetic material and they can be approximated by

$$P_{1e} = ke * s^2 * \Omega^2 * B^{1.6} \tag{5.8}$$

The other rotor losses, the air losses, depend on the geometry of the rotor and its surrounding parts. For an example the run-down curve of a magnetically suspended rotor has been measured in air and in vacuum (fig.16). The braking torques caused by the various effects are shown in fig.17.

The stator losses are dominated by the copper losses due to the resistance in the coils. According to equations (4.24, 4.25) they are proportional to the square of the current i in the coils,

$$P_{1c} = kc * i^2 \tag{5.9}$$

As a consequence, considering (4.23), the stator losses typically increase proportional with the bearing force f.
The amplifier losses are the main source for the electrical losses. In the conventional analogue amplifiers the losses depend on the maximal output voltage, that has to be kept available for a good dynamic performance even when it is not required by the quiescent load. Switched amplifiers, however, work much more economically as their losses are dependent on the actual load only. In this case the amplifier loss increases proportional to the load.

6. APPLICATIONS

Magnetic bearings have been used for rotating machines in vacuum-techniques because there will be no contamination by wear or lubricants. One of the first industrial applications was for a turbomolecular pump. An extremly low vibration level (o.o5 mm/s at the pump), an operational speed of 5oo Hz, and an ultrahigh vacuum of 10 E-10 mbar free of any hydrocarbon contamination are outstanding specifications. Now several companies are developing and building such pumps.

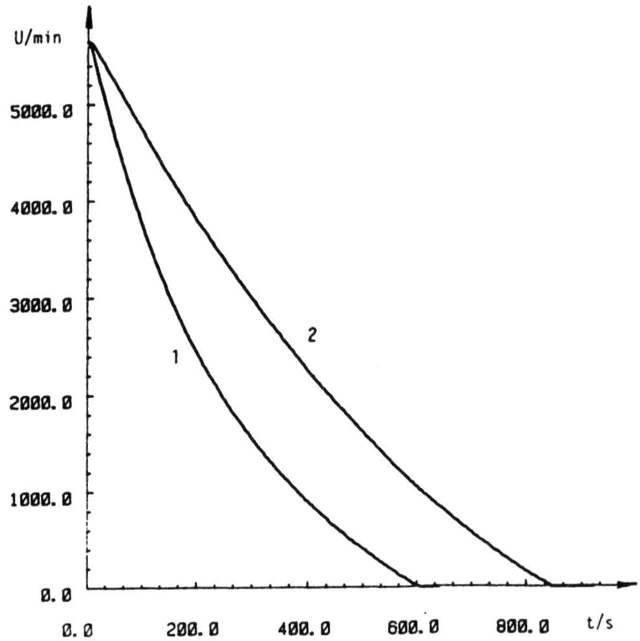

Fig. 16: Run-down curve for a test-rotor
1 - in air, 2 - in vacuum)

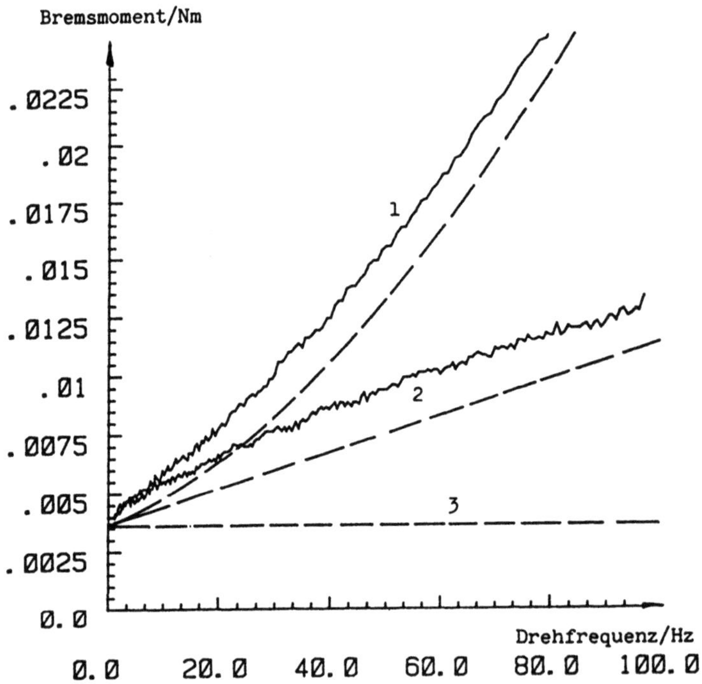

Fig. 17: Braking torques (—— measured,--- calculated,
1 - in air, 2 - in vaccum, 3 - only hysteresis losses)

A centrifuge for epitaxy experiments on semiconductor material under UHV-conditions is using magnetic bearings. The bearings are outside a vacuum chamber for suspending the rotor within the vacuum /ST 83/. The set-up of the centrifuge is shown in fig. 18.

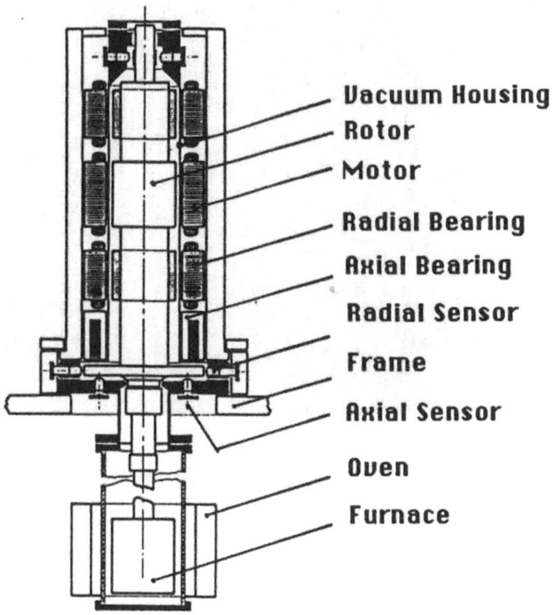

Fig. 18: Set-up of the magnetically supported epitaxy-centrifuge

An eight stage high speed centrifugal compressor on magnetic bearings has been built and tested recently /HK 85/. The rotor weight is about 37oo N, the bearing span 1.3 m, and three critical speeeds are below the operating speed of 13,ooo rpm. Helium compressors, compressors for submarines, and centrifuges for chemical industry are other examples.

Machine tools, too, have been designed with magnetic bearings: a lathe, replacing a polishing machine for heavy cylinders in the graphical industry, high speed milling machines in aerospace industry, and spindles for high speed grinding of threads.

A future application could be the vibration control of elastic rotors with magnetic bearings or dampers. For a rotor, modeled in fig. 19 and shown on a test-rig in fig. 20, the vibration response is given in fig. 21. The reduction of the resonance amplitude at the critical speed by using an active damper is obvious.

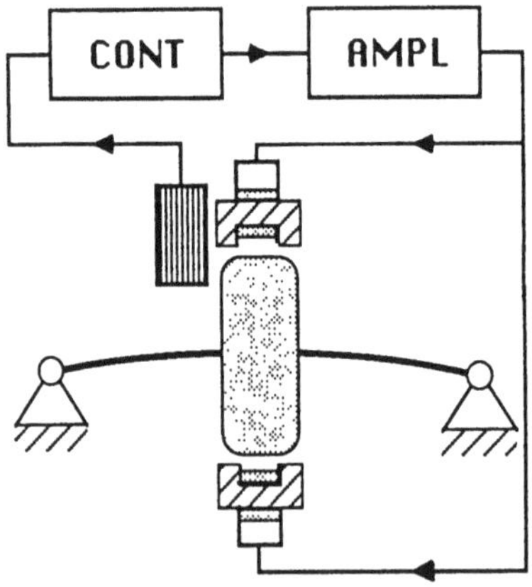

Fig. 19: Magnetic damper for a rotor on an elastic shaft

Fig. 20: Test rig

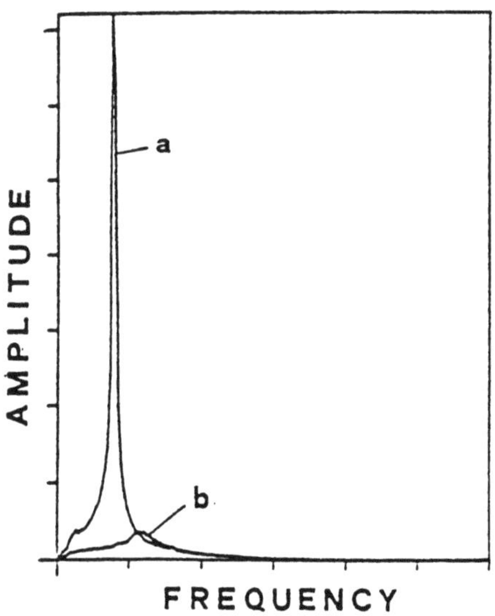

Fig. 21: Frequency response
a - undamped,
b - with active damper

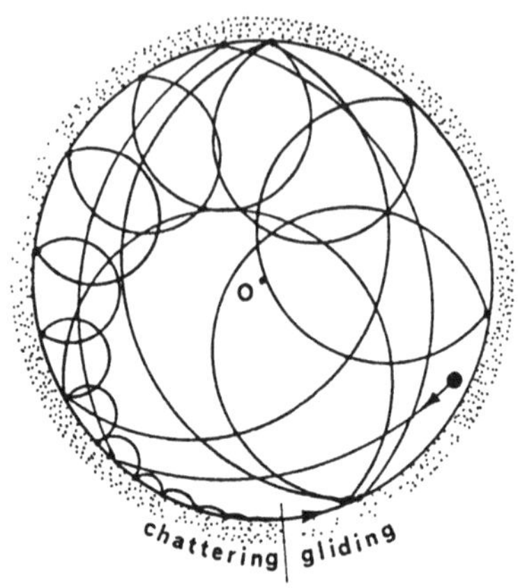

Fig. 22: Trajectory of rotor shaft center for a rotor touching down on the bearing

7. CONCLUSIONS

Magnetic bearings offer some features that allow to tackle some of the problems of classical rotor dynamics in a new way: no mechanical wear, no lubrication, low maintenance effort, low losses, adjustable dynamic performance. Of course there are some drawbacks, too, at least up to now: the bearing works within a closed loop system , and thus for its design it typically requires system engineering and mechatronics. This knowledge is not yet widely spread, and therefore the acceptance level in industry is still low, except for some advanced machinery. Part of the current scepticism comes from the question of reliability. It has not yet been solved in a sytematic way even though for space applications and other ones the specified reliability has been demonstrated. Further effort therefore will have to go into this direction. For example, the dynamic behaviour of a rotor touching down on an emergency bearing will have to be investigated. Fig. 22 shows a typical trajectory of the shaft center of such a rotor /SC 85/. Another area of interest is the active control of rotor vibrations, where future theoretical and practical results can be expected.

Research activities of various companies indicate that these specific features of magnetic bearings are going to lead to new designs and a new generation of rotating machinery.

References

/BL 84/ Bleuler, H.: Decentralized control of magnetic rotor-bearing systems. Diss. ETH Zurich No. 7573, 1984.

/BR 39/ Braunknecht, W.: Freischwebende Körper im elektrischen und magnetischen Feld. Zeitschr. f. Physik, 1939, S. 753-763.

/BU 85/ Bucher, Ch.: Contributions to the modeling of flexible structures for vibration control. Diss. ETH Zurich, No. 7700, 1985.

/BY 46/ Beams, J.W., Young, J.L., and J.W. Moore: The production of high centrifugal fields. J. Appl. Phys., 1946, pp. 886-890.

/GH 84/ Gondhalekar, V., and R. Holmes: Design of electromagnetic bearing for vibration control of flexible transmission shaft. NASA Conf. Publ. 2338, 1984.

/GO 84/ Gottzein, E.: Das "Magnetische Rad" als autonome Funktionseinheit modularer Trag- und Führsysteme für Magnetbahnen. Fortschr.Ber. VDI-Z., R.B. Nr. 68, Düsseldorf VDI-Verlag, 1984.

/HB 85/ Habermann, M., and M. Brunet: The active magnetic bearing enables optimum control of machine vibrations. ASME Paper 85-GT-22, Gas Turbine Conf. Houston, March 1985.

/HK 85/ Hustak, J., Kirk, G.R., and K.A. Schoeneck: Active magnetic bearings for optimum turbomachinery design. Symp. on Instability in Rotating Machinery, Carson City, June 1985, NASA Conf. Publ. 2409.

/MA 82/ MADYN-Handbuch, Krämer/Klement, TH-Darmstadt, 1982.

/PW 82/ Pietruszka, W.D., and N. Wagner: Aktive Beeinflussung des Schwingungsverhaltens eines magnetisch gelagerten Rotors. VDI-Bericht Nr. 456, 1982.

/SA 84/ Salm, J., and G. Schweitzer: Modeling and control of a flexible rotor with magnetic bearings. Conf. on Vibrations in Rotating Machinery, Inst. of Mech. Eng., York, 1984, pp. 553-561.

/SC 85/ Szczygielski, W., and G. Schweitzer: Dynamics of a high-speed rotor touching a boundary. IUTAM/IFToMM-Symp. on Multibody Dynamics, CISM Udine, Sept. 1985, Springer-Verlag, 1986.

/SL 76/ Schweitzer, G., and R. Lange: Characteristics of a Magnetic Rotor Bearing for Active Vibration Control. Conf. on Vibrations in Rotating Machinery, Inst. of Mech. Eng., Cambridge, U.K., 1976.

/ST 83/ Schweitzer, G., Traxler, A., Bleuler, H., Bauser, E., und P. Koroknay: Magnetische Lagerung einer Epitaxiezentrifuge bei Hochvakuumbedingungen. Vakuumtechnik 32 (1983), S. 70-74.

/TM 84/ Traxler, A., Meyer, F., and H.P. Murbach: Fast digital control of a magnetic bearing with a microprocessor. 11. Internat. Congr. Microelectronics, Munich, Nov. 1984.

/TR 85/ Traxler, A.: Eigenschaften und Auslegung berührungsfreier elektromagnetischer Lager. Diss. ETH Zürich No. 7851, 1985.

CHAPTER 12.2

VIBRATIONS IN VARIABLE SPEED MACHINES

H. Irretier
University of Kassel, Kassel, FRG

ABSTRACT

In a wide range of dynamically loaded machines variable speeds occur which lead to time-dependent excitation spectra of forced vibrations, passages through resonances and other phenomena. Some examples of the basics of these instationary vibrations are discussed and the most important mathematical relations for instationary forced vibrations of linear systems are presented.

1. INTRODUCTION

An important field in practical mechanical engineering is the construction of dynamically loaded machines e.g. compressors, combustion engines, gas and steam turbines and other machines in which rotating or accelerated components are used /1/, /2/, /3/. These components are loaded by the forces of the machine process and/or their own inertia forces. Thus, high stress levels may occur in particular when the frequency spectrum of the load excites the component or the machine in resonance.

In a lot of practical cases, these exciting frequency spectra are time-independent. This is valid for instance for those machines which run with constant angular velocity and with no change in the machine process. However, important cases exist where the exciting frequencies change with time /4/. This non-stationary case occurs for instance in turbines which run up or run down, in compressors with external load increase or decrease or in electro-motor-driven systems. In these cases, it is mostly unavoidable to protect the machine from resonance vibrations.

The purpose of this paper is to consider the special effects which occur in linear systems which are subjected to resonance vibrations by excitations with time-dependent spectra. In particular the run-up and run-down of various machines and their components are considered. All of them belong to the class of mechanical problems which can be described by the equation of motion

$$\underline{M}\,\ddot{\underline{w}} + \underline{D}\,\dot{\underline{w}} + \underline{K}\,\underline{w} = \underline{p} \tag{1.1}$$

for a discretized system. In this equation $\underline{w}$ denotes the displacement vector of the system, $\underline{M}$, $\underline{D}$ and $\underline{K}$ are the mass, damping and stiffness matrix, respectively, and $\underline{p}$ is the vector of the external forces. In a lot of practical cases, for instance in rotating machinery, these forces are periodic in time and can be described by the Fourier-series

$$\underline{p} = \sum_{\nu=-\infty}^{\infty} \hat{\underline{p}}_{\nu} \exp(j\nu\Omega t) \tag{1.2}$$

in which Ω is the angular speed of the machine, t is the time and $\hat{\underline{p}}_{\nu}$ are the Fourier-coefficients. The stiffness matrix $\underline{K}$ in equation (1.1) often is of the form

$$\underline{K} = \underline{K}_0 + \Omega^2 \underline{K}_{\Omega} + \Delta\underline{K}(t) \tag{1.3}$$

in which $\underline{K}_0$ is a time-independent part which describes the stiffness properties of the system itself, $\Omega^2\underline{K}_{\Omega}$ is an additional stiffness term which comes into consideration in those cases where the stiffness properties of the machine change with rotational speed. Very important examples for this effect are rotating turbine blades and discs. Finally, $\Delta\underline{K}(t)$ describes a time-dependent part of the stiffness matrix which must be included when parametric excitation occurs. This is true when the stiffness properties of the system change periodically. Important examples for such types of parametric excitations are the vibrations of cracked shafts where the bending stiffness changes periodically with rotation and beam-like structures which are loaded in longitudinal direction by periodic forces.

The solution of the equations (1.1) to (1.3) has been an important subject in the past. Many results are reported in the literature. However, in the most cases they are restricted to constant rotational speeds Ω of the machine which yields spectra with time-independent frequencies. Time varying excitations as they occur in variable speed machines were considered much less and - because of mathematical difficulty - mostly for some principal studies.

In the following chapters some examples are discussed in which instationary vibrations and passages through resonances occur due to changing excitation frequencies. The examples stand for the large amount of related problems in practical mechanical engineering.

2. EXAMPLES, MECHANICAL MODELS AND MATHEMATICAL DESCRIPTION

2.1 SPRING-MASS-SYSTEM EXCITED BY FORCE

The simplest model of a mechanical system which is able to perform vibrations is the single-degree-of-freedom system shown in Fig. 1 which consists of a spring with a spring-constant k, a viscous damper with the coefficient d and a mass m. If this system is excited by a harmonic force $F(t)=\hat{F}\cos\Omega_0 t$ its behaviour is described by the differential equation

$$m\ddot{w} + d\dot{w} + kw = \hat{F}\cos\Omega_0 t. \quad (2.1)$$

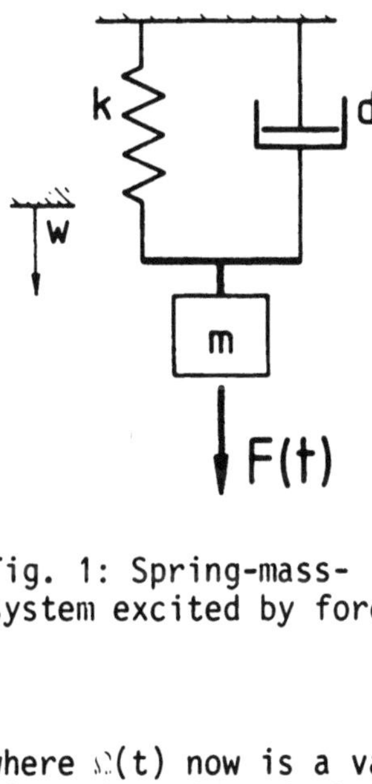

Fig. 1: Spring-mass-system excited by force

If $\hat{F}$, the exciting force amplitude, and Ω_0, the exciting frequency, are constant, equation (2.1) is a standard problem of vibration theory and there is no need to repeat the results here. However, when instationary conditions, for instance by an increasing or decreasing exciting frequency, are considered, $\Omega_0 t$ must be replaced by the phase

$$\varphi(t) = \int_0^t \Omega(t)dt \quad (2.2)$$

where $\Omega(t)$ now is a variable exciting frequency. Thus, the equation of motion of the spring-mass-system reads

$$m\ddot{w} + d\dot{w} + kw = \hat{F}\cos\varphi(t) \quad (2.3)$$

where $\varphi(t)$ is a more or less complicated function of time. This differential equation will be solved in chapter 3 for a linear decreasing and increasing excitation frequency.

2.2 SPRING-MASS-SYSTEM EXCITED BY UNBALANCE

A classical problem of mechanical engineering is the study of the behaviour of a spring-mass-system which is excited by a rotating unbalance. A corresponding mechanical model is shown in Fig. 2. If time-varying angular speeds and, thus, non-zero angular accelerations

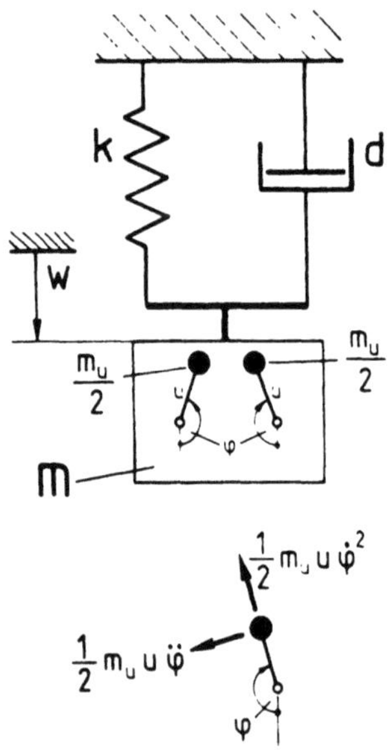

Fig. 2: Spring-mass system excited by unbalance

$$\dot{\varphi} = \Omega(t) \rightarrow \ddot{\varphi} = \dot{\Omega}(t) = \alpha(t) \quad (2.4)$$

are included in the consideration, the equation of motion is (/3/, /5/)

$$(m+m_u)\ddot{w}+d\dot{w}+kw=m_u u(\dot{\varphi}^2\cos\varphi+\ddot{\varphi}\sin\varphi) \quad (2.5)$$

in which φ, $\dot{\varphi}$ and $\ddot{\varphi}$ are functions of time. We notice from this equation of motion that now besides the phase φ also the excitation amplitude can be a complicated time-dependent function.

2.3 UNBALANCED FLEXIBLE ROTOR

The understanding of the dynamics of a flexible rotor, which is excited by its rotating unbalance, is a classical problem of mechanical engineering sciences. In the most cases such problems have been considered for a constant rotational speed of the rotor. However, from the practical point of view for some types of rotating machines it is of importance to study the behaviour of the rotor also for varying rotational speeds e.g. a run-up or a shut-down.

The most simple model of a rotor is the well-known Laval-rotor which is shown in Fig. 3. If y_C and z_C denote the coordinates of the centre of mass of the disc, m its mass, d a viscous damping coefficient, k the spring constant of the shaft, ε the unbalance, J the polar mass moment of the disc, and T the torque driving the rotor, the equations of motion of the deflected Laval-rotor are(/5/, /6/, /7/)

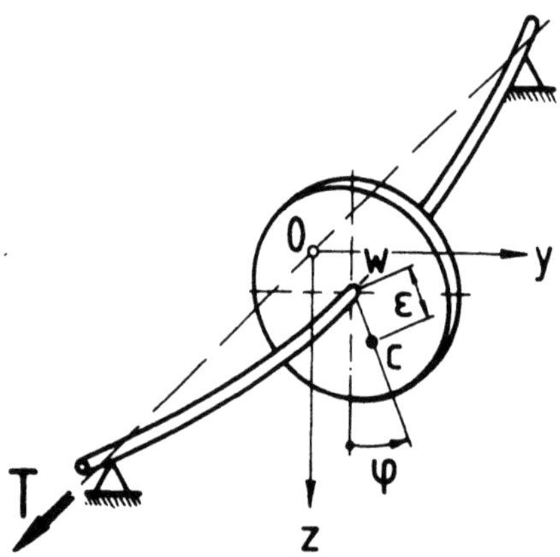

Fig. 3: Deflected Laval-rotor

$$m\ddot{y}_C+d\dot{y}_C+ky_C=k\varepsilon\sin\varphi$$

$$m\ddot{z}_C+d\dot{z}_C+kz_C=k\varepsilon\cos\varphi \quad (2.6)$$

$$J\ddot{\varphi}-k\varepsilon(y_C\cos\varphi-z_C\sin\varphi)=T.$$

We notice from these equations that generally the rotor vibrations are described by non-linear differential equations. However, for the most practical cases for rotors with small unbalance in comparison to the mass radius of gyration of the disc, the non-linear term in equation (2.6) can be neglected, the problem is linear and the acceleration $\ddot{\varphi}$ is proportional to the driving torque T. Thus, an increasing or decreasing torque yields a proportional increasing or decreasing angular velocity Ω and again the angle φ which describes the angular position of the disc, is a time-dependent function according to equation (2.2). In this way, for this instationary case, the differential equations (2.6) are of similar type as equation (2.3) for the spring-mass-system excited by an external force.

2.4 ROTATING TURBINE BLADE

In gas and steam turbines the blades are one of the most significant components from the structural mechanics point of view. In Fig. 4 a typical turbine rotor with its shaft, its discs and some blades is shown. In view of the fact that the rotor blades run

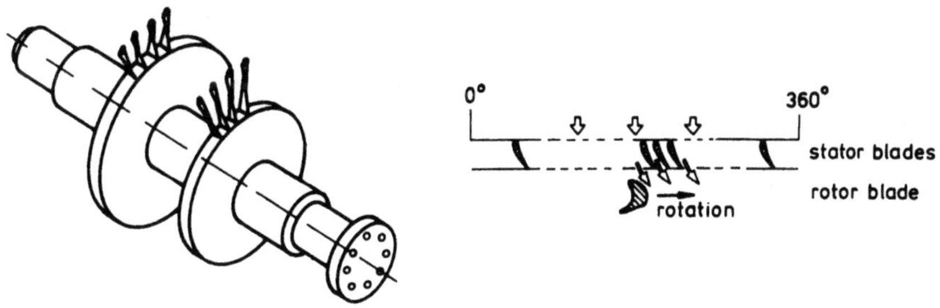

Fig. 4: Turbine rotor and nozzle excitation of the rotor blades

behind several stages of fixed stator blades, they are loaded periodically by aerodynamic forces which can be described by the Fourier-series

$$p = \sum_{\nu=0}^{\infty} \hat{p}_\nu \cos(\nu z \Omega t - \psi_\nu) \qquad (2.7)$$

where $\hat{p}_\nu$ are the amplitudes of the various harmonics, z is the number of nozzles in the stage, and Ω is the angular speed of the blade. If the very simple model of an Euler-Bernoulli-beam is introduced to describe the behaviour of the blade, the corresponding equation of motion reads

$$\frac{\partial^2}{\partial x^2}\left(EI \frac{\partial^2 w}{\partial x^2}\right) + \rho A \frac{\partial^2 w}{\partial t^2} = p \qquad (2.8)$$

where EI is the bending stiffness of the blade, ρA its mass density per unit length and $w = w(x,t)$ describes the deflection. We notice from this equation of motion that - for varying angular speeds Ω of the turbine - again a problem of instationary excitation of a mechanical system must be solved to find the response of the rotating blade during a run-up or shut-down.

In general, the rotating blade vibration problem is more complicated than described by the equation of motion (2.8). Refined models have been developed in the past /8/, /9/. All these models can be set up in the form of equation (1.1) where it is particularly important to mention that the stiffness matrix $\underline{K}$ depends on the angular speed of the blade because the centrifugal force field increases the stiffness of the rotating blade considerably. Thus, the stiffness matrix follows the equation (1.3) with $\Delta\underline{K} = \underline{0}$, so that an additional effect occurs that is the fact of changing eigenfrequencies of the system during a run-up or run-down of the blade, /10/, /11/.

2.5 TORSIONAL VIBRATIONS OF ELECTRO-MOTOR-DRIVEN SYSTEMS

A final example for a mechanical system in which variable speed problems occur is an electro-motor-driven compressor system. A sufficient mechanical model for such types of systems is a discretised spring-mass-model of several degrees of freedom as shown in Fig. 5. During the run-up or the shut-down of this motor-compressor-system one or more torsional critical speeds of the machine can be passed which yield strongly increasing stress levels. A typical plot of the shaft torque is shown in Fig. 6 for a motor start-up /12/.

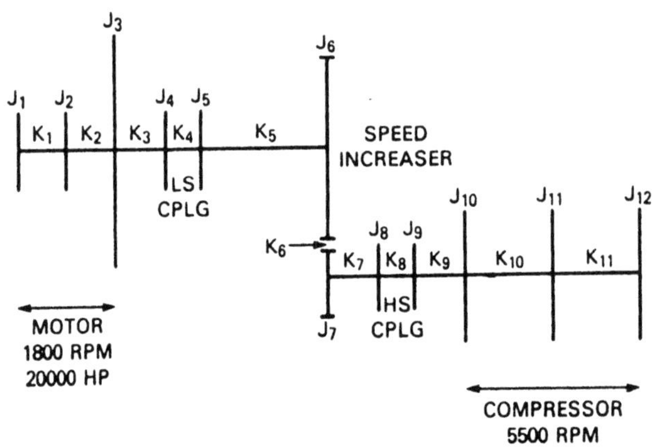

Fig. 5: Typical spring-mass-model for a torsional vibrating system /12/

This plot of shaft torque shows - and by this means represents a transition to the following chapters - the typical response of a linear mechanical system running through resonance. Consequently, the next considerations deal with these phenomena in application to the examples discussed here.

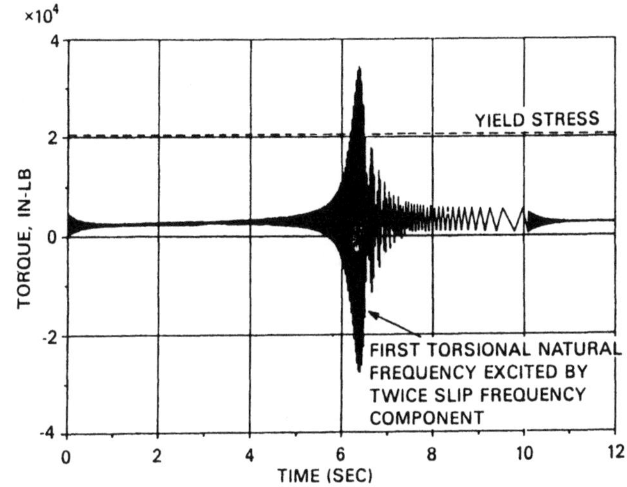

Fig. 6: Shaft torque as function of time during motor start-up /12/

3. BASIC PHENOMENA IN MECHANICAL SYSTEMS RUNNING THROUGH RESONANCES

The examples in the previous chapter stand for all similar problems in machines with variable speed. Next, the vibrational behaviour of the systems during passages through resonances is considered.

A typical response of a linear system running through a resonance is plotted in Fig. 7 which shows the deflection of a Laval-rotor

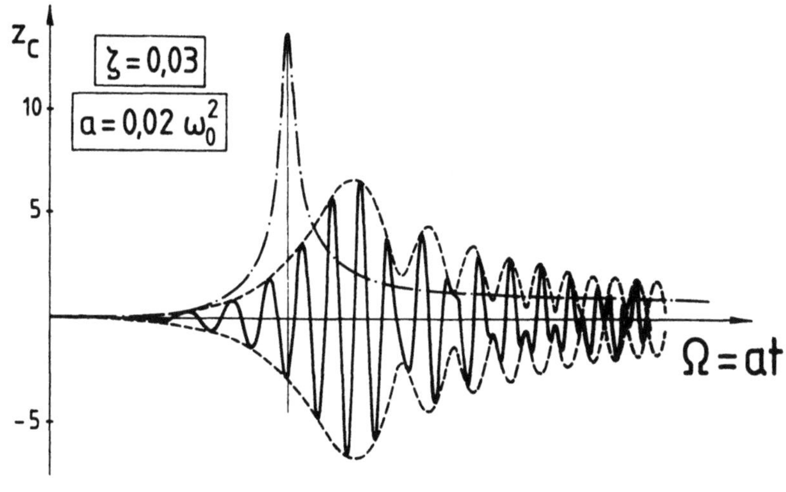

Fig. 7: Deflection of a Laval-rotor running through resonance as function of time /7/

(Damping ratio $\zeta = 0{,}03$) as function of time during a run-up compared with the response curve for stationary resonance which is synonymous with an infinite-slow run-up. All other comparible examples and problems show a similar behaviour during the passage through resonance. This behaviour is characterized by the following phenomena:

i) The maximum of the amplitude of the instationary response did not occur when the exciting frequency equals the eigenfrequency of the system but shifts to later times during a run-up and to earlier times during a run-down of the system, respectively. This shifting increases with increasing velocity of changing of the exciting frequency and vice versa.

ii) The magnitude of the amplitude maximum depends on the velocity of resonance passing and proves to be smaller than the amplitude during a stationary resonance. The more the velocity of frequency changing is, the less is the maximum of the response amplitude.

iii) After running through resonance typical beat vibrations occur because the response of the system consists of the excited and the natural motion, and shortly after resonance both are vibrations with frequencies closed to each other.

The next step of consideration is to discuss numerical results for the various examples considered before to find the dependence of the different parameters of the system and its response during a passage through resonance. In view of the shortness of the present paper, it is only possible to give the fundamental results; for details on the calculations the reader is referred to the literature.

4. EXAMPLES AND NUMERICAL RESULTS OF SYSTEMS RUNNING THROUGH RESONANCES

The most considered problem with respect to non-stationary resonance vibration are the spring-mass-systems shown in the Figures 1 and 2. A lot of numerical and even analytical results had been given in the literature. The most extensive reference is /4/ where the mathematics of solution is discussed in detail for a lot of problems. Recent papers like /5/ and /11/ to /14/ deal to some extend with applications to special problems and only a few results can be quoted here.

4.1 SPRING-MASS-SYSTEMS

The spring-mass-system of Fig. 1 is considered first. Introducing the eigenfrequency, the damping ratio and the dimensionless time

$$\omega_0 = \frac{k}{m} \quad ; \quad \zeta = \frac{d}{2\sqrt{mk}} \quad ; \quad \tau = \omega_0 t \tag{4.1}$$

the equation of motion (2.3) can be transformed into

$$w'' + 2\zeta w' + w = \cos\varphi(\tau) \tag{4.2}$$

where a unit force for which $\hat{F}/k$ = "1" is assumed to simplify matters. If now a linear increasing ($a > 0$) or decreasing ($a < 0$) exciting circular frequency according to the equation

$$\Omega(t) = \Omega_0 + at \tag{4.3}$$

is assumed, it is possible to solve equation (4.2) analytically. For this case, the phase $\varphi(t)$ in equation (4.2) follows from equation (2.2) to

$$\varphi(t) = \Omega_0 t + \frac{1}{2}at^2 \quad \text{or} \quad \varphi(\tau) = \eta_0\tau + \frac{1}{2}\alpha\tau^2, \tag{4.4}$$

where

$$\eta_0 = \frac{\Omega_0}{\omega_0} \quad \text{and} \quad \alpha = \frac{a}{\omega_0^2} \tag{4.5}$$

are the exciting circular frequency for t=0 related to the eigenfrequency and the parameter which relates the angular acceleration $\dot{\Omega} = a$ to the square of the eigenfrequency. Introducing equation (4.4) into (4.2) the mathematical model of a force excited linear single-degree -of-freedom system with linear increasing or decreasing exciting frequency becomes

$$w'' + 2\zeta w' + w = \cos(\eta_0\tau + \frac{1}{2}\alpha\tau^2) \quad . \tag{4.6}$$

As shown in /4/ the solution of the equation (4.6 e found in the form

$$w = |\hat{w}(\tau)|\cos\{\varphi(\tau) - \psi(\tau)\} \tag{4.7}$$

where $\hat{w}(\tau)$ describes the time-dependent amplitude of the response of the system and $\psi(\tau)$ is a time-dependent phase angle. The amplitude function $\hat{w}(\tau)$ is a complex function

$$\hat{w}(\tau) = |\hat{w}(\tau)|\exp\{-j\psi(\tau)\} \tag{4.8}$$

which can be calculated with the help of probability integrals/4/.

The physical interpretation of the equation (4.7) is very simple if the dynamical behaviour of a mechanical system excited under stationary conditions with time-dependent exciting frequency is considered for comparison. First, the system follows the phase φ given in equation (4.4) but with a phase difference ψ which is time-dependent. Second, the amplitude of the system is time-dependent, too, so that an amplitude-modulated vibration occurs which depends on the damping ratio and the actual ratio of the exciting frequency to the eigenfrequency of the system.

The amplitude function $|\hat{w}(\tau)|$ and the phase angle $\psi(\tau)$ are calculated in the references /4/, /5/, /13/ and /14/ for spring-mass-systems subjected to various types of excitation. Here, the response of a spring-mass-system shown in Fig. 1 during a passage through its resonance frequency is considered first. Fig. 8 shows the amplitude function $|\hat{w}(\tau)|$ as a function of the dimensionless angular velocity $\varphi' = \eta_0 + \alpha\tau$ for a damping ratio $\zeta = 0{,}02$. Parameter on the curves is the quantity α given in equation (4.5) which is the most important parameter to decide the behaviour of the system during the passage through resonance in comparison to the stationary case. Both cases, a run-up ($\alpha > 0$) as well as a run-down ($\alpha < 0$), are plotted in the diagram in Fig. 8. We notice

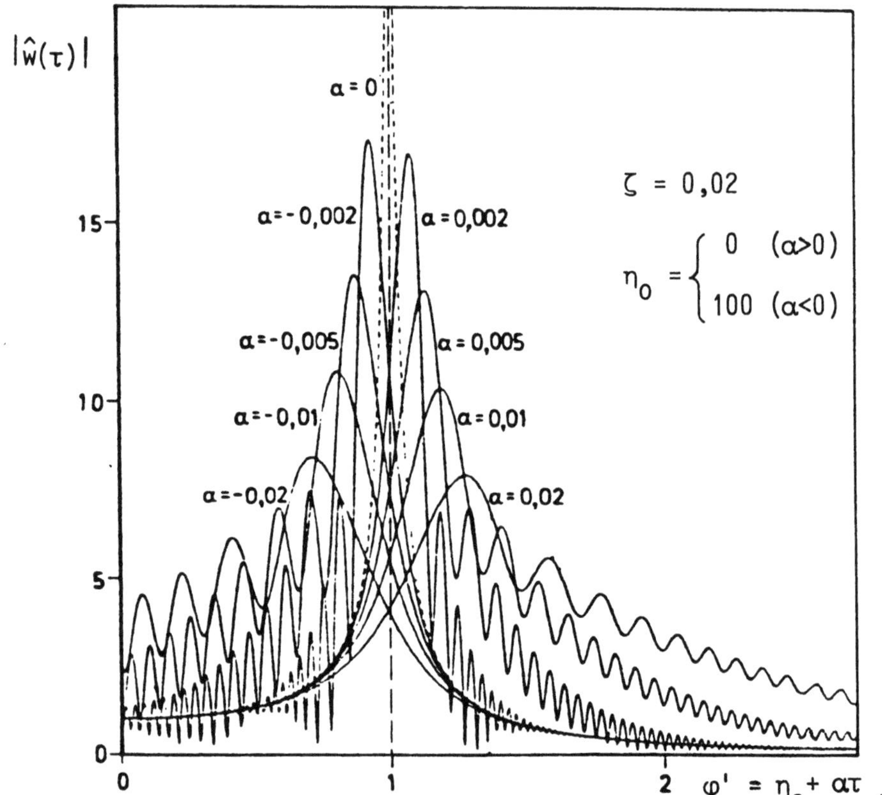

Fig. 8: Amplitude functions for a spring-mass-system excited by force during a passage through resonance by linear increasing and decreasing exciting frequency /14/

from the curves that the more α shifts from 0 to positive or negative values the more the amplitude maximum shifts from the stationary resonance point and the more its magnitude decreases in comparison to the stationary case.

As an second example, in Fig. 9 a running through resonance is plotted for the spring-mass-system under unbalance excitation (Fig. 2). The curves show the amplitude of the force at the foundation of the system related to the centrifugal force of the unbalance. Again, the typical behaviour of the passage through resonance is visible. In addition, we notice that the maxima of amplitudes which occur are different for the run-up and the run-down. More details and parameter studies are given in /5/ and further investigations are presented in /4/, /13/ and /14/.

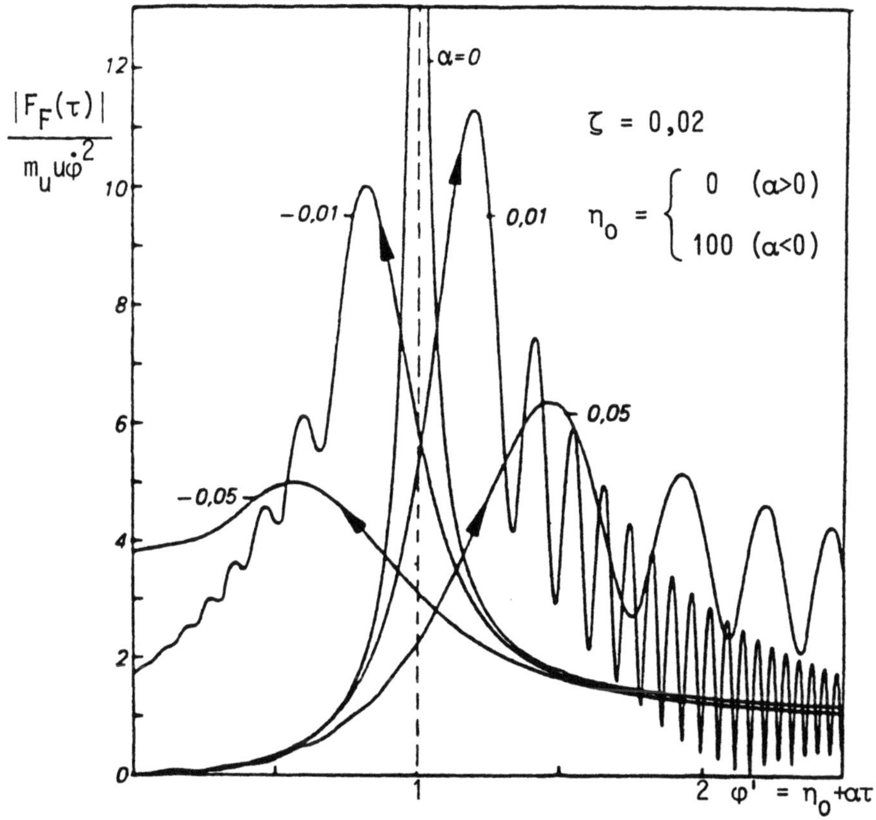

Fig. 9: Force at the foundation of a spring-mass-system excited by unbalance during a passage through resonance by linear increasing and decreasing exciting frequency /5/

4.2 RUN-UP AND SHUT-DOWN OF A FLEXIBLE ROTOR

To continue our consideration of an unbalanced flexible rotor in chapter 2.3, now its response is discussed for the case of a passage through resonance due to a constant angular acceleration. The describing, linearized equations of motion (2.6) are of similar type as for a force excited spring-mass-system. Thus, the solution is possible in a way comparable to that one described in the previous chapter. Fig. 10 shows the amplitude of a Laval-rotor running through its critical speed during a run-up ($\alpha > 0$) and a shut-down ($\alpha < 0$). The damping ratio is $\zeta = 0,02$ and curves are plotted for various values of the constant angular acceleration α. Again, the typical behaviour of a linear system running through resonance is visible. The maximum of response occurs after the critical speed for stationary conditions is reached, and this maximum is less than in the case of stationary rotation of the rotor. For acceleration $|\alpha| > 0,1$ no typical resonance response is observed even for the small damping ratio of 0,02.

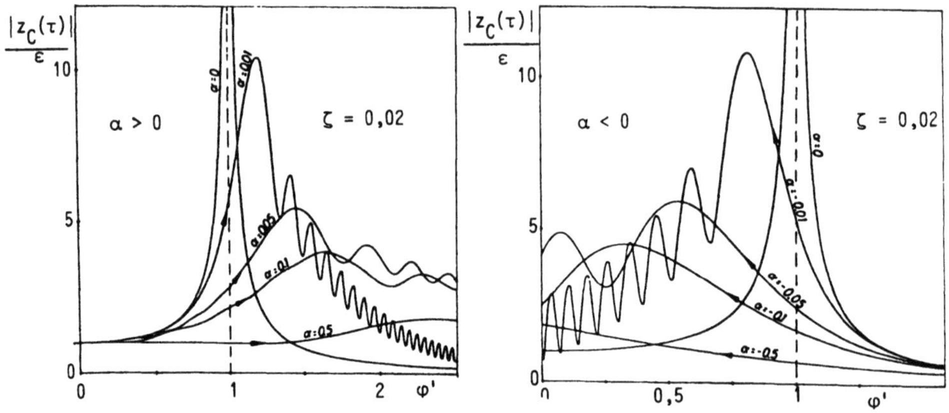

Fig. 10: Response amplitude during a run-up and a shut-down with constant angular acceleration /7/

In the discussion of the dynamical behaviour of a flexible rotor running through critical speeds, an important phenomena may not be ignored. The plots shown in Fig. 10 are only valid, if the driving torque T in equation (2.6) exceeds a certain value to insure that the system is able to overcome the critical resonance rotation. Details on that are reported in /4/ and /7/. Here,only some basic considerations are performed.

If the problem of sufficient driving torque is considered mathematically, the complete non-linear equation (2.6) must be solved. For a constant driving torque T and a large excentricity ε it proves during the passage through resonance, where y_C and z_C increase to large values, that the deviation of the angular velocity φ from a linear time-dependent function can increase to a considerable amount. This effect is shown in the left diagram of Fig. 11 which shows the angular velocity as a function of time during the passage through resonance for the case of a sufficient driving torque to exceed the resonance vibration. The response of the rotor shows the typical form. In contrast, if the driving torque does not suffice to accelerate the rotor through the critical speed, the angular velocity reaches the critical value and then oscillates below this value as shown in the right diagram of Fig. 11. The system does not pass the critical resonance vibration and the amplitude rests on the resonance level prescribed by the value of damping.

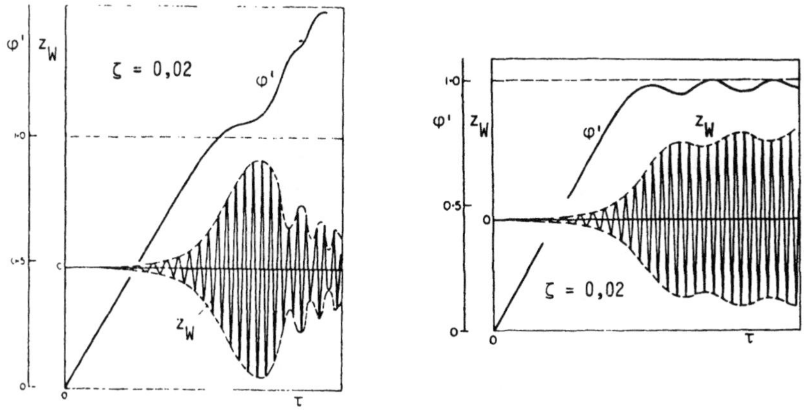

Fig. 11: Angular velocity and deflection of a Laval-rotor accelerated through the critical speed and blocked at the critical speed /7/

4.3 RUN-UP OF A TURBINE BLADE LOADED BY PARTIAL ADMISSION

An outlook on more complicated problems, in which transient vibration caused by a variable speed of the machine can occur, is discussed now. A rotating turbine blade is considered which runs up from one rotational speed to another while it passes several resonance points. This problem differs from the example regarded in the previous chapters in two sorts of respect . The first point is, that the turbine blade must be described by a continuous mechanical model which basically requires a formulation in a partial differential equation as it was shown for a simple model by equation (2.8). Thus, more than one eigenfrequency must be considered. The second fact is that, due to the rotation of the blade, its spectrum of eigenfrequencies increases with rotational speed. In this way, complicating effects occur when the acceleration of the blade through critical speeds is considered.

As one example, which is discussed in detail in /11/ a rotating blade is considered which is loaded by partial admission. This partial admission is modelled by two 90°-sectors (Fig. 12) while it is unloaded in the two other sectors. The blade force is distributed over the blade length as shown in Fig. 12, too. During a run-up of the blade with such type of partial admission it is mostly unavoidable to pass through several resonance points as indicated in the Campbell-diagram in Fig. 13 which is taken from /11/. The corresponding response of the blade is simulated by a computer program which is described in /11/, too. A typical result is shown in Fig. 14. The simulation starts with a rotational frequency of f_D = 10 Hz and ends at a rotational frequency of 60 Hz which is reached within 25 s (Fig. 13). During this run-up the blade passes several resonance points as indicated in the

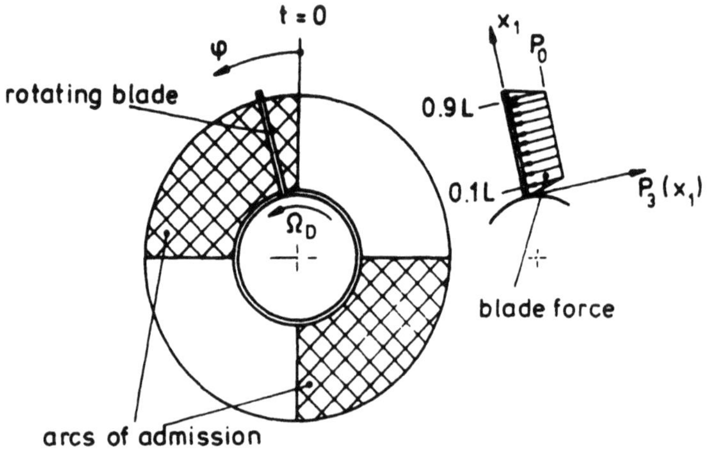

Fig. 12: Rotating blade and load due to two arcs of partial admission /11/

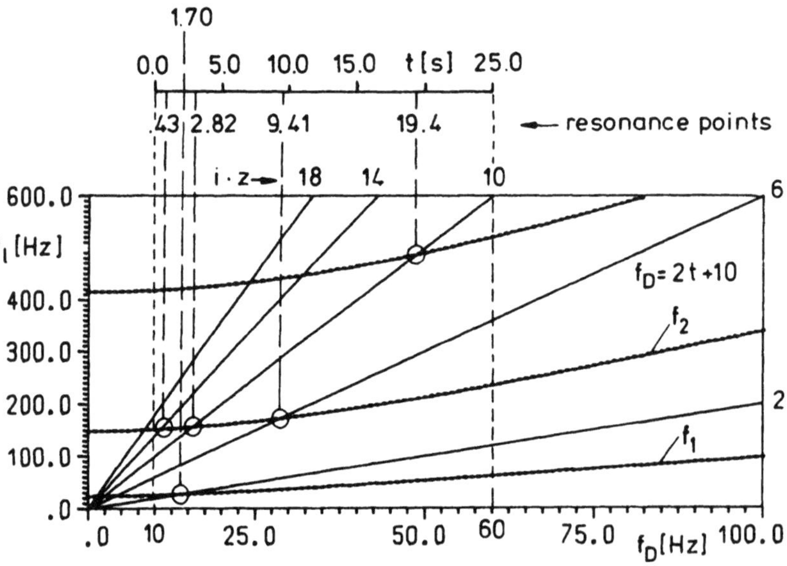

Fig. 13: Campbell-diagram of a simulated run-up of a turbine blade loaded by partial admission /11/

Campbell-diagram in Fig. 13.

The response of the blade is shown in Fig. 14 where the blade root stress is plotted as function of time during the run-up. We notice again the more or less time difference due to the non-stationary conditions between the resonance points indicated in the Campbell-diagram and the maxima of stress in Fig. 14.

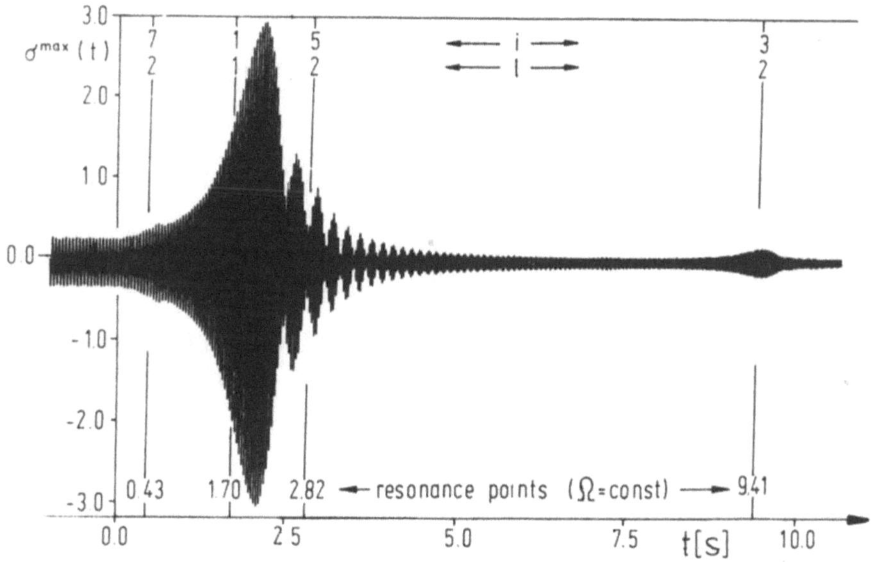

Fig. 14: Response of a turbine blade loaded by partial admission during a run-up /11/

Moreover, the plot shows that for the type of load distribution along the blade only the resonances between the first engine order and the first eigenfrequency and third engine order and second eigenfrequency are of significance. This effect occurs because of decreasing intensity of the exciting spectrum for increasing engine order.

5. CONCLUSIONS

The presented examples of dynamical problems occuring in machines with variable speed show that a lot of additional phenomena exist in comparison to stationary conditions. The most important results are the decrease of the resonant amplitude and the shifting of the resonance frequency. As to more complicated effects for instance the influences of non-linearities or parametric resonances the reader is referred to the literature in particular to /4/ in the following list of references.

6. REFERENCES

/1/ IMechE: Proc. 3rd. Int. Conf. on Vibrations in Rotating Machinery, York (England), Sept. 1984

/2/ IFToMM: Proc. Int. Conf. on Rotordynamic Problems in Power Plants, Rom (Italien), Sept./Oct. 1982

/3/ Holzweissig, R.; Dresig, H.: Lehrbuch der Maschinendynamik. Springer-Verlag Wien/New York, 1979

/4/ Goloskokow, E. G.; Filippow, A. P.: Instationäre Schwingungen mechanischer Systeme. Akademie-Verlag, Berlin, 1971 (translated from Russian)

/5/ Markert, R.; Pfützner, H.: An- und Auslaufvorgänge einfacher Schwinger. Forsch. Ing.-Wes. 47,4 (1981),117-125

/6/ Gasch, R.; Pfützner, H.: Rotordynamik. Springer-Verlag Berlin/Heidelberg/New York, 1975

/7/ Markert, R.: Biegeschwingungsverhalten unwuchtiger elastischer Rotoren bei der Resonanzdurchfahrt. VDI-Bericht 381 (1980),155-160

/8/ Traupel, W.: Thermische Turbomaschinen. Springer-Verlag Berlin/Heidelberg/New York, 1968

/9/ Rao, J.S.: Turbine Blade Vibration. Shock and Vibration Digest 12 (1980), 19-26

/10/ Irretier, H.: Turbine blade dynamics. CISM-Courses and Lectures - Rotor Dynamics II, Udine (Italien), Oct. 1985

/11/ Irretier, H.: Computer Simulation of the Run-up of a Turbine Blade Subjected to Partial Admission. ASME Design Eng. Techn. Conf.; Cincinnati (USA), Sept. 1985; ASME Publication 85-DET-128 (1985), 1-12

/12/ Evans, B.F.; Smalley, A.J.: Simmons, H.R.: Start-up of Synchronous Motor Driven Trains: The Application of Transient Torsional Analysis to Cumulative Fatigue Assessment. ASME Design Eng. Techn. Conf., Cincinnati (USA), Sept. 1985; ASME Publication 85-DET-122 (1985), 1-9

/13/ Dittrich, G.; Sommer, H.: Das instationäre Verhalten des linearen Schwingers mit einem Freiheitsgrad. VDI-Zeitschrift 118, 8 u. 10 (1976) 375-378 u. 477-482

/14/ Markert, R.: An- und Auslaufvorgänge einfacher Schwinger im Kriechgrenzfall D = 1. Forsch. Ing.-Wes. 48,1 (1982), 11-14

MIX
Papier aus verantwortungsvollen Quellen
Paper from responsible sources
FSC® C105338

If you have any concerns about our products,
you can contact us on
ProductSafety@springernature.com

In case Publisher is established outside the EU,
the EU authorized representative is:
Springer Nature Customer Service Center GmbH
Europaplatz 3, 69115 Heidelberg, Germany

Printed by Libri Plureos GmbH
in Hamburg, Germany